Practical RF Circuit Design for Modern Wireless Systems

Volume I

Passive Circuits and Systems

For a listing of recent titles in the *Artech House Microwave Library*,
turn to the back of this book.

Practical RF Circuit Design for Modern Wireless Systems

Volume I

Passive Circuits and Systems

Les Besser
Rowan Gilmore

Artech House
Boston • London
www.artechhouse.com

Library of Congress Cataloging-in-Publication Data
A catalog record of this book is available from the Library of Congress.

British Library Cataloguing in Publication Data
Besser, Les
Practical RF circuit design for modern wireless systems
Vol. 1: Passive circuits and systems.—(Artech House microwave library)
1. Radio circuits—Design 2. Wireless communication systems
I. Title II. Gilmore, Rowan
621.3'8412

ISBN 1-58053-521-6

Cover design by Yekaterina Ratner
Text design by Darrell Judd

Cover image courtesy of Agilent, Inc.

685 Canton Street
Norwood, MA 02062

International Standard Book Number: 1-58053-521-6
A Library of Congress Catalog Card Number is available from the Library of Congress.

10 9 8 7 6 5 4 3 2 1

To our wives, Susan and Nicole

Contents

Preface

This text is intended to be a populist book.

With so many complex equation-filled engineering books lining the shelves of our bookstores, perhaps you might be wondering whether the science of microwaves and RF is ready for a text that can be understood by those who do not speak Latin or wear black robes. We believe so. The goal of a populist book is to appeal as much to the academic at a highbrow university as to the practitioner working in today's frantic production environment. We hope you will find this text as relevant to your work of teaching others as to improving your own skills.

This two-volume book is written for practicing engineers and for those who would like to become one. And these days, who can afford not to keep learning? Whether you are a student at your final year of university, an engineer in industry who has just been assigned your first RF design project, or a seasoned veteran of the magic of microwave design, we hope that you will all find something useful in these pages. Even if you are a microwave or RF industry guru with most of the answers already, our experience in writing this has been that there is still a thing or two out there that needs explaining. If you cannot find anything that seems inexplicable, then at least you will have the satisfaction of reassuring yourself that you have indeed been right all these (long!) years.

We do not suggest you throw away your other excellent textbooks that explain semiconductor transport equations, Green's functions, or the complex mathematics of filter design; just that this effort might make those paperweights all the more relevant. Don't misunderstand us—we do not imply that anyone can become a high-grade RF circuit and system designer without using any complex algebra. We feel strongly however that you *do not need as much of it* as some of the courses you have taken before may have included.

These two volumes are the culmination of more than 40 joint years of teaching these topics to *thousands of practicing electrical engineers* from around the world. Little by little, we have extended the scope of our courses and learned the simplest ways to convey basic ideas to our audience. We have often been surprised and have found for the most part that our audience is generally not interested in obtaining guru status or academic knowledge, but in gaining an understanding of microwave and RF circuits, *in gaining intuitive insight*, and in applying that to their work. We hope we have captured that spirit herein.

This book is not written for the expert. If anything, we have omitted specialist material (the text is long enough as it is!). We often begin our courses by telling our students that if they have spent the past year characterizing the intermodulation properties of a device to design a predistorter circuit, they are probably already one of just a handful of experts in the world in that area—and that they can probably teach us something! Although we hope this book will convey the background and insight to set you on the road to becoming an expert, it will not take you down the narrow and winding lanes that make you one. We have focused on discrete circuits and discrete circuit design rather than IC design, believing that only when discrete design is mastered can those techniques be applied to integrated circuits. In consciously stopping short of IC design, we have not considered many worthy topics, such as RC or AGC oscillators or complex biasing techniques. Nor have we considered integrated systems such as phase-locked loops. All these topics are worthily covered elsewhere in expert texts of their own, and rightly so! Perhaps a third volume of this series will one day attempt to simplify those topics as well, should our wives ever let us back near our computers again.

The book can be used as a final-year text in applied RF engineering towards a bachelor's degree in electrical engineering, or as part of a master's degree coursework material, or as a reference by the engineer who has already reached that level. In the university context, it is suited for a two-semester course. We assume that the student already has an understanding of basic topics such as phasors, electromagnetics, Fourier transforms, circuit analysis, and semiconductors. To be on the safe side, in the first two chapters we have summarized most of the fundamentals needed. From that background, we recommend that the text be taught in the order in which we have presented the chapters. Our experience is that after some initial preliminaries, the systems material starting in Chapter 3 of this volume can motivate the rest. It contains simple applications of radio technology so that the student can feel a worthwhile accomplishment early on, and will see good reason to pursue his or her subsequent detailed work. We attempt to close the circle at the end of Volume II by returning to the radio systems aspects started at the beginning, but now armed with a more detailed understanding of the technology. Prototyping a radio system with some of the integrated circuits in that chapter would be a worthwhile student project that could proceed throughout the year, building on the self-discovery process in parallel with the formal learning. In the middle, we cover all the important techniques of RF, such as impedance matching, device characterization and modeling, amplifiers, oscillators, and mixers. Knowing how to build high-speed blocks for gain, loss, frequency conversion, and oscillation enables the student to go on to build almost any RF component.

This text differs from others in that we focus on the systems aspects of a design. To use a metaphor, we look at the forest rather than the trees, although we have included plenty of different greenery and spend ample time examining the leaves and branches as well. We assume the student comes with a basic knowledge of agriculture, understanding the soil, rain, sun, and so on. We have not focused on any one particular topic, although because the property of amplification is so fundamental, it is covered in rather more detail. We use amplification to learn about devices, simulation, distributed elements, characterization, impedance matching, stability, gain and power, and nonlinear behavior. We have also been generous in the use of the simulator to illustrate each tree with many examples, and encourage students to develop their own. Our goal is that by the end of the text, students will be able to plan and seed their own forest with enough interest to make it grow!

Throughout the book, we emphasize computer-aided design (CAD) techniques and encourage you to use them as much as possible in your daily work. At the same time, we abhor the idea of *blindfold optimization* without first obtaining a reasonable initial estimate and intuitive feel for the outcome. Although today's CAD tools are powerful enough to reach a solution at times for simple problems, relying on optimization without understanding the underlying circuit or system fundamentals is a poor practice that inevitably leads to failure. Combining CAD techniques with a thorough understanding of RF fundamentals and use of traditional engineering tools is the best way to be successful.

The text contains material that is both mature and state-of-the-art, although we have been inclined to retain mature material if it is still current and where it can provide more fundamental understanding than a result just published in a recent journal. We believe great textbooks are written to last for years. They should teach fundamental principles that can be applied to each recent technological advance as it comes along, and not become obsolete in the process. Our courses have attempted to do just that, and in encapsulating the core of what we have taught, we hope we will achieve that here as well.

Acknowledgments

Like many others, this book has grown out of course notes. In particular, the Applied RF Techniques I, Applied RF Techniques II, and Applied Wireless and Microwave Techniques courses we have taught at Besser Associates and CEI Europe have been particularly fruitful in this regard. However, the book does differ from similar books in that we have taught these courses to the rather more demanding audience of graduate, practicing engineers, which we hope will make it quite relevant. Accompanying us in our teaching of these courses have been other notables in the industry, who have helped shape these notes, and therefore this book. In particular they include Ed Niehenke and Alan Podell, who have helped to develop, structure, and formulate these presentations over the years. Their contributions to this book are sometimes implicit, but nevertheless manifold, especially in Volume II. Other Besser instructors, Rick Fornes, Steve Hamilton, and Lynne Olsen, have also provided many helpful suggestions and improvements. The comments of Giora Goldberg, Bob Morrow, and Irving Kalet in their particular areas of expertise have also enriched the text.

We want to express our thanks to Istvan Novak for contributing Chapter 9 on high-speed design considerations to provide RF professionals a glimpse of the "other technology."

We are also indebted to the numerous CAD vendors whose products we have liberally used both in this text and our coursework. In particular, the efforts of the highly professional and responsive staff at Applied Wave Research have really enabled much of this work. Dane Collins, the vice-president of engineering at AWR, has frequently gone out of his way to assist us along with their support group. Steve Maas, the chief technology officer at AWR, has also assisted us with technical issues.

We are also thankful to Bill Bridges, Joe Civello, and Danielle Flint at Agilent Technologies for providing circuit and device information, as well as software, so we could keep the focus of this book practical.

The contribution of Bob Stengel of Motorola and Bill Eisenstadt of the University of Florida on mixed-mode *S*-parameter techniques was particularly helpful since there is little published material on the subject. Guillermo (Bill) Gonzalez of the University of Miami provided numerous helpful suggestions and corrections to the books. Dan Swanson, Bob Wenzel, and George Szentirmai gave us input on EM simulation and filter synthesis and also provided a copy of Filsyn.

Others who assisted with design and CAD information include Rich Carlson at Motorola, Akifumi Nakatani at Ansoft, Mike Meehan at Agilent Technologies, Shawn Carpenter, Jim Rautio, and Volker Muehlhaus at Sonnet, and former colleague Peter Sturzu. We acknowledge and appreciate their help.

Thanks should also go to the many staff at the semiconductor manufacturers who helped us with device information and drawings. Their companies include Agilent, Analog Devices, Anadigics, California Eastern, IBM, Infineon, M/A-COM, Maxim, Motorola, Peregrine, and Philips. We also received support from the staff of ATC, Coilcraft, Maury Microwave, Modelithics, and Murata who helped us with their data sheets and device data—sometimes hot off the press.

Colleagues at Besser Associates, Jeff Lange and Annie Wong have also assisted. Rex Frobenius worked long hours to make corrections and provided ideas as well as support with many of the illustrations. Les' daughter Daphne also lent her graphics expertise to generate many of our illustrations, like the "water pipes" that convey complex ideas in simple forms. Thanks should also go to Les' son Kent and younger daughter Nanci for their proofreading and helpful suggestions.

The staff at Artech House who produced this book has also worked tirelessly. Mark Walsh, Barbara Lovenvirth, Rebecca Allendorf, Judi Stone, and Darrell Judd have all managed the idiosyncrasies and extra effort of working with two authors on opposite sides of the globe. We thank them all that through their efforts we managed to finally make it to print! Our reviewers also helped to reshape sections and clarify points when we had strayed, and encouraged us when we were on the right path. We thank them for keeping us honest.

To former students from various countries we express our gratitude and appreciation for providing criticism and suggestions about the content of the courses we have presented. We appreciated the feedback that enabled us to fine-tune our delivery, tweak our examples, and stay focused on a practical theme.

Finally and most importantly we thank our families who missed us for weeks on end as we stayed glued to our computers. There must be a fundamental law defining how much time it takes to write a two-volume book, but instead of listening to those who have done it before, we set out to prove that we could do it in 6 months. More or less, we thought. Much, much more as it turned out! We eventually stopped keeping track of the 18-hours workdays of thinking, writing, talking, simulating, and drawing that required withdrawal and silence. We are indebted to our families for putting up with all that. Without their understanding that "what a man's got to do, he's got to do," this would never have been accomplished.

CHAPTER 1

Introduction

When you opened this book for the first time, perhaps your first motivation was to discover exactly what *Practical RF Circuit Design for Modern Wireless Systems* really means. In this chapter, we explore those words to lay some common ground before we start our studies.

1.1 Defining RF

Perhaps the distinguishing feature of this book when compared with other books on analog circuit design is captured in two letters: RF. Arbitrarily, the *radio frequency* portion of the spectrum is considered to extend from 3 to 1,000 MHz (1 GHz). This book is certainly applicable to circuits operating within those frequencies, but it is more than that. Although frequencies above 1 GHz are considered to be within the microwave domain, we cannot neglect them. Except for certain distributed circuits embodied in microwave applications such as waveguides or parabolic dish antennas, we will include many microwave techniques in our treatment of RF design. At the opposite end of the spectrum, below the *high-frequency* (HF) band (3–30 MHz), the AM radio band extends from 650 to 1,650 kHz. At these relatively low frequencies the effects of parasitic inductance and capacitance on circuit designs are minimal. This marks the lower boundary at which we still need to apply RF techniques.

We suggest, however, that RF is less defined by a frequency demarcation than it is by its characteristics. In [1], M. Gupta relates the amusing story of being caught off-guard by his son, and later by his university students, when asked what "RF" implies. At first, he went through the usual historical definitions but found them all lacking. For example, RF circuits are normally tuned circuits, and by definition, narrowband when compared to the carrier frequency. Literally, they are "radio frequency" circuits. They occupy the *radio* bands from HF through UHF; they are frequently used for carrier signals rather than baseband; and they tend to be small in size, comparable with a wavelength of the carrier frequency, and thus distributed in nature. For each of these aspects, however, there are numerous exceptions, and the historical basis of the terms is inexact. Each definition proved too restrictive to ultimately convey what an RF circuit designer actually works on.

Although there is no doubt that while RF circuit design is a subset of analog circuit design, the RF designer needs to be mindful of several additional distinguishing characteristics. These can then provide a more useful basis for characterizing when RF techniques are needed. Gupta suggested an RF circuit or system is one where:

- The *phase shift* of the signal is significant over the extent of the component, because its size is comparable with a wavelength.
- The *reactances* of the circuit must be accounted for, particularly those associated with the parasitics of active devices.
- Circuit *losses* cause degradation of Q, reduction of frequency selectivity, and thermal noise.
- *Noise*, especially arising from within the circuit, can be significant and its effect frequently needs to be modeled.
- Electromagnetic *radiation* and capacitive *coupling* among elements cause unintentional loss and may also significantly alter the performance of the circuit.
- *Reflections* between elements occur because circuit size is of the order of a wavelength. Circuit design requires special treatment to ensure reflections do not cause loss of gain, power, or failure of components.
- *Nonlinearity,* which causes distortion and unwanted frequency components, can be undesired but at the same time is an essential part of circuit operation, as in mixing or local oscillators.

Consequently, RF circuit design can be defined by a set of techniques that account for these effects. Conversely, even if the operating frequency is within the defined RF frequency range, integrated circuits are much smaller than a wavelength and reflections are not always an issue, so standard analog techniques can still be appropriate. RF integrated circuits may not require certain RF techniques such as *S*-parameters or impedance matching, since existing design techniques like *Y*- or *Z*-parameters are still appropriate.

Depending on the application, RF and microwave techniques can be used from near dc up to optical frequencies. Microwave circuit design is not too dissimilar, except that the transmission line almost exclusively replaces lumped elements like the resistor, capacitor, and inductor, except perhaps in device models. In this text then, we present a range of techniques that help the designer to account for such RF effects, and to comfortably accommodate them.

1.2 Circuits and systems

Most RF communication systems can be represented by the block diagram shown in Figure 1.1. The application starts with the need to convey some information to a remote location. To do so the information is used to modulate a higher frequency which may itself then be mixed with a still higher signal used as a *carrier*. The carrier signal with its information modulation is radiated from the transmit antenna to the receive antenna. This radio path may be only a few feet or thousands of miles, and in the case of spacecraft, as much as a billion or more miles.

As can be seen in the figure, upon reception the frequency translation process is reversed. The final IF signal is demodulated and the information extracted as the best estimate of the sent information. The RF power level ratio between transmit and receive may differ by 10 to 15 orders of magnitude. The frequency difference between the information content and the carrier frequency is generally less, but still several orders of magnitude. Clearly, for such systems, measurements such as power, frequency, and information throughput will be important defining system parameters.

1.2.1 System specification

We consider a system to be a collection of components, each built up of a circuit or a number of circuits. The system is usually specified by a top-level system specification that characterizes its desired performance and ensures interoperability between systems from different manufacturers.

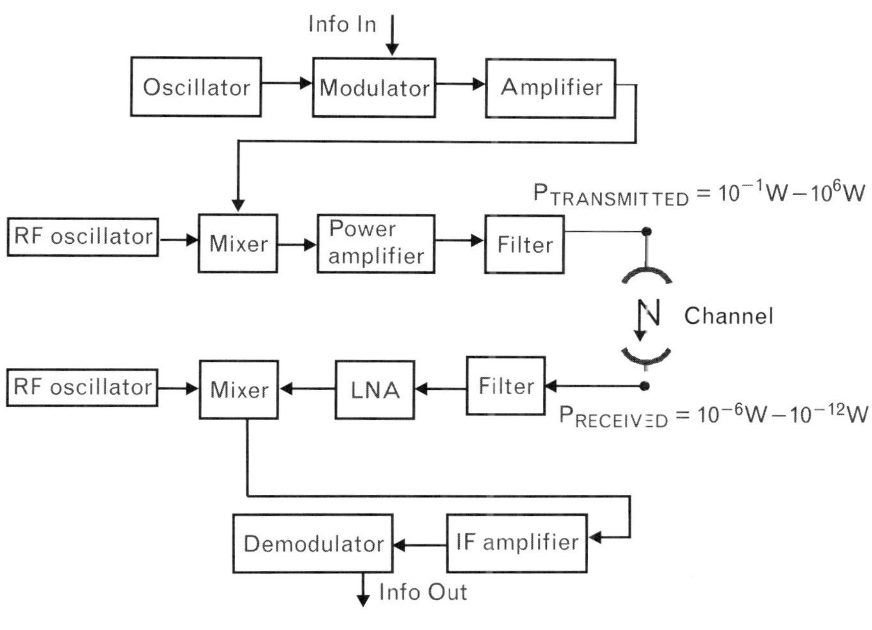

FIGURE 1.1
A representative block diagram for an RF communication system.

The design of wireless and other microwave and RF systems begins here. For example, modern digital radio systems, and especially high bit rate, third generation wireless systems, contain specifications on parameters like the byte and channel timing, signaling, and the air-interface with which both base stations and mobile handsets must comply. The air-interface is of most interest to the RF designer, since it typically specifies parameters such as the transmit power, allowed adjacent channel interference, spurious signal levels, and spectral allocation. Understanding how to specify a wireless system will be our starting point in Chapter 3.

1.2.2 System design

From the system specification, we can begin to characterize each component in the system. For an amplifier in the receiver, this might be its noise figure, gain, and linearity; for an amplifier in the transmitter, critical parameters could be its power, efficiency, and linearity. Similarly, for a mixer, there will be requirements on noise figure, conversion gain, and power handling capability that must be met in order for the overall collection of components to meet the system specification.

Proper setting of these important parameters determines whether or not the subsystem, or system, meets the system specification. If not, either the topology of the system or the component specifications must be changed for the overall system design to conform with the requirements. A behavioral simulator can be used to model system performance. In this type of simulator, the interconnected components are each modeled as a black box. The input-output behavior of each component is described by a functional specification for parameters such as its power or gain compression point.

Once the required component performance has been established, we can begin to drill down to a more detailed level and begin the task of circuit design.

1.2.3 Circuit design

The bulk of this book will focus on the design of components such as low-noise and power amplifiers, oscillators, and mixers that will meet given specifications. We make extensive use of linear and nonlinear *computer-aided design* (CAD) tools to do this. We will develop an understanding of such models and tools to enable us to design and analyze RF components, and in turn, RF systems.

Accurate modeling requires a detailed knowledge of:

- The devices used in the component (i.e., device models);

- The (linear) components used to match the device or to generate the required behavior from the device (i.e., a circuit model).

These device and circuit models can be used in a circuit simulator to simulate the input-output behavior of a component at a variety of frequencies and at a variety of input power levels. Such simulators can predict whether a given device/circuit combination will meet the required power, gain, frequency, intermodulation, and other specifications. The ability to predict such parameters *quantitatively* is a necessary design capability since many of the specifications for wireless RF or microwave systems can be met only by careful control of the circuit layout, parasitics, matching circuits, and losses, which can all be modeled. However, a *qualitative* understanding is equally important since this can help explain why circuits do not always work the way in which they are intended. We will study how to predict the performance of circuits well before they are committed to a circuit simulator for optimization.

1.3 Wireless

Radio transmission is nearly 100 years old, and the term *wireless* was originally used by Marconi. Yet we see it applied today as if it were a contemporary invention, as in *"modern" wireless systems*. To some people, a wireless is an old, polished wood case containing valves that gathered dust in the corner of an old house when people would gather around it to tune in to an early AM or SW broadcast. Yet within the last decade we have experienced an explosive growth in information transmission, especially for communications and control, largely because of the perfection of the integrated circuit. This has enabled the low-cost production of cellular telephones, garage door openers, television and radio circuits and their remote controllers, pagers, and numerous other devices—a veritable grocery list of radio applications. Many of these applications were previously either served exclusively by wired control or, as with the remote garage door opener, were altogether impractical until implemented via radio link. These new devices were termed *wire-less* to distinguish them from the old way of doing things, and the terminology has stuck.

The available electromagnetic or *radio* spectrum depicted in Figure 1.2 used by these devices has changed little. The bands from about 1 MHz to 30 GHz comprise nearly all of the available bandwidth for wireless devices, neglecting the infrared used for numerous television and VCR controllers. However, way beyond anything Marconi may have dreamed of, the electromagnetic spectrum has now become such a limited and scarce resource that telecommunications companies are willing to pay massive amounts of money in license fees to use it.

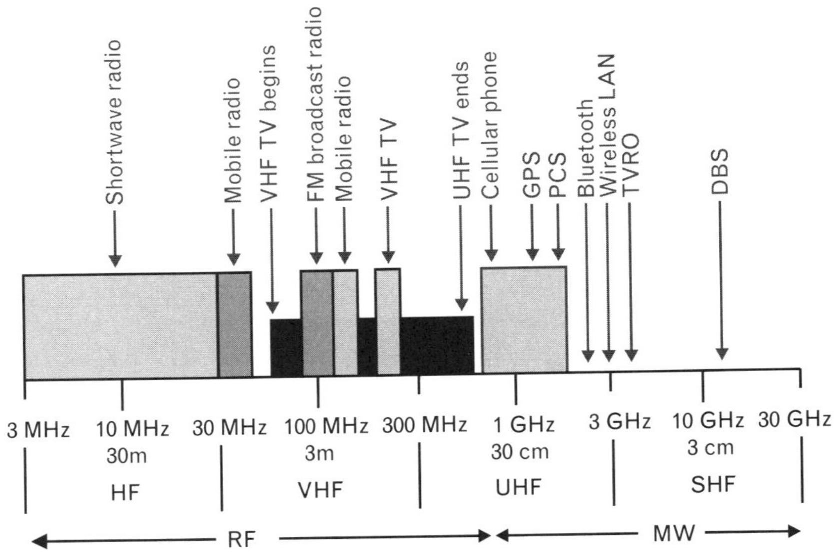

FIGURE 1.2 *The electromagnetic spectrum employed for RF and microwave applications.*

The RF portion of this spectrum is the "sweet spot" of radio communications, providing good radio transmission properties with the least engineering effort and device resources. Circuits are easier to model and construct. Integrated circuits and transistors function more efficiently at RF than at higher frequencies. Antennas are simple and effective. It is no surprise, therefore, that this part of the spectrum is the most utilized.

As a result, numerous applications have experienced crowding. For example, cellular telephones began operation near 800 MHz and have moved to extended bands around 1,900 MHz. Wireless LANs, Bluetooth, and cordless phones operate at even higher frequencies, up to 5,000 MHz. We can expect further reallocations within the wireless bands as newer applications are judged to be more important than existing ones. As a result of this spectral crowding, emphasis is on pushing the boundaries of signal power and coding techniques to maximize information throughput while occupying less spectrum. What will not change is the importance of this RF band, home of wireless engineering.

In this text, it does not matter whether or not *wireless* systems are actually used for *wire-less* propagation. The techniques you will learn are useful for any system in which we observe RF effects.

1.4 Conclusion

The *Practical* of the book's title means built with tools you can use and easily obtain, illustrated with examples you can recreate and use as design templates, and circuits that you can build. This is backed up with theory that is kept simple enough so you actually believe it is necessary, see its value, and feel that it is appropriate.

There you have it—*Practical RF Circuit Design for Modern Wireless Systems*. Our first step is to start at the top, with systems, and then to work our way down to the detail of circuit design. We hope you enjoy the journey as you learn the practical RF techniques to do so.

Reference

[1] Gupta, M., "What Is RF?" *IEEE Microwaves Magazine*, December 2001, pp. 12–16.

CHAPTER 2

RF circuit fundamentals

2.1 Introduction

In this chapter we review some basic, yet important definitions that you probably had (or should have had) in undergraduate engineering courses. In addition, many practical considerations that are generally not covered in college are also included. Even if you are an experienced engineer, consider scanning through the review material so that you have a common knowledge base before progressing to more in-depth topics and their applications.

Since many of the students attending our continuing education courses come with a digital background and have not used complex numbers for some time, we have added a brief review on that subject. Although we do not expect you to do most of the complex algebra manually, an understanding of complex numbers and vectors is essential to follow the material.

2.2 The decibel scale

In Chapter 1 we examined the block diagram of a typical RF communication system. At the transmitter, signal levels may be in the watt, kilowatt (10^3W), or in some special cases even in the megawatt (10^6W) range. At the receiver, the signal incoming to the antenna may be only a few picowatts (10^{-12}W). The power ratio of the transmit/receive signals may be 1,000,000,000,000,000:1 or greater. This can be very difficult to keep straight in one's head. We need to use a better method to squeeze these large ratios into more manageable forms.

Rather than deal with such wide power ratios, we can use their 10 based logarithms. The *Bel* scale was originally defined as the logarithm of the base 10 of the power ratio.[1] One Bel is a factor of 10, two Bels a factor of 100, and so on. Using Bels simplifies the expression of large ratios. For

1. A lesser-used similar unit based on natural log is called Neper = 8.686 dB.

example, the large ratio shown in the previous paragraph is equal to 15 Bels, or the exponent of 10^{15}.

As useful as it is, the Bel scale has too small steps and we generally use a quantity that is 10 times smaller than the Bel, called the deciBel (dB). To compute decibel we take 10 times the power ratio's logarithm. Accordingly, 10 dB is a power ratio of 10; 20 dB is a ratio of 100; and so on. For smaller ratio increments, 1 dB is a power ratio of about 1.26, while 3 dB is a ratio of about 2. Thus, the ratio of two power levels P_2 and P_1 is defined in the decibel scale as

$$\mathrm{dB}(\mathrm{Ratio}) = 10 \log \frac{P_2}{P_1} \tag{2.1}$$

To translate a power ratio to the decibel scale, we need only count the exponents of 10 in the ratio and multiply by 10. Thus, if in a test system an attenuator reduces 100W (10^2W) to 10 mW (10^{-2}W), the power ratio converted to decibels is

$$\mathrm{dB}(\mathrm{Ratio}) = 10 \log \frac{10^{-2}\ \mathrm{W}}{10^{2}\ \mathrm{W}} = 10(-4)\,\mathrm{dB} = -40\,\mathrm{dB}$$

Since the signal is smaller at the output, we understand the minus sign is implicit when we say the *loss* is 40 dB. This implicit sign usually causes no problems in human comprehension, but keep this in mind when using RF simulators *wherein loss in decibels must be expressed as a quantity with a negative sign, referring to negative gain.*

Filter designers generally specify desired loss with positive numbers in their computations, and they get erroneous results from RF simulators that expect negative gain for input. Back in the 1970s when I, Les Besser, sold computing time with the Compact[2] software program, one of my customers called after long sessions of unsuccessful optimizations. He told me that he started with an ideal (lossless) transmission line filter, selecting element values from filter tables. Analyzing the ideal circuit with electrical parameters gave him the expected results. Next, he converted to a physical layout with Compact's conversion subprogram. The physical circuit, of course, gave different results, due to transmission line discontinuities and losses. Optimizing the circuit made the performance much worse—in fact, the optimized circuit did not behave like a filter any more.

Without going through all the details, the solution to his problem was simple. In the stopband he wanted *40-dB loss, but he did not use the negative sign to indicate loss.* The program, interpreting his request as 40-dB gain,

2. Compact (Computerized Optimization of Microwave Passive and Active CircuiTs) was first used through commercial timeshare systems in 1973.

continuously wiped out the initial stopband response. Since it could not get gain from a passive circuit, the next best thing was to minimize the loss by leveling off the frequency response. (When the engineer heard my explanation, his reaction was similar to how I feel these days when Microsoft Windows crashes on me and I lose hours' worth of work because I forgot to save my changes.)

Often, we may refer to the voltages of two signals, v_2 and v_1. *Assuming that they both are referred to the same impedance level,* their power ratio is easily computed in decibels by taking 20 times the logarithm of their ratio; thus,

$$\mathrm{dB}(\mathrm{Ratio}) = 20 \log \frac{v_2}{v_1} \tag{2.2}$$

Since decibel measure is 10 times the logarithm of the power ratio, to convert from decibels back to an absolute power ratio, one simply raises 10 to the exponent (dB/10); thus,

$$\frac{P_2}{P_1} = 10^{\mathrm{dB}/10} \tag{2.3}$$

$$\frac{v_2}{v_1} = 10^{\mathrm{dB}/20} \tag{2.4}$$

Table 2.1 compares representative voltage and power ratios and their decibel values. A convenient calculator scale, relating decibels to power and voltage ratios, is shown in Figure 2.1.

The decibel scale has proved to be extraordinarily useful for representing power ratios and for computing the overall loss or gain of transmission paths. When individual signal losses and gains are expressed in decibels, the overall signal path loss or gain is simply the algebraic sum of the individual decibel values. We will explore this in greater detail shortly.

Gain slope is often expressed by a ratio that shows how much the gain increases or decreases for a given change in frequency. For example, 20-dB/decade (equal to 6-dB/octave) voltage gain roll-off specifies 20 dB gain drop for every decade of frequency increase (equal to 6 dB decrease every doubling of frequency.) To prove the two ratios are the same, we convert the two decibel values to voltage ratios from Table 2.1.

- Voltage ratio "Decade" for 20 dB/decade = 10
- Voltage ratio "Octave" for 6 dB/octave = 2

TABLE 2.1 VARIOUS POWER AND VOLTAGE RATIOS AND THEIR DECIBEL REPRESENTATIONS

Power Ratio			Voltage Ratio	
Ratio	Exponent	Decibel Value	Exponent	Ratio
1	10^{0}	0	10^{0}	1
2	$10^{0.3}$	3c	$10^{0.15}$	1.41
3	$10^{0.477}$	4.77	$10^{0.24}$	1.73
4	$10^{0.6}$	6	$10^{0.3}$	2
10	10^{1}	10	$10^{0.5}$	3.16
50	$10^{1.7}$	17	$10^{0.85}$	7.07
100	10^{2}	20	10^{1}	10
0.5	$10^{-0.3}$	−3	$10^{-0.15}$	0.707
0.25	$10^{-0.6}$	−6	$10^{-0.3}$	0.5
0.1	10^{-1}	−10	$10^{-0.5}$	0.316

Taking the ratios of Decade/Octave, as $10/2 = 5$, we confirm the relationship between decade and octave ranges, 20 dB − 6 dB = 14 dB.

2.2.1 Illustrative example: decibel calculations

Measuring the output power of a high-power amplifier we have a 36-dB attenuator connected between the amplifier and the power meter. When the power meter's reading (P_{MEAS}) is 0.5W, what is the actual power (P_{ACT}) at the output of the amplifier? Assume that all components are matched[3] to 50Ω.

We could use (2.3) to compute the power ratio corresponding to 36 dB attenuation, but instead let us get the answer from Table 2.1. Since the

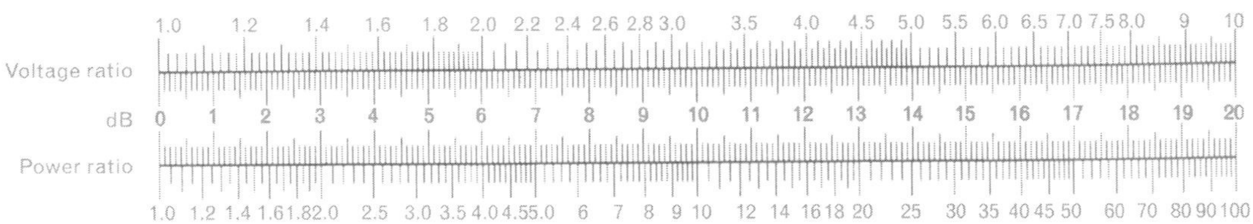

FIGURE 2.1 *Graphical conversion between power or voltage ratios and decibels. Draw a vertical line across the known parameter value and read the desired quantity on the appropriate scale.*

3. In a real-life situation, none of the components are perfectly matched and in addition we also need to compute the mismatch uncertainties.

table does not include 36 dB, we can get the power ratios corresponding to 20, 10, and 6 dB first, and multiply the ratios to get the final answer.

- Splitting decibel values: 20 dB + 10 dB + 6 dB = 36 dB
- Equivalent power ratios: 100 × 10 × 4 = 4,000

The 36 dB is equal to a power ratio of 4,000. Multiplying the P_{MEAS} reading with 4,000 gives us P_{ACT} = 2,000W, or 2 kW of power (better not touch that attenuator).

Alternatively, using the scale of Figure 2.1, we can do the same in two steps. First find the power ratio corresponding to 20 dB and multiply it with the power ratio of 16 dB.

- Splitting decibel values: 20 dB + 16 dB = 36 dB
- Equivalent power ratios: 100 × 40 = 4,000

That, of course, also verifies our result found from the decibel table.

2.2.2 Absolute power level reference

Since decibel values denote ratios, they are dimensionless. If an amplifier has 20-dB gain, we only know that the output signal is 20 dB higher than the input. A filter may have –30-dB gain; that is 30-dB loss. None of these quantities refer to absolute values.

It is useful to set standard reference values for power and voltage specifications. For low-power applications the dBm scale is used where everything is referenced to 0 dBm being equal to 1-mW absolute power (Figure 2.2). For example, if an amplifier is capable of delivering 10-mW output power, that is, 10 times 1 mW. On the dBm scale the output power is 10 dB above the reference power level, or +10 dBm. If power level is given as 1 μW = (1/1,000) W the equivalent dBm value is 30 dB *below* the reference level, or –30 dBm.

For high-power applications it is more convenient to use dBW scale with respect to 1W reference. Conversions between dBW and dBm are easy, since there is a 1,000:1 ratio difference. Thus, 30 dBW (1,000W) is 60 dBm (also 1,000W).

To find dBm and dBW values,

$$\text{dBm} = 10\log\left(P_{\text{mW}}\right) \tag{2.5}$$

$$\text{dBW} = 10\log\left(P_{\text{W}}\right) \tag{2.6}$$

and to convert dBm and dBW values back to absolute power ratings,

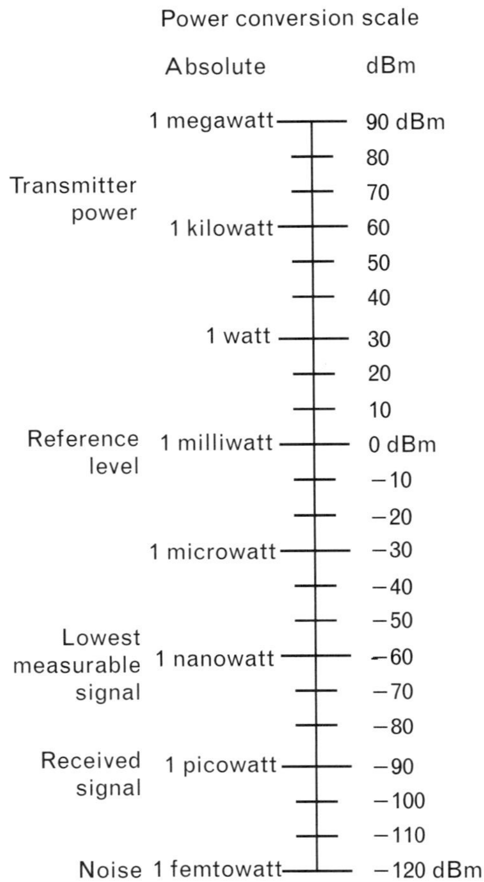

FIGURE 2.2
The dBm scale shows absolute power levels, referenced to 1 mW equal to 0 dBm. Absolute power levels less than 1 mW are expressed in negative dBm values.

$$P_{\mathrm{mW}} = 10^{\mathrm{dBm}/10} \tag{2.7}$$

$$P_{\mathrm{W}} = 10^{\mathrm{dBW}/10} \tag{2.8}$$

Communication receiver sensitivities are frequently referenced to 1V (dBV), or 1 mV (dBmV). Another commonly used term is dBc, which references readings to modulated signal carrier level.

2.2.3 Illustrative exercise: power conversions

The power output (P_T) transmitted from a cellular phone is +30 dBm. At the receiver the signal power (P_R) is down to 5 pW. What is the attenuation A_{dB} of the signal path between the transmitter and receiver?

Solution 1

Find the absolute power ratio first and convert it to decibels.

1. Convert +30 dBm to absolute power using (2.7):

$$P_T = 10^{\text{dBm}/10} = 10^{30/10} = 1{,}000 \text{ mW}$$

2. Compute the attenuation power ratio:
$A = P_R/P_T = 5 \text{ pW} / 1{,}000 \text{ mW} = 5(10^{-9}) \text{ mW}/10^3 \text{ mW} = 5(10^{-12})$

3. Convert the attenuation power ratio to decibels:
$$A_{\text{dB}} = 10\log\left[5\left(10^{-12}\right)\right] = 10\log 5 + 10\log 10^{-12}$$
$$= (7 - 120) \text{ dB} = -113 \text{ dB}$$

Solution 2

Express the powers in dBm first and then find their difference in decibels.

1. Convert $P_R = 5$ pW to dBm with (2.5):
$$P_R(\text{dBm}) = 10\log P_{R\text{mW}} = 10\log\left[5\left(10^{-9}\right)\right] = 10\log 5 + 10\log 10^{-9} = (7 - 90) = -83 \text{ dBm}$$

2. Subtract the transmitted power level from the received power level:
$$A_{\text{dB}} = P_R(\text{dBm}) - P_T(\text{dBm})$$
$$= -83 \text{ dBm} - (+30 \text{ dBm}) = -113 \text{ dB}$$

Notice that subtracting two quantities in dBm corresponds to taking their ratio, hence their dimensions (milliwatts) cancel and we are left with their ratio in decibels, which is dimensionless. Also note that Solution 2 required less computational effort. When the result sought will be in decibels, converting to decibel format early in the solution generally requires less computational effort.

2.3 Complex number review

We shall use complex numbers in the linear ac analysis employed throughout this text to analyze circuits. A brief review of complex number algebra is presented here.

A complex quantity (number) consists of two parts. It can be thought of as a vector having magnitude and direction, called the *polar form*, or in *rectangular form* as a location in the x–y coordinate plane having *real* (x-axis) and *imaginary* (y-axis) components.

In rectangular coordinates the complex quantity consists of *real* and *imaginary* parts. The term *imaginary* is a mathematical definition for the

y-axis component to distinguish it from a customary *real* concept. For electrical engineering applications, it is more insightful to think of the imaginary part as being in phase quadrature with the real part. A complex number,

$$A = 2 + j1$$

specifies 2 units out along the horizontal (real) axis and 1 unit up along the vertical (imaginary) axis. The operator of the imaginary number j is the square root of minus one, or $j = \sqrt{-1}$. For example, $A = 2 + j1$ really means $A = 2 + \left(\sqrt{-1}\right)1$ as shown in Figure 2.3.

In polar coordinates a complex number is expressed as a *magnitude* (M) and *angle* (θ). Thus the same vector $A = 2 + j1$ becomes $A = M \angle \theta° = 2.24 \angle 26.6°$ in polar representation.

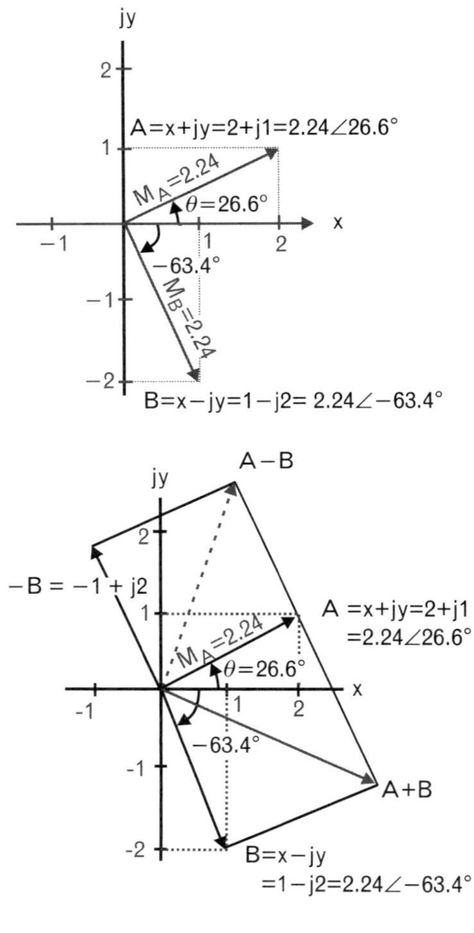

FIGURE 2.3 *Complex numbers expressed in both rectangular and polar form. Right-side plot illustrates complex number manipulation through vector addition and subtraction.*

A single polar form converts the phase angle, θ, to exponential form. The same complex quantity, A, becomes

$$A = Me^{-j\theta}$$

Most of the measured and published data for RF components comes in complex form. Test equipment generally defaults data to polar coordinates, although the option is usually there to have rectangular form also.

It is useful, especially in RF/MW engineering, to define the complex conjugate of a complex number. The complex conjugate of a number has the *same real part but the negative of the imaginary part*. Thus, if $A = x + jy$, then $A^* = x - jy$, where the * superscript denotes the complex conjugate. If the complex quantity is expressed in polar coordinates, its complex conjugate has the *same magnitude and we negate the phase angle*. For $A = 2.24 \angle 26.6°$, $A^* = 2.24 \angle -26.6°$.

Vector addition and subtraction of complex numbers are more conveniently performed in rectangular coordinates, by *adding* or *subtracting* their respective real and imaginary parts.

In Figure 2.3 we show $A = 2 + j1$ and $B = 1 - j2$:

- The sum: $A + B = (2 + 1) + j(1 - 2) = 3 - j1$;
- The difference: $A - B = (2 - 1) + j(1 - [-2]) = 1 + j3$.

For vector multiplication and division use the polar form. For multiplication, the magnitudes (M) are multiplied together while the angles (θ) are added together. For division, take the ratio of the magnitudes and the difference of the angles:

$$\textit{Multiplying:} \quad AB = \left(2.24\angle 26.6°\right)/\left(2.24\angle -63.4°\right) = 5\angle -36.8°$$

$$\textit{Dividing:} \quad A/B = \left(2.24\angle 26.6°\right)/\left(2.24\angle -63.4°\right)$$
$$= \left(2.24/2.24\right)\angle\left(2.26 - [-63.4]\right)° = 1\angle 90°$$

2.4 Normalization

The simple operation of normalization (dividing by a reference quantity) takes on considerable importance in RF and microwave engineering. In the design of filters it is found that once a design for, say, a 50-Ω characteristic impedance, 1-GHz filter is completed, the design of a similar performance 75Ω, 2-GHz filter can be scaled directly from it. To do so one first *normalizes* the frequency response and impedance level of the known design (usually to 1Ω and 1 Hz). Next, one *unnormalizes* the result at the new conditions to yield the circuit parameters for the new application frequency and characteristic impedance. In another application, the transformation of impedances through a transmission line is easily computed on the Smith chart once the load impedance has been normalized to the line's characteristic impedance.

Normalized parameters, denoted throughout this text by lowercase characters, are computed by dividing the original parameter by the reference quantity. Unnormalized parameters and reference values are shown by capital letters. Exceptions are: frequency symbol, *f*, because its generally used form is the lowercase letter, and ac current and voltage *i* and *v*. Matrix quantities will always be shown with capital letters.

If the reference quantity:	$Z_0 = 50\Omega$	$Z_0 = 75\Omega$
and the unnormalized parameter:	$Z_L = 150\Omega$	$Z_L = 150\Omega$
then, the normalized parameter:	$z_L = 3$	$z_L = 2$

To unnormalize a normalized parameter, multiply by the reference quantity.

If the normalized parameter:	$z_L = 2 + j1$	$z_L = 2 + j1$
the reference quantity:	$Z_0 = 50\Omega$	$Z_0 = 75\Omega$
the unnormalized parameter:		

$$Z_L = z_L Z_0 = (2 + j1)50\Omega = (100 + j50)\Omega \qquad Z_L = z_L Z_0 = (2 + j1)75\Omega = (150 + j75)\Omega$$

2.5 R-L-C voltage-current relationships

Throughout the next five sections we describe the characteristics of two-terminal passive networks, and we will make the customary assumptions about them. Specifically, unless noted otherwise, we shall assume they are:

1. *Lumped* (their dimensions are small compared to the wavelength of the operating frequency);
2. *Linear* (they satisfy Ohm's law in that the current through the component is proportional to the applied voltage);
3. *Finite* (their component *R*, *L*, and *C* values are described by finite values);
4. *Passive* (there is no possibility for gain in the network);

5. *Bilateral* (current flow through the component is accomplished with the same facility in both directions).

Networks formed by components that satisfy these conditions are called lumped, linear, finite, passive, and bilateral networks, and we focus on these circuits first. Later, while *de-embedding* test-fixture–related parameters, and modeling active circuits (Chapter 2), we will refer to *negative elements* also.

In basic ac analysis there are only three kinds of passive two-terminal circuit elements: resistors, inductors, and capacitors (*R-L-C*). The entire theory of steady-state ac analysis rests on the fact that when a sinusoidal voltage is impressed across one of these three ideal[4] elements, the resulting current through it is also sinusoidal in form (Figure 2.4) and either:

1. Has a phase that is the same as that of the applied voltage for a resistor;
2. Lags by 90° for an inductor;
3. Leads by 90° for a capacitor.

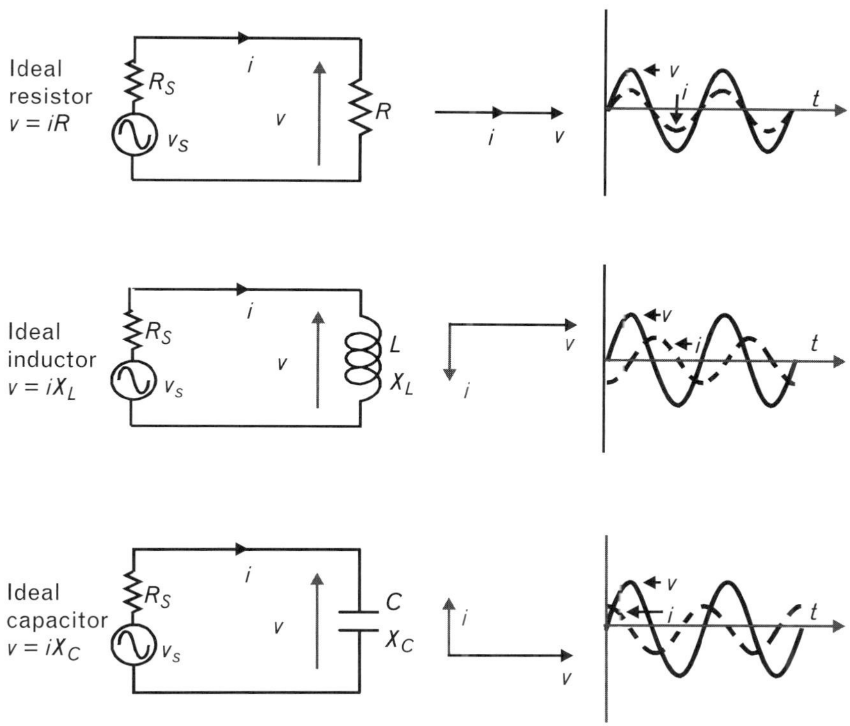

FIGURE 2.4
Voltage-current relationships of ideal R-L-C components. Impedance of the resistor is a real number because the voltage and current are in phase. In the reactive elements the voltage and current are 90° apart.

4. In our books an "ideal" component does not have any unwanted capacitance, inductance, loss, radiation, or coupling to unrelated elements. Ideal components follow their strict mathematical definitions.

It turns out that the solution for the amplitude and phase of the current response is greatly facilitated by the use of complex numbers. Ohm's law can be written as

$$i = v/Z$$

where i, v, and Z are complex quantities representing the current response, the applied voltage excitation, and the total impedance of the circuit. The impedance of an ideal inductance or capacitance is called reactance, X_L and X_C, because these elements react to an applied voltage with a current that is always 90° out of phase with the applied voltage, hence they do not dissipate any power. The impedance of an ideal resistor is called resistance, with the voltage across and current through that are always in phase. Out of the three elements, *only resistances can dissipate power*. Reactance expressions are given in Section 2.8.

The above conditions describe steady-state operation, where the voltage excitation to the component is assumed to have been applied long enough ago (theoretically an infinitely long time ago) and all transients resulting from its application have died out. This restriction results in an enormous simplification to the circuit analysis.

In our analysis we also assume that the network is linear; that is, the response at the output is directly proportional to magnitude of the single frequency excitation. Linearity also requires that the application of multiple frequencies results in the superposition (simple addition) of the responses to each applied sinusoid.

2.6 Complex impedance and admittance systems

Passive circuits may be described by impedances or admittances with positive real parts. When described in the rectangular impedance coordinate system their impedances (Z) or admittances (Y) all lie in the right-half plane. Mathematically,

$$\mathrm{Re}(Z) \geq 0 \text{ and } \mathrm{Re}(Y) \geq 0$$

Although at RF we are not able to measure voltages and currents, components are frequently defined by their terminal impedances or admittances in the complex rectangular coordinate system.

Complex impedance is the vector sum of resistance and reactance shown in Figure 2.5.

$$\text{Impedance} = \text{Resistance} \pm j\text{Reactance}$$

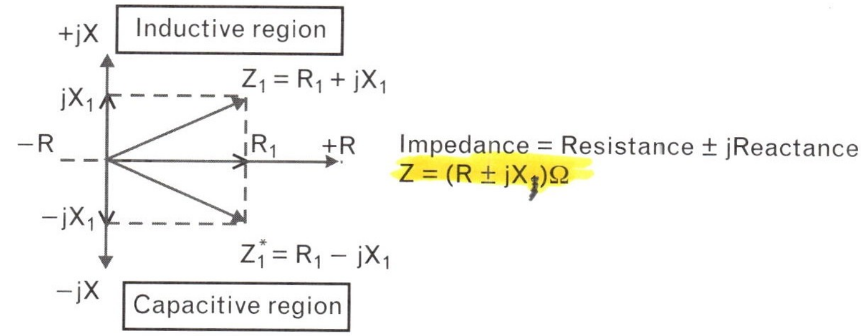

FIGURE 2.5 *Impedance plotted in the complex plane. A short-circuit is at zero impedance and an open-circuit is at infinity. The top half is inductive while the lower half is capacitive. Impedances of passive circuits always lie in the right-hand plane.*

$$Z = (R \pm jX)\Omega \tag{2.9}$$

Series connections of *lumped* two-terminal components are handled conveniently in the rectangular impedance system [1]. The total impedance of the series connection is equal to the sum of the individual impedances. The same can be extended to two-ports characterized by their impedance parameters.

On the other hand, a parallel combination of elements is better handled in a complex admittance system (Figure 2.6). Admittance, $Y = 1/Z$, measured in Siemens, is the reciprocal of impedance.

$$\text{Admittance} = \text{Conductance} \pm j\text{Ssceptance}$$

$$Y = (G \pm jB)\text{Siemens} \tag{2.10}$$

Conductance of an ideal resistor, R, is equal to the reciprocal of resistance, $G = 1/R$. For an ideal inductive reactance, jX, the corresponding susceptance $1/jX = -jB$. However, if given a complex impedance, $Z = R +$

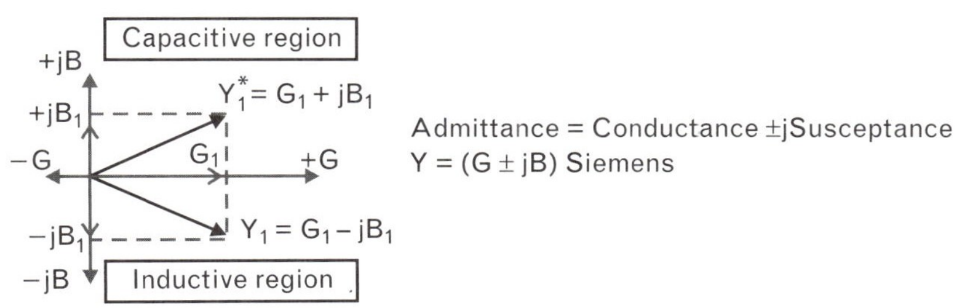

FIGURE 2.6 *In the complex admittance plane an open-circuit has zero admittance and a short-circuit is at infinity. The top half is capacitive and inductive susceptances are on the lower half.*

jX, you must use complex algebra to find the equivalent parallel admittance, Y,

$$\begin{aligned} Y &= \frac{1}{Z} = \frac{1}{R + jX} = \frac{1}{R + jX}\frac{R - jX}{R - jX} = \\ &= \frac{R}{R^2 + X^2} - \frac{jX}{R^2 + X^2} \\ &= G - jB \end{aligned} \tag{2.11}$$

Similarly, converting admittance $Y = G + jB$ to impedance yields

$$\begin{aligned} Z &= \frac{1}{Y} = \frac{1}{G + jB} = \frac{1}{G + jB}\frac{G - jB}{G - jB} = \\ &= \frac{G}{G^2 + B^2} - \frac{jB}{G^2 + B^2} \\ &= R - jX \end{aligned} \tag{2.12}$$

Notice that in both equations we multiplied numerator and denominator by the complex conjugate of the denominator to remove the imaginary numbers from the denominator. This technique, called *rationalization,* is a convenient means of evaluating the reciprocal of a complex number. Later we will use this technique as a basis for impedance matching.

Equations (2.11) and (2.12) show the change of sign of the imaginary part when converting between impedance and admittance. Keep in mind however that the nature of the complex quantity does not change due to a conversion. Inductive reactance, $+jX$, remains inductive after being converted to susceptance, even though the sign of the imaginary operator is changed to $-jB$.

2.7 Unloaded and loaded Q definitions

Losses in practical components affect their RF behavior. We can define various forms of *quality factors* that relate the energy loss in reactive components to the energy they store during a signal cycle [2]. The unloaded quality-factor, Q_U, of a physical (nonideal) reactive component is defined as

$$Q_{\text{UNLOADED}} = Q_U = \frac{\text{Energy stored in the component}}{\text{Energy dissipated in the component}} \tag{2.13}$$

In the next section we define the frequency-dependent Q in terms of measurable circuit parameters and show how they affect the loss and bandwidth of RF circuits. Ideal reactors (capacitors and inductors) have no losses—they do not dissipate energy, therefore they have infinite Qs.

When the component is used in a circuit, a loaded Q definition is useful:

$$Q_{\text{LOADED}} = Q_L = \frac{\text{Energy stored in the component}}{\text{Total energy dissipated in the component and external circuit}} \quad (2.14)$$

When a reactive component is terminated resistively, $Q_L < Q_U$ since the terminations increase the dissipated energy.

In L-C resonators energy bounces between the two components at resonance; therefore, we only compute the energy stored in one of them (see Chapters 5 and 7). Using the loaded Q-factor at resonant frequency we can compute the 3-dB bandwidth of the resonator's frequency response, as f_0/Q_L, where f_0 is the resonant center frequency. Another type of Q definition, Q_N, called the nodal-Q, will also be covered in Chapter 5.

In Volume II, Chapter 6, we apply the loaded Q concept to investigate its effect on oscillator performance.

2.8 Complex series impedance of RF components

At RF and microwave frequencies reactance calculations are more convenient if we specify the frequency in gigahertz (10^9 Hz), inductance in nanohenries (10^{-9} H), and capacitance in picofarads (10^{-12} F). Inductive reactance is

$$X_L = 2\pi f_{\text{Hz}} L_H = 6.283 f_{\text{GHz}} L_{\text{nH}} \quad (2.15)$$

If the reactance is normalized to 50Ω,

$$x_L = 0.1257 f_{\text{GHz}} L_{\text{nH}} \quad (2.16)$$

Unnormalized and normalized capacitive reactances are given as

$$X_C \approx \frac{1}{2\pi f_{\text{Hz}} C_F} \approx \frac{159}{f_{\text{GHz}} C_{\text{pF}}} \quad (2.17)$$

$$x_C = \frac{3.183}{f_{\text{GHz}} C_{\text{pF}}} \quad (2.18)$$

Reactance of an inductor or capacitor is frequency dependent. It is a very good idea to memorize the unnormalized value of reactance of an inductor and that of a capacitor at some common frequency, such as 1 GHz. Then the reactance of either of these elements at any other frequency and component value can be estimated quickly by a simple scaling. For example, using (2.15) and (2.17),

- 1-nH inductance at 1 GHz has inductive reactance $jX_L \approx j6.28\Omega$;
- 1-pF capacitance at 1 GHz has capacitive reactance $-jX_c \approx -j159\Omega$.

If you remember these results at 1 GHz, you can easily scale the reactance value of an inductor or capacitor at any other frequency. Notice that the scaling requires remembering which way the reactance goes with the parameter value and frequency:

- Directly proportional to L and f for inductors;
- Inversely proportional to C and f for capacitors.

With a little practice you can estimate mentally the value of reactance of any L or C at any frequency.

The terminal impedance of a two-terminal inductor is a complex quantity that may be reduced to a series-connected resistance and reactance, R_S and X_S (Figure 2.7). The unloaded quality factor of the component is computed from the ratio of the stored and dissipated energy (2.13). Since the reactive portion stores and resistance dissipates energy, we can also compute the Q-factor[5] of the component shown in series equivalent form from its impedance parameters,

$$Q_S = \frac{X_S}{R_S} \tag{2.19}$$

For ideal (lossless) components, the terminal impedance is equal to the series reactance and the unloaded quality factor is infinite. Low series resistance leads to high Q.

2.9 Complex parallel admittance of RF components

The terminal admittance of a lossy inductive or capacitive element is a complex quantity that may also be represented as a parallel-connected

5. When we just show Q, we refer to unloaded Q-factor. QS is used for series circuits and QP is used for parallel forms.

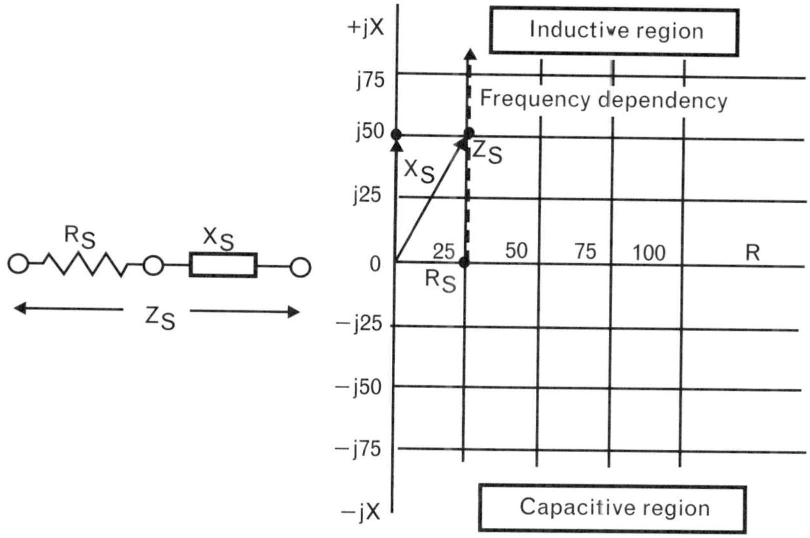

FIGURE 2.7
The terminal impedance of a lossy inductor, Z_S, shows $j50\Omega$ inductive reactance in series with 25Ω. A very low Q of 2 at this specific frequency is caused by the large series resistance. At dc, Z is equal to R_S. As the frequency increases, a varying amount of inductive reactance is added to R_S. In real physical circuits R_S may also be frequency dependent.

conductance and susceptance, G_p and B_p (Figure 2.8). This two-element form may not be a broadband equivalent circuit of the component, but it describes the admittance at a single frequency. Use of admittance parameters is recommended for parallel circuit component connections, because the total admittance is the sum of the individual admittances. The related mathematics is very simple, only additions and subtractions are required.

The conductance is computed as $G_p = 1/R_p$. Using a reference impedance of 50Ω, the unnormalized and normalized inductive and capacitive susceptances are computed as

$$B_L = \frac{1}{2\pi f_{\text{GHz}} L_{\text{nH}}} = \frac{0.159}{f_{\text{GHz}} L_{\text{nH}}} \tag{2.20}$$

$$b_L = \frac{7.96}{f_{\text{GHz}} L_{\text{nH}}} \tag{2.21}$$

$$B_C = \frac{f_{\text{GHz}} C_{\text{pF}}}{159} = 0.006283 f_{\text{GHz}} C_{\text{pF}} \tag{2.22}$$

$$b_C = 0.314 f_{\text{GHz}} C_{\text{pF}} \tag{2.23}$$

The unloaded component quality factor of the parallel circuit of Figure 2.8 is computed from the ratio of parallel susceptance to conductance.

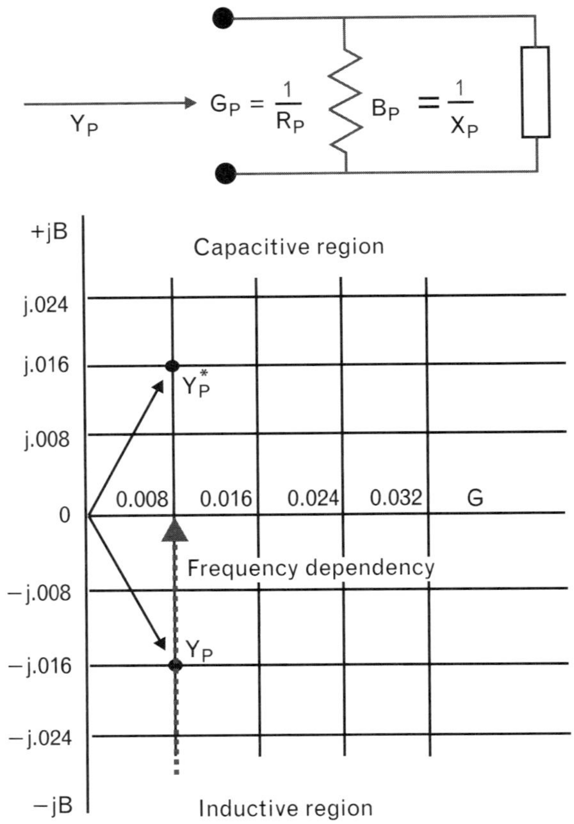

FIGURE 2.8
Admittances of RF components are very convenient for parallel connections. Admittance plots always move upward with the increase of frequency for physical components.

Since most of us relate better to impedance parameters, we can also express Q_p in terms of R_p and X_p.

$$Q_P = \frac{B_P}{G_p} = \frac{1/X_p}{1/R_p} = \frac{R_P}{X_p} \tag{2.24}$$

For ideal (lossless) parallel elements,

$$G_p = 0 \text{ and } R_p = \infty, \text{ therefore } Q_p = \infty$$

Comparing the impedance parameters of (2.19) and (2.24) shows that in the series form *small resistance* leads to high Q. In the parallel form, as we stated above, *large parallel resistance*—little loading effect—gives high Q.

For convenience, the formulas to compute inductance and capacitance when either the reactance or susceptance is given are summarized here. The normalized reactance and susceptance formulas are also included in Table 2.2, since we will use a normalized Smith chart in Chapter 4.

TABLE 2.2 EXPRESSIONS TO COMPUTE INDUCTANCE AND CAPACITANCE FROM NORMALIZED (x AND b) AND UNNORMALIZED (X AND B) REACTANCE AND SUSCEPTANCE (SUMMARIZED IN APPENDIX)

	FROM UNNORMALIZED REACTANCE AND SUSCEPTANCE	FROM NORMALIZED REACTANCE AND SUSCEPTANCE
INDUCTOR VALUE	$L_{nH} = \dfrac{0.159X_L}{f_{GHz}} = \dfrac{0.159}{f_{GHz}B_L}$	$L_{nH} = \dfrac{7.96x_L}{f_{GHz}} = \dfrac{7.96}{f_{GHz}b_L}$
CAPACITOR VALUE	$C_{pF} = \dfrac{159}{f_{GHz}X_C} = \dfrac{159B_C}{f_{GHz}}$	$C_{pF} = \dfrac{3.183}{f_{GHz}x_C} = \dfrac{3.183b_C}{f_{GHz}}$

Note: The 50-Ω reference is used for all normalizations.

2.9.1 Illustrative exercise: computing elements from admittance specifications

Compute the simplest parallel lumped element form for the admittance shown in Figure 2.8. Assume the data was given at $f = 1$ GHz.

The admittance shown on the plot is

$$Y_P = G_P - jB_P = (0.008 - j0.016)S$$

The parallel resistance is equal to the reciprocal of admittance, $1/G_p$. Since the susceptance has a $-j$ operator, it represents an inductor. Using the expressions of Table 2.2, we compute the inductor value representing the imaginary part:

$$R_P = \frac{1}{G_P} = \frac{1}{0.008} = 125\Omega$$

$$L_P = \frac{0.159}{f_{GHz}B_P} = \frac{0.159}{1(0.016)} = 9.94 \text{ nH}$$

The true equivalent circuit may include many components, and our solution is the simplest one based on single-frequency admittance data.

2.10 Series and parallel L-C resonant circuits

In the two circuits of Figure 2.9, at zero frequency the reactance of an inductor is zero while that of a capacitor is infinite. As frequency increases the inductor's reactance increases while that of the capacitor decreases. Furthermore, the reactances of *Ls* and *Cs* are opposite in sign. It follows

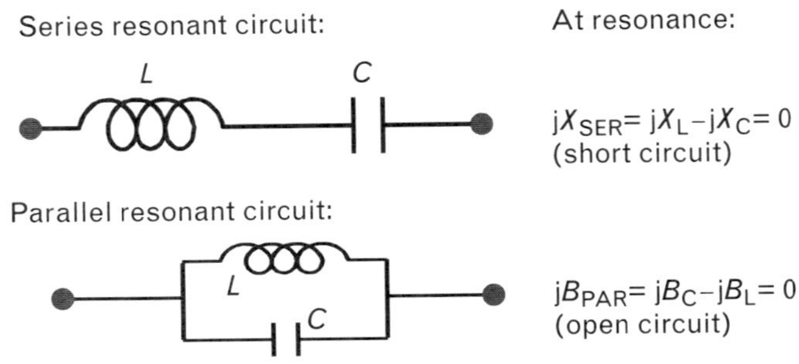

FIGURE 2.9 *Series and parallel combination of a given set of inductor and capacitor resonate at the same frequency, f_R. If the components are ideal, at resonance the series circuit represents a short-circuit while the parallel circuit looks like an open-circuit. For real-life components, a small series resistance is added to the series circuit and a large parallel resistor to the parallel network.*

that at some *resonant frequency*, f_R, the two reactances will have exactly the same magnitudes and, because they are of opposite sign, will cancel when the two elements are in series. That is, at a unique frequency, f_R, the series combination of an L-C circuit has zero reactance. The circuit is said to be *series resonant*. Neglecting any resistance they might have, it presents an ac short circuit at f_R.

If, instead, the same two elements are connected in parallel, as shown in the same figure, their net susceptance is zero at f_R, and the circuit is said to be *parallel resonant*. Again, neglecting resistive effects, a parallel resonant L-C presents an open circuit at f_R. The resonant frequency is only a function of the two component values,

$$f_{R\mathrm{GHz}} = 5.033\sqrt{\frac{1}{L_{\mathrm{nH}}C_{\mathrm{pF}}}} \tag{2.25}$$

It is important to note that both series and parallel resonances occur at the same frequency, where the reactance (or susceptance) magnitudes are equal. The behavior of L-C circuits near resonance has profound implications throughout electrical engineering, and certainly in wireless applications. The L-C resonators form the basis of numerous useful filters. Less useful, the parasitic reactances of practical inductors and capacitors change the behavior of these elements and must be taken into account if designs are to be modeled accurately.

A *reactance chart* (Figure 2.10) proves useful to visualize the frequency dependencies of inductors and capacitors. It also helps us to get a quick estimate of resonant frequencies. Because of its logarithmic scales, reactances follow straight line plots with frequency. We illustrate the use this chart with a simple component.

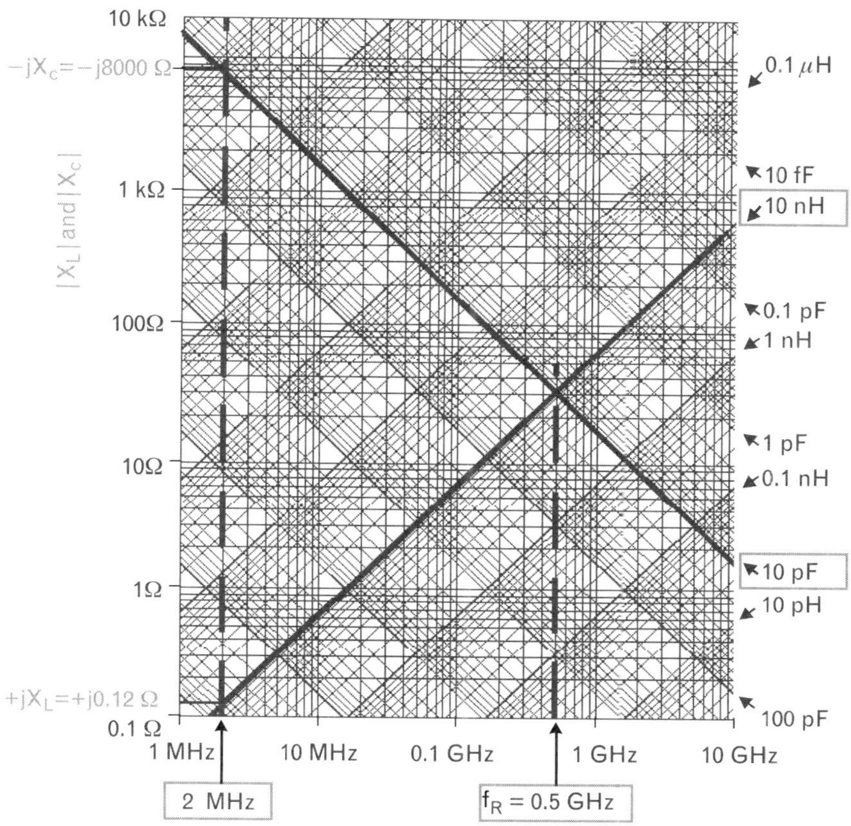

FIGURE 2.10
A reactance chart provides a ready measure of the frequency-dependent reactances of Ls and Cs.

Suppose we have a 10-pF discrete capacitor. Being a practical element, it has wire leads, and these leads offer series inductance—for our example suppose it is 10 nH. Because this reactance was not intended, it is called *parasitic inductance*. If we neglect the losses of the component, we can get a quick idea about the behavior of our capacitor at various frequencies.

The 10-pF capacitor has a reactance (magnitude) of 8 kΩ at 2 MHz, which drops to only 1.6Ω at 10 GHz. The magnitude of the 10-nH inductor's reactance increases from 0.12Ω to 628Ω for the same frequency range. Their magnitudes are equal around 500 MHz, indicating resonance. Clearly, the series circuit is dominated by the capacitor at low frequencies. At 2 MHz, the $+j0.12\Omega$ inductive reactance is completely negligible to the $-j8k\Omega$ capacitive reactance. As we approach 500 MHz, the inductive reactance gets more noticeable and reduces the effective capacitive reactance. The result is, contrary to popular beliefs, an *increase* in the effective capacitance. Above self-resonance, the component is obviously controlled by the inductor at the high frequencies. In Chapter 7 we look at the effects of parasitic inductance and capacitance.

2.11 Series and parallel conversions of lumped R-L-C networks

At any single frequency a given *series equivalent impedance can be converted into its parallel equivalent admittance.* In Section 2.6 we showed the complex arithmetic necessary to convert between impedance and admittance. We now point out that the conversion at a single frequency does not effect the Q. Whether we use the series Q_S calculation with (2.19) or the parallel Q_P form of (2.24) for a given set of components, the results are the same. Because the Q is unchanged in converting between the series impedance and the shunt admittance formats, we just use the symbol Q, instead of Q_S and Q_P. The equality can be used as the basis for a very convenient method to make the conversion mathematically.

To convert the series *R-L* circuit of Table 2.3 to its parallel equivalent at a particular frequency *f*, let us recall the Q of the series circuit from (2.19) and express X_S from it.

$$Q = \frac{X_S}{R_S} \quad \Rightarrow \quad X_S = QR_S$$

The terminal impedance of the series circuit can now be modified as

$$Z_S = R_S + jX_S = R_S + jQR_S = R_S\left(1 + jQ\right)$$

Converting series impedance to parallel admittance

$$Y_P = \frac{1}{Z_S} = \frac{1}{R_S\left(1 + jQ\right)} = \frac{1}{R_S\left(1 + jQ\right)}\frac{\left(1 - jQ\right)}{\left(1 - jQ\right)}$$

which can be simplified to

$$\begin{aligned} Y_P &= \frac{1}{R_S\left(1+Q^2\right)} - j\frac{Q}{R_S\left(1+Q^2\right)} \\ &= \frac{1}{R_S\left(1+Q^2\right)} - j\frac{Q}{\left(X_S/Q\right)\left(1+Q^2\right)} \\ &= G_P - jB_P \\ &= \frac{1}{R_P} - j\frac{1}{X_P} \end{aligned}$$

Equating the real and imaginary parts of the second and fourth lines of our last equation gives us the parallel component values in terms of the series circuit.

$$R_P = R_S\left(1+Q^2\right) \tag{2.26a}$$

$$X_P = X_S\left(\frac{1}{Q^2}+1\right) \tag{2.26b}$$

Multiplying through both sides of (2.26b) by $2\pi f$ gives us the solution for inductances of an R-L network.

$$L_P = L_S\left(\frac{1}{Q^2}+1\right) \tag{2.27}$$

Since capacitive reactance is inversely proportional to capacitance, the Q-function of (2.27) is inverted to give the capacitor value of R-C networks. The resistive part is computed from (2.26), and the capacitance is

$$C_P = C_S\left(\frac{Q^2}{1+Q^2}\right) \tag{2.28}$$

Equations (2.26) through (2.28) hold whether we convert from the series to the parallel equivalent circuits or vice versa. Thus we will refer to either process as a *series-parallel* conversion. The following are some points to remember:

1. The parallel resistance is greater than the series resistance by the factor $(1 + Q^2)$.
2. The parallel reactance is $1/Q$ times the parallel resistance.
3. The sign of the parallel reactance is the same as that of the series reactance. Thus, an R-L series circuit has an R-L parallel equivalent (but the sign of the susceptance is opposite that of the reactance for a given reactor type such as, for example, an inductor).

We have emphasized the series-parallel conversion due to its useful applications. Conversions are required throughout impedance matching steps. In fact, the basis for resistive transformations at a single frequency depends upon it, as we will show in Chapter 5. We also use the transforms in component models to combine different kinds of losses.

Note that the series-parallel conversion works no matter what the value of Q is. The conversion is a mathematical identity and does function regardless of the value of Q, so long as it is finite. However, viewing (2.27) and (2.28), if the Q is reasonably high (i.e., greater than 10) the reactive elements of both circuits are practically the same. The parallel resistance becomes Q^2 times greater than the series resistance.

The specific relationships for the series-parallel conversions are given in Table 2.3. The specific values for resistance, inductance, and inductive reactance apply at a single frequency only where the Q is calculated. Depending on the Q, the series and parallel circuits may be similar enough for narrowband applications. The lower the Q the less difference we see moving away from the frequency at which the equivalence was established.

2.11.1 Illustrative example: converting a series parallel equivalence

The complex input impedance of a transistor is given as a series *R-C* circuit in Figure 2.11. For a specific application it is more desirable to view the input in parallel form. Find the component values of the equivalent parallel circuit at 1 GHz. Plot the input impedance responses of both circuits versus frequency, and measure the frequency range through which the magnitudes are nearly equal (±5%).

Solution

1. Calculate the Q of the series circuit at 1 GHz, using (2.19). Since the two circuits must have the same Q, we will label it just as Q.

$$Q_S = Q = \frac{X_S}{R_S} = \frac{\frac{159}{f_{\text{GHz}} C_{SpF}}}{R_S} = \frac{\frac{159}{1(7.95)}}{10} = 2.0$$

2. Using (2.26) and (2.28), calculate the equivalent parallel resistor and capacitor values.

$$R_P = R_S\left(1+Q^2\right) = 10\left(1+2^2\right) = 50\Omega$$

TABLE 2.3 THE SPECIFIC RELATIONSHIPS USED IN THE CONVERSIONS BETWEEN SERIES AND PARALLEL *R-L* AND *R-C* EQUIVALENT CIRCUITS

	R-L CIRCUITS		*R-C* CIRCUITS	
	INDUCTANCE	REACTANCE	CAPACITANCE	REACTANCE
PARALLEL TO SERIES	$L_S = \frac{Q^2}{1+Q^2} L_P$	$X_S = \frac{R_P^2 X_P}{R_P^2 + X_P^2}$	$C_S = \left(1 + \frac{1}{Q^2}\right) C_P$	$X_S = \frac{R_P^2 X_P}{R_P^2 + X_P^2}$
	$R_S = \frac{1}{1+Q^2} R_P$	$R_S = \frac{R_P X_P^2}{R_P^2 + X_P^2}$	$R_S = \frac{1}{1+Q^2} R_P$	$R_S = \frac{R_P X_P^2}{R_P^2 + X_P^2}$
SERIES TO PARALLEL	$L_P = \left(\frac{1}{Q^2} + 1\right) L_S$	$X_P = \frac{R_S^2 + X_S^2}{X_S}$	$C_P = \frac{Q^2}{1+Q^2} C_S$	$X_P = \frac{R_S^2 + X_S^2}{X_S}$
	$R_P = \left(1+Q^2\right) R_S$	$R_P = \frac{R_S^2 + X_S^2}{R_S}$	$R_P = \left(1+Q^2\right) R_S$	$R_P = \frac{R_S^2 + X_S^2}{R_S}$

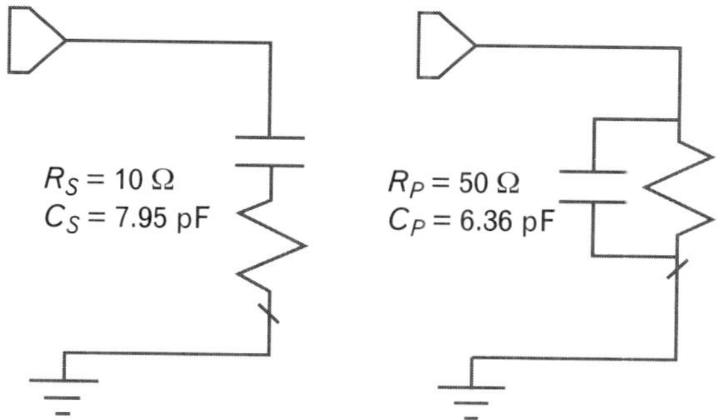

FIGURE 2.11
We can always find a parallel circuit to electrically replace a series R-C circuit at a single frequency. The closeness of the two circuits' terminal impedances depends on their Qs.

$$C_P = C_S\left(\frac{Q^2}{1+Q^2}\right) = 7.95\left(\frac{2^2}{1+2^2}\right) = 6.36 \text{ pF}$$

3. Plotting the input impedance magnitudes of the two circuits with the Microwave Office circuit simulator shows the same 22.4Ω for both circuits at 1 GHz [Figure 2.12(a)]. Between 942 and 1,062 MHz the two magnitudes are within ±5% of the 1-GHz value. For additional information, we also plotted the same parameters in polar form [Figure 2.12(b)], which is a better way to compare two *complex* impedances. A disadvantage of polar plots, however, is that they do not show frequencies without added markers.

2.12 One-port and multiport networks

As frequencies increase into the RF range, conventional voltage and current measurements become impractical and we need to use traveling-wave based types of measurements. Connections to circuits are done through various numbers of *ports*. Each port has two terminals, one of which is grounded. All circuits, systems, and test equipment with ports are called single-ended, or unbalanced types. Balanced (differential) circuits have terminal pairs referenced to each other instead of ground, and we cover those in Section 2.19.

A one-port circuit, as its name implies has only one port. No matter how many elements may be connected within a one-port network, it can always be completely described at one frequency by a simple complex impedance or admittance. These are the *equivalent impedance* and *equivalent admittance*, respectively, of the one port at that frequency. However, impedances and admittances are typically computed from voltage and current measurements that cannot practically be performed at RF, so why do we even talk about them in this book?

In our introductory engineering courses we learned about impedances and admittances and are familiar with them. We already showed in this

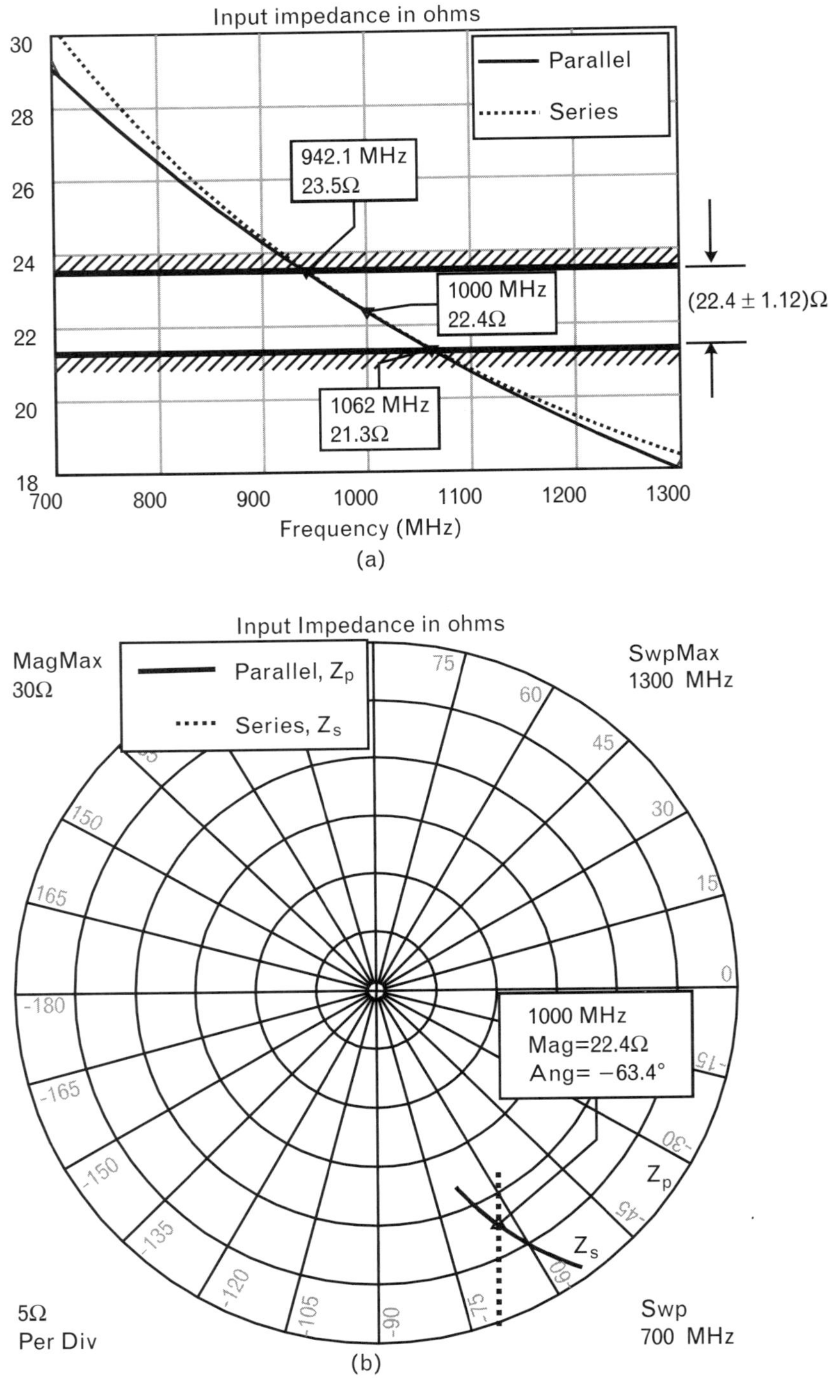

FIGURE 2.12
Plotting the magnitudes of (a) terminal impedances and (b) complex impedances shows exact equality of the two circuits at 1 GHz. Impedance variation between the circuits is almost negligible for ±20-MHz bandwidth and less than ±5% through 120 MHz. The closeness of the two responses is inversely related to the Qs of the circuits. Lower Q leads to near equivalence through a wider frequency range. Plots are shown for the 700–1,300-MHz frequency range.

chapter that impedance and admittance concepts help in visualizing simple equivalent circuits to represent our components. They are also convenient to make series and parallel connections of lumped components. Since they are worthwhile parameters we want to use them. Therefore, we find ways later to convert the measurable RF parameters to impedance and admittance.

Multiport circuits need a more detailed description because applied signals can be transferred among ports. Impedance or admittance matrices, in addition to port impedances and admittances, also have *transfer* impedances and transfer admittances. The order of the parameter matrix is equal to the number of ports.

An example of a one-port may be a precision 50-Ω termination. A conventional lowpass *L-C* filter is a two-port device. A two-way power-divider acts as a three-port device. A dual-directional coupler, to be covered in Section 2.16.8, is an example for a four-port circuit.

2.13 Importance of power transfer when cascading system components

Low-frequency circuit design courses use voltage-gain and current-gain concepts for cascading system blocks. Active devices are either voltage or current driven, and in low-frequency design we are interested in their voltage- or current-gain capabilities. Cascading one amplifier stage to another, we wanted to see the stage being terminated in a high load to maximize the output voltage. At RF however, we generally focus on *power transfer* between stages and system blocks. Since it is easy to understand the need for power transfer when we cascade two transistors, let us use that example here. Later, we can expand the principle to more complex circuits.

At low frequencies, a voltage applied to the input terminals of a device's package is effectively the same as the voltage appearing inside the package at the transistor's gain-cell level. The same is true at the output side—a voltage produced by the chip is effectively the same as the voltage at the output terminal of the package. Therefore, when we cascade two devices, the output voltage at the chip of the first stage is effectively the same as the input voltage of the second stage, even though the devices are embedded into packages.

At RF and microwave frequencies, the voltage applied to the input terminal of the package is *not* what the transistor cell sees inside the chip. You can clearly see the difference even in the highly simplified transistor equivalent circuit of Figure 2.13. The device's current-controlled current source is driven by the current flowing through resistor *R*. Parts of the input current, i_{IN}, branch into parasitic capacitances instead of resistor *R*. In a voltage-driven device, such as an FET, parts of the input voltage, v_{IN}, are wasted across the parasitic inductances instead of generating amplification.

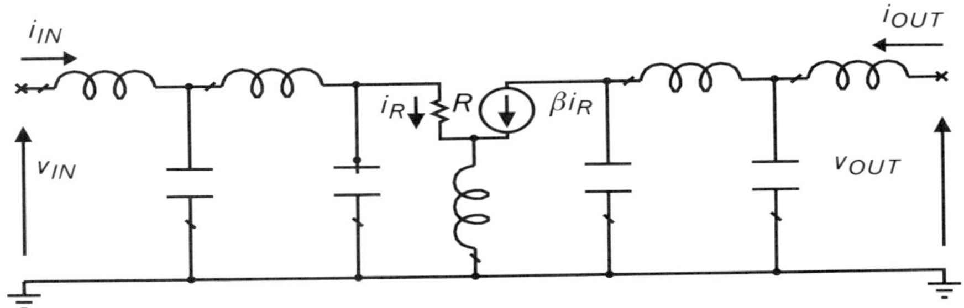

FIGURE 2.13 *Simplified equivalent circuit of a bipolar RF transistor, showing some of the package, as well as internal parasitic inductances and capacitances. The device is driven by the current (i_R) passing through resistor R, which is quite different from the input current (i_{IN}) at the base terminal. Similar differences exist in voltage-driven devices also.*

Either way, without an exact knowledge of the device's equivalent circuit, the input current or voltage does not tell us what the device is capable of doing if the parasitic elements were tuned out.

If instead of the input current or voltage, we know the amount of *power delivered* to the input port, the problem is solved since reactive elements do not dissipate power. (By *delivered* we mean the *difference between applied and reflected* power.) At the output side, if the true available power[6] gain of the device is given, we know exactly what to expect when the device is properly terminated, assuming no power is dissipated in the parasitic elements.

For the above reasons, RF and microwave circuits are almost always designed with power flow, or power transfer considerations, as will be shown in Chapters 4 and 5. At low frequencies, even though the same parasitic inductance and capacitances do exist, their effects are generally negligible. Therefore, voltage or current-based designs are feasible and practical.

2.14 Importance of impedance matching

At RF we need to maximize power transfer among the successive stages of a system. The *maximum power transfer* between a generator and load occurs *when their respective impedances are complex conjugates* of one another. Consider first a generator and load having only real impedances R_S and R_L (only resistances), as shown in Figure 2.14(a). This situation is analogous to the flow of water between two pipes [Figure 2.14(b)], having circular cross-sections with perpendicular ends. For our illustration, let us imagine that the diameter of a pipe represents the resistive part of an electrical port. It is reasonable to expect that the most efficient water flow would occur when

6. The maximum power available under conjugate matched condition (see Chapter 4).

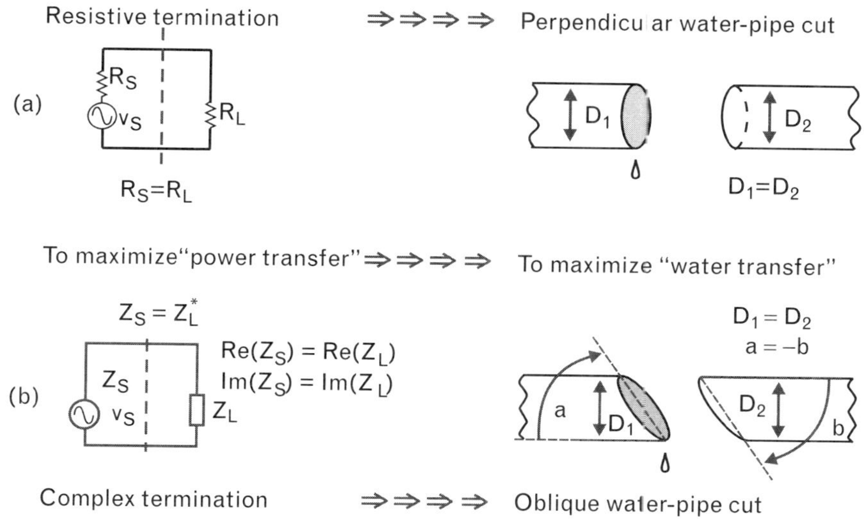

FIGURE 2.14
(a) Maximum power transfer occurs when generator and load have conjugate impedances, (b) analogous to water flow in pipes having the same diameter and complementary angles at the splice joint.

the two pipes have the same diameters. One might suggest that if the receiving pipe were of a larger diameter, the transfer would be even better; however, a diameter change at the interface would introduce turbulence, actually slowing the flow.

Next consider that the generator has a complex impedance, Z_S. Then the load should have impedance $Z_L = Z_S^*$, the complex conjugate of the source impedance. Again, using the water pipe analogy we note that when the source pipe is cut at an angle, the receiving pipe should have the same diameter (meaning equal resistance to the source), and it should be cut at the complementary angle so that the joined pipes match smoothly. We could view the slope of the cut as analogous to the reactive portion of the complex impedance. Positive slope may be considered "inductive" and negative slope as "capacitive." Electrically, when the impedances are complex conjugates the *net reactance is zero,*[7] *and the real parts are equal.* We cover impedance matching in Chapter 5 with more scientific justification.

2.15 RF components and related issues

Components behave quite differently at RF from expectations based on their nominal values. Parasitic inductance and capacitance associated with components can change their behavior radically, even to the extent of changing their reactance type (i.e., a capacitor behaves inductively above its self-resonant frequency). Modeling is needed because parasitics,

7. A common mistake is to have two identical complex terminations facing each other. In that case the reactances add, not cancel (see Section 5.1).

coupling, and grounding effects cannot be ignored at the high frequencies found in the RF band.

2.15.1 Parasitic inductances and capacitances

The unavoidable presence of parasitic inductances and capacitances of components introduces frequency-dependent changes to the performance of the circuit. Although we may expect some variation after a simple physical inspection of the component, the true effective change can be startling. For example, knowing that a discrete (leaded) capacitor has wires attached to both ends, we may think that at RF the effective capacitance is reduced by the presence of the series self-inductance of the wires. Far from being true—the effective capacitance actually *increases*. The increase is true even for surface mount (chip) components. In inductors the parallel self-capacitance of the winding has the same effect, increasing the nominal inductance.

The *primary parasitic* of a component, like the series inductance of a capacitor, leads to *primary resonance*, above which the component is controlled by the primary parasitic portion. Modeling the primary parasitics is a relatively easy task. As the frequency increases, we may also witness *secondary resonances*, which are more difficult to model, and their effects can yield strange results. There will be more on this subject in Chapter 7.

2.15.2 Limited range of practical element values

As we will see in Chapter 7, when inductors and capacitors are used above 20% to 25% of their primary self-resonant frequencies, the effective inductance and capacitance increases in a nonlinear fashion with frequency. In lumped components, the resonant frequency is a function of the nominal value and parasitic reactance. Even with chip components, we face an upper limit of nominal values that can be safely used to avoid significant changes with increased frequency.

Tolerances limit low element values. For example, when we read capacitor manufacturers' specifications, they may offer components with ±1% tolerances—that sounds very good. Reading the fine print, however, we see that it states ±1% or ±0.1 pF, *whichever is greater*. Buying a 0.6-pF capacitor with such a specification really means (0.6 ± 0.1) pF, which represents a possible total change of 33% of the nominal value.

Considering the high and low value limitations, the useful lumped component value range is usually no more than 4:1, perhaps as high as 5:1. We see later that even with transmission lines we face limits. Having a characteristic impedance for microstrip lines (see Section 2.16) below 25Ω to 30Ω requires line widths that causes propagation problems. Above 100Ω to 120Ω, the lines are too narrow[8] and become very lossy. Etch tolerances[9] and trapezoidal etching effects [3] make things even worse for narrow lines. Once again, the ratio of practical range of high-to-low impedance is rather limited.

2.15.3 Measurement and test-fixture considerations

Most components used in RF circuits come in forms that cannot be attached directly to the commonly available RF test equipment for characterization, and we need to use special test fixtures [4] for the measurements. Although specialized, more expensive test fixtures come with calibration standards that allow us to extend the plane of measurements; frequently the measured results include data for the component and fixture. Deembedding removes the effects of the test fixture and provides the true parameters of the component. A subsequent process leads to component equivalent circuits, called *models*.

If the test fixture contains only a segment of 50-Ω transmission line, deembedding is a very simple process. Many times however, the fixture also includes dc biasing or RF stabilizing elements, and the necessary procedure to remove their effect takes more work. We discuss this topic along with component modeling in Chapter 7.

2.15.4 Grounding and coupling effects

Single-ended (unbalanced) circuit components require access to RF ground that is often not where we need it. A surface-mount capacitor mounted on top of the PC board may get its grounding through a plated (via) hole with nonzero self-inductance. Although the inductance of the ground-path may only be a few tenths of a nanohenry, the effective inductive reactance can easily be several ohms. We see in Chapters 4 through 9, and in Volume II, Chapters 1 and 2, that seemingly small inductance can completely change the expected performance, and in active circuits may even lead to parasitic oscillation.

As our circuits get smaller, the capacitive and inductive coupling among components increases. Coupling takes place through air and also through the dielectric media we use. Finding the exact causes of coupling often require the use of *electromagnetic* (EM) simulation (see Chapter 6).

8. With special techniques like suspended substrates the upper limit of characteristic impedance can be increased. Twin-lead lines, used with TV antennas, go up to 300Ω. Additional discussion on transmission lines is given in Chapter 7.

9. Typically equal to $\pm t$, where t is the thickness of the metal layer.

2.16 Lumped elements versus transmission lines

Interestingly, nearly all analog electronic devices are designed using alternating current (ac) theory, where input signals used to activate the circuit are assumed to be sinusoidal in form (see Figure 2.15) and to have been applied an infinitely long time ago. The resulting circuit response is then the *steady-state* solution, which prevails after any transient effects due to the initial application of the excitations have died out.

For a sinusoidal wave of frequency f, propagating in free air, the wavelength λ_0 is defined as

$$\lambda_0 = \frac{C_L}{f} = \frac{2.998 \times 10^8 \text{ m/sec}}{f_{\text{Hz}}} \approx \frac{30 \text{ cm/sec}}{f_{\text{GHz}}} \approx \frac{12 \text{ in/sec}}{f_{\text{GHz}}} \tag{2.29}$$

where C_L is the velocity of light (and of all other electromagnetic waves) in a vacuum. Using (2.29), we see that a 1-GHz wave traveling through air would have a wavelength of about 30 cm, or approximately 12 inches.

When the electromagnetic wave enters a nonmagnetic, homogenous[10] dielectric medium other than air, its propagation slows by the square root of the *relative dielectric constant* of the medium [5]. The velocity of propagation in such lossless media is

$$v_P = \frac{C_L}{\sqrt{\varepsilon_R}} \tag{2.30}$$

The relative dielectric constant, ε_R, is the ratio of electric field storage-capability of the material compared to that of free space, and it is a

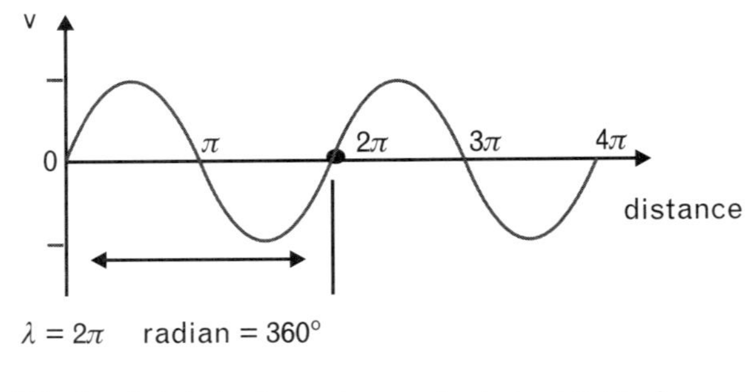

FIGURE 2.15 *Wavelength, l, is the distance between the start and finish of a complete sinusoidal cycle. At 60 Hz the free space wavelength is over 3,000 miles (5,000 km), but at 2 GHz it is only 6 inches (15 cm). When the waves propagate in dielectric media, other than free space, the propagation slows down and wavelengths become shorter.*

10. Same in all directions.

dimensionless number greater than unity. Practical dielectrics vary considerably, having relative dielectric constants of about 2.2 for Teflon, 4.5 for FR-4, and 13.1 for GaAs. When the EM waves propagate in uniform, homogenous media, the *effective dielectric constant*, ε_{EFF}, is equal to ε_r. For nonhomogenous media, such as microstrip transmission lines, ε_{EFF} is less than ε_r because part of the waves propagate in air [6]. Thus the actual *guide* wavelength, λ_G, sometime referred to as effective wavelength, λ_{EFF}, within any kind of dielectric medium is given by

$$\lambda_G = \lambda_{EFF} = \frac{\lambda_0}{\sqrt{\varepsilon_{EFF}}} \approx \frac{30 \text{ cm}}{f_{\text{GHz}} \sqrt{\varepsilon_{EFF}}} \approx \frac{12 \text{ in}}{f_{\text{GHz}} \sqrt{\varepsilon_{EFF}}} \tag{2.31}$$

Closed-form expression are available to compute ε_{EFF} for different forms of transmission line realization [7]. As we mentioned above, for transmission lines having homogenous dielectric filling, such a coaxial lines, $\varepsilon_R = \varepsilon_{EFF}$.

Fractional wavelength is the percentage of the physical length of the line, ℓ, compared to the effective wavelength.

$$FW(\%) = \frac{\ell}{\lambda_{EFF}} 100\% \tag{2.32}$$

It may also be expressed in degrees, as

$$\theta(\text{degrees}) = \frac{\ell}{\lambda_{EFF}} 360° \tag{2.33}$$

At low frequencies a wire used to interconnect two points in a circuit is assumed to be at one voltage throughout its entire length, at any particular instant of time. This is a satisfactory assumption when the length of the lead is short, less than about 5% of the operating wavelength. For discrete circuits, such as printed circuit boards, a practical limit of the lumped approximations is typically between 500 and 1,000 MHz. Due to the small dimensions in microcircuits employing chip elements and in semiconductor integrated circuits, the same lumped element approximations may be used up to 5 or even 10 GHz.

A 1-GHz signal has a wavelength of 30 cm (11.8 inches) in air, which may be reduced to about half that in dielectric medium. Five percent of 15 cm is 7.5 mm, (295 mil) only, so conductors exceeding that length should be treated as *distributed elements*. Of course, the 5% figure is just a guideline. Although a distributed model is always more accurate for any frequency above dc, experience shows that the 5% guideline is a good point at which to transition from lumped to distributed analysis. If the conductor has uniform dimensions and ground underneath, we call it a transmission line instead of a wire, or a printed-board connection. We must then employ the distributed analysis techniques characterized as microwave design.

2.16.1 Illustrative example: fractional wavelength calculations

To appreciate the importance of fractional wavelength, consider an ℓ = 1.27 cm = 0.5 in long conductor laid out on a PC board, with $\varepsilon_{EFF} = 2.25$. How should this conductor be treated at 10 MHz, 100 MHz, and 1 GHz?

1. At 10 MHz the effective wavelength from (2.31) is

$$\lambda_G \approx \frac{30 \text{ cm}}{10^{-2}\sqrt{2.25}} = 2{,}000 \text{ cm} = 20 \text{ m} \approx 800 \text{ in}$$

The 1.27-cm conductor represents a very small fraction of the 2,000-cm wavelength,

$$FW(\%) = \frac{\ell}{\lambda_{EFF}} 100\% = \frac{1.27}{2{,}000} 100\% = 0.063\%$$

The conductor is only 0.063% of the wavelength. Clearly it is safe to assume that the signal voltage will not change significantly at any instance within such a short portion of the wavelength, and the conductor may be treated as a short circuit.

2. At 100 MHz, where the fractional length is 0.63%, the conductor is no longer treated as a short-circuit, yet we have not reached the upper limit of lumped models. In Chapter 7 we look at the actual circuit models of passive RF components and see that a short conductive line on a PC board may be considered as a lumped inductance—*yes, we say inductance*—instead of a short circuit. If there is a ground plane at the other side of the PC board, we may also need to add shunt capacitance to our model, as shown in Figure 2.16.

3. The fractional wavelength of the 1.27-cm conductor at 1 GHz is 6.35%. Accordingly, at that frequency the conductor requires full-distributed circuit modeling.

Our consideration must also include all components, not just simple conductors. For example, if the lengths of conductors within a capacitor are less than the above stated limit, the total capacitance of the plates can be lumped together at the two contacts of the component. Otherwise, the capacitor should also be modeled in distributed form, as an open-circuited transmission line stub, covered in Section 2.16.4.

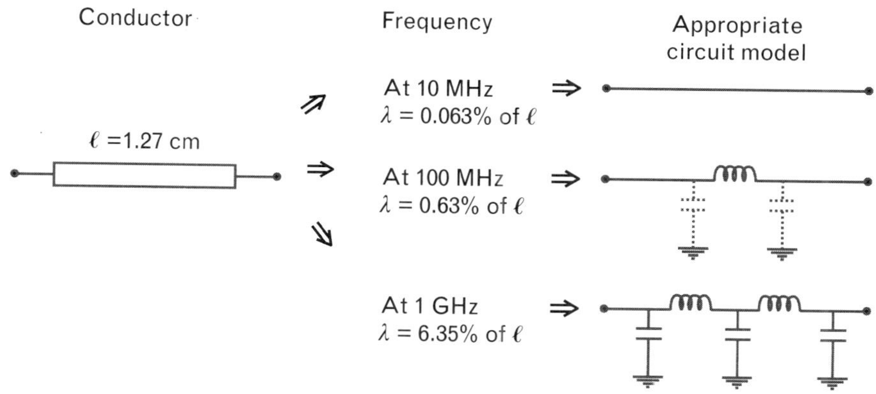

FIGURE 2.16 *The behavior of a 1.27-cm long PC board conductor depends on the relationship between its physical length and the effective wavelength at different frequencies: nearly a short-circuit at 10 MHz; a small inductance and perhaps some stray capacitance at 100 MHz; and distributed L-C at 1 GHz.*

2.16.2 Two-conductor transmission lines

When parallel wires or printed circuit traces are used to convey RF signals, it is appropriate to treat them as *transmission lines*. Those having two separate metallic conductors can propagate signals from dc up to any desired frequency, although practically the undesirability of propagation in higher order modes [8] limits the applicability of each line's geometry to some upper frequency limit.

When an ac voltage is applied to the two-conductor transmission line of Figure 2.17, the instantaneous voltage along the line varies at any particular instance. If we record the voltages at time t_1 and remeasure them again a very short time later at t_2, we get different readings. At another time interval t_3, we get a third set of voltages. Superimposing the three sets of voltages on the same plot, we get the effect that the *voltage waves are traveling* along the transmission line, giving the name to *traveling waves*. We can visualize EM wave propagation in this form.

We mentioned that higher order modes are created at higher frequencies, and let me, Les Besser, share my favorite explanation presented by one of my mentors and former colleague, Steve Adam.

As the electromagnetic waves propagate on a transmission line, the associated fields vary in periodic manner. As the signal frequency increases, the effective wavelength is reduced and eventually becomes compatible to the transmission line's cross-section, creating additional EM field configurations. Propagation velocities of the higher order modes are different from that of the dominant (or principal) mode and require more complex analytical treatment. Here is a simple physical analogy to illustrate their concept.

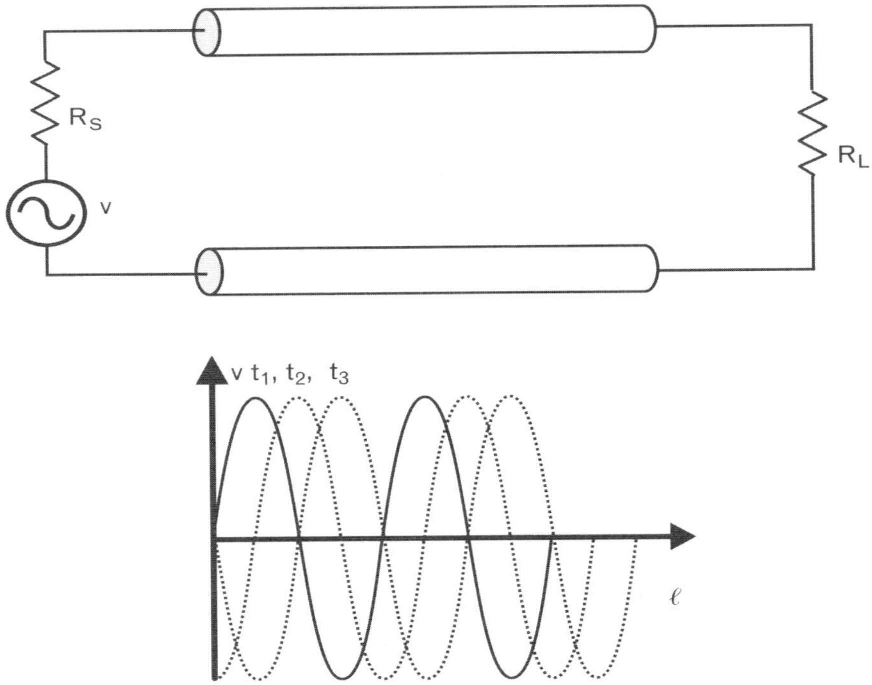

FIGURE 2.17 *When the transmission line's physical length is long enough compared to the wavelength, the applied RF voltage varies along the transmission line in form of a sine wave. Measurements taken at a fixed location at various times t_1, t_2, and t_3 give the impression that waves "travel" from the input toward the load.*

Figure 2.18(a) shows a blowgun with rice pellets being moved by air pressure. Due to the dimensions of the inner diameter of the blowgun compared to the rice, the traveling rice can only propagate in one form. Knowing all the details, such as air pressure and coefficient of friction, the flow of rice can easily be monitored. However, if the inner diameter of the tube is increased [Figure 2.18(b)], the rice can tumble into different

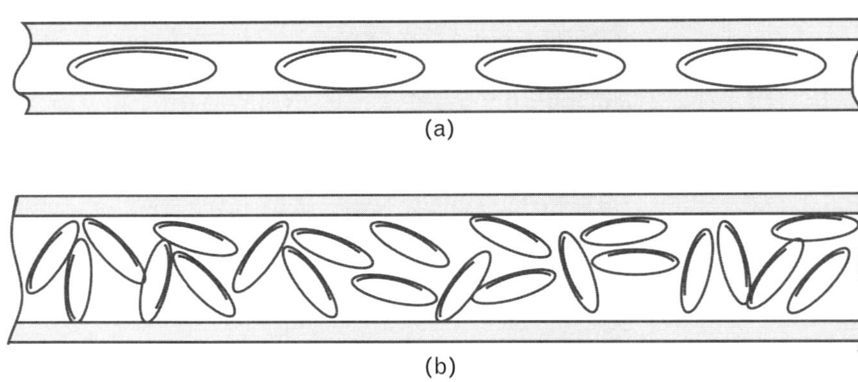

FIGURE 2.18 *Rice sent through a blowgun (a) is restricted to one form of travel when the inner diameter of the tube is small, and (b) can travel into different directions in a larger tube. The pictures represent some analogy to EM wave propagation in transmission lines. (From: [8]. © 1992 Adam Microwave Consulting. Reprinted with permission.)*

directions, perhaps even spin around, and slow down propagation. Modeling the rice flow in this second case is much more difficult, and gives us an impression about the existence of higher order modes in transmission lines.

The electric and magnetic field patterns for two conductor lines operated in the dominant *transverse electric and magnetic* (TEM) mode is shown in Figure 2.19. The mode is called transverse because both the electric (E) and magnetic (H) fields surrounding the wires are contained in the plane perpendicular to the direction of energy propagation, and neither has a component in the direction of propagation. Transmission lines used in most RF circuits fall into this category [8].

Transmission lines that conduct at direct current necessarily are TEM mode. Waveguides that only operate in *transverse electric* (TE) or *transverse magnetic* (TM) mode have a cutoff frequency below which attenuation-free propagation is not possible. For a TE waveguide mode, all of the electric field lines lie in the plane transverse to the direction of propagation while

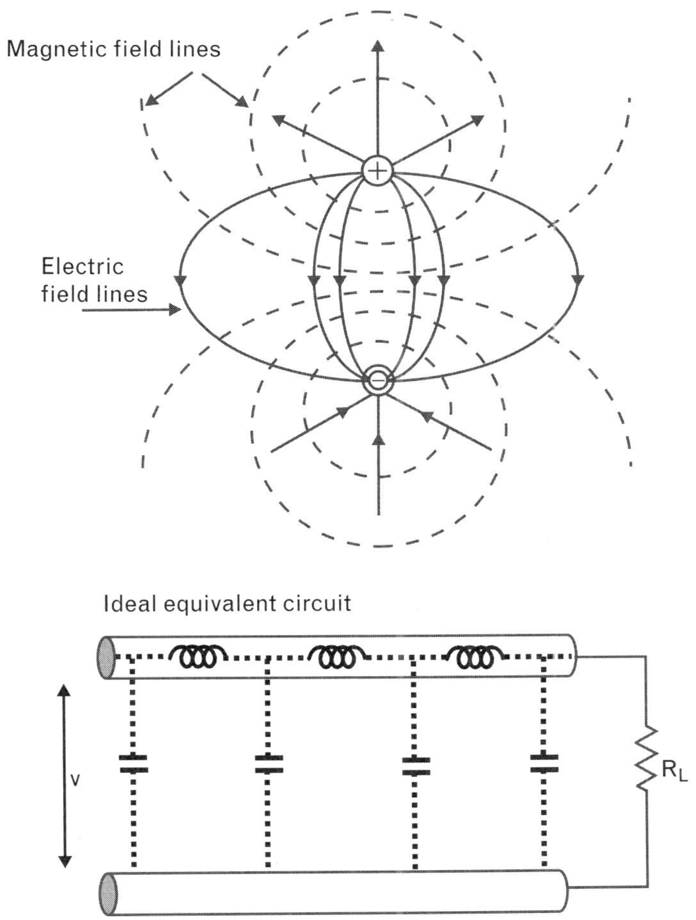

FIGURE 2.19
An ideal uniform transmission line can be segmented into cascaded L-C sections. Cross-sectional view on the top side shows the transverse electric and magnetic field lines caused by and applied voltage and resultant current. Physical transmission lines are modeled with lossy inductors and capacitors to represent conductive and dielectric losses. (After: [8].)

there are M (magnetic) fields in the direction of propagation. The reverse is true for TM waveguide modes.

The effect of propagation along a line is more complex than the mere addition of some parasitic inductance and capacitance to account for the wire length. To deal with this situation we resort to the use of uniform transmission lines with *constant series impedance and shunt admittance* per unit length in RF circuits. In this manner we can estimate the effects of reflections on the lines and the corresponding impedance changes seen by a generator to which the line is connected. Later we will also use a graphical tool, called *Smith chart,* to do this efficiently.

2.16.3 Transmission line characterizations

There are two ways to describe transmission lines: physical dimensions and electrical parameters. In physical form we need to provide the length and cross-sectional dimensions, in addition to all the applicable physical dimensions and electrical parameters of the dielectric medium. For electrical characterization the *characteristic impedance, electrical length,* and *attenuation constant* are required.

For all TEM mode transmission lines a simplified theoretical analysis may begin with the equivalent circuit in which each incrementally small length of line is modeled as a series impedance and a shunt admittance (Figure 2.20). A wave traveling along the line has voltage, v_T, and current, i_T, related by the characteristic impedance of the line segment, Z_{TL}, by [7]

$$Z_{TL} = \frac{v_T}{i_T} = \sqrt{\frac{Z_s}{Y_p}} = \sqrt{\frac{R + j\omega L}{G + j\omega C}} \tag{2.34}$$

The *propagation constant* of a real physical lossy transmission line is

$$\gamma = \alpha + j\beta = \sqrt{Z_S Y_P} = \sqrt{(R + j\omega L)(G + j\omega C)}$$

where α is the attenuation constant in Neper/unit length, and β is the phase constant in radians/unit length.

The phase velocity is

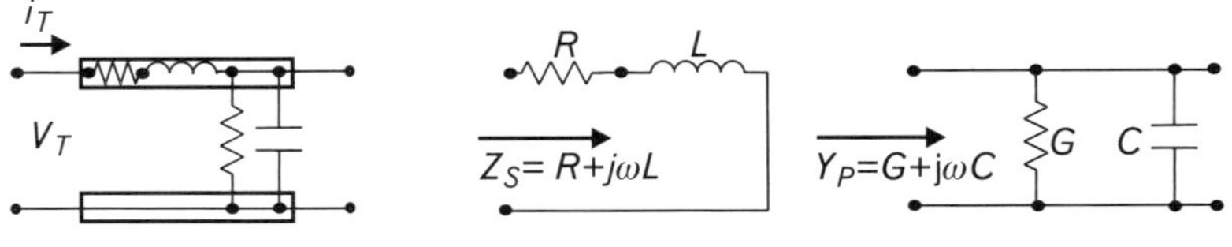

FIGURE 2.20 *Characteristic impedance of a transmission line can be derived from the short-circuited series impedance and open-circuited shunt admittance of its incremental segment.*

$$v_p = \frac{\omega}{\beta}$$

The value Z_{TL} is called the complex characteristic impedance of the transmission line. If the loss elements R and G are small compared to the associated inductive reactance and capacitive susceptance, then the impedance simplifies to a real number

$$Z_{TL} = \sqrt{\frac{L}{C}} \quad (2.35)$$

Thus for a *lossless* transmission line the characteristic impedance, Z_{TL}, is a real number and independent of frequency. The result of this is that a voltage applied to the properly terminated line will propagate at the speed of v_p in the line's dielectric medium without reflection or absorption. On reaching the end of the line, if the terminating impedance, Z_L, is a pure resistance equal to Z_{TL}, there will be no reflection. If Z_L is not equal to Z_{TL}, then a reflection occurs, which we shall describe shortly.

Of course, a real transmission line is not lossless; it has a small series resistance and a small shunt conductance due to metallic and dielectric losses. These losses are also frequency dependent and the exact value of the characteristic impedance may be a complex number. Still, (2.35) provides a very good approximation for practical transmission lines used in RF circuits.

Note that we refer to the characteristic impedance of transmission lines as Z_{TL}, while in many textbooks the authors use Z_0. We reserve the use of Z_0 for the characteristic (reference) impedance of the system, which is generally 50Ω, because in impedance matching circuits we often use transmission lines with characteristic impedances other than 50Ω. Later we will introduce open-circuited and short-circuited transmission line stubs, and use the terms Z_{OS} and Z_{SS} for their characteristic impedances, which may also be some arbitrary value.

In case you wonder why 50Ω is used in most cases as characteristic impedance in RF and microwave systems, there are two explanations. One is more practical and the other is more scientific.

In microwave transmission systems of the 1930s, coaxial transmission lines were initially fabricated in England with a standard British plumbing pipe. Using a commonly available center conductor to this pipe led to a 50-Ω characteristic impedance. According to some, that was the birthplace of the 50-Ω standard. Another equally reasonable explanation is that for minimum signal attenuation, coaxial transmission characteristic impedance is around 77Ω. For maximum power handling capability the optimum impedance is around 30Ω. Standardizing the characteristic impedance at

50Ω seems like a good compromise for the both performance considerations. However, there are exceptions where using a different characteristic impedance makes more sense. For example, in long-haul cable television systems, where the signal needs to be amplified repeatedly to overcome cable losses, attenuation is more important than power handling capability. Accordingly, the cable-TV industry uses 75Ω as a standard, requiring special test equipment, cables, connectors, and other components for their operation.

The term *electric length* (θ) refers to the ratio of the physical length (ℓ) of the transmission line to the wavelength (λ_G) in the applicable dielectric. Expressions are given for lengths in centimeters and inches.

$$\theta = \beta\ell = \frac{\ell}{\lambda_G}360^\circ \approx \frac{360^\circ\, f_{\mathrm{GHz}}\, \ell_{\mathrm{cm}}\sqrt{\varepsilon_{EFF}}}{30_{\mathrm{cm}}} \approx \frac{360^\circ\, f_{\mathrm{GHz}}\, \ell_{\mathrm{in}}\sqrt{\varepsilon_{EFF}}}{11.8_{\mathrm{in}}} \quad (2.36)$$

If a complex signal is passing through a transmission line, all frequencies making up the signal must experience the same phase velocities and attenuation to avoid any distortion. This is a difficult requirement because the attenuation increases with frequency, and the phase velocity can only be constant in perfectly homogeneous dielectric medium.

2.16.4 Various TEM transmission line configurations

We provide here a brief summary of some of the most commonly used two-conductor transmission line types (Figure 2.21). Each geometry offers its own advantage and disadvantage. More detailed discussion follows in Chapter 7.

- The *twin-lead* line was probably the earliest low-cost attempt at a uniform transmission system. Twin-lead is a balanced configuration with high radiation losses, not practical above the VHF range. Characteristic impedance may be relatively high, up to several hundred ohms. Shielded form is also available at higher cost.
- *Coaxial* line provides perfect shielding (if the outer conductor is solid) and its propagating energy does not radiate or is susceptible to radiation. This structure, however, is called an *unbalanced line* because its two conductors are not equidistant in impedance from ground. It lends itself to flexible cable but is not easily employed for complex circuits requiring transmission line branches and other component installations.

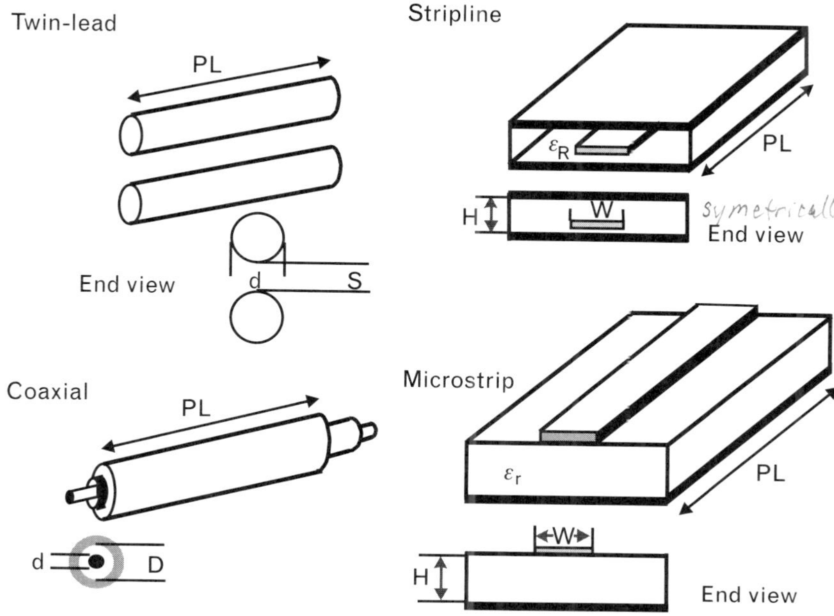

FIGURE 2.21 *Various transmission line configurations. Twin-lead is a balanced transmission line since neither conductor is grounded. Coaxial, stripline, and microstrip lines are unbalanced types.*

- An evolution of coaxial line led to a flattened geometry, called *stripline*, which can be realized in printed circuit board form. Stripline has the total shielding potential (if ground straps are used to interconnect the two ground planes), and it provides a much more suitable manufacturing medium for passive circuits such as filters and directional couplers. The flat center-conductor is etched chemically using a photographic process, allowing nearly identical reproduction at low cost.

- A further variation of stripline formed an open center conductor trace, which is the *microstrip transmission line* geometry. This circuit realization allows us to add other components and perform "tuning" when necessary. (In contrast, the stripline needs to be taken apart for adjustments.) Microstrip is the most frequently used transmission line form in RF circuits.

We mentioned earlier that if the dielectric medium is not homogeneous, the phase velocity becomes frequency dependent. For example, in microstrip lines the EM waves propagate in two different dielectric media with different phase velocities, changing the characteristic impedance. The effect, called dispersion, is already noticeable in the RF range and requires computational corrections.

It is possible and very useful to design circuits having series and shunt combinations of transmission line sections (Figure 2.22). A line directly connected into the main signal path is called a *cascade transmission line*,

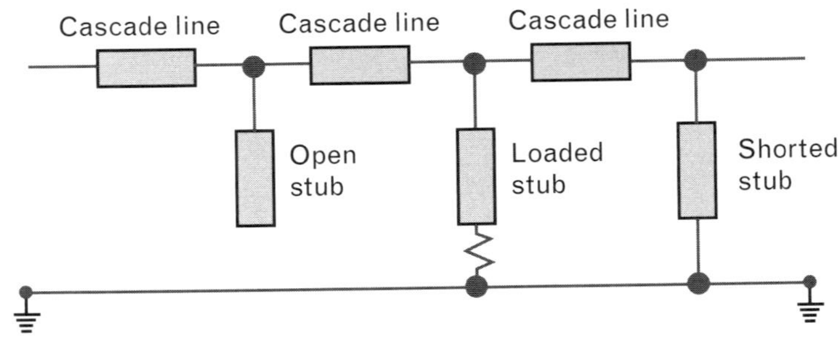

FIGURE 2.22 *Six-element transmission line network with cascade lines and parallel stubs. Depending on the far-end termination of the parallel branch, it is classified as a open-circuited, loaded, or short-circuited stub.*

although in other publications it may be referred to as a *series line*[11] or *unit element*. Conductive traces branching away from the main path are called *transmission line stubs*. There are series or parallel stubs, but the former require more complex technologies, which we do not cover in this book. Parallel stubs on the other hand are very practical in PC board applications. They can be used to replace parallel capacitors and inductors at the higher frequencies where lumped element realization is no longer possible.

Complex transmission line circuits, if realized in coaxial lines or waveguides, would require fabrication similar to household plumbing, an expensive and laborious task. However, with microstrip configurations such intricate connections merely require suitable etching of the "center conductor" pattern, as shown in Figure 2.23. Since this is done photographically, the design labor is confined to producing a suitable etching negative, from which countless duplicate circuits can be reproduced quite economically. Ground connections between top and bottom sides may be done with conductive strips or plated through holes, called *via holes*.

When two or more transmission lines of different forms or parameters are connected together, we experience discontinuities. Abrupt changes,

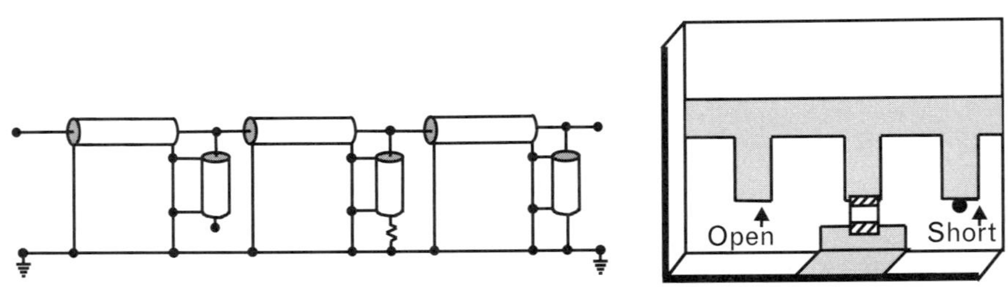

FIGURE 2.23 *Coaxial and printed-circuit forms of the transmission line circuit shown in Figure 2.22. Microstrip realization offers low cost at high degree of reproducibility. Since microstrips radiate and are susceptible to radiation, some form of shielding is recommended.*

11. Since there are also series stubs in certain transmission line forms, we prefer to use the "series" term there.

such as the open end of a parallel stub, also require special consideration and we cover these topics in Chapter 7.

2.16.5 Reflected waves on transmission lines

Earlier we stated that the impedance parameters are very convenient to add components in series, because the total impedance of the series connection is simply the sum of the individual impedances. Our theory, however, does not apply to distributed elements because those are made up by an infinite number of cascaded *series L – parallel C* sections, as shown in Figure 2.24. For transmission line elements we need to use a more complex approach [9]. To illustrate the need, let us take a simple example of an ideal transmission line terminated with a resistive load. Assume that at a specified frequency the reactance of each series inductor is $j5\Omega$, and each capacitor represents $-j200\Omega$. Our goal is to compute the input impedance, Z_{IN}, using a step-by-step approach.

We start by using a low value of 1Ω for R_L and look at the effect of C_1. Clearly, since the capacitive reactance is 200 times larger than R_L, we can neglect the component and move to L_2. Adding the reactance of the inductor increases the total impedance to $(1 + j5)\Omega$. The reactance of capacitor C_3 is still much larger than $(1 + j5)\Omega$, so we can move to inductor L_4 and add its series reactance. Now we have a new total impedance of $(1 + j10)\Omega$, which is still much smaller than the reactance of the next parallel capacitor, so we could ignore C_5 also. We could repeat this process for a few more inductors, but eventually the parallel capacitors must also be considered. From this exercise we could conclude that a *short transmission line behaves like a series inductor, if the termination is low impedance.*

If we repeat the calculations with a 250-Ω termination, the transmission line acts in a different way. Now the parallel capacitors are the dominant parasitics and the series inductors are signifcant—at least at the beginning. After going through 5 to 10 sections, the series inductors can no longer be neglected. We could then also conclude that a *short segment of transmission line terminated with high impedance behaves capacitively.* Is it

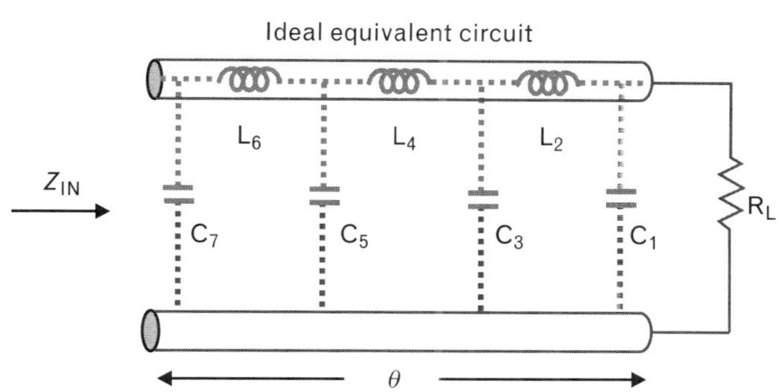

FIGURE 2.24 *Behavior of a short transmission line segment depends on the load. For low-impedance loads the series inductors dominate and with high loads shunt capacitances have more effect.*

possible there is a certain load for which the inductors and capacitors have equal effects? Let us look at the exact mathematical solution.

If an ideal transmission line of characteristic impedance Z_{TL} and electrical length of $\theta = \beta\ell$ is terminated with a complex impedance Z_L, the new input impedance is

$$Z_{IN} = Z_{TL}\frac{Z_L + jZ_{TL}\tan\theta}{Z_{TL} + jZ_L\tan\theta} \tag{2.37}$$

For lossy transmission lines, the input impedance is a function that involves hyperbolic tangent,

$$Z_{IN} = Z_{TL}\frac{Z_L + jZ_{TL}\tanh(\gamma\ell)}{Z_{TL} + jZ_L\tan(\gamma\ell)} \tag{2.38}$$

where the propagation constant $\gamma = \alpha + j\beta$, and ℓ is the physical length of the line.

We notice that if in (2.37) and (2.38) Z_L is equal to the characteristic impedance of the transmission line ($Z_L = Z_{TL}$) the input impedance is simply Z_{TL}, *regardless of the length of the line*. This is one of the most important properties of transmission lines, and it is true for lossy and lossless lines. Restating it, if a transmission line is terminated in a pure resistance equal to Z_{TL}, it is said to be *matched* or the load is a *matched load*. Under this condition, a wave applied to the input is completely dissipated in the termination. On the other hand, when the line is terminated by any impedance other than Z_{TL}:

- A portion, or all of the traveling waves are reflected, and therefore, not all the power is dissipated in the termination;
- The input impedance is analytically determined by solving either (2.37) or (2.38). Alternatively, we show in Chapter 4 how it can be done with the Smith chart.

The wave reflection can be described in terms of an incident voltage (v_F) and an incident current (i_F), which upon reaching the improper load give rise to a reflected voltage (v_R) and reflected current (i_R), as shown in Figure 2.25. Some authors prefer v^+ for v_F and v^- for v_R.

The total load current is equal to the difference of i_F and i_R, while the total voltage is the sum of v_F and v_R,

$$i_L = i_F - i_R \tag{2.39}$$

$$v_T = v_F + v_R \tag{2.40}$$

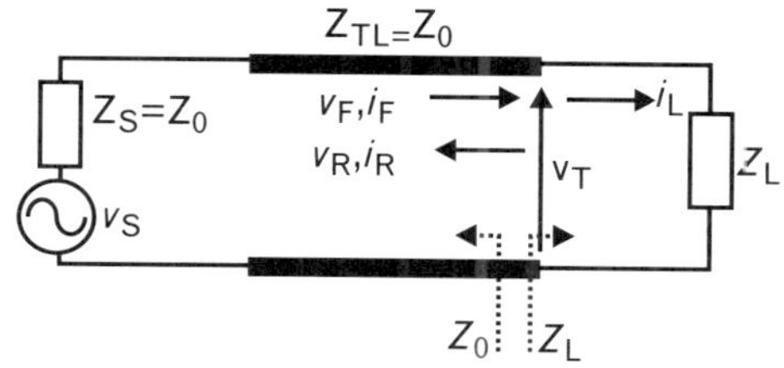

FIGURE 2.25 *If the transmission line is terminated at the output with an impedance $Z_L = Z_{TL}$, there is no wave reflection from the load and the input impedance of the line is always equal to Z_{TL}. When $Z_L \neq Z_{TL}$, a portion of the incident voltage and current are reflected. The magnitude and phase of these reflected waves are functions of Z_L and Z_{TL}.*

It should be noted that while we describe the process as an incident wave, which upon reaching the load, reflects back toward the generator, our analysis is a steady state analysis instead of a transient type. The wave reflected from the load traveled back to the generator (then gets rereflected from the generator if Z_S is not equal to Z_{TL}). Our analysis assumes that an infinite number of such journeys were completed and the voltages and currents everywhere on the line are now in the steady state condition. There is a standing wave pattern on the line resulting from the relationship of the load impedance Z_L in relation to the characteristic impedance Z_{TL}.

Perhaps the easiest way to visualize the concept of reflection is to use a short circuit for the load. Then, the total instantaneous voltage at the load must be zero at all times. From (2.40) we can show that the reflected voltage becomes the negative of the incident voltage.

If

$$v_T = 0 = v_F + v_R$$

then

$$v_R = -v_F$$

Since v_F is fluctuating between its positive and negative peak values, v_R must also follow in the opposite form to maintain zero voltage at the short-circuited load. Power is not dissipated in a load without resistance, so there is a full reflection back toward the source. The interaction of the two voltage waves flowing in a long line in opposite directions creates the standing-wave pattern shown in Figure 2.26. Standing waves are only shown for the last wavelength of the line.

An interesting observation from Figure 2.26 is that as we move away from the load, at every half-wavelength interval the total voltage of the line drops to zero, which means that those locations are at ground potential.

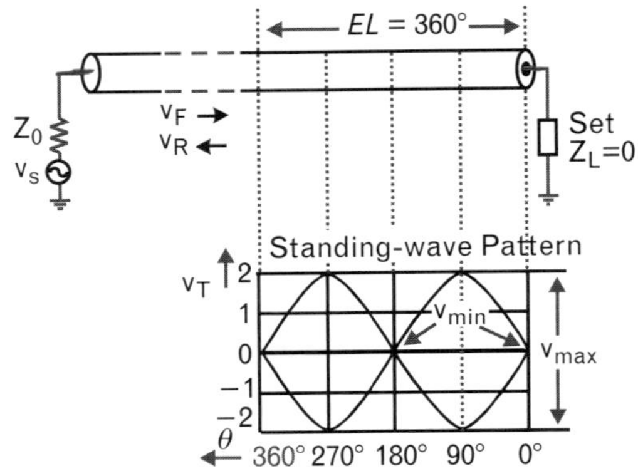

FIGURE 2.26 *Standing wave pattern caused by sending a $v_F = 1V$ peak voltage wave into a short-circuit terminated transmission line. The characteristic impedance of the line is the same as the source impedance, so there are no rereflections from the source. Voltage peaks as well as nulls are one half-wavelength apart, referenced to the load.*

These are the points where the voltages of the two traveling waves cancel each other. A quarter-wavelength away from the short-circuit load termination, the voltage is twice the magnitude of v_F—that is where the two voltage-waves are in-phase with equal magnitudes, doubling the voltage. These cycles repeat throughout the line.

Until the mid-1960s we were not able to measure separately the incident and reflected voltage waves. All we could do was to detect the maximum and minimum of the composite standing wave pattern.[12] The introduction of the vector voltmeter and the network analyzer enabled us to measure them, even through a range of swept frequencies. The transmission line structure that enables us to separate the incident and reflected waves from each other is the *directional coupler.*

2.16.6 Transmission line stubs

We showed in Figures 2.22 and 2.23 that at RF, transmission line type parallel branches can be used instead of lumped elements. We can replace lumped parallel capacitors with open-circuited parallel stubs and parallel inductors with parallel short-circuited stubs, frequently referred to as *open stubs* and *shorted stubs*. We also mentioned in Figure 2.24 how sensitive transmission line elements are to their applied terminations; and let us repeat the extreme cases for terminations:

- Very short transmission line segments *terminated with low impedances behave like inductors*;

12. The ratio of those two voltages, called voltage standing wave ratio, is covered in Section 2.17.3.

- Very short transmission line segments *terminated with high impedances behave like capacitors.*

Figure 2.27 compares the circuit symbols used to represent the lumped element and parallel stubs.

A short-circuited parallel stub is terminated with the lowest possible impedance—a short circuit. Therefore, it is reasonable to expect inductive behavior, if the length is short compared to the wavelength (i.e., less than 5–10%). What happens when the stub is longer? Let us find out how a short-circuited stub behaves with any arbitrary length.

Rewriting (2.37) for the input impedance of a transmission line terminated with an arbitrary load, Z_L,

$$Z_{IN} = Z_{TL} \frac{Z_L + jZ_{TL} \tan\theta}{Z_{TL} + jZ_L \tan\theta}$$

Setting the termination to a short-circuit, $Z_L = 0$, and simplifying the equation gives us a purely reactive form for the input impedance of a *lossless* short-circuited stub,

$$Z_{IN} = R_{SS} + jX_{SS} = 0 + jZ_{SS} \tan\theta_{SS}$$

where Z_{SS} the characteristic impedance of the shorted stub, and θ_{SS} is its electrical length in degrees.

Extracting the imaginary part gives us the input reactance of a short-circuited parallel stub, X_{SS},

$$X_{SS} = Z_{SS} \tan\theta_{SS} \tag{2.41}$$

The input susceptance of the short-circuited stub is just the reciprocal of X_{SS},

$$B_{SS} = \frac{1}{Z_{SS} \tan\theta_{SS}} \tag{2.42}$$

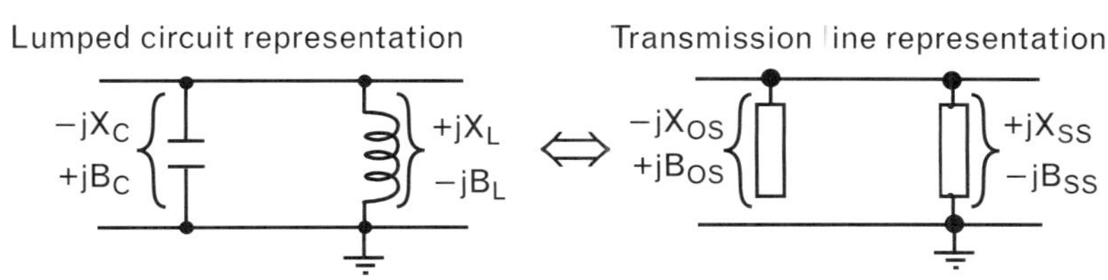

FIGURE 2.27 *A parallel lumped capacitor and inductor may be characterized by its terminal reactance or susceptance. The same can be done for parallel open and shorted stubs.*

Similarly, applying an open-circuit termination instead of the short, we obtain the input impedance of an open-circuited stub, $Z_{IN} = -jX_{OS}$, where

$$X_{OS} = \frac{Z_{OS}}{\tan\theta_{OS}} \tag{2.43}$$

where Z_{OS} is the characteristic impedance and θ_{OS} is the electrical length of the open stub. The corresponding input susceptance is

$$B_{OS} = \frac{\tan\theta_{OS}}{Z_{OS}} \tag{2.44}$$

When the electrical lengths of the shorted and open stubs are less than 90°, they behave like parallel inductors and capacitors, respectively. When the lengths exceed 90°, the sign of tangent θ changes repetitively through every 90°. Thus, both the short-circuited and open-circuited stub may look like an inductor or a capacitor, depending on their electrical lengths. Although we keep saying that the stubs behave like inductors and capacitors, we must remember to use (2.41) through (2.44) to compute their input impedances and admittances for accurate characterization.

In actual circuit design parallel elements are generally treated as admittances, as shown in Figure 2.28. Remember that even though at a single frequency we can always find electrical equivalence between lumped and distributed form, the reactances and susceptances of the two component types are computed differently. The equivalence can only be established at single frequencies, although for a narrow frequency range they can be reasonably close to each other.

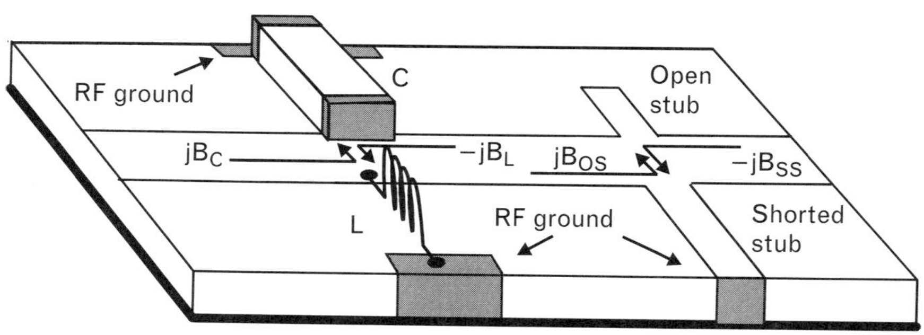

FIGURE 2.28 *Printed circuit board layout showing lumped parallel inductor and capacitor with parallel open and shorted stubs. Transmission line representation is always more accurate, assuming all discontinuities, such as ground parasitics, are properly modeled.*

2.16.7 Illustrative example: open-circuited parallel stub computation

The two ideal lumped elements of a resonant circuit of Figure 2.28 are $C = 3.16$ pF and $L = 2$ nH. Find their electrical equivalent open and shorted stubs at 2 GHz, using characteristic impedances of $Z_{OS} = Z_{SS} = 50\Omega$ for both stubs.

Solution

Using (2.20) and (2.22), compute the susceptance magnitudes of the inductor and capacitor.

$$B_L = \frac{0.159}{f_{\mathrm{GHz}} L_{\mathrm{nH}}} = \frac{0.159}{2(2)} = 0.0397S = 39.7 \text{ mS}$$

$$B_C = 0.006283 f_{\mathrm{GHz}} C_{\mathrm{pF}} = 0.006283(2)3.16 = 0.0397S = 39.7 \text{ mS}$$

Since the inductive and capacitive susceptances are of the same magnitudes, 2 GHz is the resonant frequency of the parallel L-C circuit. The total admittance at resonance, Y_{RES}, assuming lossless components ($G_{LOSS} = 0$),

$$Y_{RES} = G_{LOSS} + jB_C - jB_L = 0 + j0.0397 - j0.0397 = (0 + j0)S$$

which of course is an open circuit.

Next, we need to find the parameters of the equivalent parallel stubs. Generally, since a stub has two variable parameters, they cannot both be determined from the input susceptance.[13] Since we already specified the characteristic impedances to be 50Ω, we solve (2.42) and (2.44) for the electrical lengths of the two stubs by setting $B_{SS} = B_L$ and $B_{OS} = B_C$.

$$\theta_{SS} = \tan^{-1}\left(\frac{1}{Z_{SS} B_{SS}}\right) = \tan^{-1}\left(\frac{1}{50(0.0397)}\right) = 26.7°$$

$$\theta_{OS} = \tan^{-1}(Z_{OS} B_{OS}) = \tan^{-1}(50(0.0397)) = 63.6°$$

At 2 GHz, the parallel resonant L-C circuit may be replaced with a parallel shorted stub of 26.7° and an open stub of 63.6° electrical lengths. Characteristic impedances are set to 50Ω for both stubs. Of course, we could use other impedances and recalculate the corresponding electrical lengths. Changing impedances affects the frequency response.

13. This fact is major roadblock for the synthesis of distributed components. Since we cannot solve one equation with two unknowns, we must set the value of either the electrical length or the characteristic impedance (see Chapter 6).

Frequency responses of the two ideal resonators are shown in Figure 2.29. The *L-C* circuit has smooth monotonic impedance roll-off above and below 2 GHz, while the parallel stub combination shows cyclical changes between high and low impedances. If we add the unavoidable parasitic elements to the inductor and capacitor, their effects lead to additional resonances.

2.16.8 Directional couplers

The description of voltages and currents on the transmission line in terms of incident and reflected waves is important at RF and microwave frequencies because it is impractical to measure actual instantaneous or rms values of voltages and currents at these frequencies. Instead we separate and measure the incident and the reflected power, and deduce the values of the voltages, currents, reflection coefficients, and impedances. This is possible through the use of *directional couplers*.

There are many different ways of forming a directional coupler [10–12]. They can be constructed in stripline, microstrip, coax, and even waveguide. One of the most common forms is shown in Figure 2.30. It consists of two transmission lines placed close together along the axes of propagation. It turns out that the maximum coupling occurs when the coupled lines are an odd number of quarter-wavelengths long. In some cases the coupling is backward. Thus, if a signal is incident at Port 1 the coupled wave exits at Port 3. With proper design and proper terminations, no signal exits at Port 4 (the isolated port). The power not coupled to Port 3, of course, exits at Port 2, the far end of the input line. Because of this coupling characteristic, this type of coupler shown in Figure 2.30 is sometimes called a *backward-wave* coupler.

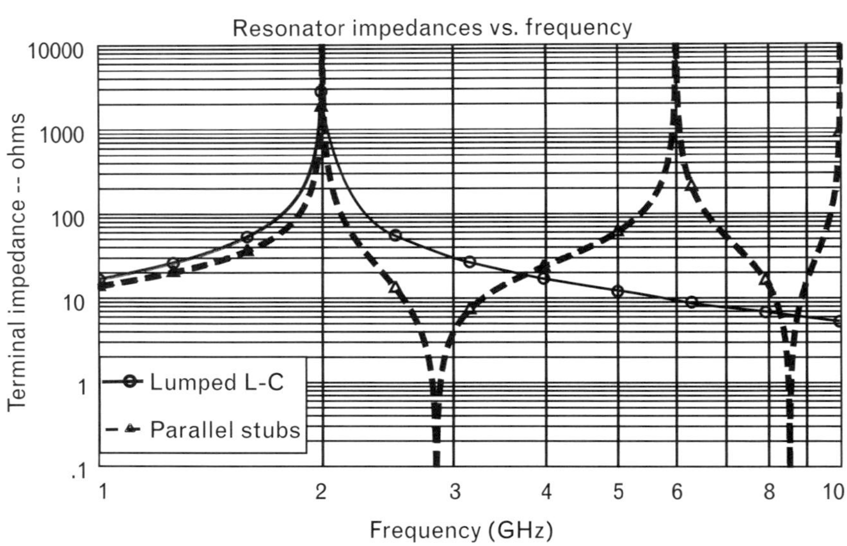

FIGURE 2.29
Since we designed an electrical equivalent of the LC circuit with transmission line stubs at 2 GHz, both circuits show resonance at that frequency. Additional parallel stub resonances occur at 6 GHz and 10 GHz. The open-circuited stub behaves like a short circuit at 2.84 GHz. Repeated resonances also occur above 10 GHz.

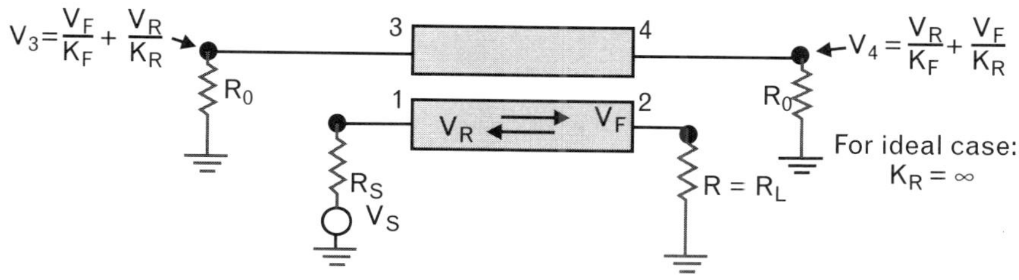

FIGURE 2.30 *A directional coupler allows us to separate incident and reflected waves. Waves entering at Port 1 travel directly to Port 2, which is called the "thru port." Determined by the forward coupling factor, K_F, part of the forward wave, v_F, is coupled to Port 3. In ideal couplers $K_R = \infty$ and v_F is not coupled to Port 4. However, a wave reflected from Port 2, labeled v_R, couples to Port 4, but not to Port 3.*

Suppose that the load on Port 2 is not matched to the characteristic impedance of the coupler. Then there is a reflected wave, which reenters the coupler at Port 2. Since the coupler is symmetric, this new wave couples with Port 4 by the same K_F factor that applies between Ports 1 and 3.

In actual practice the directional coupler is usually more complex in its implementation, particularly if employed in a broadband network analyzer. Two separate line sections might be employed for sampling the incident and reflected waves (Figure 2.31) in order to achieve better isolation of the forward and reverse going waves. This is because slight discontinuities at the entrances and exits of the coupled line sections deteriorate the theoretically perfect isolation in a practical coupler. Also, while a single quarter-wave section for the coupled region might yield up to an octave bandwidth of useful operation, additional cascaded sections having different degrees of coupling can with proper design yield several octaves of bandwidth.

2.17 Circuit parameters using wave relations

As frequencies reach 100 MHz, the voltages and currents are difficult to measure. A more practical set of parameters can be defined in terms of traveling waves. Two of those parameters are used in this section, as follows.

2.17.1 Reflection coefficient definitions

A new parameter Γ, called the *reflection coefficient,* is defined to show what fraction of an applied signal is reflected when a Z_0 source drives a load of Z_L. By the use a directional couplers, forward and reflected traveling waves inside a transmission line can be separated and measured. A simplified network analyzer setup for reflection coefficient measurements is shown in

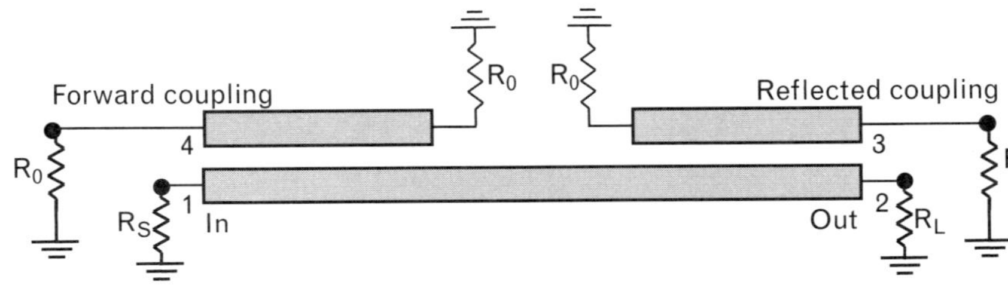

FIGURE 2.31 *Practical couplers may use more than one coupled section to achieve better isolation between incident and reflected wave sampling. The coupler shown is called a dual directional coupler.*

Figure 2.32. Reflection coefficient is generally displayed in the polar coordinate system, providing magnitude, ρ, and phase angle, θ.

$$\Gamma_L = \frac{v_R}{v_F} = \frac{b_1}{a_1} = |\Gamma_L|\angle\theta = \rho_L e^{-j\theta} \tag{2.45}$$

where a_1 and b_1 are *normalized* traveling voltage waves [1], defined at the one-port load termination as

$$a_1 = \frac{v_F}{\sqrt{Z_0}} \tag{2.46}$$

$$b_1 = \frac{v_R}{\sqrt{Z_0}} \tag{2.47}$$

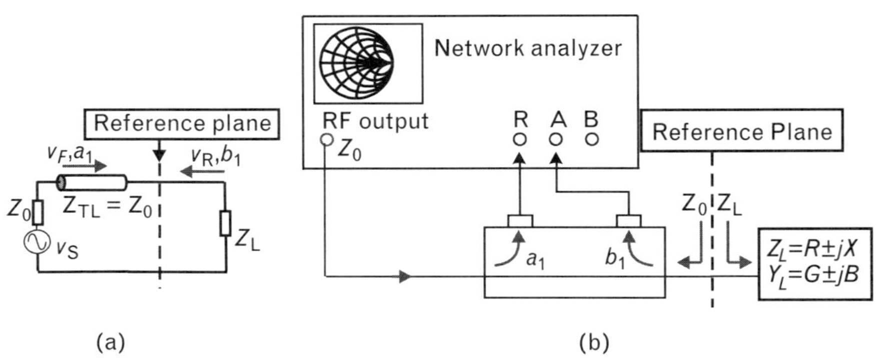

FIGURE 2.32 *(a) Connecting a signal source to an arbitrary load through a transmission line may cause a signal reflection. (b) Impedance measurement with a network analyzer system is based on this principle. An arbitrary load impedance Z_L or admittance Y_L can be determined by measuring the incident a_1 and reflected b_1 voltage waves with a calibrated dual-directional coupler.*

In terms of two terminations facing each other, the voltage reflection coefficient is also computed as,

$$\Gamma_L = \frac{Z_L - Z_0^*}{Z_L + Z_0} \tag{2.48}$$

We can see that when the load is equal to the complex conjugate of the source impedance, the reflection coefficient is zero. However, most of the literature assumes that the reference impedance is real, in which case $Z_0^* = Z_0$. Therefore, the commonly used form of (2.48) does not include the complex conjugate symbol.

For *passive* Z_L terminations $0 \leq |\Gamma_L| \leq 1$. Perfect match is when $|\Gamma_L| = 0$. Unity $|\Gamma_L|$ means 100% reflection, caused by a short- or open-circuited load, as well as any ideal reactive element (remember that only resistance dissipates power). Active circuits may display reflection coefficient magnitudes greater than unity.

The reflection coefficient is measured by either a *scalar or* a *vector network analyzers* [13, 14]. The former measures magnitude only (in decibel form equal to return loss, covered in Section 2.17.2), while the latter provides both magnitude and phase information.

In impedance measurements, we know the reference impedance and measure the reflection coefficient of the unknown load, Γ_L. If Z_0 is real, we solve (2.48) for Z_L, which allows us to compute the unknown impedance.

$$Z_L = Z_0 \frac{1 + \Gamma_L}{1 - \Gamma_L} \tag{2.49}$$

For practice, let us evaluate the impedance associated with two load reflection coefficients measured in $Z_0 = 50\Omega$ system as

$$\Gamma_{L1} = 0.33\angle 0^\circ$$

and

$$\Gamma_{L2} = 0.33\angle 180^\circ$$

Substituting the measured load parameters into (2.49) we obtain $Z_{L1} = 100\Omega$ and $Z_{L2} = 25\Omega$, pointing out the importance of phase information. Clearly, without knowing the phase angle, we could not differentiate between the two loads.

Power is reflected to the source from any load having improper impedance, even if the two components are directly connected to each other, as shown in Figure 2.33. The magnitude of reflection coefficient,

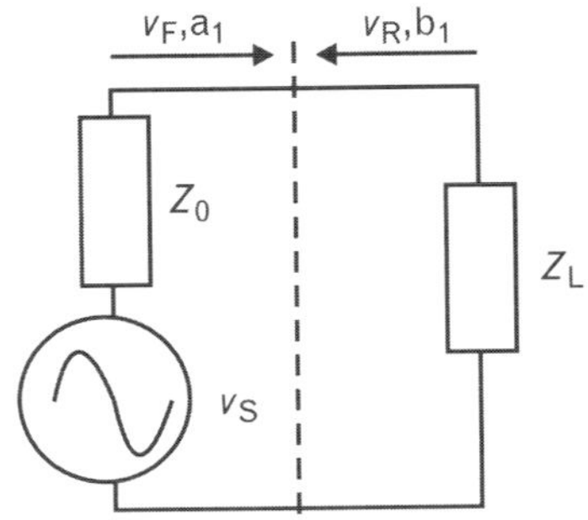

FIGURE 2.33 *Connecting a source of Z_0 impedance to a load with arbitrary impedance may cause power reflection even if the two terminations are connected directly to each other. Traveling voltage waves v_F and a_1 move toward the load while v_R and b_1 represent reflection coming back from the load.*

$|\Gamma_L|$, indicates a voltage ratio, while $|\Gamma_L|^2$ represents the power reflected. Connecting elements between the source and load generally makes the reflection more frequency dependent

2.17.2 Return loss

The reflection coefficient is a *ratio* of two normalized traveling voltage waves. Its magnitude squared represents the reflected power. Converted to decibels, it is called the *return loss,*

$$RL_{dB} = 10\left(\log|\Gamma|^2\right) = 20\left(\log|\Gamma|\right) \tag{2.50}$$

The return loss evaluates the *difference* between the reflected wave and the incident wave in decibels.

Return loss does not have phase information and it is primarily used in production testing where the goal is to reject parts that do not meet a specified minimum limit. For design purposes reflection coefficient is more useful, but it takes a more expensive test equipment to get the phase information also.

For passive circuits, $0 \text{ dB} \leq R_L \leq \infty$ dB. Active circuits may have *return gain* instead of *return loss*, meaning that the reflected signal is greater then the incident wave. Before transistors had useful gain at microwave frequencies, we used special diodes[14] that were capable of generating reflection coefficients with greater than unity magnitude to amplify signals.

14. Tunnel diodes, mentioned in Volume II, Chapter 1, fall into this category.

2.17.3 Voltage standing wave ratio

In the early years of transmission line measurements the determination of a load impedance proceeded by rather crude methods. Initially, sufficiently directive directional couplers were not generally available, and network analyzers had not yet been invented. The test instrument of choice, called a *slotted line*, consisted of a carefully fabricated transmission line having a thin longitudinal slot cut in its outer conductor and fitted with an exterior carriage. Within the carriage was mounted a small wire probe connected to a rectifying semiconductor crystal. The crystal's rectified voltage output was connected to an oscilloscope.

When EM waves propagate in two directions inside a transmission line, a "standing-wave" pattern is formed. *Voltage standing wave ratio* (*VSWR*) is by definition the ratio of the maximum to the minimum voltages in the line. During measurements, the probe was slid along the line through a distance of at least a quarter-wavelength to record the maximum and minimum voltages of the standing wave pattern. The maximum was proportional to $(1+\rho)$ and the minimum was proportional to $(1-\rho)$.

VSWR is a "leftover parameter" from decades ago, rarely measured these days. However, it still appears on many data sheets and specifications and we need to be familiar with it. *VSWR* provides magnitude only without any phase information. Some of the literature refers to it simply as *SWR*.

$$VSWR = \frac{v_{MAX}}{v_{MIN}} = \frac{v_F + v_R}{v_F - v_R} = \frac{1+|\Gamma|}{1-|\Gamma|} = \frac{1+\rho}{1-\rho} \tag{2.51}$$

Ideal reactive terminations, including short and open circuits, reflect voltage waves with magnitudes equal to that of the incident waves. The result is perfect cancellation at half-wavelength intervals, and doubled voltage values between the nulls, as we saw in Figure 2.26.

For passive circuits, $1 \le VSWR \le \infty$. Reactive terminations have infinite *VSWR*s. Matched terminations (zero reflection) leads to unity *VSWR*.

Solving (2.51) for the reflection coefficient magnitude gives us

$$|\Gamma| = \frac{VSWR - 1}{VSWR + 1} \tag{2.52}$$

In the case of a circuit where two terminations facing each other are purely resistive, the *VSWR* between them is calculated by the *ratio of the larger resistor to the smaller one*. For example, between a 100-Ω source and a 50-Ω load the *VSWR* is equal to 2.0. Note, however, that if the source and the load resistor values are interchanged, the *VSWR* remains the same. A 50-Ω source and 100-Ω load also lead to a $VSWR = 2$!

2.17.4 Mismatch loss

Mismatch loss (*ML*) is a computed parameter that tells us the power wasted between two interconnected ports, as shown in Figure 2.32, due to mismatch. It is a useful parameter since it tells us how much gain improvement we get by proper impedance matching. Using a one-unit normalized signal power available from the source, there are two ways to compute mismatch.

1. *Arbitrary load and Z_0 source* ($\Gamma_L \neq 0, \Gamma_S = 0$). If the source impedance is equal to Z_0, any power reflected from the load is dissipated in the source without any rereflection toward the load. In that case, the normalized voltage reflected from the load is $|\Gamma_L|$ and the normalized power reflected from the load is $|\Gamma_L|^2$.

   The normalized power delivered to the load is the difference of what is available from the source and what is reflected, $(1 - |\Gamma_L|^2)$. The reciprocal of this amount is the fractional increase of power delivered to the load without reflection, called *mismatch loss,*

$$ML = \frac{1}{1 - |\Gamma_L|^2} \tag{2.53}$$

   Converting (2.53) to decibels,

$$ML_{\text{dB}} = -10\left(\log\left(1 - |\Gamma_L|^2\right)\right) \tag{2.54}$$

2. *When both port impedances are other than Z_0* ($\Gamma_S \neq 0$, $\Gamma_L \neq 0$), the mismatch loss is more complicated because the signal bounces back and forth between the two terminations. The fractional mismatch loss is

$$ML = \frac{|1 - \Gamma_S \Gamma_L|^2}{\left(1 - |\Gamma_S|^2\right)\left(1 - |\Gamma_L|^2\right)} \tag{2.55}$$

   Finally, converting (2.55) to decibels gives us

$$ML_{\text{dB}} = -10\left[\log\left(\frac{\left(1-|\Gamma_S|^2\right)\left(1-|\Gamma_L|^2\right)}{|1-\Gamma_S\Gamma_L|^2}\right)\right] \tag{2.56}$$

Equations (2.55) and (2.56) tell us that without having phase information for Γ_S and Γ_L we cannot compute the exact amount of mismatch. Since the phase of is generally not specified for purchased components, from the magnitudes we can only compute the *maximum and minimum amount of the mismatch*, sometimes called *mismatch uncertainty* (MU).

$$MU_{\text{dB}} = 20\left(\log\left(1 \pm |\Gamma_S\Gamma_L|\right)\right) \tag{2.57}$$

Note that (2.54) and (2.56) had negative signs at their fronts to change the negative values of the expressions to positive numbers, showing loss as a positive quantity. Equation (2.57), however, does not have a negative sign and we can get either a negative or positive decibel value as a result. We will explain this further in Chapter 4 and see that cascading two amplifiers, where each one has 10-dB gain, could give us less or more than 20-dB overall gain.

Between passive terminations the reflected signal is eventually attenuated to noise level since the magnitudes of the reflection coefficients are less than unity. In active circuits however, large reflection coefficients may lead to oscillation, as we see in Volume II, Chapter 1.

The four circuit parameters (Γ, *RL*, *VSWR*, and *ML*) are interrelated. Knowing one, the magnitudes of the others can be computed.

2.17.5 Difference between mismatch loss and return loss

Return loss and mismatch loss are frequently interpreted incorrectly, so let us review them one more time.

Return loss represents the difference between the reflected and incident power. A *large* return loss value indicates a good match.

Mismatch loss represents the maximum possible power gain improvement when the mismatched source is matched to the load with a lossless matching network. A *small* mismatch loss results in a well-matched case.

2.18 Impedance transformation and matching

In circuit designs it is important to distinguish between two similar and therefore often confused activities, impedance *transformation* and impedance *matching*. Since we use these terms frequently throughout our books, let us illustrate their differences (Figure 2.34).

Impedance transformation is used to transform one impedance to a different value. Impedance transformation of Z_1 to Z_2, as the term implies, changes Z_1 to Z_2. At the output of the transformer network, the new output impedance, $Z_{OUT} = Z_2$. We are not concerned about what termination is connected to the output port. For example, we may want to transform a $Z_1 = 50\Omega$ source to $Z_2 = (80 + j65)\Omega$ because that is the source impedance for minimum noise operation of a transistor. The output termination for this circuit, which is the input impedance of the transistor, does not enter into our impedance transformation computation.

Impedance matching is always performed *between two specified terminations*. Impedance matching between Z_1 and Z_2 changes Z_1 to the complex conjugate of Z_2. At the output of the transformer network, the new output impedance, $Z_{OUT} = Z_2^*$.

If Z_2 is real ($X_2 = 0$) impedance transformation and matching have the same results. When Z_2 is complex, the outcome of the two processes is always different.

2.19 Single-ended versus differential circuits

Growth in digital integrated circuits has been driven by a doubling of complexity and speed every 18 months through the past decades. Reductions into deep submicron dimensions benefit RF integrated circuits with increased clock speed and increased maximum frequency performance. Wireless RF communication circuits and integrated circuits are also becoming more complex and packing functionality and signals into an ever-closer space. While the reduced size offers many advantages, it also introduces new problems. Increased level and density of integration lead to

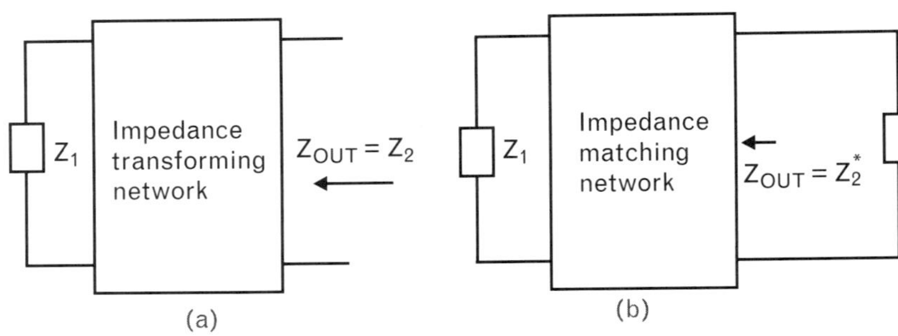

FIGURE 2.34 *Illustrating the difference between (a) transforming an impedance to another value and (b) matching a given impedance to a second one. Impedance transformation is not concerned about the second termination, while impedance matching always takes place between two specified terminations.*

a high-level of electromagnetic interaction between circuit nodes. Low voltages, which are required for dense digital functionality, make it troublesome for common RF circuits to meet performance parameters such as linearity, dynamic range and output power.

Mixed-signal designs with complex digital and analog functionality, called *system-on-a-chip* (SoC) designs, are now becoming commonplace. Isolation of circuit modules such as analog and digital sections becomes extremely difficult in these SoC designs. Using conventional RF circuit design and existing isolation techniques, it may not be possible to achieve electromagnetic compatibility between noisy digital and sensitive analog sections.

We already mentioned that RF circuits are typically single-ended (unbalanced), while most of the low-frequency analog integrated circuits are differential (balanced) type. Since more and more of the RF communication systems are being integrated, we examine the fundamental differences between the two approaches, and in Chapter 4 we provide more details of modern differential design technique for RF applications.

2.19.1 Single-ended RF circuits

In single-ended (unbalanced) circuits dc and signals currents flow through a common—generally ground—terminal as shown in Figure 2.35. Grounded conductive structures are also used for shielding, to protect the upper conductors from interfering signals and prevent radiation.

In transmission lines parallel capacitance is distributed together with series inductance. In many other cases however, the presence of shielding introduces additional capacitance, which can alter the performance of a circuit. Finite, nonzero, ground impedance leads to signal loss, noise, and other interference.

Benefits of the single-ended circuitry are as follows:

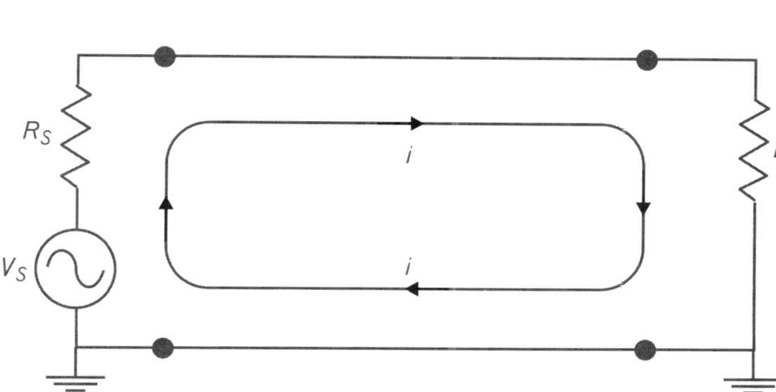

FIGURE 2.35 *In unbalanced circuits a good ground connection is required between the signal source and the load, because the ground is in the path of the signal current. Ideally this ground should represent zero impedance.*

- Well-established design capability using traveling wave–based techniques, such as scattering- (*S*-) parameters.
- Wide range of available RF/MW test equipment: network analyzers, spectrum analyzers, and so forth.

Although the development of RF test equipment and design techniques enabled us to improve the performance of single-ended circuits and systems during the past decades, unbalanced operation has limitations, as follows:

- Susceptibility to noise and interference through magnetic and capacitive coupling, power supply, and ground connection. Any electrical interference, presented to the input in the form of coupling, radiation, or hum, is amplified and attached to the signal at the output of the circuit, as shown in the simplified amplifier of Figure 2.36. At the output it can require excessive filtering or signal processing to eliminate such unwanted interference.
- Limited input-output isolation during test, particularly at chip level where the test probes are physically very close to each other.
- Need for good RF ground which is almost always not located where needed.

2.19.2 Differential RF circuits

Differential signal transmission and processing has been used at lower frequencies in the telephone industry for a very long time. The basic wired telephone circuit in the residence or the commercial workplace is a differential transmission and reception device with twisted pair wire connections that help reject common mode noise. Now differential techniques have found their way into RF transceiver systems throughout the analog and

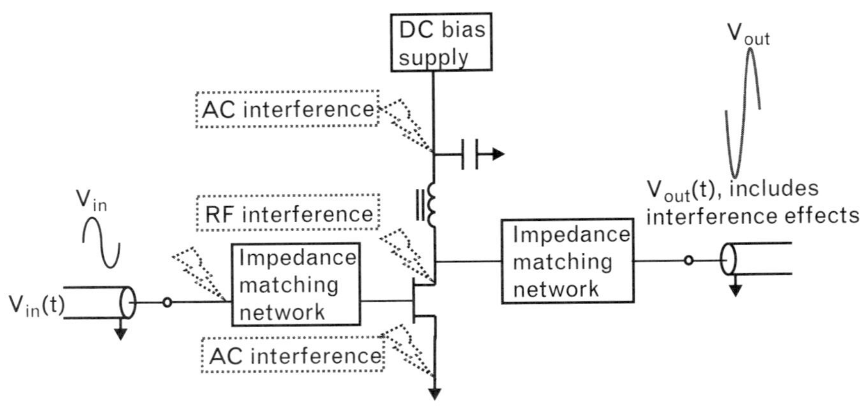

FIGURE 2.36 *Block diagram of a typical single-ended two-port RF amplifier, showing several points where electrical noise and interference may be added to the signal.*

high-frequency signal processing sections. This includes the transceiver integrated circuits in the transmitter, the receiver, and right up to the antenna.

In a differential circuit, signals with equal magnitude and opposite phase are transmitted and processed across a matched two-conductor system. The ground or dc supply path, ideally, is not a path for the signal currents or the signal return currents to the source.

Differential technology is the processing of signals along a two-conductor system without ground or dc supply as a path for the signal currents. A simple balanced differential circuit is displayed in Figure 2.37, which is comprised of two equal amplitude and opposite-phase single-ended signal sources, v_{S1} and v_{S2}, with source impedances of R_{S1} and R_{S2}. The sources are terminated with loads R_{L1} and R_{L2}. Both sources and loads are connected to a common ground. The differential, or balanced current flow, i_D, that does not include the ground reference path is represented with dashed lines. The single-ended unbalanced currents, i_1 and i_2, generated by the two signal sources, are marked with dotted lines.

In Figure 2.37 the source signal is applied differentially with amplitude equal to two times v_{S1} (or v_{S2}) while the load is also applied differentially. When the input signals are equal amplitude and opposite in phase, R_{S1} and R_{S2} as well as R_{L1} and R_{L2} are also equal, the resulting currents i_{c1} and i_{c2} cancel each other in the ground path. No ground currents are expected to flow into or out of the ground nodes. This means that balanced differential circuit does not have any component applied to the ground reference nodes.

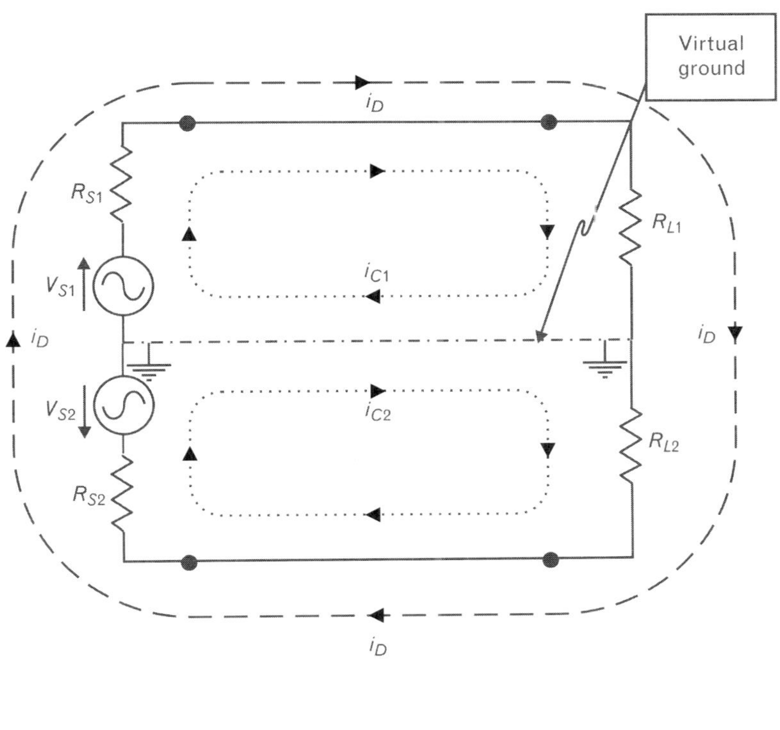

FIGURE 2.37
A balanced circuit operates equally well with or without a common ground connection since the net current flow in the ground conductor is zero.

Even when the balanced conditions are not exactly fulfilled, the signal currents on the ground or reference path are much less than a single-ended circuit, where the entire signal flows through the ground reference path.

Differential amplifiers have been used by integrated semiconductor industries for a long time (Figure 2.38), offering several benefits over single-ended circuitry. As the useful frequency range of RFICs reached and passed the gigahertz range, researchers began to focus on developing new design and test capabilities. Still, since most RFIC designers have low-frequency analog background, it takes time to adopt the *S*-parameter based design techniques used by RF and microwave designers for some time.

Advantages of differential over single-ended circuitry are as follows:

- *Better immunity to common-mode interference*[15] *and noise:* The coupled noise or other interfering signal couples (nearly) equally and in phase to both channels in the balanced circuit. As a result it induces the same common mode voltage on the pair. But since only differential mode (180° out of phase) signals on the pair constitute a signal, the induced *common mode* voltage is rejected and does not constitute interference.

  Of course, this presupposes that the two sides of the differential, are perfectly symmetrical with respect to each other and to ground circuits in order that the common mode induced voltages on both sides of the circuit are exactly equal in magnitude and phase. Asymmetry in either the coupling or the circuit compromises this interference reduction property of the balanced circuit.

FIGURE 2.38
Simplified circuit of a differential amplifier

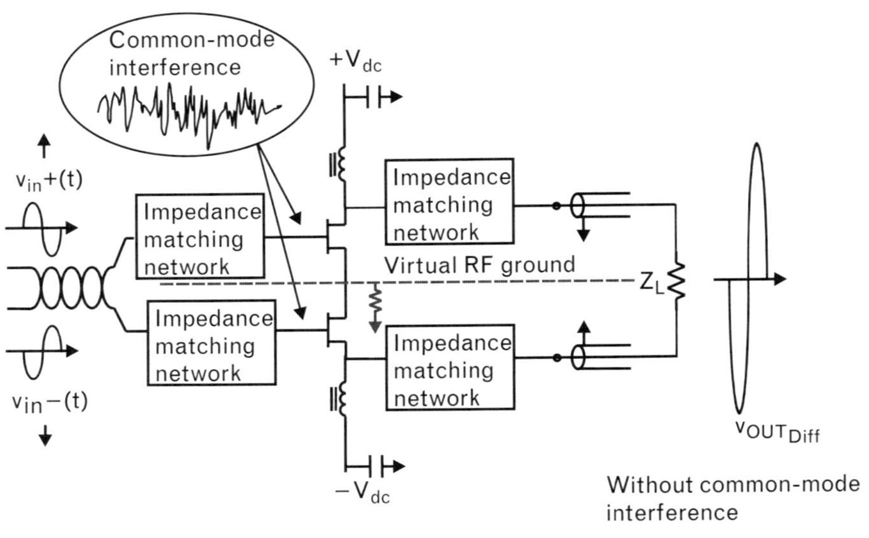

15. How much of a common-mode input signal shows up across the differential output.

- *Low even-order distortion:* Balanced operation eliminates even-order distortion products caused by the nonlinearities of active circuits. Since there are two (nearly) identical amplifiers in a balanced circuit, at their respective outputs the phases of a second harmonic distortion products are identical, thus they cancel each other in the differential mode. The same argument holds for all even harmonics.

- *Tolerance for nonideal RF grounds:* In a differential circuit there is no net current going to ground because both of the balanced sides of the circuit have oppositely directed ground currents. As a result, in a perfectly symmetric differential circuit the ground connection is *virtual*; that is, there is no differential current carried by it and ideally no need for a ground contact at all. Even if the symmetry is less than perfect the extent to which it exists minimizes the need for a ground. Since maintaining a low impedance connection to ground is often a problem in unbalanced circuits, the lack of a need for ground in the balanced circuit conveys a considerable practical advantage.

- *Power supply noise suppression:* To the extent that there is power supply noise, such as ac hum, on a common dc supply for both amplifiers of a balanced circuit, this noise will appear as a common mode voltage and will not interfere with the differential mode. (Note: This suppression may not apply to power supply *voltage bounce* by which the power supply voltage falls as the transistor turns on.)

- *Improved isolation for on-wafer RF probing:* Many wireless circuits are realized as integrated circuits on a semiconductor wafer. Probing the wafer for, say, the gain of a given amplifier is complicated by the fact that the input and output of the circuit may be separated by only a few mils (one thousandth of an inch). Consequently, the probe used to sample an unbalanced circuit may, when connected to the input, pick up an induced signal from the output. But when probing a balanced circuit, each of the probe leads tends to pick up the same voltage from both sides of the output (a common mode interference) and this is inherently isolated from the differential mode voltage being measured.

- *Improved crosstalk*[16] *rejection:* Crosstalk from an input pad to both output terminals tends to be symmetric; it therefore appears as a common mode and is suppressed in a balanced measurement.

- *Higher output power capability:* In a single-ended power amplifier the output voltage swings between a fixed maximum, v_{MAX}, and ground. In the differential output the voltage swing doubles since it swings

16. We refer to the coupling from one input to both output terminals.

between $\pm v_{MAX}$, giving 6 dB more power. As a result, dynamic range[17] also increases.

That all sounds great. Why do we bother then with single-ended circuits? Well, even differential circuits are not perfect—they have the following shortcomings compared with the single-ended approach:

- Twice as many components and more complex circuitry, although with integrated circuits it only increases the chip size;
- Twice as much current drain or voltage;
- More input and output terminals that may reduce the benefits of reduced crosstalk;
- Low passive component Qs in RFIC realization [*microelectromechanical* (MEM) [15] technology promises help here by lowering component losses];
- Requires a true balance for the circuitry, which is difficult to achieve with discrete circuits;
- Last but not least, until recently accurate characterization did not exist for balanced circuits. The new test and design methodologies are *very slowly* finding their ways into RFIC design and test departments. Mixed-mode *S*-parameter techniques [16] are covered in Chapter 4.

2.19.3 Electromagnetic compatibility and interference

Electromagnetic compatibility (EMC) and *electromagnetic interference* (EMI) [17] refer to the issue of remote signal crosstalk. The goal of EMC is to prevent electromagnetic interference that disturbs wireless signal reception. Interference may be caused by conduction, radiation, and induction. EMC and EMI examine crosstalk between the *victim* circuit surviving a hostile remote interference environment and the *offending* circuit propagating signals. Electromagnetic crosstalk creates a reciprocal reception-propagation relationship between the *victim* and *offending circuit*. It is the responsibility of electronic equipment manufacturers to design EMC into their products, along with the EMI requirements. Cost of first introducing EMC into a product increases significantly through the various stages of development and production. Computer-aided design (see Chapter 6), including that with EM simulation, is essential to identify potential EMC problems and to find the optimum circuit and component layout that meets requirements. In cases where shielding alone is not sufficient, differential circuit

17. Difference between lowest detectable and maximum useful signal power levels.

techniques may bring the needed improvements over single-ended designs, for both EMC and EMI considerations.

2.20 Time domain versus frequency domain

For most logic applications, transmission-line effects and termination issues can be safely ignored if the round-trip delay of the longest interconnect path is less than about 10% of the signal transition time (rise time or fall time, whichever is shorter), regardless of mismatches and reflections.

Dual-state signals, typically referred to as digital logic signals, are significantly more tolerant of noise than analog signals. In digital signals, the dual states are defined as *logical "0"* and *logical "1,"* which in standard CMOS logic gates are defined as the power rail voltages, ground or v_{ss} and supply or v_{dd}. The result is a wide range of voltages defining the two digital states, compared to an analog signal with precise floating point state values at every point along the time axis.

Analog signal integrity is degraded immediately with even the smallest variation from the original signal value. However, a digital signal can endure a great deal of signal variation before the state definitions are affected. As the speed of digital signal processing increases, there is a point where an analog signal can be represented at a sampled rate with digitized values. Adding an interference signal with less than half the rail-to-rail supply voltage will ideally have no effect on the digitized signal quality. On the other hand, as we already stated in the previous section, adding an interfering signal of *any magnitude* to an analog signal degrades its signal quality.

2.20.1 Periodic waveform definitions

Closer examination of the digital signal does reveal analog properties in both the timing and amplitude values. A finite transit time, called rise time, is required for a digital signal to change between the "0" and "1" states as shown in Figure 2.39—duration between 10% of low-state minimum voltage, v_L, and 90% of high-state maximum voltage,[18] v_H. Other digital signal parameters with analog properties are fall time and duty cycle, which is the ratio between the duration of high voltage ("1") and the sum of high and low voltages ("1" and "0"), within one period.

Definitions of the various terms shown in Figure 2.39 are as follows:

- Rise time $t_R = t_2 - t_1$

18. In some specifications the rise and fall times refer to the time difference between 20% and 80% voltage levels, which gives the impression of better performance, when compared against 10% to 90% readings.

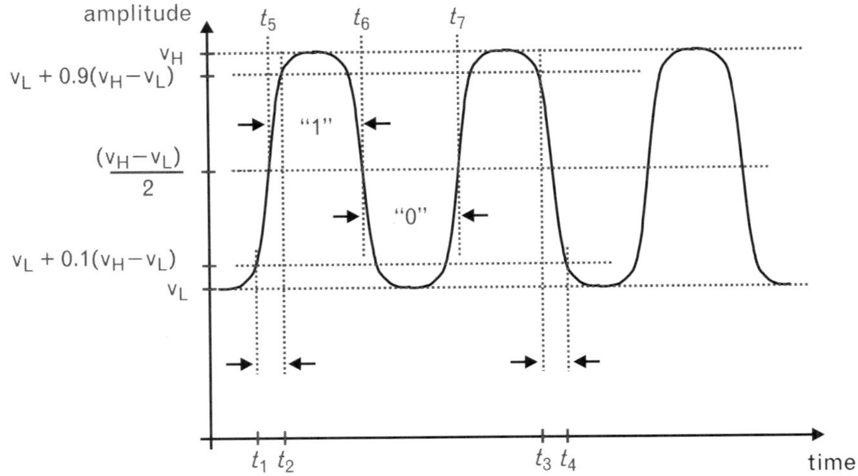

FIGURE 2.39
Time-domain definition of a periodic digital clock signal with analog definitions of rise time, fall time, and duty cycle.

- Fall time $t_F = t_4 - t_3$
- Clock period $T = t_7 - t_5 = 1/\text{clock frequency}$
- Duty cycle $D = (t_6 - t_5) \,/\, (t_7 - t_5)$
- Signal amplitude $A = v_H - v_L$
- State "1" $v > (v_H - v_L)/2$
- State "0" $v < (v_H - v_L)/2$

For the clock signal shown the "0" and "1" states are equal in time, giving a 50% clock cycle. The generalized expression for the complex spectrum of the periodic time-domain signal [17] of Figure 2.39 is given as

$$C_n = \frac{A}{2\pi n}\left(Y\sin(\beta) - X\sin(\alpha)\right) + j\frac{A}{2\pi n}\left(Y\cos(\beta) - X\cos(\alpha)\right) \tag{2.58}$$

where

$$X = \frac{\sin\left(n\pi t_R / T\right)}{n\pi t_R / T}$$

$$Y = \frac{\sin\left(n\pi t_F / T\right)}{n\pi t_F / T}$$

$$\alpha = n\pi t_R / T$$
$$\beta = n\pi\left(DT + t_R\right)/T$$

with

T: period of signal

D: duty cycle of signal

The envelope of the spectrum given by (2.58) is shown in Figure 2.40. The conversion formulas between frequency-domain and corresponding time-domain parameters are given by (2.59) through (2.61).

The spectral envelope is described by three characteristic frequencies:

1. *Fundamental* (nominal) *frequency* of signal, f_0

$$f_0 = \frac{1}{T} \tag{2.59}$$

2. *First cutoff frequency*, f_1

$$f_1 = \frac{1}{\pi TD} \tag{2.60}$$

3. *Second cutoff frequency*, also referred to as the bandwidth of signal, does not depend on f_0

$$f_2 = \frac{1}{\pi t_R} \tag{2.61}$$

Faithful transmission of a digital signal requires passing all harmonics up to *at least* the second cutoff frequency f_2, which is therefore called the *bandwidth* (BW) of the signal. The second cutoff frequency of the spectrum envelope depends only on the signal rise time: $BW = f_2 = 1/(\pi t_R)$. The spectral envelope is flat between f_0 and f_1. This flat portion is hardly noticeable on signals with close to 50% duty cycle, but gets wider as the duty cycle approaches 0 or 1. Harmonics beyond the second cutoff frequency have little effect on the waveform.

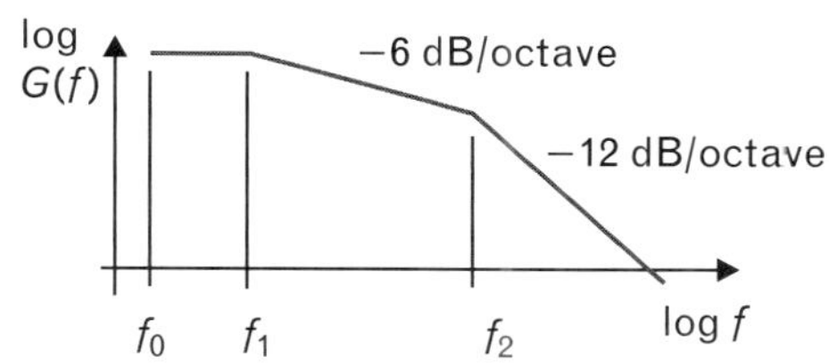

FIGURE 2.40 *Spectral envelope of a symmetrical periodic time-domain signal, showing key frequency points.*

Periodic digital signals (e.g., clock signals) have comb spectra, which have spectral lines only at integer multiples of the fundamental frequency.

Note that:

- Shorter rise-time edges result in higher second cutoff frequency and less attenuation of higher order harmonics.
- Frequency f_1 almost equals f_0 for duty cycles close to 50% (e.g., clock), and $f_1 >> f_0$ for small duty cycles (e.g., noise glitches).

In the waveform of Figure 2.39 we showed a sharp transition between 0 and 1 states at the $(v_H + v_L)/2$ voltage level. In practice a much wider margin is used [18], as shown in Figure 2.41. Setting v_L and v_H for high and low logic states, the drivers are specified to provide at least v_{HS} for high state. Logic lows are reached by dropping and holding below v_{LS}.

To minimize the effects of noise and interference, logic highs must reach at least v_{HMIN} and logic lows must remain below v_{LMAX}. For such specifications the noise margins [18] are:

$$\text{Noise Margin}_H = v_{HS} - v_{HMIN}$$

$$\text{Noise Margin}_L = v_{LMAX} - v_{LS}$$

2.20.2 Jitter

Noise or interference signals summed with a two-state digital signal modify the amplitude along the digital signal including the transit time. The two-

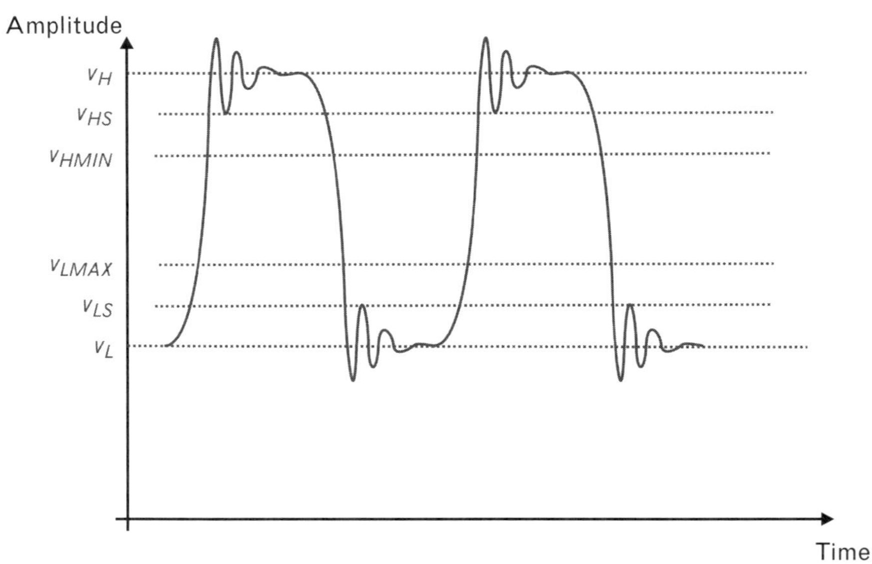

FIGURE 2.41
A distorted form of the waveform displayed in Figure 2.39, showing overshoot and undershoot in the high and low logic states. (After: [18].)

state portions of the digital signal are removed by an inverter stage. However, any added random noise during the transit portion of the digital signal moves the transition forward or backward in time. If moved far enough this time difference can affect the state value defined within a subwindow of the clock period, known as the bit period. For a clock signal source this time shift is called *random jitter* (Figure 2.42) [19] and is defined as the rising edge cycle-to-cycle period variation in statistical rms values. For low probability of bit error during detection, jitter must be much less than one-half of the bit period. More details are covered in Chapter 9.

2.20.3 Eye diagram

Single bits, also referred to as symbols, may affect the shape of following symbols if they are not settled before the clock transition. Plotting a random bit sequence of "1" and "0" states creates an eye-diagram [19], shown in Figure 2.43. A time-shifted signal holds its state for a duration equal to the sum of the maximum transit shift forward and backward, subtracted from the bit period. This time value is defined as the eye opening duration. The amplitude of the digital signal has a variation for both "1" and "0" states within this eye opening duration. Eye opening amplitude is defined as the maximum amplitude range between the lowest "1" state and highest "0" state voltages. The eye opening time duration and amplitude value defines the window where the bit state can be detected. Eye opening amplitude is compared to the $v_{HMIN} - v_{LMAX}$ difference shown in Figure 2.41 to determine possible errors in detection. The objective of a digital signal detector is to sample the signal at the center of this eye opening, with sufficient amplitude resolution to distinguish between the two digital states. *Intersymbol interference* (ISI) is when symbols affect detection of others.

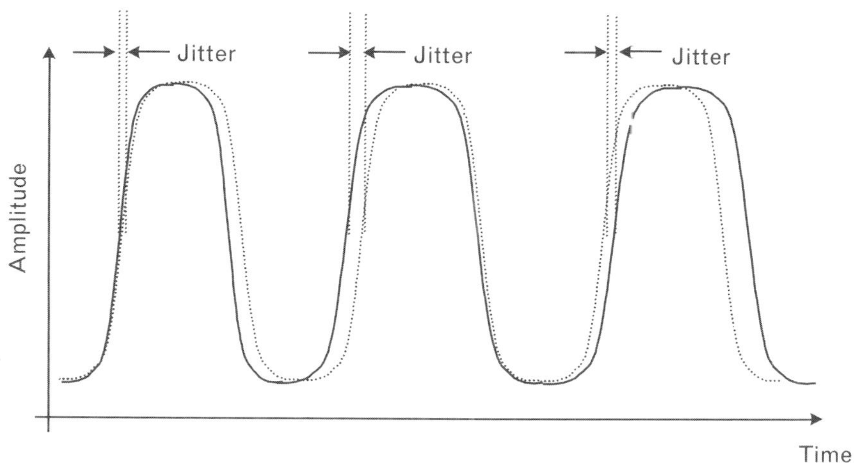

FIGURE 2.42
Clock signal showing cycle-to-cycle jitter induced by random noise, causing a time difference between ideal and actual voltages.

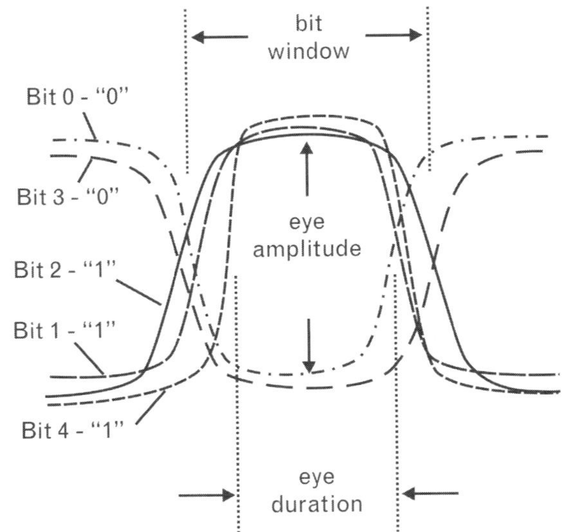

FIGURE 2.43
Digital detector output signal—eye diagram—shows the effect of random jitter. A large eye amplitude and small difference between bit window and eye duration are necessary for low bit error rate.

One measure of receiver signal quality or signal integrity is defined as the *bit error rate* (BER) [19], or the percentage of incorrectly detected digitized binary signals relative to the total number of digital signals received. A very important objective in a communications system is to minimize bit error rate. Additional bits can be added to provide bit error detection and correction, at the expense of slower signal transfer rate.

2.21 Summary

In this chapter we reviewed most of the essential background information needed to proceed with the rest of the book. Our attempt was to provide the minimum amount of necessary mathematical expressions, and only deviate from that rule to provide:

- Logical development of definitions commonly used in practical daily work;
- Answers to frequently asked questions by students in our short courses.

After teaching RF and microwave continuing education courses at various levels for several decades, we have seen a gradual decline in basic understanding of RF/MW fundamentals. If you have sailed through this chapter in a short time, then you are an exception and have a head start. Those of you who needed frequent review, and also looking up additional references, we recommend slowing down the pace and working through the examples with one of the RF circuit simulators. Change some of the circuit parameters and see if you can predict the results caused by the change.

If your interest is just to get enough out of this book to get your project going next week, our hope is that you will also find the book helpful.

References

[1] Gonzalez, G., *Microwave Transistor Amplifiers Analysis and Design,* 2nd ed., Englewood Cliffs, NJ: Prentice Hall, 1997.

[2] Kajfez, D., *Q Factor*, Oxford, England: Vector Fields, 1994.

[3] Wadell, B. *Transmission Line Design Handbook*, Norwood, MA: Artech House, 1991.

[4] Besser, L., "Skillful De-embedding," *Applied Microwave & Wireless*, Fall 1994.

[5] Laverghetta, T., *Practical Microwaves*, Englewood Cliffs, NJ: Prentice Hall, 1996.

[6] Maas, S. A., *The RF and Microwave Circuit Design Cookbook*, Norwood, MA: Artech House, 1998.

[7] Sinnema, W., *Electronic Transmission Technology*, 2nd ed., Englewood Cliffs, NJ, Prentice Hall, 1988.

[8] Adam, S. F., *Microwave Theory and Applications*, Los Altos, CA: Adom Microwave Consulting, 1992..

[9] Moreno, T., *Microwave Transmission Design Data*, Norwood, MA: Artech House, 1989.

[10] Howe, H., Jr., *Stripline Circuit Design*, Dedham, MA: Artech House, 1974.

[11] Pozar, D. M., *Microwave Engineering*, Reading, MA: Addison Wesley, 1990.

[12] Losee, F., *RF Systems, Components, and Circuits Handbook*, Norwood, MA: Artech House, 1997.

[13] Laverghetta, T. S., *Modern Microwave Measurements and Techniques*, Norwood, MA: Artech House, 1988.

[14] Wartenberg, S. A., *RF Measurements of Die and Packages*, Norwood, MA: Artech House, 2002.

[15] De Los Santos, H., *Introduction to Microelectromechanical (MEM) Microwave Systems*, Norwood, MA: Artech House, 1999.

[16] Weber, R. J., *Introduction to Microwave Circuits*, New York: IEEE Press, 2001.

[17] Paul, C., *Introduction to Electromagnetic Compatibility*, New York: Wiley, 1992.

[18] Young, B., *Digital Signal Integrity*, Upper Saddle River, NJ: Prentice Hall, 2001.

[19] Larson, L. E., (ed.), *RF and Microwave Circuit Design for Wireless Communications*, Norwood, MA: Artech House, 1996.

Selected Bibliography

Bowick, C., *RF Circuit Design*, Newton, MA: Butterworth-Heinemann, 1982.

Johnson, H. W., and M. Graham, *High-Speed Digital Design*, Englewood Cliffs, NJ: Prentice Hall, 1993.

Ludwig, R., and P. Bretchko, *RF Circuit Design, Theory, and Applications*, Upper Saddle River, NJ: Prentice Hall, 2000.

CHAPTER 3

The radio as typical RF system

The radio is a good example of a system in which most of the components we will study are used. The system is typically designed to meet a set of standards that all manufacturers must meet. Such conformance to a common specification enables units from different manufacturers to work together. From an RF point of view, the standard of most relevance is the air-interface specification, which outlines the requirements of the physical layer of the communications protocol and the wireless interface.

In this chapter, we will delve into how the radio standards are used to outline the system requirements, and from the system specification how the component specifications for each of its subsystems can be derived. We will assume some background undergraduate knowledge of communications concepts and vocabulary, which will be necessary in order to specifically understand the wireless interface of a radio.

3.1 Receiver architecture

The radio receiver is often a student's first introduction to the world of analog electronics. However, today's receivers are amazingly complex structures. The architecture of modern receivers continues to evolve to account not only for improvements in the analog performance of devices but for advances in *digital signal processing* (DSP) that permit more functions to be programmed in software rather than being hardwired in the circuits themselves. We begin by studying the receiver, and defer the transmitter until later chapters.

3.1.1 The simple detector receiver

The simple AM detector receiver is illustrated in Figure 3.1. As depicted, this is the forerunner of today's common AM radio receiver. Its basic components are:

- The *antenna*, whose function is to receive the incoming RF signal. All systems use carrier frequencies that can propagate through the media that makes up the communications channel (generally air). The antenna should ideally be matched to the media at the chosen carrier frequency to ensure good sensitivity. Modern systems will also impose requirements on its size, directional sensitivity, and possibly its sensitivity to out-of-band signals.

- An *RF filter*. In this simple receiver, the filter is tuned to select the desired frequency band (or *channel*) to be received, and to eliminate any undesired signals. These could be channels in adjacent frequency slots, or out-of-band spurious products.
- A *detector structure*. For an AM system the detector could be simply a diode that rectifies the RF signal and responds to the peak amplitude of the RF envelope. The purpose of the shunt inductor that precedes the diode in Figure 3.1 is to serve as an RF choke (i.e., to keep the input of the detector at dc ground but to preserve a high RF impedance in parallel with the diode, so as not to affect its RF performance). The RF choke needs to be able to carry any dc current, or rectified current from the RF, that the diode might generate.
- An *audio amplifier* (optional). This component is the only component that requires dc power or battery voltage, and it drives an audio output device (such as a speaker) with the demodulated waveform.

Such a system is clearly simple and cheap, with a minimum of required parts. The system shown detects AM modulation only, although detection of frequency- or phase-modulated signals could be performed by replacing the diode detector with other types of detector. However, such a receiver architecture is limited because its ability to differentiate between adjacent channels (or its *selectivity*) is determined by the ability of the input RF filter to screen out unwanted signals, and to pass ("select") only the desired signal. We will see that this is related to the quality factor Q of the filter,

FIGURE 3.1
A simple diode detector AM radio receiver

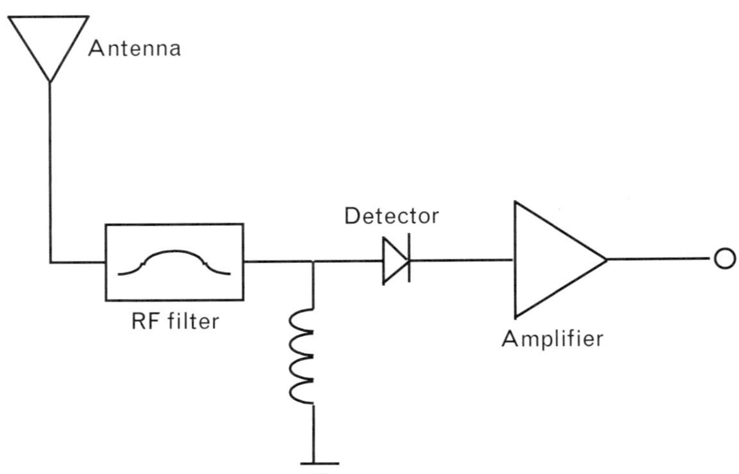

which needs to be very high to provide good differentiation between channels. This becomes more difficult to achieve at high frequencies, particularly if we want to tune the center frequency of the filter across large bandwidths. Furthermore, the ability to detect very small signals (or the receiver *sensitivity*) is limited by the noise that is detected together with the desired signal.

Later in this chapter we shall more precisely define the selectivity and sensitivity of a system, but these quantities are fundamental parameters that relate to a user's perception of the receiver quality. In a simple diode detector receiver selectivity and sensitivity are very limited and system performance is poor as a result.

3.1.2 The direct conversion (homodyne) receiver

In order to overcome the need to tune an RF filter to select the desired channel, the concept of downconverting a signal from RF to a baseband at dc can be used. In a direct conversion, or homodyne, receiver (Figure 3.2), a mixer is introduced, and the incoming signal is mixed with a carrier signal of identical frequency to produce a baseband signal prior to detection. In this way, the desired channel is selected by tuning the frequency of a local oscillator to match the incoming carrier frequency. This enables a fixed RF filter to be used after the antenna, which can typically now be built with higher Q than with a tunable RF filter. Of course, its bandwidth must now be wider to ensure coverage of the entire RF band relevant to the system specification. Such a receiver is simple, low cost, and adaptable to its environment.

The selection of the desired channel is made by tuning the local oscillator exactly to the desired RF channel frequency, and unwanted channels appearing at higher baseband frequencies, after downconversion, are

FIGURE 3.2
The direct conversion, or homodyne, receiver.

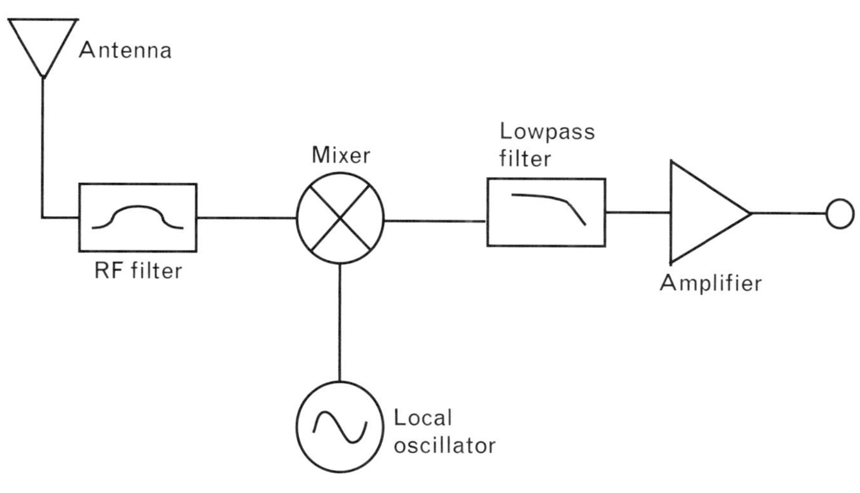

removed by a lowpass filter following the mixer. In receivers where the signal is digitally encoded, such channels can be removed by digital filtering within the digital signal processors that will also perform demodulation. In this case, an in-phase and quadrature LO signal is used to drive two separate mixers. They generate an in-phase (I) and a quadrature (Q) baseband output that retains both the amplitude and phase of the modulation, in Cartesian form. So-called quadrature downconversion is necessary because in general the two sidebands of the RF spectrum around the carrier frequency are different and a single mixer would destroy the distinction between them.

The ability to use a higher Q, fixed bandwidth RF filter preceding the mixer and the lowpass filter following it helps to prevent out-of-channel noise entering the detector and to minimize its loss, thus improving its sensitivity. With the use of DSP for detection, it also allows multiple modulation formats to be detected provided the signal phase is preserved, with I and Q baseband outputs. Such a system has a low number of components, requiring only an additional mixer and local oscillator compared with the simple detector receiver. Nor does it require many of the components in the more complicated superheterodyne system described below, such as the image reject filter or the IF stages.

However, there are several limits to the performance of the direct conversion receiver. First, there are a number of noise sources at dc that add to the desired signal. These include $1/f$ noise from the devices in the amplifier and mixer (we will see later that $1/f$ noise is low-frequency noise whose power density decays as the inverse of frequency; such noise is generated by impurities within the semiconductor lattice of the active devices). Another source of unwanted signal arises from dc offsets and dc products resulting from second-order nonlinear distortion (which we will also shortly define) in the amplifier and mixer. This problem can become worse when strong interfering signals are present, particularly since then the dc offset is no longer static. The dc offset can be particularly severe in CMOS circuits, where mismatches between devices can cause offsets to be up to two orders of magnitude higher than the signal itself. In addition, the local oscillator signal could leak through the mixer back out to the RF port, and because it has the same frequency as the incoming signal, out through the RF filter to the antenna. Even if an RF amplifier is used between the filter and mixer, its reverse isolation would need to be high to avoid this problem. Such a reradiated signal could interfere with other receivers in the vicinity, or be reflected from (possibly moving) reflectors back into the mixer and receiver, where it would mix with itself and exacerbate the dc offset problem. There is also a risk of the RF pulling the LO frequency if isolation in the mixer is imperfect, since the two frequencies are the same.

In spite of these problems, the homodyne receiver is reemerging as an attractive candidate for multiband receivers where the receiver must cover several RF bands and channels. This architecture has been widely used in GSM mobile phones since the late 1990s, because of the multiband coverage and the reduction in component cost it provides. We explore the alternatives more fully in Volume II, Chapter 8, Section 8.2.

Near-zero IF receivers manage to avoid these problems by downconverting to a center-frequency close to, but not exactly zero. The downconverted frequency is known as the *intermediate frequency* (IF). It preserves the modulation of the RF signal, since only the frequency is linearly translated to a different value. The IF might lie for instance at several hundred kilohertz, perhaps one or two times the channel bandwidth. Such a receiver can be thought of as a homodyne (zero IF) receiver whose output rotates at the near-zero IF rate. In a digital radio, where DSPs or *application-specific integrated circuits* (ASICs) are used to digitally select the channel from the low-frequency IF carrier, the advantages of a single downconversion are maintained without the dc problems. However, a new problem arises, and that is that the *image frequency*, and the baseband beat products arising from second-order distortion, can both fall in-band. Since they can be stronger than the desired signal, they require a high-resolution analog-to-digital converter to maintain dynamic range. The problems of the image and distortion are discussed later in this chapter.

3.1.3 The superheterodyne receiver—analog system

The superheterodyne, or superhet, receiver, is a well-known and almost universal receiver architecture for radio receivers. We will see that such a receiver can provide both good selectivity and sensitivity, because the noise bandwidth can be limited to the channel bandwidth without compromising the receiver's ability to tune across the entire RF band. Its basic components are shown in Figure 3.3.

Let us examine in turn the function of each of the blocks in this receiver.

- The *antenna*, whose function is identical to that described earlier.

FIGURE 3.3
Basic architecture of the analog superheterodyne receiver.

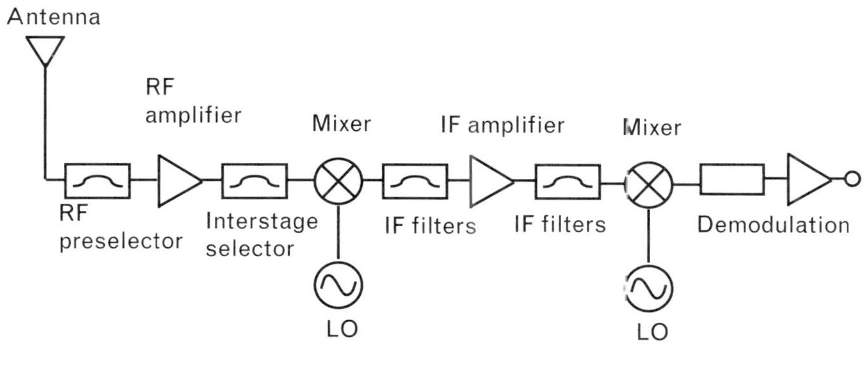

- The *RF preselector.* The purpose of this component is to filter out all unwanted signals lying outside the RF band containing the possible channels to be detected. Unwanted signals can include signals fed from the transmitter itself, which might share a common antenna. In this case the preselector is either a diplexer, a filter that allows one frequency band to pass between the antenna and receiver and another (different) band to pass between the transmitter and antenna (as, for example, in CDMA mobile systems), and/or a transmit/receive switch that turns on during the receive phase and off during the transmit phase (as, for example, in GSM systems). Such a circuit thus allows full or half-duplex operation, respectively, and prevents overload of the downstream receiver components from unwanted frequencies. The function of the preselection filter is also to suppress the undesired responses at the output that arise from incoming signals lying at *spurious* frequencies f_S that the receiver is not tuned to receive (which we will call the *tuned* frequency f_T). We will see that such spurious frequencies can include the *image* frequency as well as harmonics of the incoming tuned frequency. Ideally, the preselector will also have a good impedance match in-band to avoid bandpass ripple.

- The *RF amplifier.* The function of the input amplifier is to linearly amplify the input signal and minimize the noise added by the receiver to the signal itself. We will see that such low noise amplification can be achieved by noise matching the input of the amplifier, and is important because it can determine the overall *noise figure* of the entire receiver. It should have a good input and output match to avoid gain ripple. In addition, the input amplifier must not introduce distortion of the signal, because strong signals may be simultaneously present in adjacent (unwanted) channels, and any *nonlinear distortion* of the amplifier could swamp a weaker signal in the channel we are trying to detect. To achieve this, it will require a reasonable maximum power-handling capability indicated by its *input intercept point.* Of course, the RF amplifier can consist of multiple stages in order to provide the necessary gain.

- The *interstage selector.* The preceding amplifier will provide gain to all the channels within the RF bandwidth, and its gain is likely to roll off slowly beyond it. Furthermore, the amplifier will amplify noise across the entire band, and possibly at the image frequency as well. Therefore, this (optional) component is a filter to suppress any gain of undesired signal responses at spurious frequencies, and in particular at the image frequency. It thus maintains the system noise figure by preventing image noise from entering the mixer. It also helps to minimize LO reradiation from the RF mixer port. This component should have low in-band loss.

- The *local oscillator*. This is a strong signal that is normally generated by a frequency synthesizer, and is typically tuned across a bandwidth equal to the entire RF bandwidth, but offset from it, to choose any desired channel. Its function is to drive the devices within the mixer into a nonlinear regime for frequency translation (mixing). An important oscillator specification is its phase noise, since any phase fluctuation on the oscillator signal is directly superimposed on the mixer output signal. Its broadband noise should also be low so as not to raise the system noise floor. It will also require a good tuning range or bandwidth, and low spurious and harmonic content.
- The *first mixer*. This component translates all incoming signals in the RF frequency range into signals in some intermediate frequency range, depending on the local oscillator signal frequency. The mixer translates all frequencies linearly, preserving phase information within the new range of frequencies. Within some range of RF signal amplitudes, the amplitude of the output signal is also preserved on the IF. We will also see that the choice of mixer topology (e.g., single balanced, double balanced, and so forth) is important in rejecting certain unwanted mixer output components. It requires low LO feedthrough to the RF and IF, and a large spurious-free dynamic range for the incoming RF signal. The selection of the IF frequency is important in ensuring the receiver response to unwanted spurious responses is minimized. A historic rule of thumb for HF/VHF receivers is that signals are upconverted to an IF at twice the highest RF frequency. At microwave frequencies where the tuning range is much less, signals are typically downconverted to a lower frequency. IF frequencies from 45 to 82 MHz are common for mobile radio receivers in the 800-MHz band, and from 110 to 300 MHz for radios in the 1,800-MHz band or for both bands.
- The *IF filter*. This component rejects the unwanted signal components generated by the mixer and other components. Its bandwidth must be sufficiently wide to pass the modulation sidebands in the desired channel without distortion. It can be high Q because it is of fixed frequency. The blocking and overload characteristics of the receiver are often determined here in combination with the following IF amplifier.
- The *IF amplifier*. This component should provide adequate gain to the IF signal to drive the following stages. Because it is at a fixed frequency it can provide high gain and be well stabilized.
- The *second mixer, local oscillator*, and *second IF filter*. Sometimes known as the second *IF strip*, these components mirror the functions of the first IF strip, but across a different frequency band and with different passband characteristics. The power handling capability (intercept

point) is generally higher. The purpose of a second IF strip will unfold later in this chapter, but as a general rule, the first IF strip is designed to receive the entire RF passband and to reject spurious frequencies and the image frequency in particular, while the function of the second IF strip is to narrow in and select the desired channel from the entire passband, thereby providing additional selectivity to the receiver. For commercial AM radios, 455 kHz is a fairly standard second IF frequency, while mobile phone receivers typically use frequencies around 10 MHz. In some architectures, the second IF is omitted totally and a single downconversion from RF to a low IF is used instead. In that case, the (first) IF filter also performs channel selection.

- The *demodulator.* This component extracts the modulated signal from the IF signal and converts it to baseband. In the case of an analog system, this information will be either AM or FM; for a digital system, it will typically be symbols having multiamplitude or multiphase levels that are later decoded. (A symbol is the way a bit or combinations of bits are coded in the waveform.)
- The *baseband amplifier.* This provides output power to drive the relevant output device, which could typically be a speaker, fax output, or video screen.

One variation on the standard superheterodyne structure is to fix the (first) LO frequency and downconvert the entire RF band to a wideband IF. This enables a fixed frequency, thus high Q, low-phase noise oscillator to be used for the RF LO, and tuning and channel selection to be performed by varying the IF (second) LO [1]. This can overcome the problem of phase noise introduced by the LO, which is discussed in Volume II, Chapter 6. It can, however, result in dc offsets at baseband since the second LO will be at the same frequency as the IF, but this is not a problem unique to this architecture. It does however require a second LO with a broad percentage tuning range, and also exposes the second mixer stage to strong adjacent-channel interferers since now the RF stage has limited ability to select the desired channel.

3.1.4 The superheterodyne receiver—digital system

Although we are primarily concerned with RF circuits, it is important to understand the effect of digital circuits if they impact the overall system and RF characteristics. Modern wireless systems use digital coding on the RF carrier (i.e., the baseband information to be transmitted is digitized). For instance, the modulating waveform (if analog to begin with) is digitized into bits, and these information bits could be grouped into pairs, where each of the four possible combinations (0,0), (0,1), (1,0), and (1,1) is assigned a phase of 0°, 90°, 180°, and 270° to modulate the RF carrier. There are four possible symbols in this system, known as *quadrature phase-shift keyed* (QPSK).

A basic superheterodyne receiver for a digital signal is shown in Figure 3.4. While the input stages of this receiver are the same as shown in Figure 3.3, there are several notable differences in the output stages.

- The *matched filter*. As in the analog receiver, this is a bandpass filter that is matched for the expected modulation format. It can be shown that in a communications channel with additive Gaussian noise, the optimal detection filter is identical to that used in the transmitter channel to filter the symbols prior to upconversion. Its ideal bandpass response is the complex conjugate of the transmitted signal spectrum. Bandpass filters, such as the Nyquist filter, are covered in more detail in Chapter 8.

- The *mixer*. This is the same as for an analog signal, in that it converts the signal to a final IF frequency. However, the choice of final IF frequency is now restricted by the digital clock rates and the analog-to-digital conversion sampling frequency.

- The *IF filter and amplifier*. These also serve the same purpose as in the analog system. Following the mixer, any dc offset needs to be removed and out-of-band frequencies reduced with a lowpass filter prior to sampling.

- The *analog-to-digital converter*, including a sample and hold circuit. This digitizes the final IF signal prior to final channel selection, downconversion, and demodulation, all performed digitally.

There are in fact two such parallel output stages that map the signal into its Cartesian coordinates, and thus maintain its phase, even when the amplitude is zero. One output corresponds to an in-phase, or *I* signal, and the second corresponds to the quadrature, or *Q* signal. These allow both amplitude and phase modulation to be preserved. This now requires two

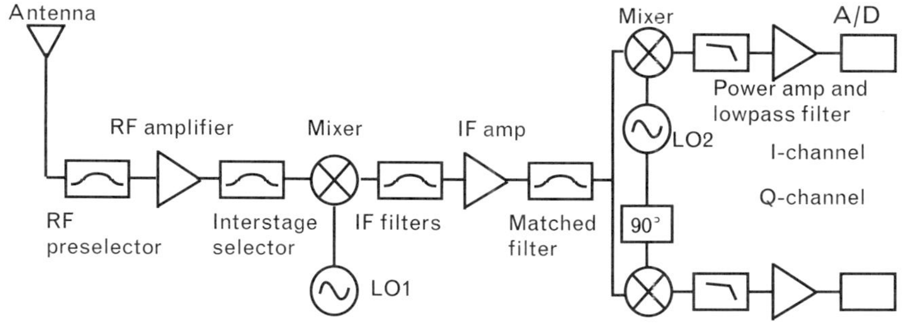

FIGURE 3.4
A basic digital superheterodyne receiver.

mixers, one driven by an LO signal of the form $\cos \omega_{LO} t$ and the other by $\sin \omega_{LO} t$. The I and Q channels ultimately correspond to the even and odd bits of the digital signal, which can be processed at half the rate of the total bit stream.

The analog and digital worlds are not totally independent. For instance, the sample-and-hold clock frequency should be set to a multiple of the LO frequency driving the second mixer *LO2*, so that any clock signal leaking into the system will not mix with the *LO2* and cause spurious signals at unexpected frequencies. Ideally, the two frequencies should be derived from the same source so that any dc offset that might arise from mixing of their harmonics is stable rather than a troublesome beat frequency. But more fundamentally, the *LO2* frequency needs to be selected in such a way so that the frequency f_{IF} of the second IF is either

$$f_{IF} \leq \frac{1}{4} f_{S\&H} \tag{3.1}$$

or

$$f_{IF} = \frac{3}{4} f_{S\&H} \tag{3.2}$$

or

$$f_{IF} = \frac{1}{4} f_{S\&H} + n f_{S\&H} \tag{3.3}$$

where $f_{S\&H}$ is the sampling frequency of the sample-and-hold circuit that precedes the analog-to-digital converter, and n is an integer.

The first of these various sampling frequencies corresponds to Nyquist sampling. The Nyquist sampling theorem states that any set of digital samples taken from an analog signal at a rate that is at least twice its highest frequency component will allow the analog signal to be completely reconstructed. Thus, if a signal such at that shown in Figure 3.5 has a total of four harmonics of the fundamental frequency $f_{IF} = 1/T$, then it requires nine time-samples, taken over one period, to be able to reproduce the original analog signal from the time samples shown. If the signal spectrum of the desired channel is equal to one-quarter of the sampling frequency, and if we ensure that there are no signal components outside the desired channel frequency at the input to the sampler, then the spectrum of the analog signal about the final IF frequency will be as shown by the solid line in Figure 3.6. Its lower band edge is shifted to dc and its upper band edge is at some frequency f_H, which is the total signal bandwidth. Sampling this

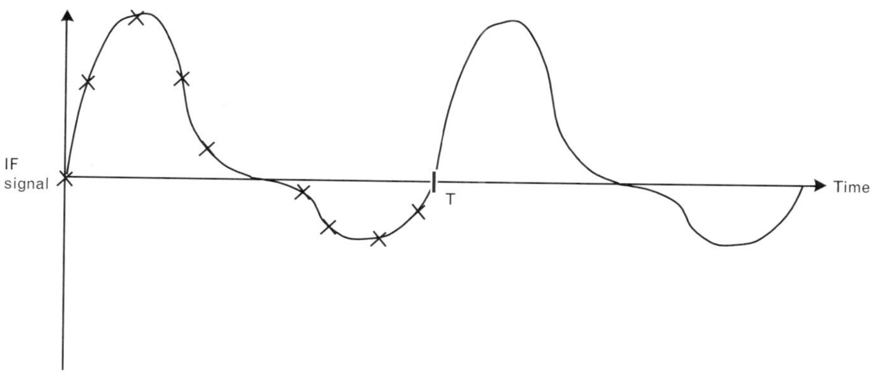

FIGURE 3.5
An analog waveform and nine time samples within a period, obtained by sampling at the Nyquist rate corresponding to the fourth harmonic of the signal.

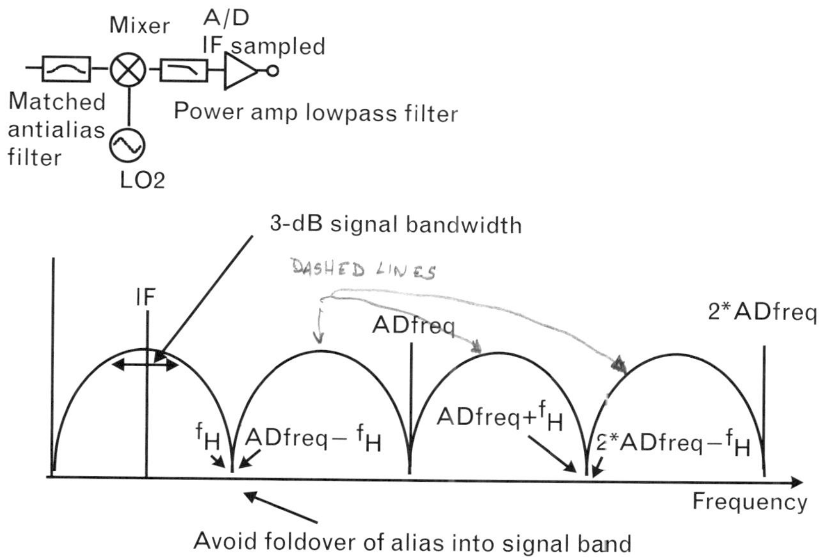

FIGURE 3.6
The spectrum of the original signal (solid line), and the spectrum of the signal (dashed line) represented by its Nyquist samples.

signal at the sample-and-hold frequency of $f_{S\&H} = ADfreq$ corresponds to Nyquist sampling, where we sample at twice the highest frequency component present in the analog waveform.

The reconstructed spectrum calculated using the finite number of sampling points that result from Nyquist sampling appears as the dashed curve in Figure 3.6. It is derived by convolution in the frequency domain, where the sampling signal and its harmonics are frequency impulses convolved with the original input signal. In effect, the original analog spectrum is translated along the frequency axis at positive and negative integral multiples of the sampling frequency *ADfreq*. The original spectrum is reconstituted at baseband; but spectral artifacts appear at harmonics outside the baseband frequency as well. This is of no concern to the digital processor since it processes the signal samples as if they contained baseband frequency components only.

Aliasing can occur when the original analog signal contains frequency components outside the assumed IF frequency range, such as out-of-band spurious signals or noise. When these are sampled at $f_{S\&H}$, translation of their spectrum along the frequency axis results in components that will fall within the baseband frequency range, and add to the (desired) signal there. They become indistinguishable from the desired signal components and are processed by the following digital circuits. This effect is known as *aliasing*. To avoid it, it is important to ensure that the analog signal is truly "clean" prior to sampling.

In general, this spurious response caused by aliasing should not exceed the level of any spurious component generated by distortion or nonlinearity within the analog-to-digital converter itself. This prevents spectral overlap from dominating the signal distortion, but avoids placing too tight a constraint on the antialiasing filter, which will have other potentially conflicting constraints such as phase linearity. Oversampling at a higher rate than the Nyquist rate relaxes the filtering requirements since it pushes out the frequency at which the spectral foldover occurs, but it also requires a faster converter to accommodate the increased digitization.

Equation (3.2) is a special case of sampling that allows a slightly lower sampling rate. In this instance, the analog signal is first mixed down to a baseband frequency that does not include dc, as shown in Figure 3.7. The

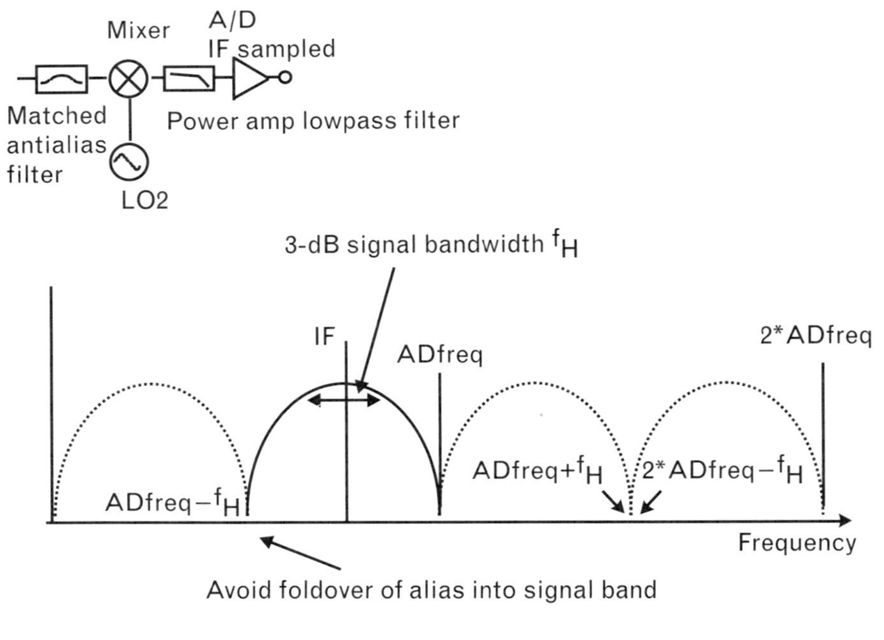

FIGURE 3.7 *Spectrum of the IF signal before and after bandpass sampling with $n = -1$.*

sampling frequency is at the upper end of the IF signal, and sampling again reproduces the original signal and harmonic artifacts at all integral multiples of the sampling rate. In this case, it is the artifact produced by the negative sideband about $-f_{IF}$ that, when convolved with the second harmonic of $f_{S\&H}$ $=ADfreq$, is interpreted as the baseband component of the digital samples. Again, aliasing can be avoided if the sampled analog signal is truly band limited about the IF.

Equation (3.3) is a more general case of the above, if we set $n = -1$. More generally, this equation represents a sampling regime known as *bandpass sampling,* and allows an analog signal to be sampled at the Nyquist rate corresponding only to the signal bandwidth, rather than to the value of its absolute center frequency. The IF *bandwidth* in this case must be less than or equal to $1/4f_{S\&H}$, although the IF center frequency itself can be much higher. Here, the analog waveform is undersampled, but since only the modulation information must be extracted, and not the carrier content, the samples are sufficiently numerous to achieve this. Mathematically, the nth harmonic of the sampling frequency creates the baseband spectral artifact, as illustrated in Figure 3.8 where a 10-MHz sampling frequency is used to sample an IF signal with 5-MHz total bandwidth centered at 102.5 MHz. By allowing a much higher IF to be used, a downconversion step can potentially be saved, and a slower clock rate used. However, remember that the sample-and-hold circuit in the analog-to-digital converter must still accept the highest frequency component present in the analog signal, which is still around the IF frequency. Analog bandpass filters with sharp rolloff must also be used prior to sampling, to prevent aliasing and consequent misinterpretation of the baseband artifact.

The terminology we have used above should not be confused with *digital direct conversion* (DDC), which is downconversion to baseband from a bandlimited IF, using a digital LO and digital mixer. Depending on speed and signal complexity, this can be performed in either DSP or with specialist DDC hardware. As before, prior to digitization, additional filtering and gain control is required to avoid aliasing and to maintain the dynamic range of the system, which would otherwise be set by the analog-to-digital converter.

FIGURE 3.8 *Spectrum of the IF signal after bandpass sampling with n = 10.*

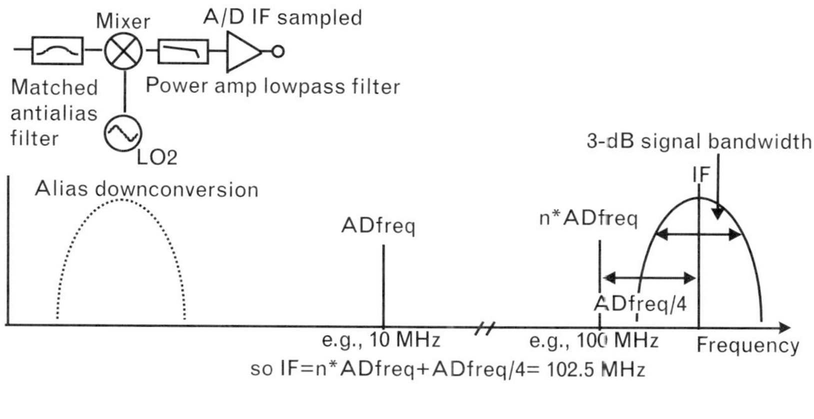

3.2 Receiver characterization

We have already used two terms—*receiver sensitivity* and *receiver selectivity*—that turn out to be the two fundamental criteria in evaluating the quality of a receiver. We will see in this section that these, and other system parameters such as the receiver dynamic range and its maximum input signal, can be determined from the system architecture and a few key parameters of each of the subsystems or components within the architecture. In particular, by configuring the gain, noise figure, power capabilities, and frequency characteristics of each stage, we can derive the overall receiver performance. In the following sections, we will derive a system budget from analysis of each of the individual components that constitute the receiver.

3.2.1 The communications channel

First, however, we will briefly consider the fundamental physical constraints and trade-offs that eventually determine the performance limits of a particular radio system, which is, after all, intended to convey information of some sort.

Information theory is a complex and well-developed field in its own right. Much emphasis has been given recently to the development of new modulation and coding schemes that conserve bandwidth, power, and minimize the effects of distortion on other systems. This is important because spectrum is a limited, scarce, and consequently expensive resource that needs to be allocated sparingly and used to the benefit of all users.

A fundamental result from this theory is Shannon's equation, which predicts the capacity or information throughput of a communications channel. It can be stated in many ways, but for our purposes

$$C = B\log_2\left(1 + SNR\right) \tag{3.4}$$

is a useful form. This states that the channel capacity C (measured as bits per second) is proportional to the channel bandwidth B and to the base-2 log of the signal-to-noise ratio SNR at the detector. This equation shows why spectrum is such an expensive resource, because as the signal occupies more bandwidth—assuming signal power and SNR remain the same—more data can be squeezed through it. In practice, modern systems use sophisticated error detection involving the transmission of redundant bits and complex coding schemes in order to approach Shannon's theoretical limit.

A result that can be derived from the above is that for a given channel with some maximum information throughput, the power (or more specifically, the signal-to-noise ratio at the detector) needed to transmit that information through it can be traded off for spectral occupancy or bandwidth B according to

$$SNR_2 = SNR_1^{(B_1/B_2)} \text{for } SNR >> 1 \tag{3.5}$$

Thus if the spectral occupancy B_2 can be increased, the signal-to-noise ratio SNR_2 can be lower for the same information transfer rate. A good example of this is to compare the GSM cellular system with CDMA. Both use the same transmission media (free space) and are subject to similar noise sources. Yet the received GSM signal must be several decibels above the noise level to be detectable, whereas the spread CDMA signal can be detected even when buried in the noise. By clever coding that smears the signal across a large bandwidth, the CDMA signal power can be reduced to a level so that the received signal appears like noise to receivers tuned to other channels.

The choice of modulation, bandwidth, and the minimum and maximum allowed power in the signal and other interfering sources are part of the standards or air-interface specification for the system, from which its RF parameters must be determined.

3.2.2 Receiver noise

We have seen already that noise plays a major role in determining the overall performance of a receiver because, with the signal power, it enters as the SNR into fundamental equations that determine the data rate of the system, and the minimum signal level that can be received.

Noise enters the receiver through a number of sources:

- The channel and into the system through the antenna, where it is usually modeled as additive, white, Gaussian noise and is thermal in nature;
- The RF preselector filter, which bandlimits the signal but has a finite insertion loss and therefore a thermal noise contribution;
- The active devices, which contribute thermal noise, shot noise, and $1/f$ noise.

These noise components produce a *noise floor* that sets the minimum signal level that can be detected. The noise is characterized by its power spectral density, and it can be a function of frequency. Power spectral density is the power contained within a given bandwidth, so has units of watts per hertz. Although we sometimes refer to noise voltage and noise current per hertz, in RF systems we typically measure power into a fixed reference impedance level (usually 50Ω). In this case, the thermal noise power is not a function of resistance and a fundamental result is

$$P_N = kTB \tag{3.6}$$

for the total noise power P_N, where k is Boltzmann's constant and T is the temperature in Kelvin. At room temperature, the thermal noise generated in a 1-Hz bandwidth is therefore

$$\begin{aligned} P_N &= \left(1.38 \star 10^{-23}\ \text{J/K}\right)\left(293\text{K}\right)\left(1\ \text{Hz}\right) \\ &= 4.057 \star 10^{-21}\ \text{W} \\ &= -174\ \text{dBm} \end{aligned} \tag{3.7}$$

or −174 dBm/Hz when expressed as a power spectral density. Clearly, the larger the bandwidth, the greater the noise power. It is for this reason that the final IF filter needs to be as narrow as possible, in order to minimize the noise power just prior to demodulation and detection. This final IF filter determines the overall noise bandwidth of the entire receiver since it will be the most narrowband component in the entire chain prior to detection.

In the radio bands below about 30 MHz, the external noise is much greater than that calculated by (3.6), due to both natural and man-made phenomenon [2]. The excess above the thermal floor varies from around 12 dB in rural areas at 30 MHz to as high as 76 dB at a noisy urban site at low megahertz frequencies. This increase in the minimum receiver noise level needs to be considered when determining the weakest signal that can be detected.

To characterize noisy systems, we measure the difference between the actual noise floor and that due solely to thermal noise power [as calculated in (3.7)]. In decibel terms, this is known as the noise figure of a component. Noise figure F_{dB} is defined as

$$F_{\text{dB}} = 10\log\left(\frac{P_{\textit{noise actual}}}{kTB}\right) = SNR_{\textit{THERMAL-INPUT}} / SNR_{\textit{OUTPUT}} \tag{3.8}$$

and compares the noise output to the noise there would be if the measured component contributed no noise. It is the degradation in SNR introduced by inserting a component.

Figure 3.9 shows a simple example. There, for an attenuator of 10 dB, the noise figure is just equal to the loss, 10 dB, because although a signal passing through the attenuator will lose 10 dB in power, the output thermal noise power is not similarly degraded because the temperature in (3.6) is the same at input and output.

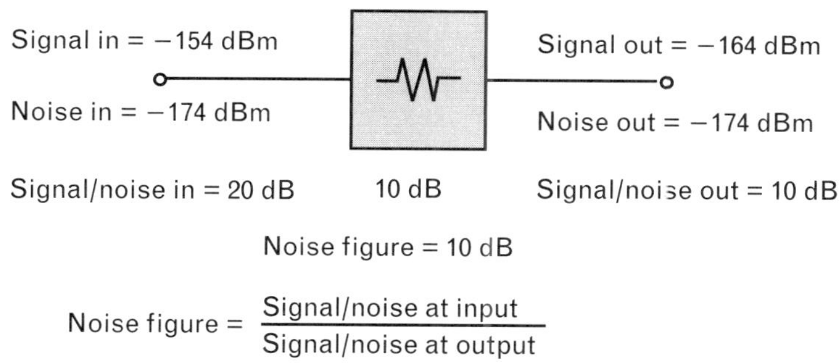

FIGURE 3.9
The noise figure of the system compares the actual noise at the output to thermal noise at the output if the component introduced no noise of its own.

An alternative definition of noise figure is sometimes made in terms of noise temperature. For this purpose, we define the noise factor F. The noise factor is proportional to the noise temperature of a component divided by a reference noise temperature.

$$F_{\text{dB}} = 10\log\left(1 + \frac{T_A}{T_R}\right) = 10\log(\text{Noise factor}) \tag{3.9}$$

The noise figure is identical to the noise factor but is expressed in decibels rather than as a pure ratio. T_A is the component effective noise temperature and is sometimes used interchangeably with noise figure or noise factor, and T_R is the effective reference or (usually) room temperature. The absolute measures of noise power are usually referred to the input of the system, while the measured noise at the output is referred back to the input by dividing the output noise power by the component gain. This enables us to compare the response to signal levels at the inputs of different systems. The desired input signal levels will stay the same, while the output responses will differ.

Figure 3.10 shows some typical noise figures and temperatures for some modern devices that we describe more fully in Volume II. Compared with room temperature of 273K, the noise temperature T_A or added noise of most devices below 20 GHz is relatively small and noise figures that are a fraction of 1 dB are common. For satellite reception however, where the receiver is pointed towards the sky and the reference temperature T_R for thermal noise falls to the low temperatures indicated by the curve marked "S" on Figure 3.10, the noise temperature of the device can become significant and its noise contribution to the thermal noise floor can become increasingly important.

The noise factor provides a useful mechanism to measure the overall added noise of a series of cascaded components. It may be calculated using

$$\begin{aligned} F = F_1 + (F_2 - 1)/G_1 + (F_3 - 1)/G_1G_2 + \ldots \\ + (F_n - 1)/G_1G_2..G_{n-1} \end{aligned} \tag{3.10}$$

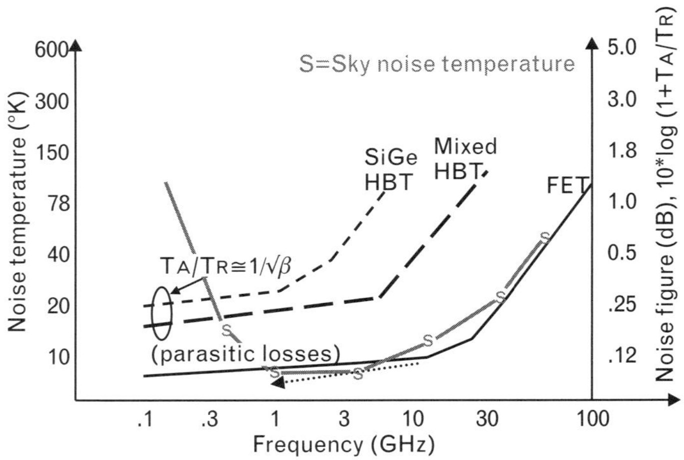

FIGURE 3.10
Noise figure and noise temperature for some typical devices versus frequency. (From: [3]. © 2003 Besser Associates.)

where the subscripts refer to consecutive stages numbered from the input to the output. Of course, in performing the addition, real ratios must be used rather than decibel quantities. This equation shows very clearly that if G_1 is high enough, the overall receiver noise factor is close to F_1 because subsequent terms are made sufficiently small. It is for this reason that a high gain *low-noise amplifier* (LNA) is usually the first component in a receiver, so that the noise factor terms from subsequent components are negligible. Losses from the antenna, filter, and cables preceding the amplifier add directly (in decibels) to the noise figure of the LNA itself.

The noise floor at the output of the receiver can now be simply measured using (3.6) and (3.10). If the gain of the total receiver is G and its final IF strip bandlimits the noise to a bandwidth B, then the noise floor is given by

$$P_N|out = kTBGF \tag{3.11}$$

3.2.3 Receiver sensitivity

The noise floor of a receiver, just prior to demodulation and detection of the signal, determines how strong the input signal must be to be correctly interpreted, either as a "1" or a "0" in a digital system, or as a high-quality analog waveform in an analog system. This minimum input signal strength needed to produce a good quality output signal is referred to as the receiver sensitivity. However, just as there are many definitions of what constitutes acceptable quality, there are equally numerous definitions of sensitivity.

The basis of sensitivity measurement comes from analog AM receivers. The noise floor prior to the AM detector is typically referred back to the antenna input by dividing the noise level there by the preceding system gain. In this way, the receiver itself can be modeled as an ideal noise-free gain block, with the noise floor at the input now accounting not only for thermal noise at the input, but all noise added by the system itself. If we now assume that an input signal must at least equal the noise level in order to be detected, then this input-referred noise floor is sometimes referred to as the *minimum detectable signal* (MDS). However, most audiophiles believe that a signal-to-noise ratio of at least 10 dB is necessary for acceptable sound quality, so that an output signal-to-noise ratio S_O/N_O of 10 dB is used to measure the system sensitivity. The more precise measurement conditions are illustrated in Figure 3.11, where the input carrier is modulated at a 1,000-Hz rate with a modulation depth of 30%.

With the receiver IF bandwidth set to 6 kHz (sufficient to pass the modulation sidebands yet reject any adjacent channel), the input carrier is increased in amplitude until the output signal-to-noise ratio of the detected AM waveform is equal to 10 dB. The resulting voltage of the carrier signal at the input, measured into an open circuit load, is defined as the receiver sensitivity.

The actual measurement is made in the presence of noise, which can be generated in the receiver by the carrier signal itself, so the carrier is always kept on and the AM modulation turned on and off. The output waveform then switches between $(S+N)_o$ and N_o, and the input carrier signal voltage is increased until the ratio of $(S+N)_o/N_o$ is 10.4 dB, corresponding to $S_o/N_o = 10$ dB. Typical sensitivity voltages of between 1 to 2 μV are obtained for modern AM systems. A further refinement to the measurement, known as *signal in noise and distortion* (SINAD), can also be made [4]. The input voltage is increased until a 10-dB SINAD ratio is achieved at the output. This measurement accounts for noise-like distortion components that exist at the output as a result of the modulation itself. To include these components in the denominator, the modulation and carrier signals must always be kept on so that the distortion terms continue to exist, but the signal is removed at the output by switching in a selective bandpass filter at the modulation frequency. This allows carrier-generated harmonics and other nonlinear distortion components to be measured with the noise, but unmasked by the signal itself.

In digital systems, it is simpler to measure the *bit-error rate* (BER) induced by noise when the signal is weak. The bit-error rate affects the data rate, so is a more useful performance measure than the signal to noise ratio,

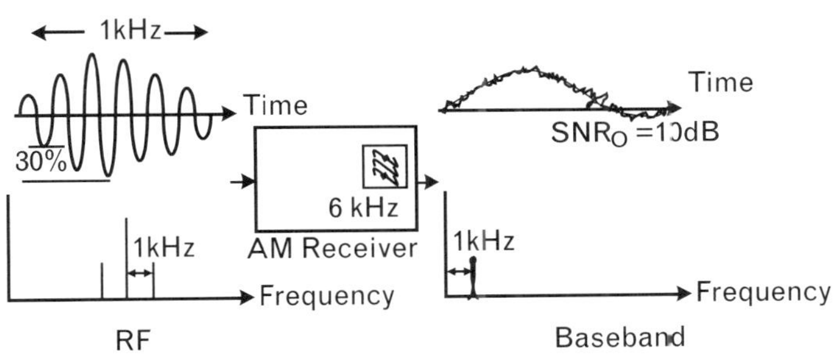

FIGURE 3.11 *Measurement conditions for determining the sensitivity of an AM receiver.*

which can differ from system to system for the same performance. Typically, a BER of 1% is specified, and the sensitivity is measured by adjusting the carrier signal strength until this BER is achieved at the output. For a given modulation scheme, this corresponds to a known signal-to-noise ratio prior to detection, and thus is conceptually similar to the AM sensitivity measurement. For instance, for QPSK modulation, the energy per bit must be 4.3 dB above the noise power density (prior to the detector) in order to be detected with a probability error of 1%. As the signal level is increased, the error rate decreases correspondingly. A number of look up tables plot the error rate against SNR for various modulation schemes, or these can be obtained from most system simulators.

3.2.4 System nonlinearity

The components in any system ultimately reach a point at which they become nonlinear. In later chapters we shall study this in much more depth, but essentially a component is nonlinear when its output amplitude or phase is no longer linearly proportional to its input amplitude or phase. In most active devices, this results predominantly from amplitude-induced changes in the device transconductance, but the variation of device resistances and capacitances with voltage amplitude also contributes to nonlinear behavior. Gross nonlinearity arises from cutoff and saturation effects that occur as the device exceeds the limits of its normal active region, even momentarily. As a result, we observe effects such as gain compression, intermodulation distortion, and cross modulation such as AM-to-PM conversion.

Figure 3.12(a) shows an example of a measurement of output power made on an amplifier. This plot shows the variation with output power at the fundamental and second, third, and fourth-harmonic frequencies as the fundamental input power is increased. Although plotted for an amplifier, such a P_{OUT} versus P_{IN} plot is quite common for many receiver components, although the measurement frequency of the output signals may not be the same as for the input (e.g., as in the case of a mixer).

It is clear in this figure that for low levels of input power, the harmonic components are negligible. Here, in this small-signal or linear region of operation, the small-signal gain is 13 dB, because at an input power of –30 dBm the output power is –17 dBm. As the input power is increased, the fundamental output power increases 1 dB for each 1-dB increase in input power. As the input power increases beyond about –15 dBm, the harmonic components start to increase: first the second harmonic, followed by the third- and fourth-harmonic components. In this weakly nonlinear region, each decibel increase in input power results in a 2-dB increase in second-harmonic output, a 3-dB increase in third-harmonic output, and so on. As the amplifier begins to compress, the harmonics increase more

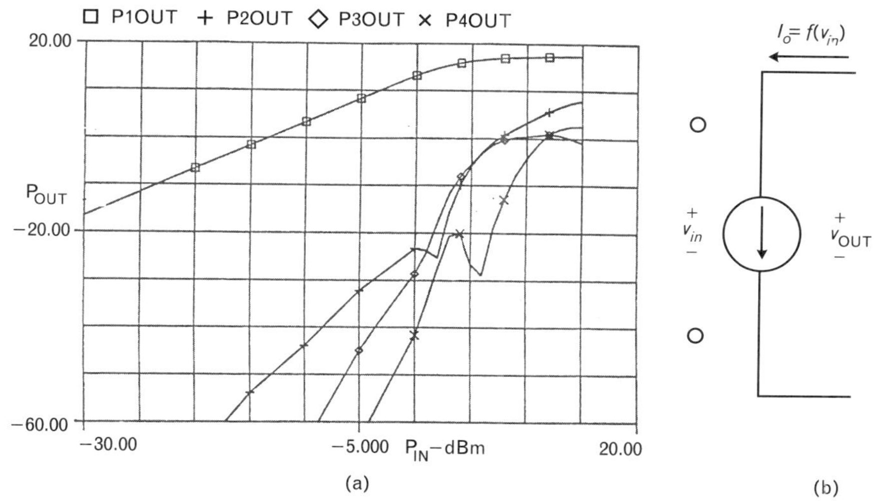

FIGURE 3.12
(a) The output response of an amplifier, showing the fundamental and harmonics generated as the input power is increased. (b) A simple transistor model to explain the observed amplifier output response.

rapidly with input power. Ultimately, the amplifier is in its strongly nonlinear region, and saturates just below 20-dBm output power. Any further increase in input power results in no further increase in the fundamental output.

The behavior observed in Figure 3.12(a) is typical of many nonlinear devices and can be illustrated mathematically by modeling the device as in Figure 3.12(b). We will show later that transistors are typically modeled as voltage controlled current sources, in which their collector or drain current I_O is some nonlinear function of the input voltage v_{IN} across the base or gate. I_O can be expanded as a polynomial series,

$$I_O = I_Q + g_m v_{IN} + g_{m2} v_{IN}^2 + g_{m3} v_{IN}^3 + \ldots \tag{3.12}$$

where g_m in the term $g_m v_{IN}$ is the linear transconductance. If the input voltage is given by $v_{IN} = E\cos\omega_o t$, then substituting into (3.12) gives

$$\begin{aligned} I_O &= I_Q + g_m E\cos\omega_o t \\ &+ g_{m2} E^2 \cos^2 \omega_o t + g_{m3} E^3 \cos^3 \omega_o t + \ldots \end{aligned} \tag{3.13}$$

By using

$$\cos^2 \omega_o t = \frac{1}{2}\left(\cos 2\omega_o t + 1\right) \tag{3.14}$$

we obtain

$$I_O = \left(I_Q + \frac{g_{m2} E^2}{2} \right) + g_m E \cos \omega_o t + \frac{g_{m2} E^2}{2} \cos 2\omega_o t + \ldots \quad (3.15)$$

where terms involving powers of E greater than 2 have been neglected. This is a good approximation when the signal strength E is not too large. Now if we let $E_{rms} = E/\sqrt{2}$ and assume resistances normalized to 1, the fundamental input power P_{IN} is simply $20\log(E_{rms})$, and the fundamental output power is $20\log(g_m E_{rms})$, so we obtain

$$P_{\text{fund}}(\text{dB}) = 20 \log(g_m E_{rms}) = 20 \log g_m + P_{IN}(\text{dB}) \quad (3.16)$$

Similarly, taking the component at $\cos 2\omega_o t$ for the second-harmonic,

$$\begin{aligned} P_{\text{second}}(\text{dB}) &= 20 \log \frac{g_{m2}}{2} + 40 \log(E_{rms}) \\ &= 20 \log \frac{g_{m2}}{2} + 2P_{IN}(\text{dB}) \end{aligned} \quad (3.17)$$

Thus, the second-harmonic power at the output rises 2 dB for every 1-dB increase in the input power. This expression breaks down once (3.15) ceases to be valid (i.e., at signal powers for which the third and higher order terms become significant). This can be observed in Figure 3.12(a) where the second-harmonic actually reduces over an interval of increasing input power, and closer to saturation as well. Similarly, one can show that the third-harmonic components increase at a 3:1 rate (30 dB/decade) with the input power, at least where the device is weakly nonlinear.

One of the most insidious forms of distortion that occurs in a radio receiver arises from the cubic term that we have so far ignored in (3.12). If the input signal is of the form

$$v_{IN} = E(\cos \omega_1 t + \cos \omega_2 t) \quad (3.18)$$

where ω_1 and ω_2 are, for instance, the frequencies of two adjacent channels in a radio system, the waveform of the two combined tones has a time-varying envelope that fluctuates between an amplitude of zero (when the two tones cancel) and $2E$ (when they add in phase). In a 1-Ω resistor, the power in the envelope of a single tone alone, and in the two tones, plots as in Figure 3.13. The two-tone envelope has a peak instantaneous power four times as high as that of a single tone, even though its average power is clearly only double. The significance of this is that the peak-to-average power ratio of such a signal is 2 (or 3 dB), and it is the peaks of what can be a relatively low-power signal that can drive a system into its strongly

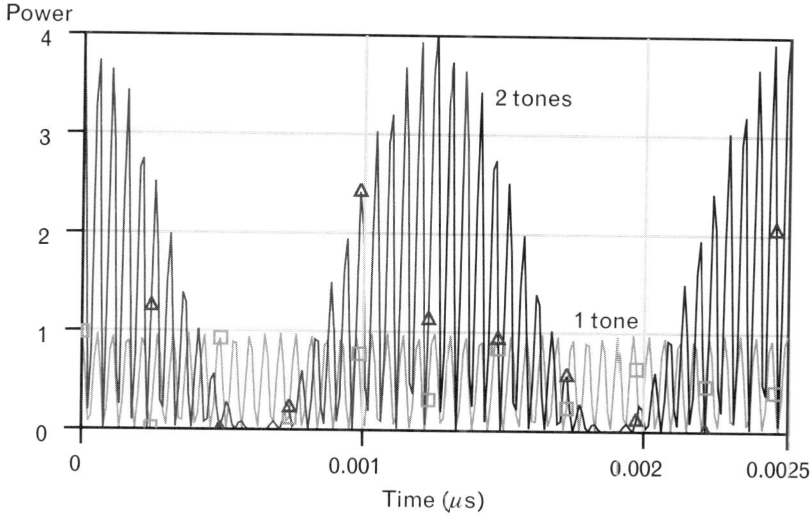

FIGURE 3.13
The instantaneous envelope power of two signals added together, each of equal amplitude and power but slightly separated in frequency.

nonlinear regime. This is important in the linearity of transmitter power amplifiers, discussed more fully in Volume II, Section 5.6.1.

Substitution of (3.18) into (3.12) results in the usual dc, sum, difference, and harmonic terms. There are also terms arising at $2\omega_2 - \omega_1$ and at $2\omega_1 - \omega_2$ that are very close in frequency to the two output tones at the input frequencies. These components have the form

$$i_{IMD3} = \frac{3}{4} g_{m3} E^3 \left[\cos(2\omega_1 - \omega_2)t + \cos(2\omega_2 - \omega_1)t\right] \qquad (3.19)$$

Because they result from the cubic term in (3.12), they are known as third-order intermodulation products, and appear as sidebands on either side of the fundamental, located as shown in Figure 3.14. We will abbreviate each of these products as *IMD3*. Because of the dependence on E^3, they rise 3 dB for every 1-dB increase in input power, or output fundamental power.

In principle, the amplitudes of these components are not functions of the tone spacing, which is related to the modulation in a real signal. In practice, there will be some dependency on the modulation bandwidth,

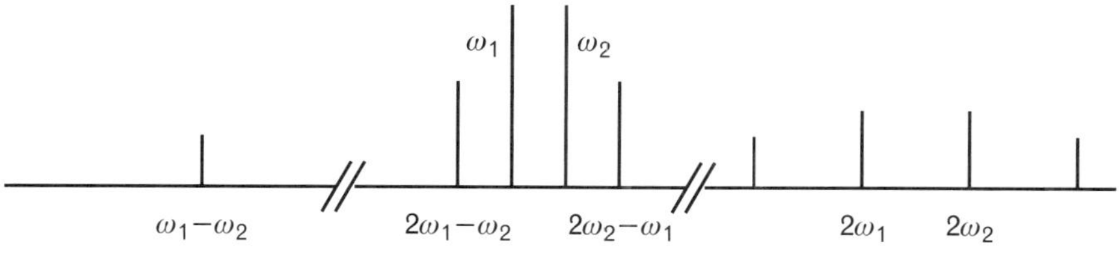

FIGURE 3.14
The output spectrum resulting from two equal tones applied to a nonlinear device.

known as *memory effect*. Second-order effects, such as the fact that the distortion in one point in a system creates distortion in a second, needs to be modeled as the cascade of two polynomials like (3.12), and can cause memory effects. The system response to the beat frequency (or envelope frequency), which in the case of a two-tone test varies from dc to several megahertz, and which is usually impacted by the bias networks of the various components, can also unintentionally add memory and create asymmetry in the intermodulation response. Thermal effects are another cause of memory in systems.

It can be interesting to observe what occurs when the two input tones are of different level—that is,when

$$v_{IN} = E_1 \cos\omega_1 t + E_2 \cos\omega_2 t \tag{3.20}$$

This test is useful in approximating the effect of different modulation formats, since it allows the envelope amplitude variation (of the beat frequency) to be adjusted by the two input tone levels. If the first tone at frequency ω_1 is the desired (weak) signal, but it is much less in amplitude than a strong interfering signal at frequency ω_2, then substitution into (3.12) gives the output current term at the fundamental ω_1 as

$$I_{\omega_1} = g_m E_1 + \frac{3g_{m_3} E_1^3}{4} + \frac{3g_{m_3} E_1 E_2^2}{2} \tag{3.21}$$

The first two terms in this expression equal the fundamental output power when just one tone at ω_1 is applied to the device. The first term is just the linear small-signal output $g_m E_1$. The second term becomes apparent at larger signal levels when the gain deviates from linear. Because a real device compresses and the gain reduces as the input signal level increases, g_{m3} must be negative. Thus, the third term is also negative, and represents the effect of the second tone on the gain of the first. If E_2 is large enough, this (negative) term can become dominant and totally suppress the gain of the first tone. This is known as small-signal gain suppression, or blocking if the effect becomes large enough. We should note as well that if the second tone is also modulated (i.e., if E_2 varies in amplitude or phase), then the small-signal gain of the weak signal is even modulated by the interferer. This effect is known as cross-modulation.

3.2.4.1 Relation between third-order distortion and gain compression

Fully expanding (3.13) into its harmonic frequency components, using all the terms shown, results in an expression for the fundamental output power that will begin to compress from its linear expected value as the input voltage is raised higher. As we can see from (3.21) with $E_2 = 0$ (i.e., with a single-tone measurement), the third-order term creates a current component $(3/4)g_{m3}E^3$ at the fundamental frequency that will subtract directly from the linear term g_mE. When this subtraction causes the fundamental output power to be 1 dB lower than the expected power using the linear term alone, the condition is known as 1-dB compression, and the output power is referred to as the 1-dB compressed output power. The power level at which 1-dB compression occurs is the most commonly used measure of the power handling capability of a device. It is also in some sense a measure of the output power at which the device ceases to be nominally linear. The 1-dB compressed power level commonly refers to the power at the output, but it can equally be referenced to the input power level as well simply by subtracting the (1-dB compressed) gain. We can see from (3.21) that when *two* tones of equal level E are present, then $E_1 = E_2 = E$, and the linear fundamental component is compressed much more quickly by the third-order term which then has magnitude $(9/4)g_{m3}E^3$. The compression is therefore three times (5 dB) greater than for a *single* tone of amplitude E, so the input 1-dB compression point is about 2 dB lower if measured on the basis of input power *per tone*. The compression point of an amplifier thus depends on the type of input signal loading, but is usually referenced to a single, sinusoidal input tone.

A secondary and equally popular measure of device power handling capability is the input third-order intercept point, IIP_3. It is measured using two input tones of equal level. This point is defined as the extrapolated input power level per tone that would cause the output third-order intermodulation products, given by (3.19), to equal the single-tone linear fundamental output power. The resulting output power at that point is the output third-order intercept point, IP_3. Graphically, it is illustrated in Figure 3.15. In this type of plot, the input power along the x-axis is the power contained in either of the two tones, so the total input power is 3 dB higher than indicated for the two-tone measurement when the third-order products are measured.

The intercept point is fictitious in that we must extrapolate the fundamental component in a linear fashion, and assume that the third-order intermodulation products increase forever with a 3:1 slope. For such a case,

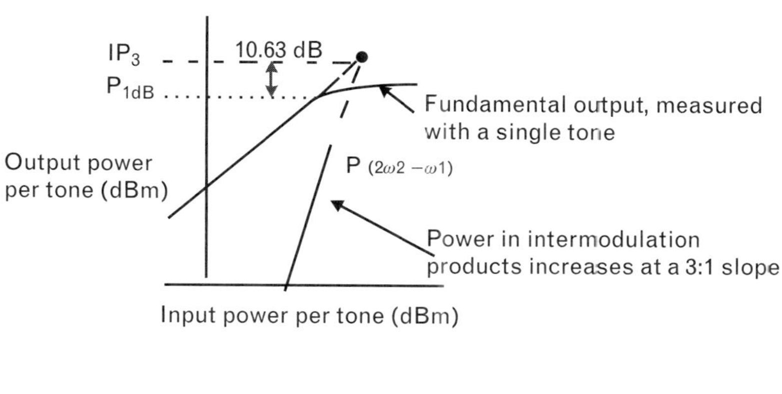

FIGURE 3.15 *Definition of the third-order intercept point IP_3.*

it can be shown from (3.19) and (3.21) and neglecting all terms beyond the cubic that

$$IP_3 = P_{1\,\mathrm{dB}} + 10.6\ \mathrm{dB} \tag{3.22}$$

where the powers are referred to the device output. (Referred to the device input, the theoretical difference is 9.6 dB, because of the 1-dB difference in gain at the compression point.) At this power level, the ratio of one of the output third-order intermodulation products to one of the fundamental output tones, or IMR_3 for short, will be approximately –21 dBc, because of the 3:1 slope. This rather idealistic situation rarely arises, principally because most devices require the equations to include terms to the fourth and fifth order for accurate modeling. In addition, there are also other nonlinear mechanisms besides the device transconductance in (3.12) that produce third-order intermodulation products, especially clipping and cutoff of the output current waveform when the device is overdriven or turned off. The phase dependence of the output signal on the amplitude of the input, AM-to-PM conversion, also significantly contributes to these products but is often neglected in analysis.

Intermodulation distortion products in some devices can rise much more steeply than the theoretical 3:1 slope predicted by the cubic model. In practical devices, the difference between $P_{1\mathrm{dB}}$ and IP_3 can range from as little as 6 dB or 7 dB to as much as 20 dB. Nonetheless, the concept of an intercept point to represent the behavior of a device, or component, to multiple signals is a useful practice. Intercept points can also be drawn for other order harmonic or intermodulation distortion components; the IP_2 point is the fictitious intersection of the second-harmonic output power with the fundamental.

3.2.4.2 Other measures of linearity

In multicarrier systems, there are many signals whose power will add, at times in phase. We saw above, for just two tones, a peak power level in the total input signal envelope that exceeded the average power by a factor of two. The peak power will be even greater as more tones are added. Although the average input power might be within the linear region of the system, the peaks might exceed it and cause nonlinear odd-order distortion, resulting in *adjacent channel power* (ACP) output. The *adjacent channel power ratio* (ACPR) is the system output power at an offset frequency normalized to the power in the channel of interest. This could be one measure of the nonlinearity of a transmitter, if it generates unwanted sidebands at an offset frequency that lie within the passband of an adjacent signal.

The two fundamental tones in Figure 3.14 could represent the sidebands of a double sideband AM signal with a suppressed carrier. In a modern wireless system, however, we might think of them as components within the modulation spectrum bordering the band edges of a modulated RF channel, with the bandwidth between the two as being occupied by the modulation of that channel. With a continuum of signals occupying the spectrum within the channel, their third-order intermodulation products will spread across the bandwidth in the adjacent channels on either side as well as within the same channel itself. The original spectrum has grown into a channel three times as broad as the original; fifth-order distortion causes such *spectral regrowth* to be five times as broad. Not only does such distortion cause growth in the amplitude of unwanted, vestigial sidebands in a modulated signal, it can also fall in-band where it is impossible to filter out.

This is illustrated in Figure 3.16, where the signal spills into adjacent channels. The requirement to minimize the spectral regrowth in order not to produce interfering signals in other channels as well as the transmitted channel itself requires tight control of the third-order intermodulation distortion output that a transmitter power amplifier can produce. Likewise, if two strong unwanted signals lie in channels adjacent to a weaker signal that is to be detected by a receiver, it is important that their third-order intermodulation distortion does not swamp the weaker signal to be detected. This places a tight requirement on the linearity of the input LNA and mixer in a receiver.

An example of the output spectra for a digitally modulated signal is shown in Figure 3.17. The linear output shows the modulation power confined about the central carrier, while the nonlinear output shows growth in spectral sidebands as the output power increases towards the compression point of the amplifier. The exact relationship between the measured two-tone C/I ratio (the ratio of power in the carrier to power in the intermodulation distortion product) or inversely, the IMR_3, and the

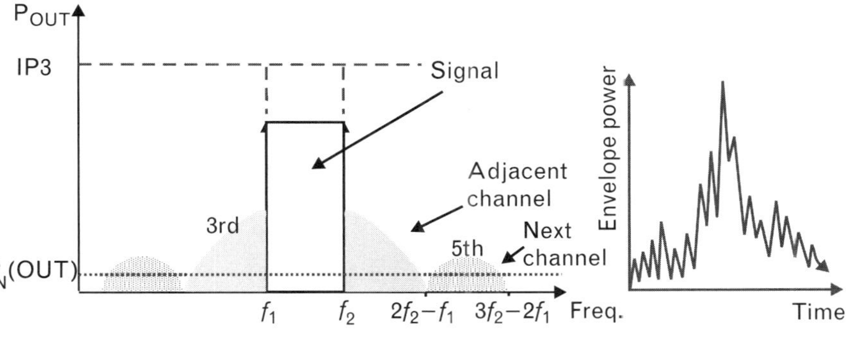

FIGURE 3.16 *Output radio spectrum showing the signal between f_1 and f_2, and the adjacent channel power created by nonlinear distortion. The envelope power fluctuates as the n multiple carriers occasionally sum in phase to a peak power level n-times the total average power. (From: [3]. © 2003 Besser Associates.)*

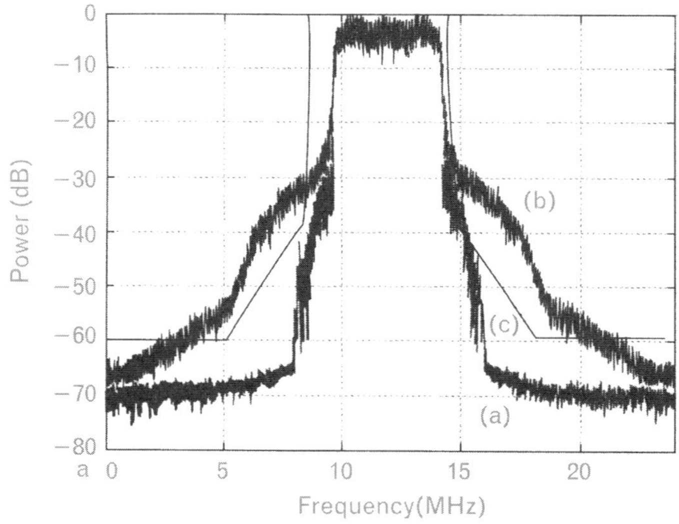

FIGURE 3.17 *Output spectra from a power amplifier for a QPSK 256-point OFDM system: (a) with infinite power backoff (ideal); (b) typical 10-dB power backoff; and (c) air-interface requirement. (From: [5]. © 2002 IEEE. Used with permission.)*

amount of spectral regrowth depends on the modulation scheme used and its peak-to-average ratio.

Spectral regrowth can be particularly severe for modulation formats in which the amplitude of the RF envelope is not constant. Then, the power amplifier will operate at different instantaneous input power levels and both AM-to-AM and AM-to-PM conversion can occur, creating intermodulation products. This variability in the envelope of digitally modulated signals can occur when amplitude is directly used as a modulating code (e.g., QAM), when multiple subcarriers of varying phase are added together (e.g., CDMA or OFDM), or when the baseband signal is filtered prior to phase-modulating a carrier (e.g., GMSK). The peak-to-average ratio of a modulation format measures the degree to which the scheme's envelope varies from a constant amplitude.

3.2.4.3 The third-order intercept point of a cascade

The concept of an intercept point enables the distortion properties of several cascaded devices to be calculated. Their individual output intercept points sum as

$$\frac{1}{IP_0} = \left[\left(\frac{1}{IP_N} \right)^q + \left(\frac{1}{G_N \, IP_{N-1}} \right)^q + \left(\frac{1}{G_N G_{N-1} \, IP_{N-2}} \right)^q + \ldots + \left(\frac{1}{G_N G_{N-1} \ldots G_2 \, IP_1} \right)^q \right]^{\frac{1}{q}} \tag{3.23}$$

where $q = (m - 1)/2$ and m is the order of the distortion ($m = 3$ for third order). The stages are numbered from 1 at the input to N at the output of the system. The G_i and IP_i ($i = 1$ to N) terms are expressed as power ratios, but the expression is somewhat simplified by recognizing that the denominator terms are simply the intercept points of each stage referred to the output of the entire system (so the total gain following each stage is simply added—in decibels—to that stage's output intercept point in decibels). In the case of third-order distortion, $q = 1$ and the expression then simply becomes

$$\frac{1}{IP_O} = \left[\left(\frac{1}{IP_N|_{out}}\right) + \left(\frac{1}{IP_{N-1}|_{out}}\right) + \left(\frac{1}{IP_{N-2}|_{out}}\right) + \ldots + \left(\frac{1}{IP_1|_{out}}\right)\right] \quad (3.24)$$

where the output reference is to the output of the whole cascade. In practice, this expression turns out to be a worst-case expression for the total output intercept power IP_O because it assumes that the distortion currents from each stage add in phase, whereas in fact they will add randomly, somewhat reducing the measured distortion power of the total cascade.

3.2.5 Receiver dynamic range

We have said that at low signal levels the minimum signal that can be detected is limited by the noise floor that results from thermal and other noise sources in passive and active devices. At large signal levels, harmonic and intermodulation distortion components arise, causing compression and interference that ultimately limit the largest signals the receiver can handle. The difference between the minimum detectable signal and the maximum signal is known as the receiver dynamic range. One interpretation of this definition is the difference between the 1-dB compressed output power and the output noise floor. But this has limited usefulness because it assumes a single channel system. Instead, the *spurious free dynamic range* (SFDR) is a more useful measure. It is defined as the range of input power levels from which the output signal just exceeds the output noise floor, and for which any distortion components remain buried below the noise floor. This is illustrated graphically in Figure 3.18, where the noise floor is shown superimposed on the P_{out} versus P_{in} distortion curves from the previous section.

Because the fundamental output power increases decibel for decibel with the input power in the range under consideration, the SFDR can also be measured along the output-power (y-) axis as shown in the figure. This enables us to establish a simple expression using the geometry of the completed triangle. Recognizing that the third-order products rise at a 3:1 slope,

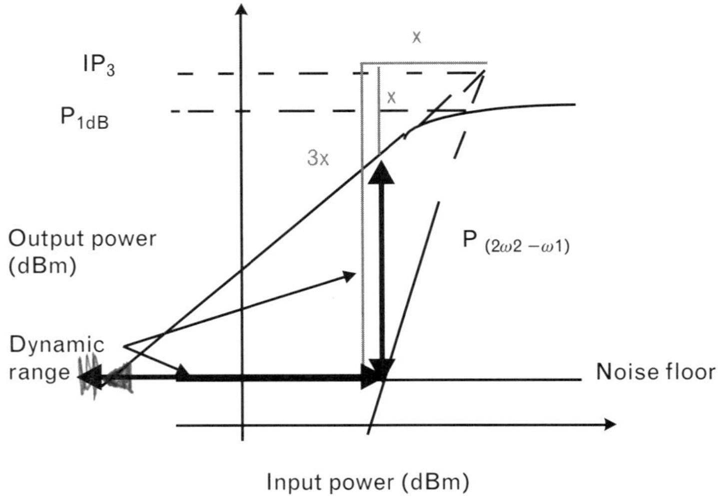

FIGURE 3.18 *Receiver output products showing the spurious-free dynamic range.*

$$3x = IP_3 - \text{Noise Floor} \tag{3.25}$$

so the spurious free dynamic range $SFDR = 2x$ is given by

$$\begin{aligned} SFDR &= \frac{2}{3}\left(IP_3 - \text{Noise Floor}\right) \\ &= \frac{2}{3}\left(IP_3 + 174 - 10\log B - G - F\right) \end{aligned} \tag{3.26}$$

using (3.11) for the output noise power and expressing all quantities in dBm or decibels. It is important to recognize in this analysis that although the noise is assumed to be bandlimited to a bandwidth of B by the final IF filter, actual distortion products will be essentially wideband and may have been eliminated by earlier filters. Thus, the distortion products shown in Figure 3.18 are not necessarily at the same frequency as the desired signal and may well have been filtered out by either the first or second IF filters. This is not always the case though, as the distortion may result from strong adjacent interferers and lie in-band (i.e., within the desired channel) where it cannot be filtered. Thus, some care is necessary in considering actual cases, since the architecture of the receiver may eliminate distortion in an earlier IF stage.

3.2.5.1 The effect of automatic gain control

In order to increase the useful range of a receiver, the gain of a system can be automatically controlled in order to decrease the gain when strong signals cause overload or distortion. However, the impact of doing this is not always obvious and will require some care in considering the signal-to-noise ratio and making a number of trade-offs. For instance, consider AGC applied to an LNA. Adding attenuation in front of the LNA will hurt noise figure, but it will help reduce the effect of large signals (which may be interferers rather than the desired signal itself). After the LNA an attenuator prevents overload of following stages but reduces the third-order intercept point. Ideally, we would like to reduce the gain, increase power handling capability, *and* minimally affect the noise figure. Thus, it is necessary to consider both the gain and *input* intercept point in tandem, since these are the principal trade-offs in setting an AGC level. The effect on system noise figure can then be derived.

With AGC, we sacrifice the system small-signal performance in order to achieve a larger signal-handling capability. Therefore, it is critical that the mechanism for reducing the gain of the variable gain element does not itself reduce its power handling capability, and thereby its intercept point. For example, this would happen if the collector current of a bipolar transistor were reduced to lower its gain. With some designs it is possible to both reduce the gain and achieve an increase in output intercept point, as for example by increasing the emitter resistor of a transistor. Dual-gate FETs and three PIN-diodes or MESFETs arranged in a pi- or T-configuration are also common voltage-controlled attenuation circuits. Of course, the ideal AGC design would also achieve this without reducing the noise figure. Unfortunately, introducing loss to reduce strong signals will inevitably degrade the noise floor. Care must also be taken that as the gain is changed, the impedance match of surrounding elements is unaffected, so that the bandpass characteristics of the receiver remain constant and frequency skew is not introduced.

Figure 3.19 shows the principle of automatic gain control. Essentially, the input power (x-) axis is shifted to the left, or the original gain curves to the right, reflecting a lowering of total receiver gain. The SFDR can be extended, as shown, provided that both strong and weak signals are equally affected. This is not always the case, since the phenomenon of small-signal

FIGURE 3.19
The SFDR can be extended by lowering the gain, effectively increasing the input intercept point.

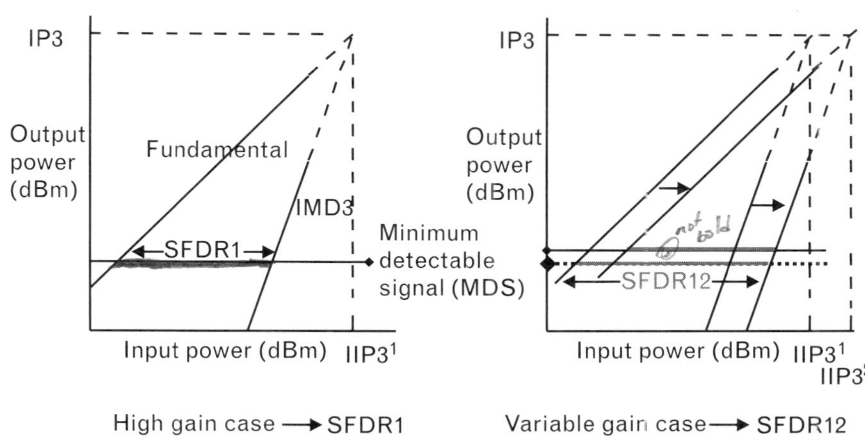

gain suppression (suggested earlier) can cause low-level signals to be suppressed when larger signals cause limiting in nonlinear amplifiers and other components.

Because lowering the gain raises the noise floor, there must be enough margin in the signal-to-noise ratio of the desired signal to allow this. The trade-off is that the input intercept point, thus the ability to handle large input signals, is increased as shown.

AGC is typically applied in an analog receiver as shown in Figure 3.20. As the desired signal increases in input power, the gain of the system is reduced in order to minimize distortion. The gain of the output stages is ideally reduced first to avoid raising the system noise figure and reducing sensitivity, and as earlier stages begin to distort, AGC is progressively introduced ahead of them. Therefore, the noise floor will eventually rise at the same rate as the input handling capability, since the noise floor is principally set by the loss in the front end.

The stronger signals may be present as interferers in other channels, and these out-of-band signals must be detected properly by the AGC circuits to sense the distortion and reduce the gain. Because such signals are present in the radio front end and are removed by the selectivity of later stages, it is important that they be detected at the front end, before they are eliminated by filtering downstream. For this reason, dual-stage AGC is sometimes used as shown to ensure proper detection of signal levels not just in the desired channel but also in interfering channels.

For example, consider two strong signals only a few kilohertz away from the desired signal. The receiver's first IF filter may not eliminate these since they are too close to the tuned, or desired signal. As a result they may cause distortion in the RF and IF front end *and* their third-order intermodulation product will lie on top of the desired signal. However, because adjacent channel signals and their harmonics will be eliminated in the much narrower second or final IF filter, they would not be detected if AGC were controlled only by the power at the output of the receiver. Such interfering signals need to be detected in the first IF stage and reduced prior to any active device in the receiver, possibly by inserting attenuation in the front end. In some older short-wave radios, switches were used to manually insert

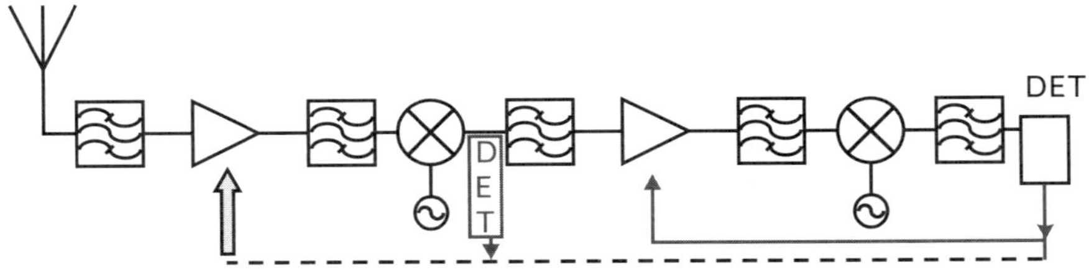

FIGURE 3.20 *The principle of AGC applied in an analog receiver, showing the points where the signal level is detected and where gain control is applied.*

RF attenuation into the signal flow to reduce strong unwanted signals from saturating the early stages of the radio. Figure 3.21 shows an example of dual-loop AGC in a short-wave radio receiver for the HF band.

In digital systems, AGC is important not only to maintain linearity in the RF and IF stages, but also in the *analog-to-digital converter* (ADC). Most modern mobile phone systems require a dynamic range of between 80 to 100 dB, and the signal level must be adjusted to lie within the linear input range of the ADC. This means the peak power is reduced by AGC to never exceed the ADC's full-scale range, while the maximum gain is sufficient to ensure that the output receiver noise floor is boosted to the quantization noise power level of the ADC. This is the input noise power to the ADC that causes the least significant bit to toggle, assuming the ADC input has the same impedance level as the receiver output.

As a general rule, selectivity should be as high as possible close to the antenna to remove large interfering signals before they enter the active devices in later stages and cause problems with distortion or overload. In practice, this is hard to achieve because filters need to cover the desired band, and high-*Q* filters are highly reactive. As a result, impedance matching can be difficult and can adversely affect the performance of other components, especially mixers.

3.2.6 Receiver selectivity

The characteristics we have so far examined in this chapter are primarily concerned with the amplitude characteristics of the receiver: its noise floor,

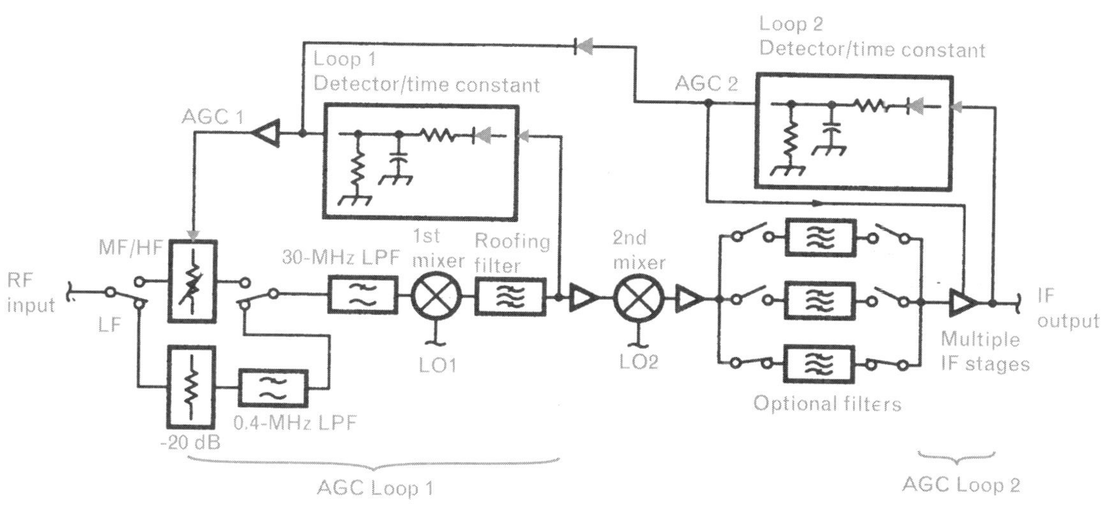

FIGURE 3.21 *An example of a dual-loop AGC system. Loop 1 is driven by the first IF signal and inserts RF attenuation, while loop 2 is driven by the final (band-limited) signal level and also controls the gain of the IF amplifiers. The loops have different time constants to prevent AGC oscillation. (From: [6]. © 1994 QST. Used with permission.)*

the minimum detectable signal, the distortion and third-order intercept point, and dynamic range. Although these are affected both by the frequency and the bandwidth of the receiver, the architecture of the receiver needs to meet a more fundamental requirement when designing its frequency characteristics—the elimination of unwanted signals at frequencies for which the receiver is not tuned.

3.2.6.1 Selectivity

The selectivity of a filter refers to its ability to reject signals outside its passband. It is defined by the attenuation of a signal at some frequency offset from its center frequency, normally measured at its passband edges. For channel selection filters in radio receivers, selectivities of 60 dB to 80 dB are typical. Selectivity is also a parameter that can be applied to a radio receiver as a whole, since selectivity is the property that enables us to separate adjacent channels. It can be measured in terms of the relative strength of an adjacent signal compared to the desired signal (e.g., 60 dBc to 80 dBc, or dB higher than the desired signal) before its detected amplitude equals the detected level of the desired signal. For instance, for the analog AMPS cellular system, the minimum adjacent channel selectivity measured in the channel 30 kHz away from the wanted signal is 16 dB; for alternate channels, two channel spacings away at 60 kHz, the minimum selectivity is 60 dB. The looser requirement on the adjacent channel is to ease the close-in attenuation requirements of the first IF filter, and also because the adjacent channel signal strength will generally be weaker than the alternate channel signal strength. This is because the frequency allocation plan for the overall system does not permit adjacent channels to be used within the same geographical cell.

A useful definition of selectivity is to specify the loss at a computed selectivity factor, which is defined as

$$\Omega = \frac{\left(f_{r2} - f_{r1}\right)}{\left(f_{p2} - f_{p1}\right)} - 1 = \frac{f_{reject}}{f_{pass}} - 1 \tag{3.27}$$

where the reject and passbands are shown in Figure 3.22. The selectivity factor is a normalized measure of the frequency band at which the skirts of the attenuation are specified, normalized to the filter passband.

3.2.6.2 Phase effects in a receiver

In calculating the distortion in a receiver, we have examined only the amplitude of the components generated by the higher order terms in (3.12). For instance, the cubic term generates the third-order distortion

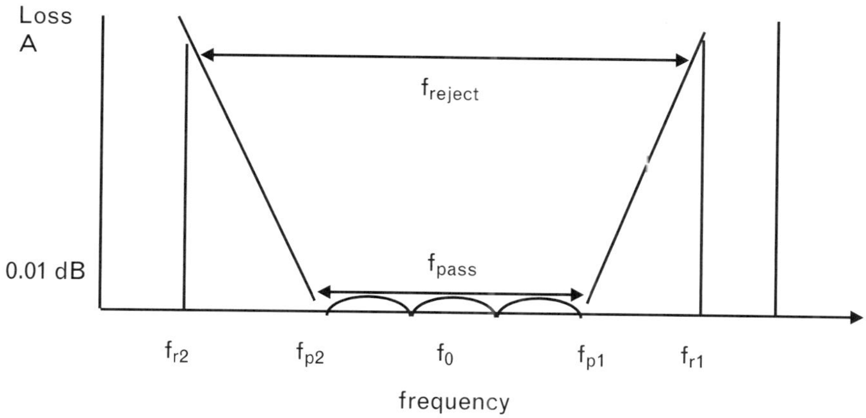

FIGURE 3.22
Terms used to define the selectivity of a filter.

that can fall close to the desired signal, and whose amplitude can be determined from a knowledge of the third-order intercept point.

However, in certain situations the output voltage will contain phase terms that are also dependent on the input voltage. This effect is known as AM-to-PM conversion, and can cause problems when a phase-modulated signal does not have constant signal amplitude, as for instance, with *quadrature amplitude modulation* (QAM). In this case, amplitude-induced changes in the output phase can be misinterpreted as phase changes due to modulation of the signal, resulting in bit errors. AM-to-PM conversion can be measured relatively simply by recording the phase of the gain in addition to its amplitude as the amplitude of a single-tone input signal is increased.

One mechanism by which AM-to-PM conversion is introduced through an active device in a receiver is illustrated in Figure 3.23, where the output of a transistor is modeled as a current source in shunt with an output resistance and output capacitance. Most semiconductor capacitances are inverse functions of voltage, so that when the reverse voltage across a PN junction is increased, more charge is pulled out from the junction and the capacitance is decreased. Clearly, a change in the output impedance seen by the current source in Figure 3.23 causes a change in the phase of the RF output voltage.

Less intuitive perhaps is that a change in the output resistor due to amplitude dependency can also change the phase of the signal, even if the

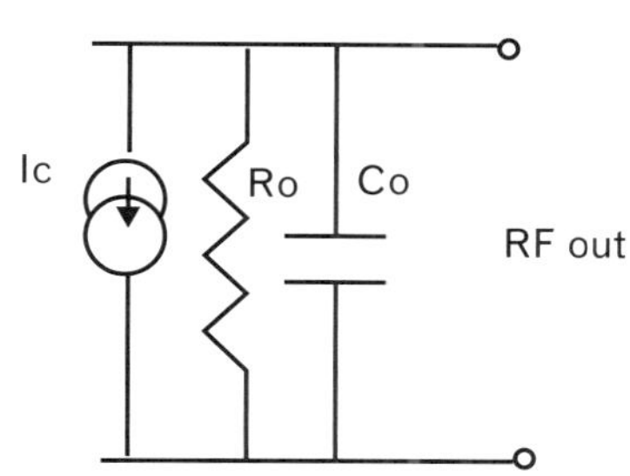

FIGURE 3.23
A model for the output of a transistor, showing a mechanism for AM-to-PM conversion.

output capacitance remains constant. This effect is common in most devices, since the output resistance of the device is usually one of the most sensitive parameters to changes in amplitude.

3.2.6.3 Spurious responses

Minimizing spurious responses in a receiver is one of the key design criteria behind selecting its frequency architecture. Spurious responses are outputs that arise from unwanted frequency components. In this context, a frequency component that is undesired is one different to that for which the receiver is tuned. For instance, if we desire to demodulate a channel whose carrier is 895 MHz, there could be a signal at 890 MHz that creates a response in the receiver that interferes with our desired channel at 895 MHz. The signal at 890 MHz is labeled a spurious frequency if it creates a spurious response. Although modern coding schemes can still detect the desired signal in the presence of interfering signals at the same frequency, spurious responses remain a problem since they reduce the sensitivity to the desired signal.

Spurious responses are caused by distortion products produced in the receiver when the receiver is excited at the spurious frequency. The responses are primarily created by nonlinearities in the mixer and amplifier, although other components such as quartz filters are also nonlinear. The responses can be generated by signals entering, or already within, the receiver. For instance, they can result from leakage paths from the transmitter or from signals in components at the back end of the receiver feeding back to the input. The excitations can occur at the IF itself, its subharmonics, the image frequency, and from higher order mixing products; they all have the property that their spurious responses fall onto the IF frequency. Although we cannot quantitatively predict their amplitude from linear analysis, we can at least predict the spurious frequencies that can cause problems, and take steps where possible to eliminate any voltages or currents at these frequencies to avoid them.

Here, we will focus on the mixer as a source of spurious responses, since by its very nature it generates nonlinear currents within it. If we label the desired signal frequency to which the receiver is tuned as f_T and the unwanted spurious frequency as f_S, then if these two frequencies enter the mixer it is inevitable that sum and difference mixing products will be generated with the local oscillator. We are normally interested in just the IF frequency at the output of the mixer, which is typically the difference (or sometimes the sum) product of the tuned frequency and the local oscillator f_0. The IF frequency f_{IF} is simply

$$\text{for a sum mixer} \quad f_0 + f_T = f_{IF} \tag{3.28}$$

$$\text{for a difference mixer (high side LO)} \quad f_0 - f_T = f_{IF} \tag{3.29}$$

$$\text{for a difference mixer (low side LO)} \quad f_0 - f_T = -f_{IF} \tag{3.30}$$

where the negative sign, of course, is simply a mathematical way of expressing a phase inversion of a positive frequency. However, mixing currents are created within the mixer at other harmonic multiples as well. These need not be present at the input terminals but are actually generated within the device itself by its higher-order terms. Some of these mixing products with the spurious signal could map onto the IF frequency f_{IF}:

$$\text{for a sum mixer} \quad nf_0 + mf_S = f_{IF} \tag{3.31}$$

$$\text{for a difference mixer (high side LO)} \quad nf_0 - mf_S = f_{IF} \tag{3.32}$$

$$\text{for a difference mixer (low side LO)} \quad nf_0 - mf_S = -f_{IF} \tag{3.33}$$

where m and n are integers. These equations can be normalized to the IF frequency by dividing through by f_{IF}, and defining $T = f_T/f_{IF}$, $S = f_S/f_{IF}$, and $O = f_0/f_{IF}$. After eliminating O from both sets of equations above, we obtain

$$\text{for a sum mixer} \quad n(1 - T) + mS = 1 \tag{3.34}$$

$$\text{for a difference mixer (high side LO)} \quad n(T + 1) - mS = \pm 1 \tag{3.35}$$

$$\text{for a difference mixer (low side LO)} \quad n(T - 1) - mS = \pm 1 \tag{3.36}$$

where the +/– has been added to allow phase inversion or mapping to the negative IF frequency as well. These three sets of equations are equations in the two variables T and S, with the harmonic integers m and n as parameters. Each of the three cases plot as a series of straight lines, in Figures 3.24 through 3.27. From substitution and as expected, the line defined by $(n,m) = 1$ in each case is just the line $S = T$, corresponding to the degenerate case when the spurious signal is the same as the tuned signal (i.e., the case of normal operation). For other (n,m) pairs, a spurious response will occur if there is a value of S that falls close to the $S = T$ line for the range of values of T over which the receiver can be tuned.

Figure 3.24 shows (3.35) plotted for values of S and T ranging between 0 and 7, and Figure 3.25 the same equation for values of T less than 1. For values of T greater than 7, the equation itself must be used to derive potential values of S that could fall within the tuning bandwidth given by the range of values T can assume.

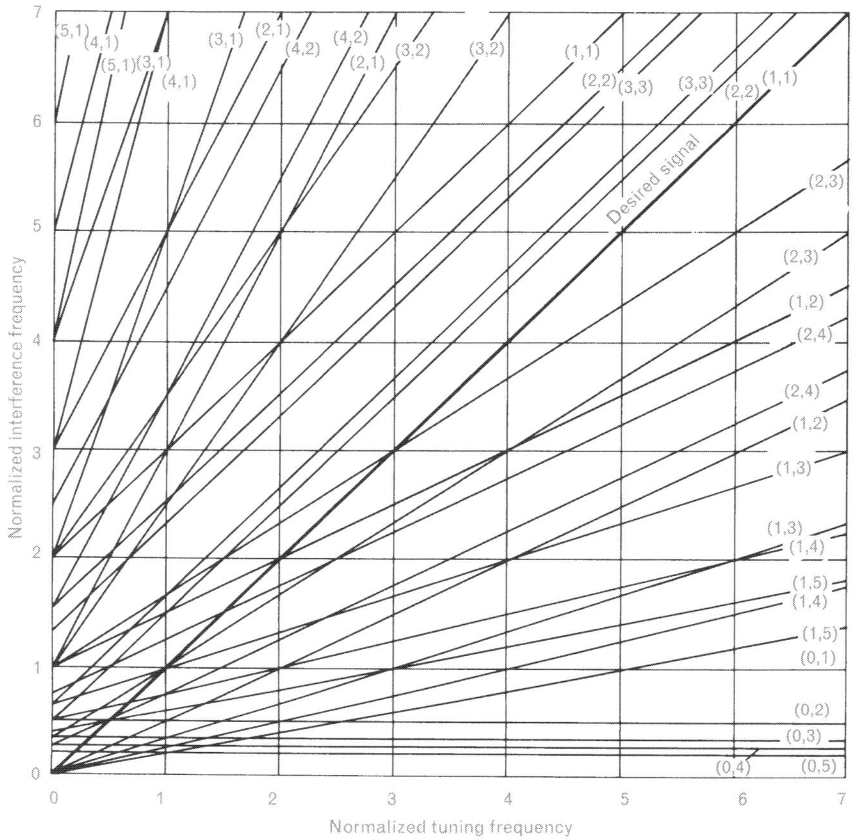

FIGURE 3.24
Values of S and T for the difference mixer, high side oscillator. The parameter is (n, m). (From: [4]. © 1988 McGraw-Hill, Inc. Used with permission.)

Even though the *n*th harmonic of the local oscillator and the *m*th harmonic of an incoming (spurious) signal may not be present at the input of the receiver, it is inevitable that currents corresponding to these harmonic multiples will be generated within the mixer itself due to its switching action. This assumes, of course, that the spurious signal has not been filtered out prior to reaching the mixer.

For example, suppose that f_{IF} is 45 MHz, and the tuned frequency f_T is 135 MHz. The local oscillator frequency must therefore be 180 MHz since Figure 3.24 is plotted for a difference mixer, high side LO. In this case, $T = 135/45 = 3$ and we can see from the figure that a spurious response will be caused by a spurious signal at the same frequency as the tuned signal, $S = 3$, generated by the (2,3) mixing product. In this case, the second-harmonic of the local oscillator ($n = 2$ at 360 MHz) will mix with the third-harmonic of the spurious signal ($m = 3$ at 405 MHz) to produce a difference signal of 45 MHz, corresponding exactly to the IF frequency and producing a signal indistinguishable in frequency from the desired IF signal, although its phase will be different.

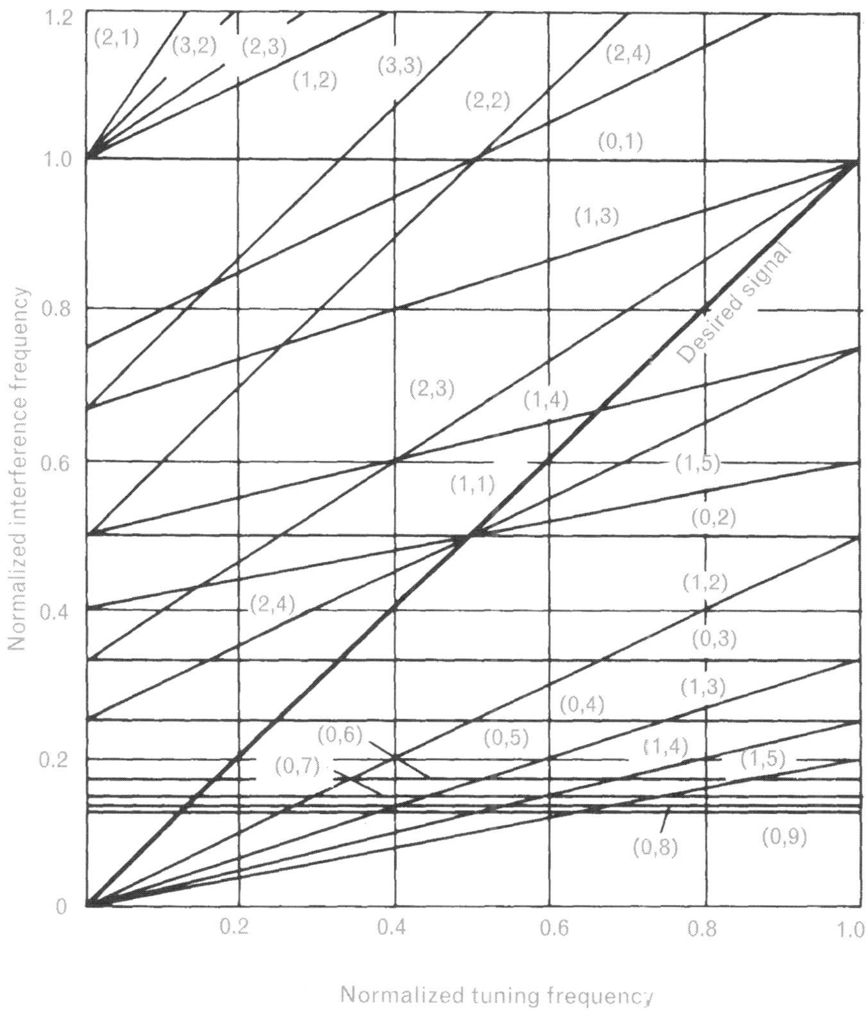

FIGURE 3.25
Values of S and T less than 1, for the difference mixer, high side oscillator. (From: [4]. © 1988 McGraw-Hill, Inc. Used with permission.)

This example is somewhat specious, since if in fact the desired signal and a spurious signal are at the same frequency, we have more than mixing to worry about! A more common problem arises when the receiver is tuned over a range of frequencies. In this case, we can draw a "box" around the possible range of values of T along the desired $S = T$ line (and hence the possible values of S if all other interferers are removed by the RF preselector) and look for any spurious frequencies that occur within the box, in the vicinity of the $S = T$ line. Thus in the previous example, if the RF frequency could be tuned to receive frequencies between 90 and 180 MHz (i.e., to values of T between 2 and 4), then within the square bounded by these values of S and T, spurious lines corresponding to (n,m) values of (2,2), (3,3), (1,2), (2,4), (1,3), (1,4), and (1,5) all exist. This implies that there are values of spurious at frequency S that fall within the bandwidth of the receiver and can potentially mix harmonically to produce an IF signal. In fact, for each of the first four (n,m) pairs listed above, there are two lines for the same (n,m) pair within the box, corresponding to the "+" and "–" values in (3.35) These simply correspond in turn to the (positive) IF frequency and its negative counterpart.

Figure 3.26 shows values of S and T for the difference mixer where the local oscillator is below the tuned frequency, and Figure 3.27 for the sum mixer where the IF is formed from the sum of the local oscillator and tuned frequency (i.e., for an upconversion). In fact, Figure 3.27 is just an image of Figure 3.26 plotted in the region where $T < 1$. This special case corresponds to the tuned frequency f_T being less than the IF frequency, and for a difference mixer, (3.30) shows that the local oscillator frequency is then also negative. Multiplying both sides of that equation through by –1 then yields (3.28), which is just that for a sum mixer.

Exercise 1

Consider the case of a cellular radio tuned between 869 and 894 MHz with an IF frequency of 220 MHz. Assume we use a difference mixer with a low

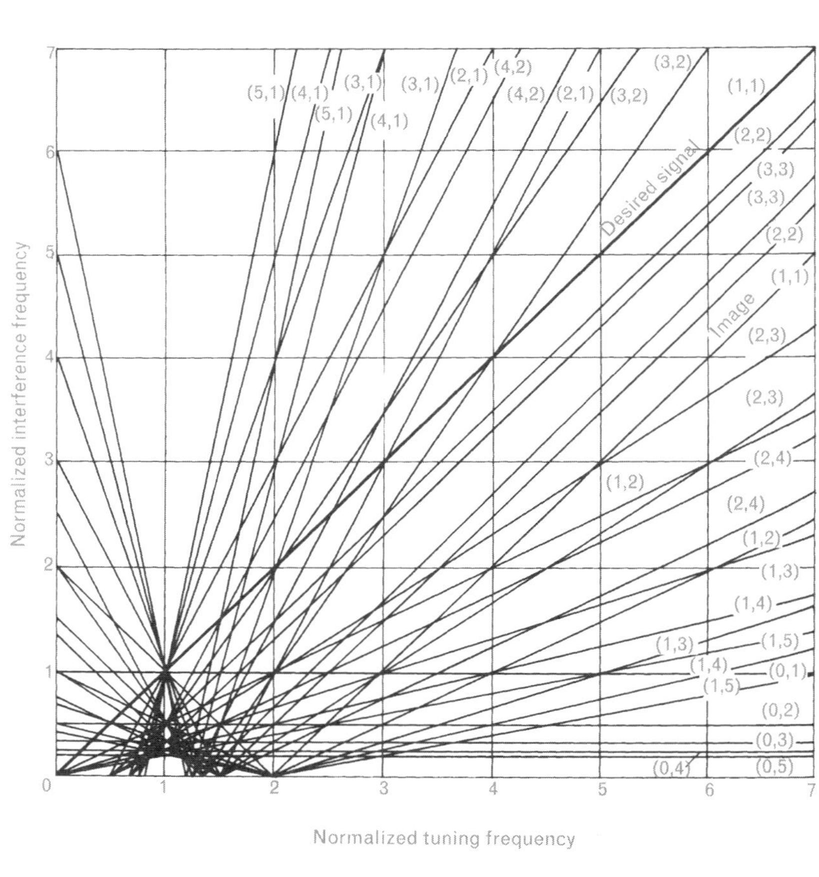

FIGURE 3.26 *Values of S and T for the difference mixer, low side oscillator. (From: [4]. © 1988 McGraw-Hill, Inc. Used with permission.)*

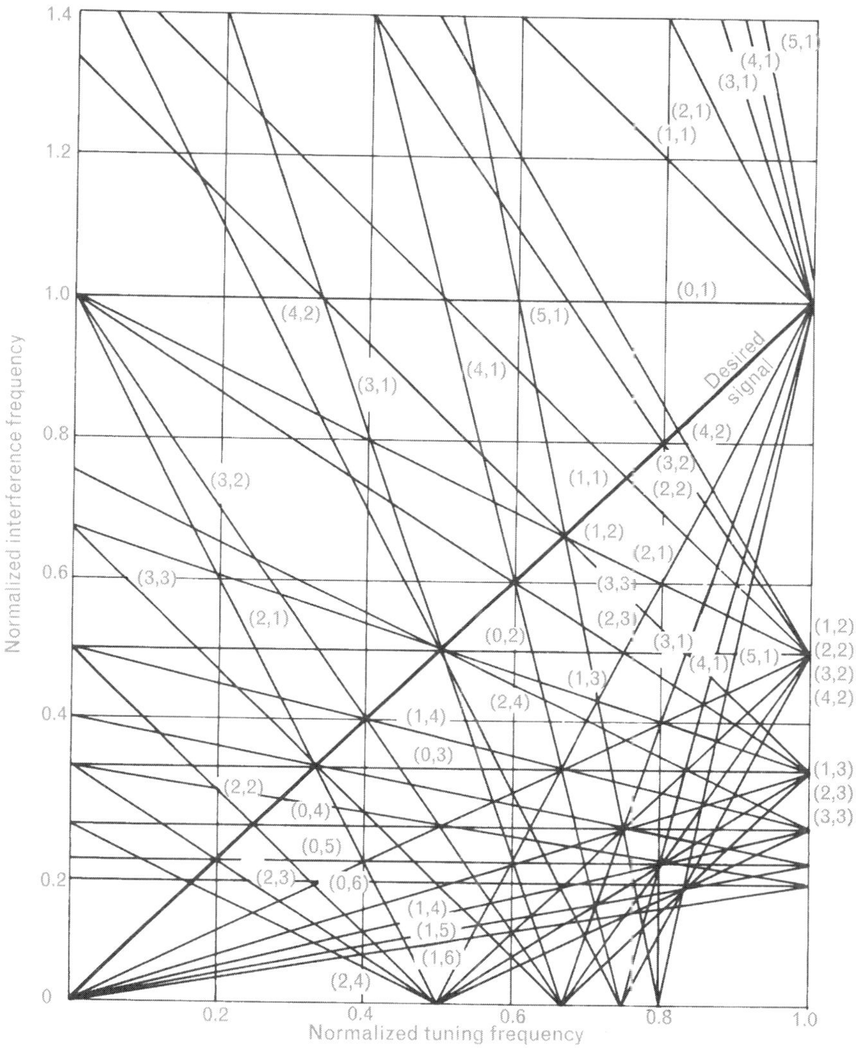

FIGURE 3.27 *Values of S and T for the sum mixer. (From: [4]. © 1988 McGraw-Hill, Inc. Used with permission.)*

side local oscillator. This corresponds to T lying between 3.95 and 4.06. From Figure 3.26, we see that the (3,2) spurious line intersects the desired signal line at $T = 4$, causing potential problems with interfering signals being downconverted to the IF of 220 MHz. This can be verified because a spurious signal at 880 MHz will produce an IF of 220 MHz when we mix its second-harmonic at 1,760 MHz with the third-harmonic of the local oscillator at 1,980 MHz.

If we choose an IF frequency higher than 220 MHz, T is reduced and the potential number of interfering spurs increases due to the density of spurious lines crossing the desired signal line in Figure 3.26. On the other hand, reducing the IF increases T and, from Figure 3.26, helps prevent spurious frequencies falling within the RF bandwidth.

We see from the exercise above the trade-off in selecting an IF frequency. As can be seen from the figures, choosing too high an IF frequency (resulting in too low a T) can cause problems with a greater number of spurious frequencies lying close to the desired $S = T$ line. However, we may wish to use a high IF frequency to reduce the size of the IF filter and possibly its cost.

Lowering the IF frequency helps to reduce the number of potential spurious frequencies. On the other hand, choosing too low an IF frequency can bring problems. First, the desired signal may fall into the flicker noise of the mixer, losing sensitivity. More critically, it makes the image frequency, which is the spurious signal that has the same frequency offset to the local oscillator as the tuned signal, but on its opposite side, more difficult to filter out. As the signal frequency offset from the LO—which is, of course, just the IF frequency itself—becomes smaller, both the local oscillator and the image frequency move closer in frequency to the tuned frequency. Eventually, they both may fall within the RF passband itself, clearly a situation we should avoid to keep the receiver passband clean!

Exercise 2

Decreasing the IF frequency in the above example to 83 MHz is an effective compromise because the image frequency will be 166 MHz away from the desired signal and still fall outside the RF passband. Furthermore, as the signal is tuned between 869 MHz and 894 MHz, T lies between 10.47 and 10.77. Although this is off the chart in Figure 3.26, we can extrapolate the lines to see that the closest spur to the desired signal line, and that falls within the tuning window, is the (2,2) spur. We need to use (3.36) directly to calculate

$$S = \left[n(T-1) \pm 1\right]/m \tag{3.37}$$

for this spur and find that for $T = 10.47$, when the radio is tuned to a signal at the low end of the RF band, the closest spur falls at $S = 9.97$. When the radio is tuned to the high end of the RF band and $T = 10.77$, the closest spur falls at $S = 10.27$. These correspond to frequencies of 827.51 MHz and 852.41 MHz, respectively (multiplying by f_{IF} or 83 MHz), and fall far enough outside the RF passband to be rejected by the RF preselector of the radio. The (2,2) spur is a common spurious frequency and can be rejected by using a double balanced mixer.

Exercise 3

Consider the RCA HF radio of Figure 3.28. Several HF bands exist at frequencies around 4, 6, 7, 9, 12, 13, 15, 17, 19, 21, and 26 MHz for broadcasting, where 5-kHz channel spacing is used but with 10-kHz nominal

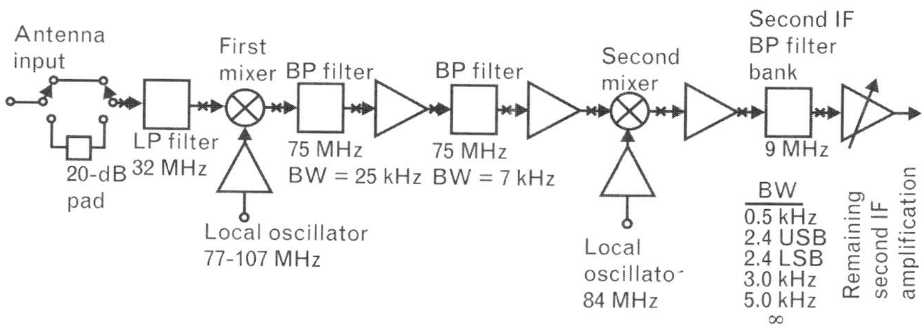

FIGURE 3.28
An RCA HF radio. (From: [4]. © 1988 McGraw-Hill, Inc. Used with permission.)

RF bandwidth. AM modulation is used. Some bands are also reserved for national broadcasters in the tropical regions around 2, 3, and 5 MHz. In this case, the RF preselector needs to filter out frequencies below about 30 MHz (32 MHz in the case of Figure 3.28). The IF filter selectivity is chosen to trade off reduction in noise and interfering signals (dictating as narrow a filter bandwidth as possible) against signal quality (dictating as wide a filter bandwidth as possible to allow the complete AM modulation sidebands to pass through).

In this radio, an upconverter is used to give a first IF at 75 MHz. The local oscillator is tuned between 77 and 107 MHz, so the mixer is a difference mixer with a high side oscillator. Tuning across the band of possible received signals from 2 to 32 MHz, results in a T between 0.02677 and 0.42667. The corresponding image frequencies lie between 152 and 182 MHz.

In this example, the first mixer *upconverts* the received signal to a higher frequency. Why could we not *downconvert* instead? Suppose we downconverted to an IF of 1 MHz. This results in three intractable problems:

1. The local oscillator would be required to tune between 3 and 33 MHz to cover the entire HF band. A 10-to-1 tuning range is impossible to achieve with a single component.

2. The isolation between the LO and RF ports of any mixer is imperfect. Local oscillator signal can leak through the first mixer to the antenna port where it would reradiate. In this case, the LO signal is in-band.

3. The image frequency now lies between 4 MHz and 34 MHz, and is impossible for the RF preselector to tune out.

Referring to Figure 3.25, we see that there are a number of possible spurious responses as the receiver is tuned to receive frequencies between $T = 0.02677$ and 0.42667. At the lower end of the HF band, the (1,2) product is closest to the desired signal line, followed by (1,3), (1,4), and higher order products. This time, we solve (3.35) for S to give

$$S = \left[n\left(T + 1\right) \pm 1\right]/m \tag{3.38}$$

yielding $S = 0.0133$, or 1 MHz. Should this spurious signal exist, then its second harmonic ($m = 2$) will fall at 2 MHz and cause interference when the LO is tuned to 77 MHz ($n = 1$) to upconvert a desired signal at 2 MHz (IF = 75 MHz). This spurious signal needs to be removed by the RF preselector.

When the signal is tuned to the top end of the HF band, the (2,4) product is closest to the desired signal line. Using $T = 0.42667$, $n = 2$, $m = 4$ and selecting the "–" sign in (3.38) results in $S = 0.4633$ or a spurious signal at 34.75 MHz when the receiver is tuned high. This can be filtered out by the RF preselector.

From Figure 3.25, other spurious lines occur close to the desired signal line at harmonic orders of (0,2), (0,3), (0,4), and so on. From (3.38) when $n = 0$, we see the spurious signals $S = 1/m$ are at the subharmonic frequencies of the IF frequency. Although the second subharmonic of the IF at $m = 2$ lies at 37.5 MHz and is outside the RF passband, higher-order subharmonics will fall within the passband and potentially create spurious responses. However, these are fixed in frequency and do not change position with tuning, so possibly could be filtered with notch filters in the RF preselector. We conclude, therefore, in this example that the upconversion architecture chosen for the first IF results in a modest LO tuning range, avoids problems with the image frequencies, and that the major spur concerns are the IF subharmonics below 0.5 and the sixth-order (2,4) product when the receiver is tuned to the top end of the RF band.

The second mixer is somewhat simpler to consider since its input frequency is fixed at the first IF of 75 MHz. By choosing an LO at 84 MHz, we can downconvert to a second intermediate frequency of 9 MHz. The second mixer is therefore also a difference mixer with high side oscillator and a tuned frequency of $T = 75/9$ or 8.333. This falls off-scale in Figure 3.24 but it is apparent that at a value of T this high there are no intersections with the desired signal line and that the closest spurious response arises from the (2,2) product. We use (3.38) to calculate S and find

$$S = \left(T + 1\right) \pm \frac{1}{2} \quad = \quad T + 1.5 \text{ or } T + 0.5 \tag{3.39}$$

which are at frequencies 4.5 and 13.5 MHz away from the tuned value. Since the first IF filter bandwidth is 25 kHz, this spurious frequency will not exist at the input to the second mixer.

3.2.7 Receiver frequency response

The *selectivity* of the receiver is dependent on its ability to filter out unwanted signals that may enter the antenna or that are present at the output of components within the receiver itself, such as mixers. Its *sensitivity* is dependent on its ability to keep the noise floor to a minimum.

It is the bandwidth of the final IF or baseband filter that determines the noise that the detector of the receiver ultimately sees and the noise bandwidth of the entire receiver. By the time the signal is this far downstream in a receiver, it is too late to filter out any in-band signals such as spurious products that fall within the desired channel bandwidth. Interfering signals that give rise to such spurious products need to be removed by the RF and higher IF frequencies preceding each mixer. The filter following the final downconversion must be sufficiently wide to pass the modulation sidebands associated with the signal, but not so wide as to pass adjacent channel information. Ideally, this filter will be as narrow as the modulation itself, to minimize the noise captured at the detector. In some radios, this final IF bandwidth is selectable to trade-off signal quality and noise.

Conversely, in the transmitter, baseband filtering of the I and Q channels can limit the spectral occupancy of the transmitted signal. However, it also lengthens the transition time between consecutive symbols, and can therefore increase the peak-to-average ratio of the modulated carrier. For instance, in BPSK, where successive symbols can alternate between −180° and +180°, the phase transition passes through zero where the instantaneous envelope amplitude is zero. Slowing down the transition (by filtering) will increase the time around zero phase and decrease the average carrier power. For this reason, modulation schemes using offset phase modulation that avoid zero transitions, or *minimum shift keyed* (MSK) schemes that vary the phase continuously at constant carrier amplitude, have more constant envelopes even with filtering, and reduce the consequent problem of spectral regrowth (explored in more depth in Volume II).

Different types of filters that can be used to set the selectivity of a radio are examined in detail in Chapter 8. There, we will see that not only is the selectivity of the filter itself important, but also its transient response. The transient response becomes more important as the bandwidth of the filter approaches that of the modulation itself (i.e., at baseband).

3.3 Analysis of a CDMA receiver handset

We have now defined a number of parameters that characterize the response of a radio receiver, and related those to the system itself. We have seen that (1) sensitivity and (2) dynamic range free from spurious responses and overload are two of the most critical parameters in choosing a receiver architecture.

The air-interface specification sets the required (minimum) overall performance, and the architecture and components that constitute the receiver and transmitter need to be chosen appropriately to meet it. In this section, we will decompose the RF and IF portions of a typical CDMA mobile receiver that meets the IS-95/IS-98 standard into its component parts, and utilize the specification of each component to determine the system-level performance. This will motivate the chapters that follow until we are better prepared for a more detailed analysis of a digital radio in Volume II, Chapter 8. The analysis will become more meaningful as the design constraints on the various constituent building blocks become apparent throughout this book.

Figure 3.29 shows a simplified block diagram of a dual-band mobile handset. The chipset used is described in more detail in Volume II, Chapter 8. This radio covers the mobile phone bands between 824 and 894 MHz (the U.S. cellular bandwidth, originally defined by the AMPS system) and 1,850 MHz to 1,990 MHz (the U.S. PCS, or personal communication system, band). The RF and IF portions of this radio can be used for the cdmaOne IS-95/IS-98 system, which is a second generation, narrowband CDMA system. With different tuning, the same receiver chips could also be used for dual-band GSM or TDMA systems in the same RF bands, since the processing and demodulation is performed digitally.

For simplicity, we will analyze here only a received CDMA signal in the PCS band, since the analysis of the cellular band receiver follows similar lines. The received CMDA signal lies in a 60-MHz spectral allocation between 1,930 MHz and 1,990 MHz. The system uses paired frequencies, in that the handset receive (downlink) frequency is 80 MHz above the handset transmit (uplink) frequency. The transmitter band is between 1,850 and 1,910 MHz, so the "guard" band separating the top edge of that transmit band and the low end of the receive band is therefore 20 MHz.

Prior to transmission, the relatively narrowband user data stream of up to 14.4 Kbps is modulated on a QPSK subcarrier at 19.2 kbaud (a baud is one modulation symbol per second). Each QPSK symbol is then multiplied by a code consisting of 64 chips per symbol, yielding a sequence with a rate of 1.2288 Mcps. A chip is basically a bit within a digital bit sequence (a code) that has zero cross-correlation with all other codes except itself. This spreads each channel over a 1.23-MHz bandwidth, consistent with the channel carrier spacing of 1.25 MHz. In the baseband demodulator, each received baseband signal is again multiplied by its own code, which is orthogonal to the codes used for all other channels. (In fact, the IS-95 standard uses the same code but shifted in time from others to maintain orthogonality.) This compresses the bandwidth of 1.23 MHz back to the original narrowband channel. Because the multiplication correlates only

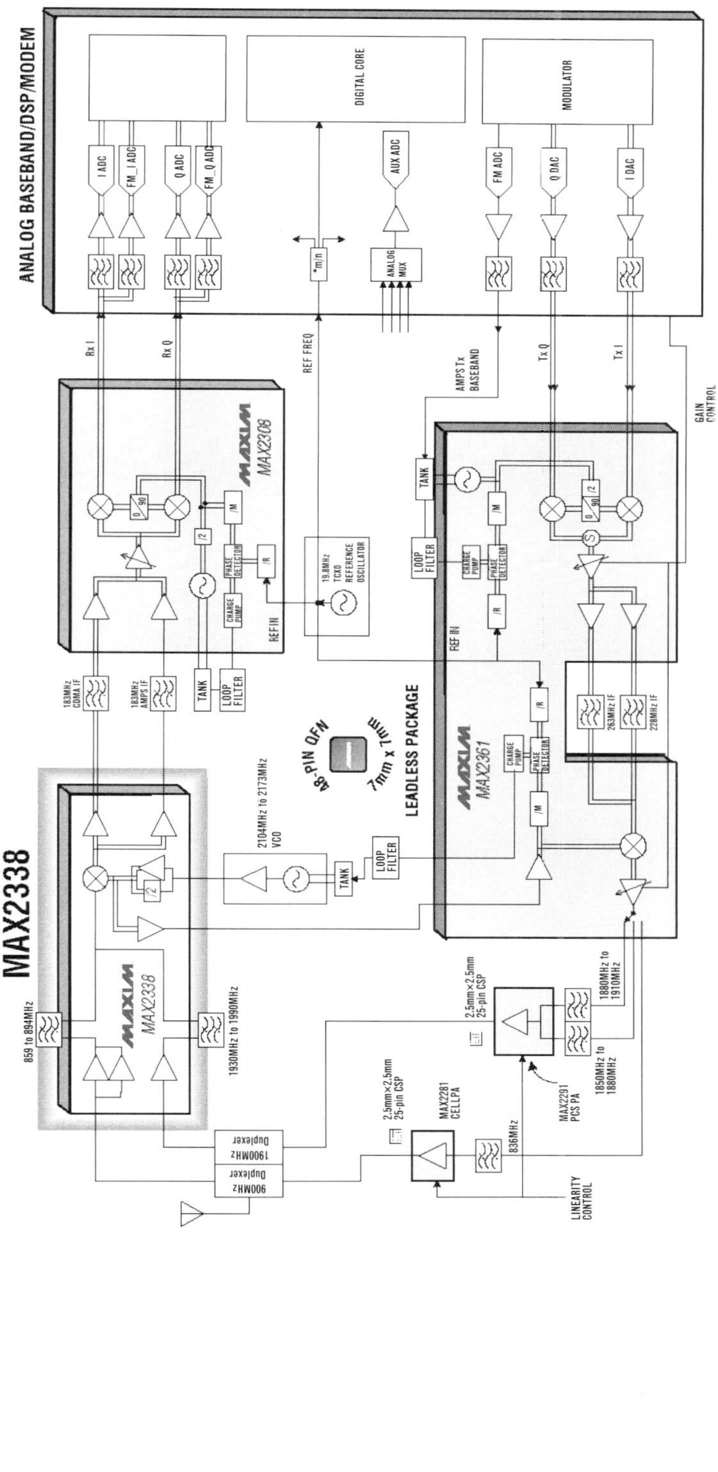

FIGURE 3.29 *Block diagram of a dual-band 900/1,900 MHz triple-mode phone, intended for the U.S. cellular and PCS bands. (Courtesy of Maxim Integrated Products.)*

signals with identical codes, the desired signal can be separated from adjacent channels, which appear within the same bandwidth as apparent noise.

The IS-95 specifications are contained in a document known as the air-interface specification. All equipment using the CDMA system within the band must be licensed to meet it. It is often a lengthy document, containing exhaustive measurement conditions to ensure conformity and interoperability between different equipment, and to ensure there is no interference to other users of the system. The relevant portions we need are summarized in Table 3.1.

Our approach here will be to analyze the system in Figure 3.29 to determine if it conforms with this specification (it does!). The handset shown is full duplex: the user can transmit and receive on a single channel at the same time (assuming, of course, that his mouth, ears, and brain work in full duplex mode). The local oscillator for both transmitter and receiver is derived from the same VCO source and multiplied appropriately to the required frequency. The transmitter path at the bottom of Figure 3.29 consists of an I-Q modulator to take the baseband I and Q channels to the IF, phase-locked loops to derive the IF and RF local oscillator frequencies, and IF amplification and filtering (one narrowband filter with appropriate bandwidth for each band). The RF section contains an RF mixer and driver amplifier, an RF filter to eliminate spurious mixing products, and separate power amplifiers for each band. Automatic gain control is used at the transmitter to ensure radiated output power according to the system

Table 3.1 Some Parameters from the IS-95/98 Air-Interface Specifications for the PCS CDMA Mobile Handset

Measurement Parameter	Specification
Receive frequency band	1,930–1,990 MHz
Transmit frequency band	1,850–1,910 MHz
Peak frequency deviation (channel bandwidth)	1,230 kHz
Maximum input power (total power at antenna connector) for error rate < 1/2%	–25 dBm
Minimum detectable signal (total power at antenna connector) for error rate < 1/2%	–104 dBm
Detection in presence of low level interference (error rate < 1%)	Detect a tone at –101 dBm in the presence of two interfering tones at –43 dBm each, offset 900 kHz and 1,700 kHz from center of channel
Transmitter output power	+23 dBm to –50 dBm
Maximum adjacent channel strength at 885-kHz offset	–42 dBc integrated over a 30-kHz bandwidth
Maximum alternate channel strength at 1,980-kHz offset	–54 dBc integrated over a 30-kHz bandwidth

requirements. The output duplexer is a split-frequency filter that allows the transmit and receive signals to share the same antenna without interfering with each other, and confines the transmitted spectrum to the transmit band. The transmitter portion is less complex and less illustrative than the receiver, so this analysis will focus on the top portion of Figure 3.29, which is the receiver part of the radio.

The receiver chain is a classic double super-heterodyne structure, with two downconversions between RF and baseband. The receiver has a duplexer and low noise amplifier to provide gain and filter the image frequency, a bandpass filter and first mixer to downconvert the signal to a first IF of 183 MHz, and a second mixer to downconvert the IF to baseband. The output I and Q channels from this quadrature mixer are digitally decoded and demodulated in the baseband DSP section.

The dual-band portion of the receiver requires parallel RF chains, one for each band. Following the IF amplifier however, the remaining functions can be shared. However, we should note that a single high side LO around 2,140 MHz is used for both RF bands, which, when divided by the on-chip divide-by-two circuit for use with the lower band, enables the RF mixer to produce a shared IF frequency around 183 MHz.

The receiver block diagram in Figure 3.30, extracted from Figure 3.29, summarizes the core RF and IF components in the receiver cascade. The first LO is at a frequency 183 MHz above the incoming RF signal (i.e., between 2,113 MHz and 2,173 MHz); the second LO is at a frequency of 183 MHz. The reason for the choice of this architecture and these mixers will become apparent during the analysis.

The component specification procedure is iterative by nature, and depends on the availability and cost of the technology available to support a given requirement. The process involves a trade-off between maintaining high power levels for best signal-to-noise ratio and low power levels for the best spurious response. Maintaining a simple spreadsheet to indicate the

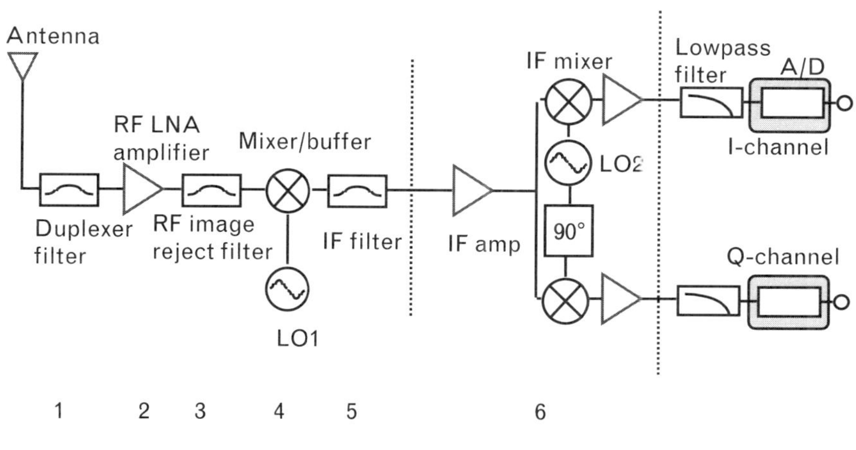

FIGURE 3.30
Core components in the receiver chain of the dual-band handset.

contribution of each stage is useful in examining these trade-offs. Let us look at each component in turn.

3.3.1 Receiver component specification

3.3.1.1 The duplexer filter specification

The input filter, variably called a duplexer (since it duplexes the transmit and receive signals), a diplexer (since it has two distinct passbands), or a preselector filter (since it selects only the receive band) has the principal function of ensuring the input amplifier is not overloaded by the transmitter. Although the selectivity of the receiver will ensure there is no output response to the transmitter signal itself since that is at a different frequency, there is a risk that a strong out-of-band signal can enter the receiver and saturate the early stages prior to such selectivity. The function of the duplexer filter is to reduce such signals below the saturation level at the input. It also reduces out-of-band noise from the input. The duplexer specifications are therefore set to minimize both transmitter leakage and out-of-band noise power.

From Table 3.1, the highest transmitter power level will be around +25 dBm. We will shortly find (after one design iteration) that the first mixer input should be less than –15 dBm to safely avoid mixer compression, and can be a little higher at the LNA input to avoid LNA compression. Thus, if we keep the leakage level at the input port below –25 dBm, there is no danger of LNA or first-mixer saturation in the wideband RF strip. This requires 50-dB attenuation from the duplexer at the upper edge of the transmitter band. Such isolation is impractical to achieve in a single package and will require careful shielding and attention to the packaging to separate the receiver and transmitter components from each other, in addition to the filtering.

Knowing the required attenuation allows us to specify the selectivity of the duplexer filter. This can be done by referring to selectivity charts or tables compiled for particular filter characteristics, but it is simpler to plot the required characteristics using the ideal bandpass filter element available in many RF linear simulators. The required characteristics are shown by the shaded lines in Figure 3.31. The passband lies between 1,930 MHz and 1,990 MHz, and the reject band, where 50-dB attenuation is required, begins 20 MHz lower than the lowest receiver frequency, at 1,910 MHz. If we allow 2.5-dB maximum passband loss, this requires a 6-pole Chebyshev filter. Duplexers that meet this requirement are commercially available.

The duplexer filter also helps to reject the image frequency. The first mixer uses a high side LO and an IF of 183 MHz, so the image is 366 MHz above the corresponding receive signal. It is apparent from Figure 3.31 that the duplexer filter *should* provide greater than 50-dB attenuation to the image. However, it is always prudent to examine the out-of-band response

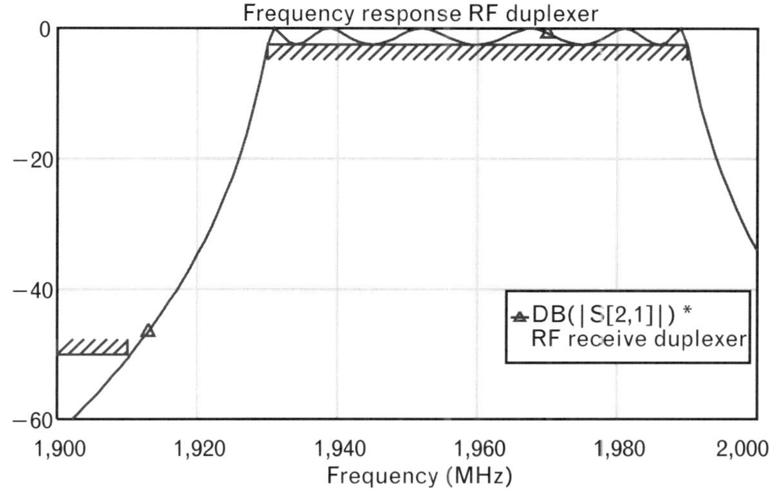

FIGURE 3.31
Duplexer filter bandpass characteristics.

of any filter, particularly if it uses transmission line elements, since their out-of-band response may contain periodic responses not evident in the idealized characteristic used for this preliminary analysis.

3.3.1.2 The LNA specification

At this stage of the design, we know that we require a LNA that will provide low-noise gain at the receiver passband. We also require that the LNA provide good linearity, because it is possible that the desired signal can be accompanied by much stronger interfering signals elsewhere in the receiver band. If the LNA is not sufficiently linear, these stronger signals can generate intermodulation responses that coincide exactly with the desired signal, and cannot be removed by any of the later components. The LNA specifications are determined by this requirement for linearity, even at maximum input power, and by the minimum detectable signal (which determines its maximum noise figure).

Since the design is iterative, our approach is to review the available components or chips. The RF front-end chip shown in Figure 3.29 provides an LNA that in its high gain, high linearity mode has a noise figure of 1.4 dB, a gain of 15.3 dB, and output third-order intercept point of 23.0 dBm. Since many ICs have variable gain RF amplifiers to handle different signal levels, in this analysis we will use the maximum gain in order to determine the receiver sensitivity, and the spurious-free dynamic range at this gain level (i.e., without AGC enabled).

3.3.1.3 The post-LNA image rejection filter specification

We have assumed already that the duplexer provides better than 50-dB image rejection. However, the LNA itself is most likely broadband and will amplify the image and/or its noise and reduce the sensitivity of the receiver. At the very least, the LNA will raise the single-sideband noise floor 3 dB because of aliasing foldover of the noise at the image frequency in the mixer, while at the worst it will possibly provide over 15 dB gain to any signal there.

The purpose of this filter is therefore to provide additional filtering at the image frequency. If we design for 20-dB loss to compensate for the potential gain of the image through the LNA, the characteristics required are shown in Figure 3.32. The selectivity is much wider than for the duplexer, since the image band is much further away than the transmitter band, lying within a 60-MHz band upwards from 1,930 + 2 (183) = 2,296 MHz. This can be achieved with a 2-pole Chebyshev filter, whose characteristics are plotted in Figure 3.32. This will provide over 40-dB attenuation at the image frequency for a maximum passband insertion loss of 2.5 dB. The total image rejection from this filter and the duplexer will therefore be at least 90 dB compared with the desired signal prior to the first mixer. This is comparable with the necessary dynamic range of the receiver, which from Table 3.1 must process input signals between −104 and −25 dBm.

3.3.1.4 LO1 frequency selection criteria

We stated that the first IF frequency is 183 MHz. This frequency was not chosen arbitrarily, but on the basis of available standard components and analysis of the spurious response. We have seen that if the IF is too low, the image frequency is too close to the receive band and too high a selectivity is required in the image rejection filter. If the IF is too high, the number of spurious responses that arise from frequencies falling within the RF passband increases dramatically.

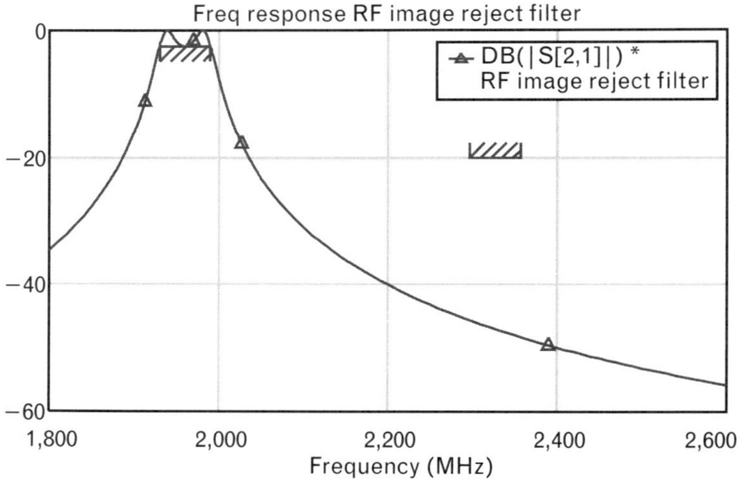

FIGURE 3.32
Characteristics of the image rejection filter.

With an IF = 183 MHz, and the desired (tuned) signal falling between 1,930 MHz and 1,990 MHz, we normalize to the IF as in Section 3.2.6.3 and obtain T = 10.55–10.87. From the spurious chart for a difference mixer, high side LO in Figure 3.24, there are no spurious responses falling within the RF passband. By way of illustration only, a bad design might choose an IF frequency of 650 MHz (admittedly unlikely in this type of application, but it is the best way to show the point), in which case we would have T hovering around 3, and from the chart the (n = 2, m = 3) spurious response would fall in-band. For instance, if the tuned (desired) signal were at 1,953 MHz, the LO would be at 2,603 MHz to create the IF = 650 MHz. Unfortunately, the (2,3) response of an in-band spurious frequency at 1,952 MHz would also fall at 2 (2,603) – 3 (1,952) = –650 MHz and be indistinguishable from the desired response. Filtering cannot remove this spurious signal since it falls in band.

With our IF of 183 MHz, the closest spur line from the chart will be the (2,2) response. For this figure, we can apply the relevant equation (3.38) for the (2,2) spur, and find spurious will occur at S = 11.05 (tuned low) or S = 11.37 (tuned high). The former falls out of band at 2,022.15 MHz and will be rejected by the duplexer; the latter falls further out of band at 2,080.7 MHz.

The rejection of this spurious signal at 2,022 MHz by the duplexer is seen from Figure 3.31 to be better than 50 dB. It is further rejected by using a double-balanced mixer, which (we will discover) nulls out the even spurious responses. We conclude that with an IF = 183 MHz spurious products will not be a problem, and we can be confident that the architecture chosen is satisfactory.

3.3.1.5 First (RF) mixer specification

The function of this mixer is to convert the incoming RF to a fixed IF of 183 MHz, selecting the desired channel by tuning the LO. We have stated above that a double-balanced mixer will reject the (2,2) spur, which is the most worrisome. With the circuit in Figure 3.29, with an integrated active mixer and buffer amplifier, we can obtain a net conversion gain of 14.5 dB, using a LO power around –3 dBm. The specified input third-order intercept point is 7.5 dBm, so the output IP3 is +22 dBm. The input 1-dB compression point is measured around –8 dBm, much lower, and the noise figure 7.8 dB. We will shortly use these numbers in determining the dynamic range of the receiver.

3.3.1.6 IF filter specification

At 183 MHz, the IF filter needs to reject the image frequency for the second mixer, the local oscillator feed-through, and all undesired mixing products. We can choose to reject the adjacent channel either here or following the second mixer. In our case, we can reject the adjacent channel here by using a narrowband SAW filter following the first mixer. This keeps the receiver selectivity as high as possible close to the antenna, and can prevent propagation of unwanted signals further downstream. Ideally its bandwidth needs to be such as to reject adjacent channel information, but still be broad enough to pass the bandwidth of the desired signal without distortion. With a peak frequency deviation of 1.23 MHz specified as part of the IS-95 requirement, the bandwidth of the first IF filter therefore must at least equal this.

To achieve a 3-dB bandwidth of 1.23 MHz its required Q will be 150, easily attainable. Its response is shown in Figure 3.33, where 5-poles have been specified in the assumed Chebyshev response and we allow for a typical SAW attenuation of around 8 dB. This filter cannot completely eliminate the adjacent channel, since its upper edge is only 1.25 MHz/2 or 625 kHz from the SAW center frequency. The remainder of the adjacent channel will need to be removed following the second IF mixer. The overall rejection requirement of the adjacent channel is determined by the maximum allowed signal strength in that channel and is part of the air-interface specification.

3.3.1.7 Second (baseband) mixer and downconverter specification

We select this component to downconvert the IF signal at 183 MHz to baseband. An integrated chip with variable gain IF amplifier, quadrature mixer, and phase-lock loop can provide all the requirements in a single package. The LO is divided by two on-chip to generate in-phase and out-

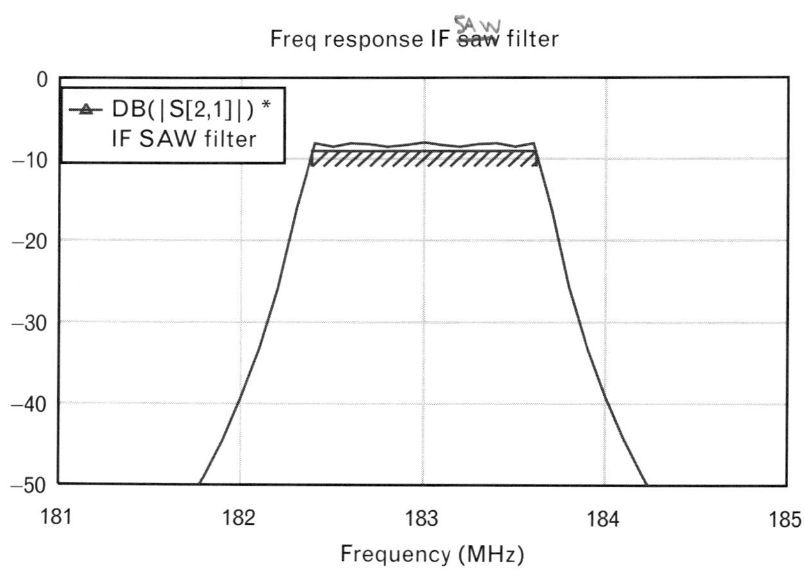

FIGURE 3.33 *Bandpass characteristics of a 5-pole Chebyshev filter centered at 183 MHz, with 8-dB insertion loss.*

of-phase signals for the mixer, so the PLL should lock the oscillator to 366 MHz. The output of the downconverter produces an I and Q baseband channel, since it uses a pair of mixers driven by the in-phase and out-of-phase LO signals. With the chip shown, at least 70 dB of IF gain control range is available, with a typical maximum gain of 35 dB. With maximum gain, the input IP3 point is –31 dBm, so the overall output IP3 is +4 dBm. The noise figure is 6.4 dB. In this example, the second IF amplifier and mixer are lumped together as a single component.

3.3.1.8 Demodulator

The I and Q baseband outputs can be filtered in a baseband chip that contains digital filters for antialiasing, sample-and-hold circuitry, analog-to-digital converters, and DSP for decoding and other demodulation functions. This chip is not analyzed here, as we want to derive the performance measures for the analog portion of the receiver.

3.3.2 Receiver response

The performance of each of the receiver components we have considered above is summarized in Table 3.2, where each stage is numbered according to Figure 3.30.

The linear components can handle, in theory at least, any input power without distortion. Of course, components such as SAW or ceramic filters will, in fact, introduce distortion and should therefore be assigned third-order intercept points. However, we will ignore such effects in these "linear" components for now and assume very large values for their output IP3 point.

The cascade response of the system can be analyzed by a tool as simple as a spreadsheet, or (preferably) either a linear systems simulator or a behavioral harmonic balance simulator. Such simulators automatically calculate the cascaded noise figure, gain, output power, and other system parameters, and determine their sensitivity to each of the cascaded components.

Table 3.2 Summary of the receiver component specifications

	Duplexer	LNA	Image-reject filter	First Mixer/Amp	IF filter	Second Mixer/Amp
Stage	1	2	3	4	5	6
NF (dB)	2.5	1.4	2.5	7.8	8	6.4
Gain (dB)	–2.5	15.3	–2.5	14.5	–8	35
IP3 out (dBm)	—	23.0	—	22.0	—	4.0

This enables the design to be iterated until a satisfactory outcome is attained.

Here are some general principles to remember in iterating:

- All the gain is there to amplify a weak signal.
- Downstream from the signal selecting filter, the signal can be considered weak.
- Ahead of a loss, dynamic range is important.
- After a loss, noise figure is important.

In a high dynamic range receiver, expect all stages to contribute to the noise figure, because strong signals will require insertion loss distributed throughout. A low noise design, on the other hand, will have high front-end gain, which dominates the cascaded noise figure.

These principles can best be examined by applying the formulae above to a typical cascade and varying the relative gains and intercept points of each component.

We will illustrate the analysis with some simple calculations first, by considering the cascade of a limited number of the components above, and compare the results using a behavioral nonlinear simulator. For simplicity of understanding, we will begin with the first five components only (i.e., the RF front end). We can use the values in Table 3.2 to derive the overall parameters of an RF/IF "supercomponent" prior to adding the final (sixth) IF-to-baseband stage.

3.3.2.1 Calculation of output noise power and SNR

The cascade noise figure given by (3.10) is repeated here for convenience:

$$F = F_1 + (F_2 - 1)/G_1 + (F_3 - 1)G_1G_2 + \ldots + (F_n - 1)/G_1G_2..G_{n-1}$$

Although G_1G_2 is high (12.8 dB or 19.05) compared with the gain of the other components, the F_3 term (following the LNA term) can not be neglected because the noise figure of the filter and mixer following the LNA totals 2.5 + 7.8 = 10.3 dB, or 10.72, which is comparably high. If we note that the first two terms of the system contribute $F = 2.5 + 1.4 = 3.9$ dB, or 2.45, to the cascade noise factor, the next term contributes ($[10.72 - 1]/G_1G_2$) = 9.72 / 19.05 = 0.51 to the total. Therefore, $F = 2.45 + 0.51 = 2.96$ or 4.7 dB.

The cascaded gain of the five components is simply the sum of the corresponding component gains—that is, $G = -2.5 + 15.3 - 2.5 + 14.5 - 8.0 = 16.8$ dB.

The output noise power is given by (3.11). At room temperature kT= –174 dBm/Hz, and because we will analyze only the analog portion of the receiver (prior to de-spreading) where the channel bandwidth is 1.23 MHz,

$$N_{OUT} = -174 + 10\log\left(1.23 \star 10^{6}\right) + 4.7 + 16.8 = -91.6 \text{ dBm}$$

The noise floor referred to the input when the cascade is modeled as ideal is –91.6 – 16.8 = –108.4 dBm. If a signal can be detected once it exceeds the noise floor, then the *minimum detectable signal* (MDS) for this system is –108.4 dBm. (From Table 3.1, the CDMA spec requires –104 dBm, if we consider, as we are here, the total channel power rather than the power in just one user component multiplexed within it.) In digital systems, the MDS is related to the probability of a bit error equaling some threshold, and this requires some additional margin (typically a few decibels) in the input power above the input-referred noise floor.

If we arbitrarily consider a "nominal" input signal level of –60 dBm, the output IF signal level is at $Pout$ = –60 + 16.8 = –43.2 dBm so the output signal-to-noise ratio is –43.2 – (–91.6) = 48.4 dB at nominal input power. The nominal output power from the simulator (which uses more complex models) is close, at –42.9 dBm.

3.3.2.2 Calculation of output third-order intercept point IP3 and distortion

To calculate the output IP3 point of the first five stages, we can substitute from Table 3.2 into (3.24), repeated here for convenience:

$$\frac{1}{IP_0} = \left[\left(\frac{1}{IP_N|_{out}}\right) + \left(\frac{1}{IP_{N-1}|_{out}}\right) + \left(\frac{1}{IP_{N-2}|_{out}}\right) + \ldots + \left(\frac{1}{IP_1|_{out}}\right)\right]$$

We can neglect the output intercept contributions from the three filters, which are assumed linear, since they contribute very large terms in the denominator of the summation. The output intercept point of the LNA referred to the cascade output is 23 dBm plus the following gain, or 23 – 2.5 + 14.5 – 8.0 = 27 dBm. The output intercept point of the RF mixer/amplifier referred to the cascade output is 22 dBm plus the following gain, or 22 – 8 = 14 dBm. We therefore calculate a system output intercept point of 13.8 dBm for the first five stages, dominated principally by the mixer.

In calculating the output intermodulation distortion $IMD3$, it is simplest to construct a diagram of the form shown in Figure 3.34. Referenced to the output, the noise floor at –91.6 dBm and third-order intercept point at 13.8 dBm can be plotted. The fundamental power and third-order

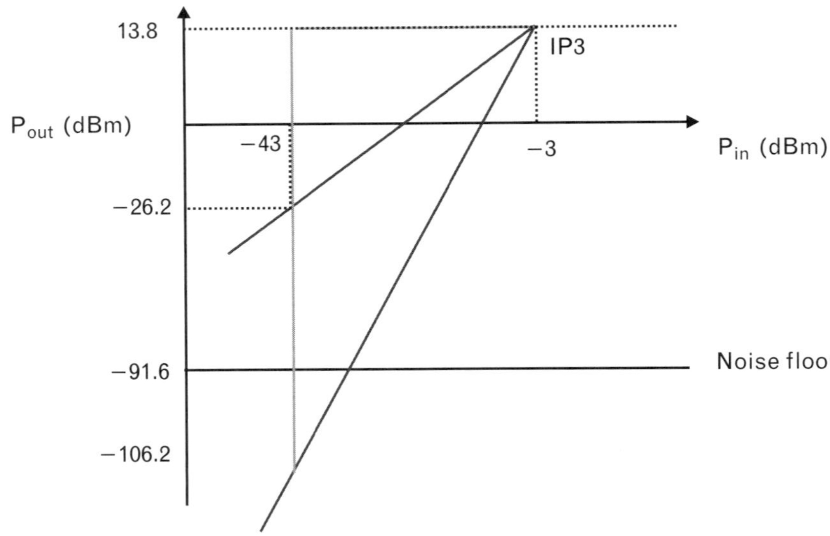

FIGURE 3.34
Calculated output fundamental and third-order distortion power versus input power of the first five components of the radio receiver of Figure 3.30.

distortion rise 1 and 3 dB, respectively, for each 1-dB increase in input power. With a cascade gain of 16.8 dB, the input power at the IP3 point corresponds to –3 dBm.

The CDMA specification for low-level intermodulation distortion uses two tones each of –43 dBm input level. These input tones turn out to be 40 dB below our calculated input third-order intercept point, so it follows that the input-referred power of their third-order distortion product is 3 (40) = 120 dB below the power of the input third-order intercept, or $-3 - 120 = -123$ dBm. Referred to the output, this distortion power is $-123 + 16.8 = -106.2$ dBm, well below the noise floor. Since the CDMA specification requires detection of a desired signal of −101 dBm input power in the presence of these third-order products, this distortion causes no loss of sensitivity as it is much lower than the thermal noise.

The CDMA intermodulation test calls for the two interfering tones to be offset 900 kHz and 1,700 kHz from the center frequency. Although they lie in an adjacent channel slot, the third-order distortion product they produce will fall into our channel, 100 kHz from the center. Our manual calculation of intercept point neglects the selectivity of the receiver filters in rejecting most interfering signals as they move downstream, so we may end up with a rather *broadband* view of the distortion power rather than a realistic *tuned* view. Depending on the relative frequencies of the distortion-producing signals, the distortion could be less than we have calculated because of the filtering provided upstream.

To simulate the manual calculation we made above, we use two RF signals of −43 dBm lying 1.25 MHz apart (i.e., situated in adjacent channels[1]). Figure 3.35 shows the resulting spectrum at the output of the IF mixer. This is the output from a behavioral harmonic-balance simulator,

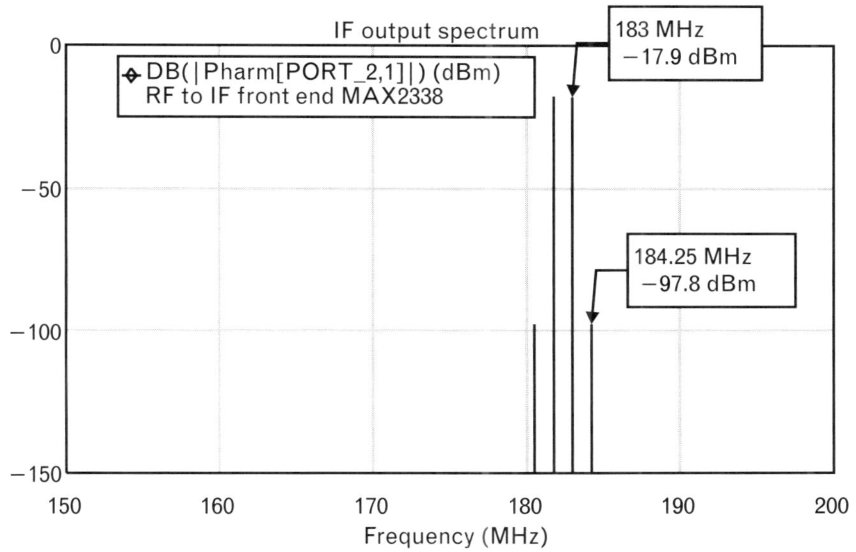

FIGURE 3.35 *The IF output spectrum of the RF front-end with two input signals in adjacent channels, prior to any IF filtering.*

such as included in the Visual System Simulator from Applied Wave Research. The signal is shown prior to the IF SAW filter (component 5) since the filter would otherwise remove all but the fundamental IF component resulting from one of the tones. The two downconverted signals are therefore both still observable at this point in the receiver, at −17.9 dBm each, as are their third-order distortion products that also fall into adjacent channels. The IMD3 power levels correspond almost exactly to those we have just calculated, after adding back in 8 dB loss from the IF SAW filter that we accounted for in the manual calculation (i.e., −97.8 – 8 = −105.8 dBm compared to −106.2 dBm).

The swept-power output of the RF-IF portion of the receiver is shown in Figure 3.36. The simulated 1-dB compression point at the input is –27 dBm, in comparison with our simple calculation of input *IP3* = –3 dBm. This 1-dB compression point is thus lower than we would have expected, but arises because the measured 1-dB compression point of the RF mixer is some 15 dB lower than its measured *IP3* point, and because of the gain ripple in the filters across the passband.

3.3.2.3 Calculation of spurious free dynamic range

The spurious free dynamic range of the RF front end can be calculated directly from (3.26) as

1. The air-interface specification calls for three tones at the input, two of them (at 900-kHz and 1,700-kHz offset) producing distortion that falls in band to interfere with the third tone at the center of the desired channel. A simpler simulation is to use just two interfering tones, one in an adjacent channel, and to calculate the distortion they produce, which will give similar results if the distortion is simulated prior to any filtering.

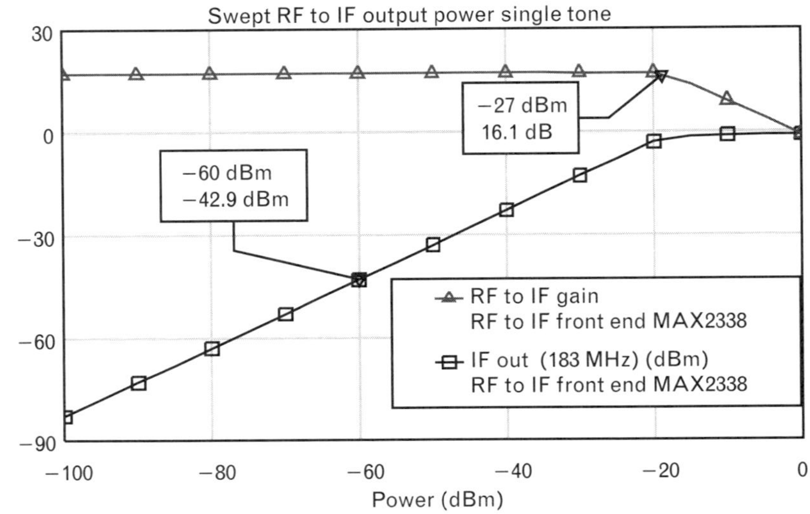

FIGURE 3.36 *Simulated swept power response parameters of the first five stages of the radio receiver of Figure 3.30, to a single-tone excitation*

$$SFDR = 2/3\left(13.8 - \left(-91.6\right)\right) = 70.3 \text{ dBm}$$

or directly from the geometry of Figure 3.34. It corresponds to the range of power levels between the minimum detectable signal (here equal to the input noise floor, −108.4 dBm) and the input power level at which the third-order distortion first appears greater than the noise. As noted earlier, the in-band distortion results from signals of equal power levels in the two adjacent channels either both above or both below our channel.

However, this calculation neglects the impact of automatic gain control. In fact, the gain is often adjustable or switchable between a few modes, so that the gain can be reduced in the presence of strong signals. This delays the onset of compression as the input power is increased and increases the range over which signals can be received before distortion components are produced. Effectively, the *x*-axis in Figure 3.34 is shifted to the left as the gain is reduced, allowing larger input powers prior to the onset of compression and an increase in the dynamic range.

3.3.2.4 Calculation of overall radio receiver response

The final (sixth) stage in Table 3.2 may now be cascaded with the composite of the first five stages just considered. This stage has a variable gain IF amplifier. In its high-gain mode, the stage provides 35-dB gain. The input *IP3* point for the stage is −31 dBm, and 1-dB compression occurs at an input IF power of −44 dBm (i.e., approximately when we apply the *nominal* −60 dBm RF input to the first stage). This is very low when compared with the (output) *IP3* point of the preceding stages, +13.8 dBm. Clearly the sixth stage is only switched into this high gain state when very low input powers are present at the input. Using these values, we estimate the output intercept point of the entire two chip-set radio receiver to be approximately that of the sixth stage alone (i.e., –31 + 35 = 4 dBm), and the cascaded gain to be 16.8 + 35 = 51.8 dB.

Although the intercept point of the final stage seems low, linearity is most important in the wideband section of the radio receiver where multiple channels are present and the possibility of intermodulation arises. Following the IF SAW filter (component 5), interfering channels have been reduced in power, and the stringent requirements on intercept point can be relaxed. In the event of strong signals that would cause the final stages to saturate, the sixth-stage gain can be reduced to below –35 dB. We will refer to this as *low-gain mode* in the following discussion. Then, the measured input power at 1-dB compression for the stage is –9 dBm and the input IP3 point rises to +3.4 dBm. With the system at this gain, 1-dB compression will now occur when the input RF power is some –9 – (–44) = 35 dB higher than before. The RF gain can also be adjusted as part of AGC in many systems, to further improve our ability to receive strong signals.

Figure 3.37(a) shows the simulated (single-tone) response of all six stages as listed in Table 3.2. The calculated total small-signal gain is 52.3 dB and with this amount of gain the 1-dB compression point occurs at –12.7 dBm output power (i.e., –64 dBm input), reasonably consistent with our manual calculation of the final output IP3 point of +4 dBm above (–48.3dBm input IP3). The dynamic range has been substantially reduced compared with that of the RF front end on its own, because the system now has excessive gain that lowers the intercept point when referred back to the input. It can be restored by switching the final stage to low-gain mode, as shown in Figure 3.37(b). There, the input 1-dB compression point increases to –28 dBm, when the final stage gain is reduced to –35 dB and the input signal is strong enough that the AGC now renders the total system gain negative (–17.7 dB). This is reasonably consistent with the CDMA air-interface specification, which requires maximum power handling capability for signals up to –25 dBm input power.

The simulator also allows a variety of modulation formats to be examined. Narrowband CDMA signals, because of the processing gain they derive from the wide spreading bandwidth, can be detected even when more than 15 dB below the noise floor, when considered prior to despreading. The RF spectrum is shown at the output of the receiver LNA in Figure 3.38, for an input power level averaging about –100 dBm. This spectrum clearly approximates a multitone intermodulation test, with a continuum of closely spaced carriers spread across the entire 1.23-MHz channel bandwidth. As a result, we would expect any third-order products generated to fall not only into adjacent channels as spectral regrowth, but also into the desired channel. The former concern is critical in transmitters, and the latter especially so in receivers.

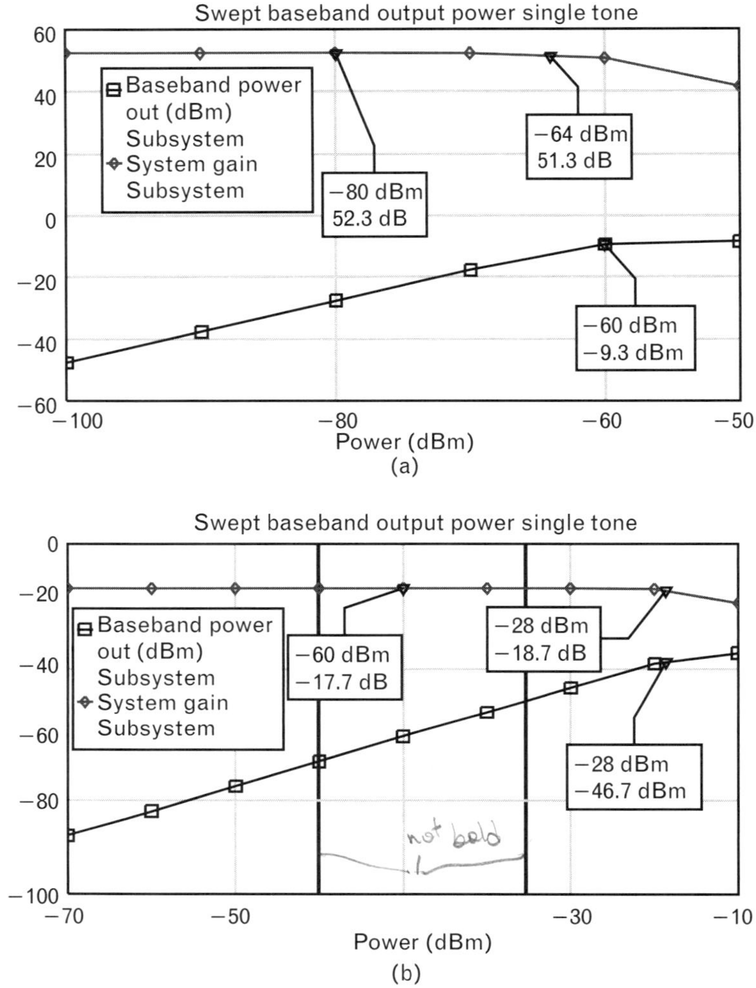

FIGURE 3.37
System diagram and cascade analysis of the six-stage radio receiver described in Table 3.2: (a) in high gain mode, when the last stage has +35 dB gain; and (b) in low gain mode, when the last stage has −35-dB gain.

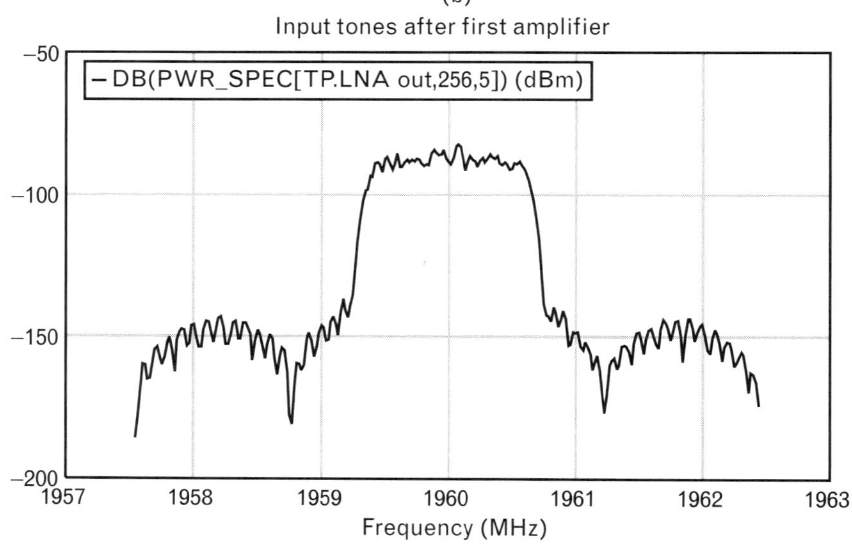

FIGURE 3.38
The RF CDMA spectrum at the LNA output.

Such spectral regrowth can clearly be seen in Figure 3.39, which shows the same CDMA spectrum at the baseband output. At a low input power level of −100 dBm [Figure 3.39(a)], the adjacent channel power ratio measured in a 30-kHz bandwidth offset the specified 885 kHz from the center frequency is only about −100 dBc. At a higher input power level of −60 dBm, however [Figure 3.39(b)], the receiver is in compression (because we have kept high-gain mode rather than lowering the system gain), and the power that spills into adjacent channels is clearly evident. In the receiver, if this signal were an interfering channel that were adjacent to the weaker, desired channel, the interfering channel would first need to be eliminated

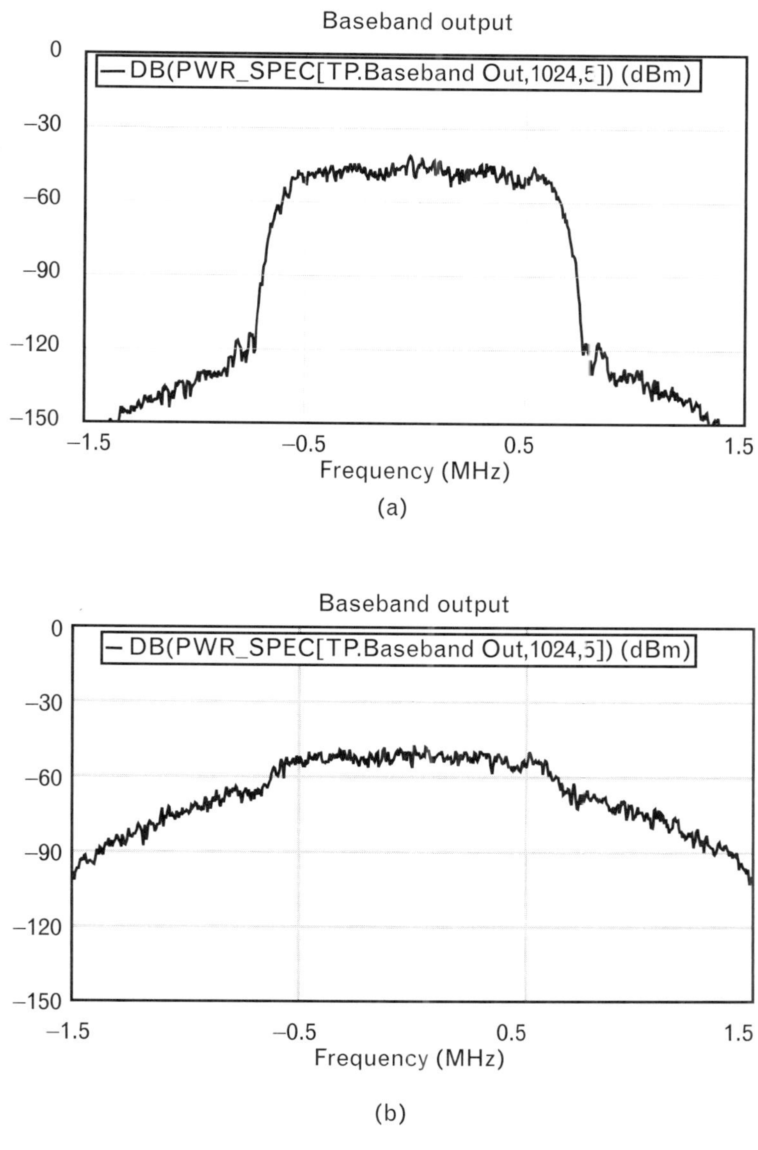

FIGURE 3.39 *The baseband CDMA spectrum at the output of the radio receiver of Figure 3.30 in high-gain mode at input power around (a) −100 dBm and (b) −60 dBm.*

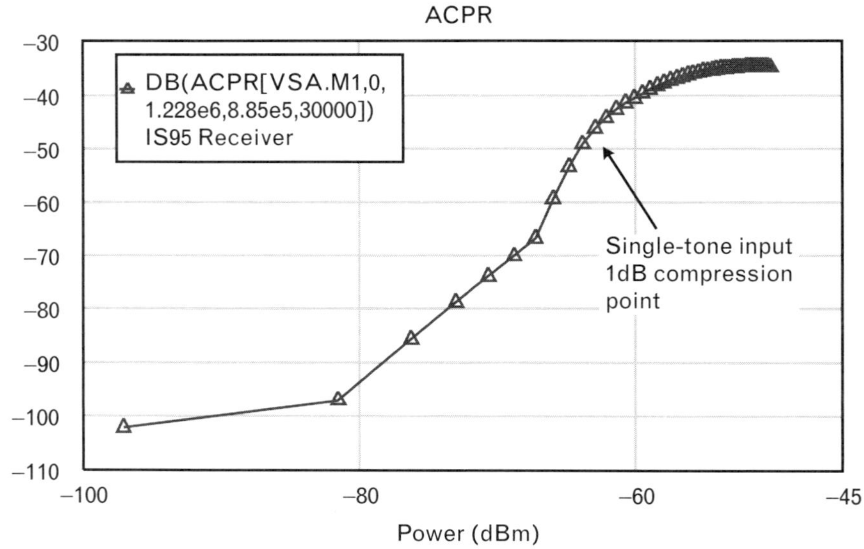

FIGURE 3.40
The ACPR of the receiver of Figure 3.30 with a CDMA spectrum, calculated in a channel bandwidth of 1.228 MHz, compared with an adjacent channel 885-kHz offset with bandwidth 30 kHz.

by IF filtering prior to the final output stages that saturate, since the spectral spill-over would otherwise swamp the desired signal.

Figure 3.40 shows the adjacent channel power ratio of the six-stage receiver in high-gain mode as a function of input power. With these gain settings, the ACPR starts to increase faster than 30 dB/decade (3:1 slope) when the input channel power exceeds about –67 dBm, and becomes quite severe around the single-tone input 1-dB compression point. In fact, the compression point with a multitone input signal is considerably lower, especially for signals with a high peak-to-average power ratio. Clearly, this receiver needs to operate with reduced gain once the signal approaches this input power level, so that the input-referred 1-dB compression point is increased.

Numerous other effects can be simulated, such as eye-diagrams, bit-error rate, and signal constellations, which assist the designer in assessing the quality of the radio in beating the air-interface specification. At this point, however, we hope that we have whetted your appetite and you are ready to begin the journey into component design.

Problems

1. In the simple diode detector of Figure 3.1, the inductor serves the purpose of an RF choke.
   (a) What is its function?
   (b) What happens to the voltage on the anode of the diode if it is not present?

(c) Does the diode become a more or less sensitive detector? Why?
(d) What might be a good definition of detection sensitivity here?

2. A signal has an SNR of 20 dB. How much can the SNR decrease if the bandwidth is doubled, assuming the same information throughput?

3. What is the theoretical maximum transmission capacity down a *plain old telephone system* (POTS) line, for which the signal to noise ratio is 30 dB? Such a line has its bandwidth restricted to 3.4 kHz.

4. Using (3.10), show that the noise figure of an attenuator preceding a low-noise amplifier is just equal to the noise figure of the amplifier (in decibels) plus the loss of the attenuator (in decibels).

5. In Figure 3.11, show that a signal plus noise to noise ratio $(S + N)_O/N_O$ of 10.4 dB corresponds to a signal to noise ratio of 10 dB.

6. Using analysis similar to (3.12) to (3.15), show that the third-order term produces a third-harmonic component that rises as the input power is increased at a rate three times as quickly as the fundamental.

7. (a) Expand (3.13), including the cubic term, and collect the terms containing the fundamental output power for a single tone input.
(b) Calculate the value of E for which the gain (or power out/power in) is 1-dB compressed from its linear value. This is the 1-dB compression point, referred to the input. (Note that this is measured with one input tone only.)
(c) Calculate the value of E for which the third-order products from (3.19) have the same value as the linear fundamental term in step (a), ignoring the higher order terms that cause the fundamental to compress. This is the third-order intercept point, referred to the input. (Note that this is measured with two input tones, and is the power in *each* tone.)
(d) Show that the third-order intercept point is 10.76 dB above the 1-dB compression point.

8. Two amplifiers are cascaded. The driver has 10 dB gain and an output intercept point of x dBm, the output amplifier has 6 dB gain and an intercept point of 30 dBm. What is the amplifier output intercept point if: (a) $x = 30$ dBm? (b) $x = 24$ dBm? (c) $x = 20$ dBm? (d) What is poor about the design with cases (a) and (c)?

9. In the two-stage amplifier of Problem 8, prove that the degradation in the output third-order intercept point for the cascade compared with that of the final stage alone is given by

$$\text{degradation(dB)} = 10\log\left(1 + 10^{\frac{IP3_{driver} - IP3_{final}}{20}}\right) \quad (3.40)$$

10. Calculate the spurious-free dynamic range of a receiver for each of the three cases of Problem 8 if the noise figures of both stages are 2 dB, and the IF bandwidth is 10 kHz.
11. In Figure 3.23, the output of a transistor is modeled by a current source shunted by an output resistance and output capacitance. If the small-signal output resistance is 500 and the output capacitance is 0.5 pF, what is the phase change at 1 GHz if as the amplifier is driven towards compression: (a) its output capacitance reduces to 0.25 pF? (b) its output resistance reduces to 250?

References

[1] Mirabbasi, S., and K. Martin, "Classical and Modern Receiver Architectures," *IEEE Communications Magazine*, November 2000, pp. 132–139.

[2] Bradley, M. J., "Wideband Receiver for Digital Radio Mondiale," *Electronics and Communication Engineering Journal*, IEE, February 2002.

[3] Podell, A., *Applied Wireless and Microwave Techniques*, Los Altos, CA: Besser Associates, 2002.

[4] Rohde, U. L., and T. T. N. Bucher, *Communications Receivers—Principles and Design*, New York: McGraw-Hill, 1988.

[5] Falconer, D., et al., "Frequency-Domain Equalization for Single-Carrier Broadband Wireless Systems," *IEEE Communications Magazine*, April 2002, pp. 58–66; originally appeared in Struhsaker, P., and K. Griffin, "Analysis of PHY Waveform Peak to Mean Ratio and Impact on RF Amplification," IEEE 802.16a cont. IEEE 802.16.3c-01/46, March 6, 2001.

[6] Rohde, U. L., "Key Components of Modern Receiver Design," *QST*, Vol. 68, No. 5, May 1994.

CHAPTER 4

The Smith chart and S-parameters

4.1 Introduction

Philip Smith, an AT&T engineer, developed a graphical tool in 1933 to simplify the task of calculating the variation of input impedance to a transmission line in passive transmission line circuits [1, 2]. Called the Smith chart in his honor, it represented one of the single most significant contributions to microwave analysis. Remarkably, the paper he wrote to describe this invention and submitted for publication to the Institute of Radio Engineers (the predecessor to today's Institute of Electrical and Electronics Engineers) was rejected for *not having any practical contribution!*

Nevertheless, the Smith chart has become one of the most popular design aids for RF and microwave engineering, with an estimated more than 70 million copies distributed throughout the world since its inception. In 1974, the IEEE MTT Society also recognized his contribution and gave him the Microwave Application Award.

Despite its wide circulation, the most fundamental aspect of the Smith chart is often overlooked, even by those who frequently use it. For many engineers who have only had a limited introduction to it, the Smith chart remains a vague concept with its intimidating collinear sets of resistance-conductance and reactance-susceptance circles. If asked, many engineers would probably respond that the Smith chart is a plot of transmission line impedances. While this is true is some sense, it overlooks the much simpler basis of the chart.

More than 30 years after the development of the Smith chart, a new line of test equipment based on traveling voltage waves and broadband resistive terminations was introduced to directly measure and display the parameters Smith was working with. Previously used voltage and current-based impedance and admittance measurements that required open and short-circuit terminations were inaccurate and impractical beyond 100 MHz. The new test equipment, called a network analyzer, could accurately characterize RF and microwave components into the gigahertz region through the use of scattering (S) parameters. Having a reliable and repeatable measurement capability revolutionized high-frequency design, changing it from an art to a science.

In this chapter we begin with graphical circuit transformations with lumped and distributed components on the Smith chart, based on reflection coefficients. In the second half we also look at signal power transmitted through circuits, using single-ended and balanced *S*-parameters.

4.2 The Smith chart: a polar plot of reflection coefficient

The Smith chart is nothing more than a polar plot of the reflection coefficient with impedance and/or admittance overlay! Philip Smith, who worked with passive transmission line circuits, saw that input impedances of transmission lines have a very broad range—they vary between zero and infinity. He also knew that reflection coefficients in his applications could not exceed unity magnitude, and they could therefore be displayed on a convenient small polar plot. In the 1930s, however, graphs with reflection coefficients did not have much meaning, even to engineers working in that field. An impedance $Z = (25 + j10)\Omega$ could immediately be connected with a physical circuit, but the equivalent reflection coefficient $\Gamma = 0.36\angle 151°$ was not well received.

In order to maintain the compact size of reflection coefficient plots and to communicate with others, in familiar terms Smith needed to create "overlays," corresponding to impedance coordinates. Being a good mathematician as well as an engineer, he used a *bilinear transformation* [3, 4] to map the normalized rectangular impedance plane for all passive circuits into a unit-radius Γ plane. The completed drawing was to become known as the Smith chart. The transformation was based on the reflection coefficient formula given in Chapter 2.

Recalling (2.48) from Chapter 2 in generalized form, and normalizing it to the reference impedance, Z_0,

$$\Gamma = \frac{Z - Z_0}{Z + Z_0} = \frac{\dfrac{Z}{Z_0} - \dfrac{Z_0}{Z_0}}{\dfrac{Z}{Z_0} + \dfrac{Z_0}{Z_0}} = \frac{z - 1}{z + 1} \tag{4.1}$$

where $z = r + jx$ is an arbitrary normalized complex impedance.

Equation (4.1) is the well-known form of bilinear transformation applied to the z- and Γ-planes. Mathematically it means that:

1. We map points in the z-plane into points in the Γ-plane on a one-for-one basis. *That is, for every z there is one and only one Γ, and vice versa.*

2. *Circles in the z-plane map into circles in the Γ-plane, and vice versa.* (For mathematicians, straight lines are special cases of circles having infinite radii.)

3. *The entire (infinite) right half of the z-plane maps inside a unit circle of the Γ-plane.*

The third property listed above for the bilinear transformation is especially useful, as we shall see, to simplify complex RF and microwave engineering problems. Mapping the entire and infinite right-half z-plane inside a unit circle allows us to perform impedance matching calculations and transmission line manipulations on a finite chart—the unit-Γ-circle Smith chart. (From now on, unless otherwise indicated, we always refer to the *unit-radius Smith chart.*)

The term "mapping" means that if $z = r + jx$ takes on a range of values in the z-plane, the contour is transformed (mapped) onto a new contour in the Γ-plane. For example, the values for which $r = 0$ and jx varies from $-\infty$ to $+\infty$ (consisting of the vertical jx-axis in the z-plane) are mapped into the unit-radius circle (circumference) of the Γ-plane. For other constant values of r, a family of circles is formed inside this unit-radius circular border, as shown in Plate 1. These circles are called constant-resistance circles, since each one represents a specific resistance value with varying amount of reactance. Note how the constant-resistance circles all have their centers along the horizontal axis (they are collinear, not concentric) and that they all have a common point, the $z = \infty$ point.

Horizontal lines in the z-plane, each representing a specific reactance with resistance varying from zero to $+\infty$, called constant-reactance lines, map into the Γ-plane as arcs of constant-reactance circles (Plate 2). Although within the unit-radius chart, we only see portions of these circles, these contours are likewise circles. Their centers lie along a vertical line drawn through the $z = \infty$ point. The full circles will be visible later in Section 4.8 when we look at Smith charts with greater than unity radius.

Combining the constant-resistance and constant-reactance circles completes the normalized impedance Smith chart (Plate 3). We should mention that the original chart developed in the 1930s was not normalized; it was referenced to 50Ω. The form of the chart was exactly the same, but all resistance and reactance values were 50 times greater. The problem with this format was that not everyone used 50Ω as a reference impedance and charts had to be customized for various needs. Developing the normalized chart helped to simplify Philip Smith's life, but it added the following two extra steps for all users:

Plate 1 and all other Plates are located at the end of Chapter 4.

1. Every impedance, resistance, or reactance must be normalized (see Section 2.4) to Z_0 before placing it on the normalized chart.

2. Readings obtained from the chart before computing the circuit element values must be normalized.

Before we begin using the chart, let's familiarize ourselves with some of its key points. We marked 10 locations with the letters A through J on Plate 3 and listed in Table 4.1 by their corresponding z and Γ values. Remember the upper half of the impedance chart is always inductive while the lower is capacitive. The region near infinity contains high impedances and on the far left we see low impedances. The reflection coefficient has finite values for all passive terminations, including open and short circuits.

You may ask why we refer to the Γ "plane" when all Γ values fall into the unit circle. Remember at this point that we are only interested in passive z values for which r is not negative. Mathematically, negative r values can also be mapped by (4.1), but the left half of the z-plane, for which $r < 0$, maps into the infinite extent Γ-plane with $\rho > 1$. We will treat the extended Smith chart later, for which r can assume negative values.

Also note in Plate 3 that the upper and lower halves of the Γ-plane are symmetric. Thus, if we define a value for Γ which corresponds to some, then the complex conjugate value of reflection coefficient, Γ^*, will correspond to z^*. Moreover, Γ^* or z^* are found on the Smith chart merely by changing the sign of the angle of Γ and the sign of the imaginary part of z^*. Put another way, the complex conjugate of Γ or z^* is found by mirroring the Γ-vector, or z-point about the horizontal axis of the Smith chart.

TABLE 4.1 INTERPRETATION OF VARIOUS POINTS AND REGIONS OF THE NORMALIZED SMITH CHART

MARKER	NAME	z-PLANE	Γ-PLANE	COMMENT
A	Reference impedance	$1 + j0$	$0.0 \angle 0°$	Single point
B	Ideal resistances	$r + j0$	$\rho \angle 0°$ or $\rho \angle 180°$	Main diagonal of the chart
C	Ideal capacitive reactances	$0 - jx$	$1.0 \angle$ any $-°$	Lower half of circumference
D	Ideal inductive reactances	$0 + jx$	$1.0 \angle$ any $+°$	Uppe half of circumference
E	Short circuit	$0 + j0$	$1.0 \angle 180°$	Single point
F	Open circuit	∞	$1.0 \angle 0°$	Single point
G	Upper half of the chart	$r + jx$	$\rho \leq 1 \angle$ any $+°$	Inductive half of chart
H	Lower half of the chart	$r - jx$	$\rho \leq 1 \angle$ any $-°$	Capacitive half of chart
I	Specific z value	$0.5 + j0.2$	$0.36 \angle 151°$	Single point
J	Complex conjugate of z	$0.5 - j0.2$	$0.36 \angle -151°$	Single point

4.2.1 Impedance manipulations on the Smith chart

Although the motive for designing the Smith chart may have been the determination of impedance transformations along a transmission line based on the variation of reflection coefficient, the Smith chart proves to be an excellent format for manipulating impedances. This is especially useful in impedance matching applications (see Chapter 5). The reason for this is that all impedances with positive real parts, including open and short circuits, conveniently fit within the Smith chart.

Series additions of components are most conveniently handled in the rectangular impedance system and the impedance Smith chart. Before we look at any specific component, let's compare side by side (see Table 4.2) the steps required to add various lumped series circuit elements to a normalized impedance, $z_L = 0.5 - j0.5$. We also summarize these movements on the Smith chart shown in Plate 4.

4.2.2 Adding series inductors on the Smith chart

Add a 8-nH series inductive reactance to a one-port termination at 1 GHz, as shown in Plate 5. (Although we use a one-port here, its input impedance could represent a port of a multiport network.) The one-port is characterized at 1 GHz by its reflection coefficient, $\Gamma_1 = 0.45 \angle -116°$. The normalized impedance Smith chart indicates that Γ_1 represents $z_1 = (0.5 - j0.5)$.

Since the impedance chart is not calibrated in nanohenries and picofarads, the first step is to convert the 8-nH inductor to normalized inductive reactance at 1 GHz. Using (2.16), calculate the normalized reactance

$$x_L = 0.1257 f_{GHz} L_{SnH} = 0.1257(1)8 = 1.0$$

The next step is to add the $j1$ unit normalized inductive reactance to our initial point z_1 and (Γ_1) on the Smith chart. Since the inductor is a

Table 4.2 Summary of Lumped Series Element Movements on the Normalized Impedance Smith Chart

Element to Be Added to z_L	Movement in the z-Plane	Movement in the Γ-Plane
Lossless series inductor L_S	Adding $+jx_L$ on the $r = 0.5$ constant-resistance (vertical)line	Adding $+jx_L$ on the $r = 0.5$ constant-resistance circle
Lossless series capacitor C_S	Adding $-jx_C$ on the $r = 0.5$ constant-resistance (vertical) line	Adding $-jx_C$ on the $r = 0.5$ constant-resistance circle
Ideal series resistor R_S	Adding r_S on the $x = -0.5$ constant-reactance (horizontal) line	Adding r_S on the $x = -0.5$ constant-reactance circle

Note: Reactive elements move on the constant-resistance circles, and resistors on the constant-reactance circles.

lossless component, we add inductive reactance by following the constant resistance circle one unit upward arriving to the point z_T (Γ_T) on the chart. This is the new input reflection coefficient of the circuit after adding the series inductor.

The reflection coefficient of the final circuit is on the horizontal axis of the chart. Therefore, the new circuit looks like a resistor at 1 GHz. (This is just a coincidence; if we use a different size of inductor, we end at a different location.) We could also say that we resonated out the internal capacitive reactance of the one-port at that frequency.

As we mentioned before, the Smith chart does not have the resolution to show small component losses (i.e., $Q > 10$). Significant inductor losses can be modeled as series R-L combinations, to be handled in two separate element movements on the chart.

4.2.3 Adding series capacitors on the Smith chart

If we add a series capacitor to the same Γ_1 starting point, the movement also takes place on the 0.5 unit constant-resistance circle. This time, however, we move into the opposite—capacitive—direction. To illustrate this movement, let us use a 3.2-pF series capacitor at 1 GHz and show the effect on Plate 6.

Once again the first step is to calculate the normalized capacitive reactance of our component. From (2.18) the capacitive reactance is

$$x_C = \frac{3.183}{f_{\mathrm{GHz}} C_{\mathrm{SpF}}} = \frac{3.183}{(1)3.2} = 1$$

Adding $-j1$ unit capacitive reactance to our initial starting point on the Smith chart, following the constant $r = 0.5$ circle, moves Γ_1 to Γ_T, which corresponds to $z_T = (0.5 - j1.5)$.

4.2.4 Adding series resistors on the Smith chart

Finally, let us see how a series resistor will move us on the Smith chart. This time we will add a 25-Ω series resistor to the same one-port used in the previous two illustrations. Once again we have to first normalize the resistor since we will be working on a normalized Smith chart.

$$r_S = \frac{R_S}{Z_0} = \frac{25\Omega}{50\Omega} = 0.5$$

Adding an $r_S = 0.5$ unit normalized resistance to z_1 will move us on the constant reactance circle toward infinite resistance to a new normalized input impedance of $z_T = (1 - j0.5)$. If we also add a 4-nH series inductor, half of what was used in Section 4.2.2, the new input impedance moves to the center of the chart. At that point the normalized impedance is $z_T = (1 + j0)$, which translates to 50-Ω unnormalized resistance.

Another way of stating this is that we can transform the initial impedance of $z_1 = (0.5 - j0.5)$ to the center of the Smith chart by adding a 25-Ω series resistor and a 4-nH series inductor. We should also point out that the order of the elements is not important. Adding inductor first and resistor next leads to the exact same result.

Note that our initial impedance of $z_1 = (0.5 - j0.5)$ cannot be transformed to the center of the Smith chart with lossless series lumped elements without including a resistor. As we will see later, transforming impedances by adding resistance is generally not the optimum solution because the added resistor dissipates some of our signal power. There are exceptions, but in most cases we want to do the transformation with reactive, lossless components. After we introduce parallel element manipulations, we will see that with series-parallel *L-C* element combinations it is possible to transform from any point to any other point on the Smith chart, without adding resistance to the circuit.

We should mention here that all three series lumped elements will eventually take us to infinity if we add a large enough reactance or resistance. The only difference is that with a series inductor we approach $+j$ infinity, with a series capacitor $-j$ infinity, and with the resistor the real infinity. On the impedance Smith chart all three of them are located on the same point: a reflection coefficient of 1.0 at 0° phase.

Let us also point out that from our initial starting point, z_1, a fourth transformation direction also exists as indicated in Plate 4, but it would take a "negative resistor." Negative resistors do not exist in real life. However, we could arrange an active circuit to behave in that manner. Negative resistance is covered in greater detail in Volume II, Chapter 1, since it requires an active circuit in order to be realized. A component displays negative resistance if an increase in the voltage applied to the component's terminals actually *decreases* the related current ($-r = \Delta v / -\Delta i$).

Let us assume that z_1 represents a capacitor fabricated on an RFIC chip. The Q-factor of the component is equal to ratio of series reactance and resistance, $x_1/r_1 = 1$. If we can add a negative resistor[1] of −0.49 unit normalized value, the initial one-port is transformed to a new impedance $z_T = (0.01 - j0.5)$, increasing the Q to 50. If instead, we add exactly −0.5 unit of negative resistance, the transformed impedance moves to the circumference of the Smith chart to a normalized impedance of $z_T = (0 - j0.5)$, that represents an ideal capacitor with infinite Q. This is a commonly used technique in integrated circuit design where the creation of negative

1. Created by an active circuit combination.

resistance makes it possible to increase the Q-factor of the components. We need to be very careful, however, because too much negative resistance transforms the impedance to a point outside the chart. In that case, the result may be unwanted oscillation instead of a high-Q capacitor.

We should also mention that in real life all components are lossy. When we add a physical capacitor there will always be a small amount of resistance representing the dissipative losses of the component. Generally this series resistance is much smaller that the series reactance of the component and is not worth showing on a graphical design, unless the quality factor of the component is less than 10. In RFIC designs where the component Qs are generally not greater than 2 to 5, we need to include the resistive portion of the component into our Smith chart manipulation.

Impedance transformations using only series reactive elements are limited movements on a specific constant resistance circle. In real life we want to be able to transform from any point on the Smith chart to any other point. This cannot be done with series reactances alone. Unfortunately, parallel elements do not move in a convenient direction on the impedance Smith chart. For example, looking at Plate 7, we see that by adding a parallel capacitor or a parallel inductor we move into totally unpredictable directions on the impedance Smith chart.

We saw in Chapter 2 that the admittance system is more convenient for parallel element manipulation. Although the first version of Phillip Smith's impedance chart could also be used for admittance manipulation, we will not discuss that technique since it is not as convenient as the one introduced next. Instead, let us develop a modified admittance Smith chart.

4.3 The admittance Smith chart

Let us go back to the initial reflection coefficient equation (4.1) and substitute $1/Y$ for Z and $1/Y_0$ for Z_0, where Y_0 is the reference admittance. If the reference impedance is 50Ω, then Y_0 is equal to 0.02 Siemens (S), or 20 milliSiemens (mS). After rearranging the algebra we have a new expression for Γ in terms of admittances:

$$\Gamma = \frac{Z - Z_0}{Z + Z_0} = \frac{\frac{1}{Y} - \frac{1}{Y_0}}{\frac{1}{Y} + \frac{1}{Y_0}} = \frac{Y_0 - Y}{Y_0 + Y} \tag{4.2}$$

Normalizing (4.2) to the reference characteristic admittance, Y_0,

$$\Gamma = \frac{Y_0 - Y}{Y_0 + Y} = \frac{\dfrac{Y_0}{Y_0} - \dfrac{Y_1}{Y_0}}{\dfrac{Y_0}{Y_0} + \dfrac{Y}{Y_0}} = \frac{1-y}{1+y} \tag{4.3}$$

As we examine forms of (4.2) and (4.3), we see that the numerator is different from (4.1) where we had $(z - 1)$ in the numerator now we have $(1 - y)$ instead. This algebraic change actually causes a 180° rotation of the impedance chart to the admittance chart shown in Figure 4.1. We are doing this to keep the inductive elements on the upper half of the admittance chart. Changing the sign of y or z is equivalent to inverting Γ.

Traditionally, in the rectangular coordinate admittance system positive susceptances are on the upper side, representing capacitors. Inductive susceptance is on the lower half, which is in contrast with the impedance system's arrangement. In addition, zero admittance is an opencircuit, which is at the same place where shortcircuits are located in the impedance system. All these can be confusing. To avoid possible misinterpretations, the admittance Smith charts in our books are all in the 180° rotated form.

In the rectangular coordinates admittance system the vertical lines represent constant conductances. Bilinear transformation converts these constant-conductance lines to constant-conductance circles on the admittance Smith chart. Horizontal lines represent constant susceptances and they become constant-susceptance semicircles in the admittance system. Once again, if we were to include the left-hand side of the admittance plane where negative conductance exists, the transformation would give us the full constant-susceptance circles on the admittance Smith chart.

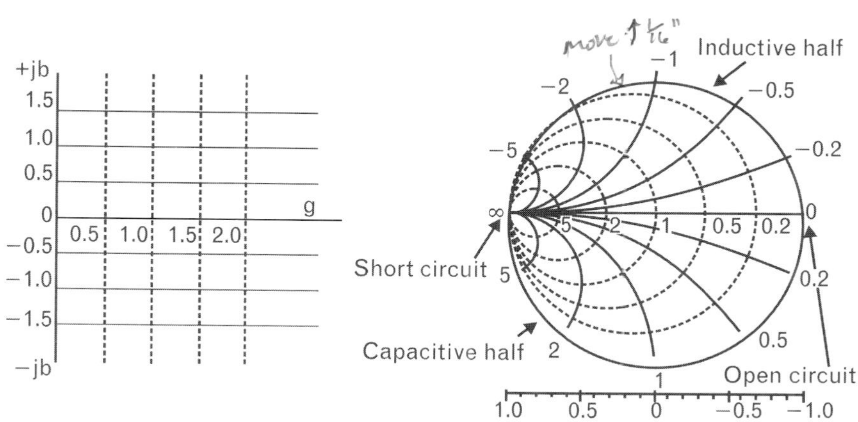

FIGURE 4.1 *The admittance chart has the same form as the impedance chart, with 180° rotation. The top half is inductive and the lower one is capacitive—same arrangement as in the impedance chart. Admittance Smith chart is made up by constant conductance circles and constant susceptance arcs.*

Examining the admittance Smith chart, we see that the top half has negative imaginary numbers and the lower half has positive. While at first it may look strange, we need to realize that inductive susceptances are negative numbers, so it makes sense that the top half of the chart has negative imaginary numbers. The opposite is true for capacitors. Capacitive susceptances are positive imaginary numbers. So once again the lower half of the chart is where capacitive susceptances are located. Infinite admittance is the same as zero impedance; therefore, the location of the shortcircuit is consistent with the impedance Smith chart. The same is true for the opencircuit that is zero admittance corresponding to infinite impedance.

Table 4.3 summarizes the movements of lumped parallel *R-L-C* components in the rectangular complex admittance plane compared to the Γ-plane. Figure 4.2 displays the lumped passive element movements on the admittance Smith chart.

Let us now practice the movement of parallel elements on the Smith chart by looking at parallel capacitors, inductors, and resistors (we call the latter conductors by their admittance names).

4.3.1 Adding parallel capacitors on the admittance Smith chart

To illustrate parallel element movements on the admittance chart, take the same one-port with $\Gamma_1 = 0.45 \angle -116°$, we used in the series element illustrations. The normalized admittance Smith chart immediately converts Γ_1 to $y_1 = 1.0 + j1.0$ (Figure 4.3).

First, let us add to this one-port a parallel 3.2-pF capacitor at 1 GHz. As we did before, we use 50Ω for normalization to take advantage of the normalized expressions listed in Chapter 2. Computing the capacitive susceptance of the component from (2.23), we can write

TABLE 4.3 SUMMARY OF LUMPED PARALLEL ELEMENT MOVEMENTS ON THE NORMALIZED IMPEDANCE SMITH CHART STARTING AT $y_1 = 1 + j1$ ($\Gamma_1 = 0.45 \angle -116°$)

ELEMENT TO BE ADDED TO y_1	MOVEMENT IN THE Y-PLANE	MOVEMENT IN THE Γ-PLANE
Lossless parallel inductor L_P	Adding $-jb_L$ on the $g = 1$ constant-conductance (vertical) line	Adding $-jb_L$ on the $g = 1$ constant-conductance circle
Lossless parallel capacitor C_P	Adding $+jb_C$ on the $g = 1$ constant-conductance (vertical) line	Adding $+jb_C$ on the $g = 1$ constant-conductance circle
Ideal parallel resistor R_P	Adding g_P on the $b = 1$ constant-susceptance (horizontal) line	Adding g_P on the $b = 1$ constant-susceptance arc

Note: Lossless elements move on the constant-conductance circles, and resistors move on the constant-susceptance circles.

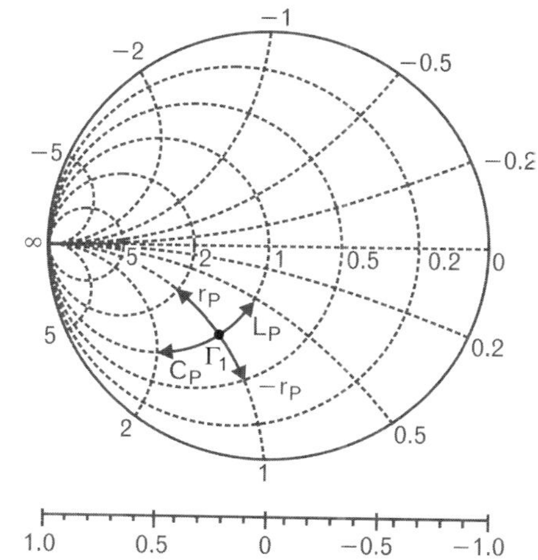

FIGURE 4.2
Parallel inductors and capacitors rotate on the constant-conductance circle while parallel resistors (conductors) move on the constant-susceptance circles. All three passive elements move the initial admittance, y_L, closer to infinite admittance (short-circuit).

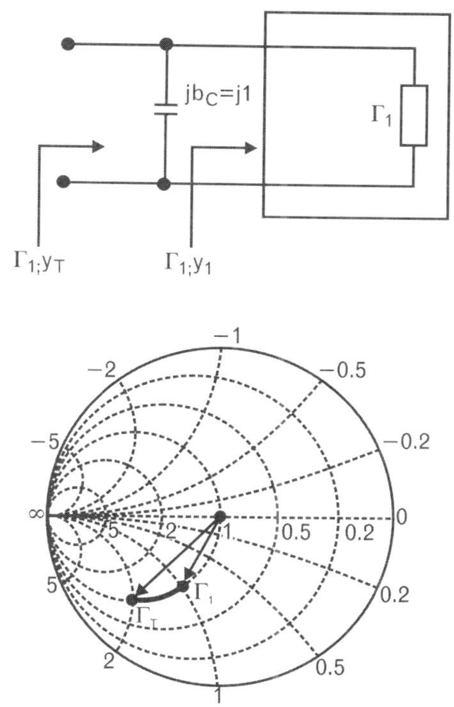

FIGURE 4.3
Admittance manipulation on the Smith chart. Adding a 3.2-pF parallel capacitor to a one-port characterized by reflection coefficient of $\Gamma_1 = 0.45 \angle -116°$.

$$b_{CP} = 0.314 f_{\text{GHz}} C_{p\text{pF}} = 0.314(1)3.2 = 1.0$$

Next, take the $+j1$ unit capacitive susceptance and move down on the constant-conductance circle from $y_1 = (1 + j1)$, adding $j1$ unit capacity susceptance. That moves us to $y_T = (1 + j2)$. The new location is labeled with Γ_T. Larger capacitor values would move Γ_T further into the capacitive region, toward the direction of a short circuit.

4.3.2 Adding parallel inductors on the admittance Smith chart

If we were to add a parallel inductor instead of a parallel capacitor, the movement would take place on the same constant conductance circle but in the opposite direction. This time we want to move toward the inductive part of the Smith chart.

Let us use the same one-port again and connect an 8-nH inductor across the input, as shown in Figure 4.4. Computing the susceptance of the 8-nH inductor at 1 GHz from (2.21) we find

$$b_{LP} = \frac{7.96}{f_{\mathrm{GHz}} L_{P\mathrm{nH}}} = \frac{7.96}{(1)8} = 1.0$$

Adding $-j1$ inductive susceptance to $y_1 = (1 + j1)$ leads to a net susceptance of zero. Since we started on a conductance circle that goes through the center of the Smith chart, the final result is $y_T = (1 + j0)$, equal to a 50-Ω unnormalized input impedance.

Parallel inductors and parallel capacitors can only move on constant-conductance circles. Their transformation capabilities are limited to specific conductance circles. Combining parallel and series elements lifts those limitations and enables us to transform to *any specific region* of the Smith chart. Before we handle the series-parallel combinations, however, let us

FIGURE 4.4
Adding parallel inductance only changes the susceptance of the circuit. Therefore, the movement must take place on the constant-conductance circle, toward the inductive region of the chart.

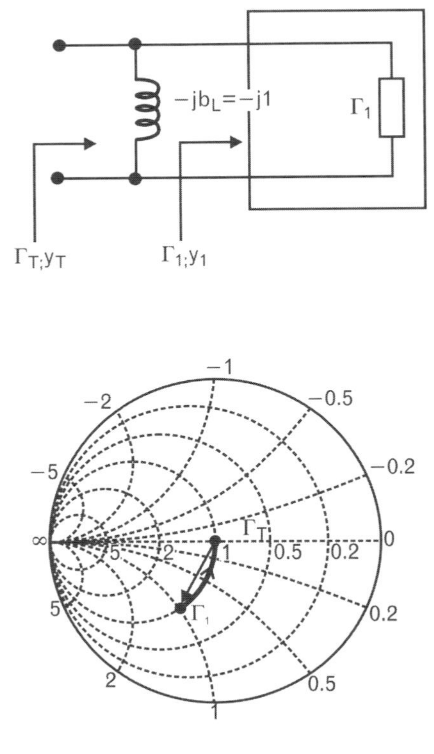

also look at parallel resistors, or as we called them earlier, parallel conductors on the admittance Smith chart.

4.3.3 Adding parallel resistors on the admittance Smith chart

Ideal parallel reactive elements moved us on the constant-conductance circles. Ideal parallel resistors add more conductance without changing susceptance. Therefore, they transform on the constant-susceptance circles. The more conductance we add, the closer we get to infinite admittance, which is a short circuit on the Smith chart.

To add a 50-Ω parallel resistor to the one-port used previously (Figure 4.5), we need to first convert the resistor to normalized conductance.

$$g_P = \frac{1}{r_p} = \frac{1}{\frac{R_p}{Z_0}} = \frac{1}{\frac{50}{50}} = 1$$

Adding one unit conductance moves us from $y_1 = (1 + j1)$ to $y_T = (2 + j1)$. We can also see from the chart that if we were to have negative conductance available we could move in the opposite direction, increasing the input Q of the one-port.

4.4 Circuit manipulations using series and parallel components

We are now ready to use both series and parallel elements for our impedance transformation. One possible way to do this is to have two separate charts: one with impedance circles and another with admittance circles.

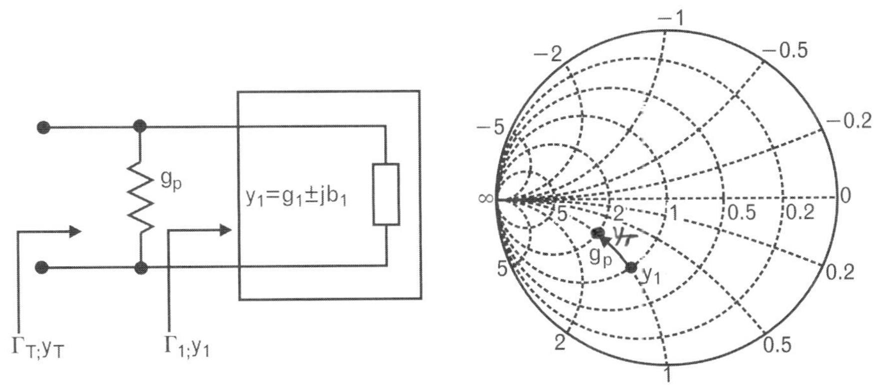

FIGURE 4.5 *Parallel resistor added to the input admittance of the one-port increases the conductance. On the Smith charts increased conductance corresponds to a movement on the $b = j1$ constant-susceptance circle toward short circuit.*

We make our movements on the chart that is more convenient for each element manipulation: impedance chart for series and admittance chart for parallel elements. This is not the simplest way, but it works. For example, if we have a four-element ladder network of Plate 8, we begin at the termination Γ_L on the impedance chart [Plate 8(a)] and add the effect of the series inductor, L_1. After the addition, carefully measure Γ_1 and transfer it to the admittance chart [Plate 8(b)]. On the admittance chart we add the effect of the parallel capacitor, C_2, which takes us to Γ_2. Measure carefully Γ_2 and transfer it back to the impedance chart to add the other series inductor, L_3. Measure the new reflection coefficient Γ_3 and transfer it to the admittance chart. Finally, add the effect of second parallel capacitor, C_4, which takes us to the new input reflection coefficient, Γ_{IN}.

Going back and forth between the two charts is a very tedious procedure. Instead, we combine the two charts and do all our manipulations on a combined Z-Y chart (sometimes called the impedance-admittance chart, or just "immitance" chart).

4.5 The immitance (Z-Y) Smith chart

Combining the impedance and admittance Smith charts provides a convenient (although not simple-looking) chart wherein both the impedance and admittance system parameters can be displayed simultaneously. With one color only, this chart can be rather intimidating and confusing, especially when many circles are used. For better visualization we use two colors for some of the charts in the next few pages: red for impedance and black for admittance (on the commercial charts the second color is generally green for admittance).

Plate 9 shows how the combined Z-Y chart simplifies the tasks shown in Plate 8. Table 4.4 provides a brief summary of the various regions on this chart. The top half is always inductive. The lower half is capacitive. Along the main (horizontal) axis we have ideal resistance and conductance with zero reactance and susceptance. The circumference of the chart is the lossless region, where $r = 0$ and $g = 0$ simultaneously.

A commercially available immitance chart [5], useful for actual computation, is shown in Plate 10. The presence of both impedance and admittance contours makes the chart initially more difficult to visualize than was the old-fashioned Smith chart. However, the benefit of having both impedance and admittance values collocated for reflection coefficient values results in less opportunity for calculation error and outweighs the added visual complexity of the chart.

TABLE 4.4 REACTANCES AND SUSCEPTANCES HAVE COMPLEMENTARY SIGNS ON THE Z-Y (IMMITANCE) SMITH CHART

Top half: Inductive	Reactance	$+jX$
	Susceptance	$-jb$
Bottom Half: Capacative	Reactance	$-jX$
	Susceptance	$+jb$

Note: The top half of the chart is inductive ($+jx$ and $-jb$), while the lower half is always capacitive ($-jx$ and $+jb$).

Better still, a variety of computer software programs are available that perform the manipulations automatically, displaying either or both impedance and admittance contours.

A review of the possible motions from the previously given reflection coefficient point on the Smith chart is summarized graphically and in tabulated form in Plate 11. Notice that each movement is along a constant resistance, conductance, reactance, or susceptance contour.

Plate 11 shows the six possible movements of series and parallel lumped elements from the same starting point. Remember that adding series resistance or reactance takes us closer to open circuit. The opposite is also true. Adding parallel conductance or susceptance moves us toward a short circuit.

Another important consideration is that once we are inside the Smith chart, only *negative resistance or negative conductance* can take us outside the chart. Extreme element values may move us to the circumference but we cannot pass that boundary without using negative resistance or conductance. The opposite is also true. If we find ourselves outside the chart, which may be the case in Volume II where we discuss active circuits, only positive resistance or conductance can bring us back inside the chart.

On the circumference of the chart in Plate 11 the reactances are always equal to the reciprocal of the susceptances. On the main axis, conductance is always equal to the reciprocal of resistance. At any particular point the impedance is equal the reciprocal of corresponding admittance. Since we have two sets of numbers related to the circles of the impedance and admittance systems, we must be very careful about which one to use. A common mistake made by first-time users is to work with a resistance circle but read the conductance value from the chart, and vice versa. On a two-color chart such mistakes rarely happen, but if we only work with black and white versions we have to be extremely careful.

4.5.1 Series R-L-C contours on the immitance Smith chart

Now we are ready to extend our examinations to a broad frequency range. Network analyzers display reflection coefficient data versus frequency directly on the Smith chart, and it is helpful if we can recognize certain contours as being produced by specific circuits.

Smith chart plots of passive lossless circuits (containing only inductors and capacitors) must necessarily fall on the periphery of the Smith chart, at which $r = 0$. If the circuit has losses, at some frequency the magnitude of the reflection coefficient must be less than unity. Ideal *R-L-C* components in series configuration always follow a constant-resistance circle, while parallel circuits stay on constant-conductance circles. For example, the reflection coefficients of three series combinations of Plate 12(a–c) stay on the $r = 1$ circle from $f = 0$ through $f = \infty$. The fourth circuit is lossless; therefore, its trace remains on the circumference of the chart. As the frequency increases, the reflection coefficients of all real physical circuit impedances follow a *clockwise* direction on the Smith chart.

Series resonant circuits of Plate 12(c, d) are capacitive up to series resonance, $f = f_R$. Above f_R, the inductor dominates the circuit.

4.5.2 Shunt R-L-C contours on the immitance Smith chart

When the same reasoning is applied to parallel connected elements, the contours shown in Plate 13 are obtained. Note that the results are the same except that admittance rather than impedance coordinates are used to describe the circuit behavior, due to the parallel combination of the elements. Parallel inductors and capacitors move on the constant-conductance circles, determined by the value of g_p, set to one unit in Plate 13(a–c). In the last circuit [Plate 13(d)], $g_p = 0$.

4.5.3 Lowpass L-C transformers

Frequently, it is desirable to transform one impedance to another for various reasons. Series reactance can only increase the impedance of the circuit because series components move towards infinite impedance (open circuit). Parallel susceptance does just the opposite since shunt elements navigate towards a short circuit. A single inductor or capacitor limits transformation possibilities because it can only move on the specific constant-resistance or constant-conductance circle that passes through the starting impedance. However, with the appropriate choice of a pair of reactive elements, one in series and the other in parallel, it is possible to transform any impedance to any other impedance. The same is true for admittances.

When the circuit is more complex than a series or parallel *R-L-C*, the impedance response follows neither the constant resistance or conductance circles. Depending on the nature of the circuit (lowpass, highpass, bandpass, and so forth), the impedance response may start at a short circuit, infinite impedance, or finite impedance at dc. Regardless of where it starts, it will describe a clockwise rotating curve as frequency is increased.

When reactive elements are used in combinations with each other for impedance transformations, they can cover only certain parts of the Smith chart. A two-element lowpass *L-C* section's impedance transformation range depends on the configuration of the first component. Transforming to lower resistance, the first element of the *L-C* section must be a shunt element as shown in Plate 14(b). To transform to lower conductance requires the opposite [Plate 14(a)]. Both circuit sections shown pass dc current and they are useful when dc feed-through is needed between two terminations. The region of possible coverage depends on the value of $r_L = \text{Re}(z_L)$. Low values of r_L lead to a wide range of up-transformations but a limited range of transforming to lower resistances.

A combination of series inductor and shunt capacitor, sometimes called a lowpass section because it passes dc and eventually rejects high frequencies, is capable of transforming an impedance Z_L as shown on Plate 14 to a portion of the Smith chart. The point at which we start affects the area of possible coverage. Our three sets of illustrative examples (Plates 14 through 16) always show the transformations from one unit resistance (r_L = 1, g_L = 1), which is at the center of our normalized chart. One of the two available topologies of Plate 14 can always be used to transform from any point of the chart to any other point.

4.5.4 Highpass L-C transformers

Similarly, given one of two configurations, highpass *L-C* sections can also transform a specific termination to any other impedance on the Smith chart (Plate 15). If our starting location is again at the center, at $r_L = 1$, each configuration transforms to its own half of the area of chart. Highpass sections block dc and severely attenuate low frequencies. As the name implies, above a certain corner frequency the effective transformation gradually diminishes.

4.5.5 Bandpass transformer sections

Using two capacitors or two inductors, impedance transformation is possible only to limited sections of the Smith chart, as shown in Plate 16. Once again, we use the center of the chart for our starting point. Bandpass sections reject both low and high frequencies, but they can serve effectively through a specified passband.

The last few sections of this chapter have been devoted to developing a familiarity with the Smith chart, specifically how to navigate various impedance and admittance movements on it. Next we work two examples to illustrate lumped element manipulations on the immitance Smith chart.

4.5.6 Illustrative exercise: series-to-parallel circuit conversions

Use the immitance Smith chart (normalized to 50Ω) to verify the results of Section 2.11.1, where we previously showed that the series *R-C* of 10Ω resistor and 7.95 pF is equivalent to the parallel *R-C* of 50Ω and 6.36 pF at 1 GHz.

Solution

Normalizing the two resistors gives us $r_S = 10\Omega/50\Omega = 0.2$, and $g_P = 1/r_P = 1/(50\Omega/50\Omega) = 1.0$. Computing the series reactance of the 7.95-pF capacitor from (2.18),

$$x_{CS} = \frac{3.183}{f_{\text{GHz}} C_{\text{SpF}}} = \frac{3.183}{(1)7.95} = 0.4$$

Computing the normalized parallel susceptance of the 6.36-pF capacitor from (2.23),

$$b_{CP} = 0.314 f_{\text{GHz}} C_{\text{PpF}} = 0.314(1)6.36 = 2.0$$

Series impedance, parallel admittance, and Q-factor of the two circuits at 1 GHz are

$$z_S = (r_S - jx_{CS}) = (0.2 - j0.4) \quad \Rightarrow Q_S = x_{CS}/r_S = 2.0$$
$$y_P = (g_P - jb_{CP}) = (1 + j2.0) \quad \Rightarrow Q_P = b_{CP}/g_P = 2.0$$

Placing z_S and y_P on the *Z-Y* Smith chart show exact equivalence (Plate 17). Note that both circuits have the same $Q = 2.0$, which is one of the requirements for series-parallel equality.

In our example we used the immitance chart to verify the computed equivalence between series and parallel *R-C* circuits. The chart can also be used to find required equivalent circuits of series and parallel forms for any two resistors. For example, if the series *R-C* circuit's resistance is 25Ω and $R_p = 50\Omega$, we can see in Plate 17 the necessary series capacitive reactance is $x_{CS} = 0.5$ unit, so now $C_S = 6.36$ pF. The 25-Ω series resistor also requires a different parallel capacitor with $b_{CP} = 1.0$, which gives us $C_P = 3.18$ pF at 1 GHz. The Q-factors of these new circuits are lower, $Q_P = Q_S = 1$. Due to the lower Qs, the two circuits ($R_S = 25\Omega$, $C_S = 6.36$ pF, and $R_P = 50\Omega$, $C_P =$

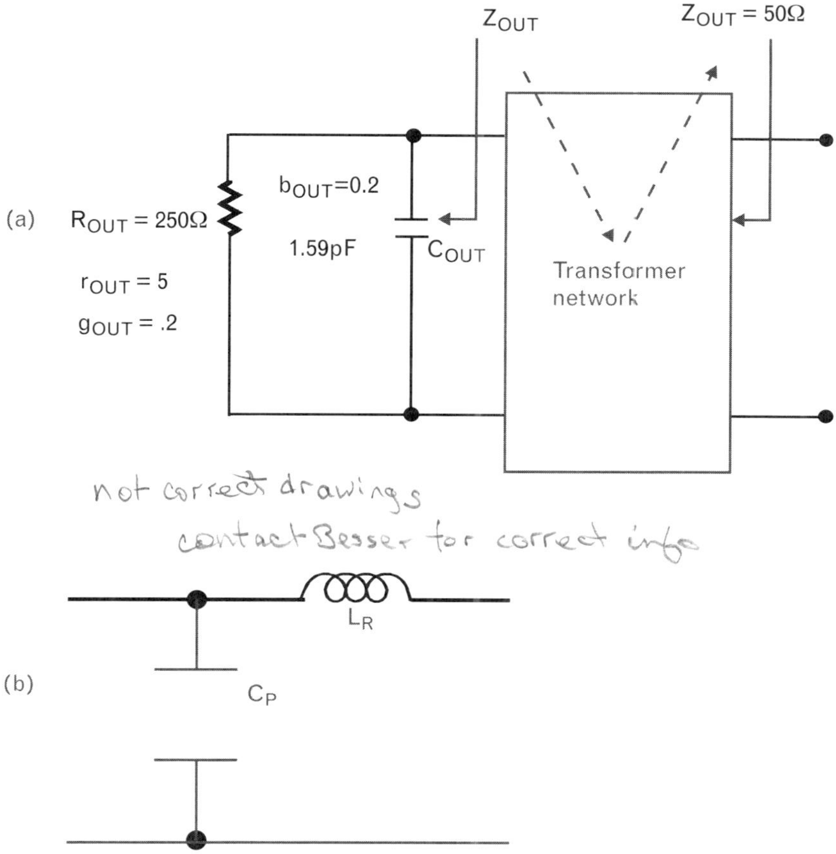

FIGURE 4.6
(a) The output admittance of a bipolar RF transistor, modeled as a parallel R-C network, needs to be transformed to 50Ω at 400 MHz with two different circuits described in (b).

3.18 pF) maintain nearly equal terminal impedances through a wider frequency range than the previous circuits with $Q_P = Q_S = 2$.

4.5.7 Illustrative exercise: impedance transformations

A bipolar transistor's output impedance, Z_{OUT}, is modeled with a parallel *R-C* combination of $R_{OUT} = 250\Omega$, $C_{OUT} = 1.6$ pF, as shown in Figure 4.6(a). Design two output networks with the Smith chart to transform the output impedance to $Z'_{OUT} = 50\Omega$ at 400 MHz with the following topologies:

- Use a two-element lossless *L-C* network in lowpass configuration. Calculate the component values needed for 400 MHz operation with the normalized formulas given in (2.15) through (2.23) and Table 2.2.
- Repeat the transformation by using a parallel inductor first to resonate out the parallel capacitance, and then find the appropriate parallel resistor for second element.

- In addition to 400 MHz, find the new output reflection coefficients at 300 and 500 MHz for both networks. Which has the better broadband response through the 200-MHz bandwidth?

Solution

Compute the normalized resistance and conductance of the 250-Ω resistor.

$$r_{OUT} = \frac{R_{OUT}}{Z_0} = \frac{250\Omega}{50\Omega} = 5.0$$

$$g_{OUT} = \frac{1}{r_{OUT}} = \frac{1}{5} = 0.2$$

Compute the normalized capacitive susceptance of the 1.6-pF capacitance at $f = 0.4$ GHz from (2.23).

$$b_{OUT} = 0.314 f_{\text{GHz}} C_{OUT\text{pF}} = 0.314(0.4)1.6 = 0.2$$

Using the admittance portion of the Z-Y chart [Figure 4.6(b)], start at the $g = g_{OUT} = 0.2$ location, labeled as g_{OUT}, corresponding to the 250-Ω resistor. Add the $b = b_{COUT} = 0.2$ unit capacitive susceptance and label the total admittance of the parallel R-C circuit, as y_{OUT} as shown in Plate 18(a). Our goal is to transform y_{OUT} to the center of the chart, and there are two paths by which we can get there with L-C elements:

- From the constant resistance circle of 1.0 by using a series L or C;
- From the constant conductance circle of 1.0 by using a parallel L or C.

From y_{OUT} we can transform to the constant resistance circle of 1.0, with either a parallel L or parallel C. Our specification states a lowpass topology that excludes the parallel inductor, leaving the parallel capacitor only. Our first move is then to add capacitive susceptance to move from y_{OUT} to the intersection of $g = 0.2$ and $r = 1.0$ circles. Label this intermediate point as y_{INT} and z_{INT}.

The susceptance difference between y_{OUT} and y_{INT} is 0.2 normalized unit, which is the amount of external capacitive susceptance we need to add to y_{OUT}. Finding the equivalent parallel capacitance from Table 2.2,

$$C_{P\text{pF}} = \frac{3.18 b_{CP}}{f_{\text{GHz}}} = \frac{3.18(0.2)}{0.4} = 1.6 \text{ pF}$$

We used the admittance portion of the chart for our parallel circuit components. It is time to switch to the impedance circles and complete the transformation with a series inductor. At the intersection of $g = 0.2$ and $r = 1.0$ circles, the reactance is capacitive, $x = -2.0$. A series inductor of $x_{LS} = +2.0$ can move us to the center of the chart. Computing the corresponding inductance at 0.4 GHz, using Table 2.2,

$$L_{\text{SnH}} = \frac{7.96 x_{LS}}{f_{\text{GHz}}} = \frac{7.96(2.0)}{0.4} = 39.8 \text{ nH}$$

Our first impedance transformer circuit then has a 1.6-pF parallel capacitor and a 39.8-nH series inductor.[2]

The second circuit starts with a parallel inductor and takes an equal amount of inductive susceptance to resonate the $b_{OUT} = 0.2$ capacitive susceptance. Finding the parallel inductor value from Table 2.2,

$$L_{\text{PnH}} = \frac{7.96}{f_{\text{GHz}} b_{LP}} = \frac{7.96}{0.4(0.2)} = 99.5 \text{ nH}$$

Adding the 99.5-nH parallel inductor moves us back to $y = 0.2 + j0$ location of the admittance chart, already marked as g_{OUT} [Plate 18(b)]. From there, we get to the center of the chart ($y = 1.0 + j0$) by adding $g_P = 0.8$ unit additional parallel conductance. Converting the 0.8 unit normalized conductance to unnormalized resistance,

$$r_P = \frac{1}{g_P} = \frac{1}{0.8} = 1.25$$

$$R_P = r_P Z_0 = 1.25(50) = 62.5\Omega$$

The second circuit includes a 99.5-nH parallel inductor and a 62.5-Ω parallel resistor (see footnote 2). Both circuits of Figure 4.6 transform the initial Z to $Z'_{OUT} = 50\Omega$ at 400 MHz. To show the new output at 300 MHz and 500 MHz, we need to compute the corresponding reactance and susceptances at those frequencies. Inductive reactance and capacitive susceptance is directly proportional, while inductive susceptance is inversely proportional to frequency changes. Table 4.5 show the reactances and susceptances at all three frequencies.

2. In this chapter we use exact computed values to illustrate the mathematical procedure. In Chapter 7 we look at practical element values.

TABLE 4.5 FREQUENCY EFFECTS ON THE REACTANCE AND SUSCEPTANCE VALUES.

FREQUENCY (MHz)	C_{OUT} = 1.6 pF b_{OUT}	C_P = 1.6 pF b_{CP}	L_S = 39.8 nH x_{LS}	L_P = 99.5 nH $-b_{LP}$	R_P = 62.5Ω g_P
300	0.15	0.15	1.5	−0.27	0.8
400	0.2	0.2	2.0	−0.2	0.8
500	0.25	0.25	2.5	−0.16	0.8

Note: Conductance of the parallel resistor is assumed to be frequency independent.

Output reflection coefficients of the two circuits are displayed in Plate 18, showing superior broadband performance for the parallel *R-L* circuit. At 400 MHz, where we computed the element values, the reflection coefficient is zero for both circuits. The ρ of the *L-C* network increases to 0.46 at 500 MHz, which translates to 1.0-dB mismatch loss into a 50-Ω load. The highest ρ for the *R-L* circuit is only 0.04, which is a negligible mismatch loss. Based on this information alone, we may conclude that the resistive circuit is the better one for this task.

Looking only at the reflection coefficient may be deceiving, since we also need to examine the dissipative loss caused by the parallel resistor. Although we have not yet covered transmission coefficients, it is helpful to compare the power transfer characteristics of the two circuits (Figure 4.7). The frequency response of the *L-C* network shows (∇-markers) zero loss at 400 MHz and gradually rolls off at both band edges, with a 1-dB

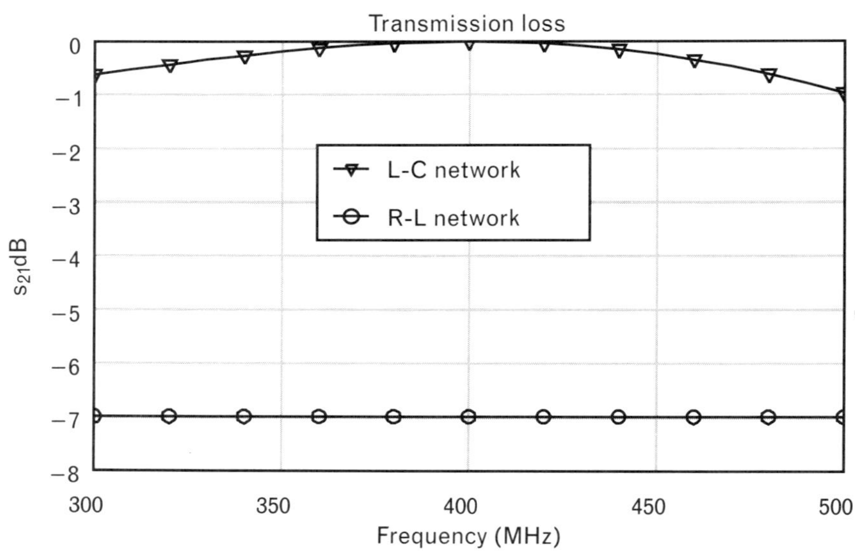

FIGURE 4.7 *Using ideal components, the L-C circuit has 1-dB maximum loss at 500 MHz, while the R-L circuit's parallel resistor dissipates 7 dB, which represents an 80% power loss.*

maximum value at 500 MHz. The response of the R-L network is impressively flat—at the cost of 7-dB loss (O-markers). The 3-dB loss means 50% of our power is wasted; 7 dB translates to 80% waste, and only 20% of the available signal power is delivered to a 50-Ω load. We will look into this issue more deeply in the next chapter.

4.6 Constant Q curves on the Smith chart

In Sections 2.8 and 2.9 we described the unloaded Q of a series or parallel network, and we summarize them here in normalized forms (Figure 4.8). Since Q calculations only use the magnitudes of the reactances and susceptances, Q is always a positive quantity.

We can easily identify both real and imaginary parts of either z or y on the Smith chart. Using the impedance chart, we can create constant Q curves by connecting all points with same Qs together. For example, at z values of $(0.1 + j0.1)$, $(0.2 + j0.2)$, and $(0.5 + j0.5)$, Q is equal to 1. At z values of $(0.1 + j0.2)$, $(0.5 + j1.00)$, and $(1.0 + j2.0)$, Q is equal to 2.0, and so on. A mirror image set of curves can also be created on the bottom half of the chart (Plate 19).

Using an immitance chart we can see that the reciprocal $y = 1/z$ values lying at the above listed points have the identical Q values. These represent the parallel equivalent circuits of the series z values, and since they are equivalent must necessarily have identical Q values. For example, where $z_S = (r_S + x_S) = (0.1 + j0.1)$, the $Q_S = x_S / r_S = 1.0$. From the corresponding $y_P = (g_P - jb_P) = (5 - j5)$ we can compute $Q_P = (b_P/g_P) = 1.0$ for the parallel equivalent circuit also.

Plotting constant Q curves is useful for predicting circuit bandwidths. The bandwidth of matching and other circuits decreases rapidly with increasing Q. Because the Smith chart already contains numerous lines to show z and y coordinates, it is not practical for the printed version of the commercial Smith chart to show Q curves as well. However, the software available to produce a computer version of the Smith chart allows automatic introduction of the Q curves as an option, adding insight into the bandwidth of various circuit matching options. There will be more on this in Chapter 5.

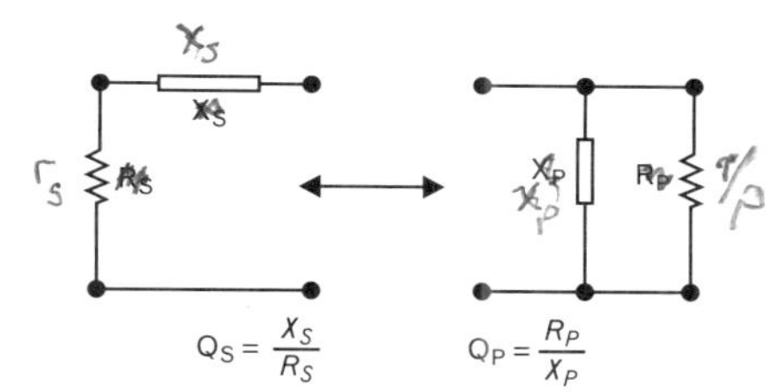

FIGURE 4.8
Q-factors for series and parallel circuits.

4.7 Negative reactive elements

A *mathematical* negative inductor's reactance is given by

$$X_L = 2\pi f(-L)$$

An erroneous belief is that a negative inductor is equivalent to a capacitor. While it is true that at a *single frequency* one can find a capacitor having the same reactance as a specified negative inductor, the equivalence does not apply as frequency is varied. This can be seen since the reactance of a capacitor is inversely proportional to frequency. Consequently, for given $-L$ and C values,

$$-j2\pi fL = -j\frac{1}{2\pi fC}$$

can only be true at one frequency.

All positive L and C elements produce reactances, which move clockwise on the Smith chart as frequency is increased. Negative elements' reactances (produced by $-L$ and $-C$ elements) move counterclockwise. Therefore, a negative reactive element has a mirror image frequency response of its positive counterpart about the horizontal diameter of the Smith chart.

Negative L and C values do not exist in physical embodiment, but they are used frequently for deembedding purposes (see Chapter 7). A simple illustration here may be helpful to emphasize their usefulness.

4.7.1 Illustrative example: removing the effect of parasitic inductance

The measured output inpedance coefficient of an RF power transistor chip (from 1 GHz to 4 GHz, in 100-MHz steps) is shown on the Smith chart of Figure 4.9. Since the power transistor is a low-impedance device, we normalized the Smith chart to $Z_0 = 5\Omega$ (note, not 50Ω) for better resolution. The measured data includes the effect of a 0.5-nH series bond-wire inductance. Find the true output impedance of the transistor chip for the complete frequency range.

Solution

Resonating out the effect of the inductor with a variable series capacitor would be a painfully long process that needs to be repeated for each frequency. A much simpler approach is to cascade a –0.5-nH inductor to the output port of the measured data. The negative inductor cancels (negates) the effect of the bond-wire for all frequencies.

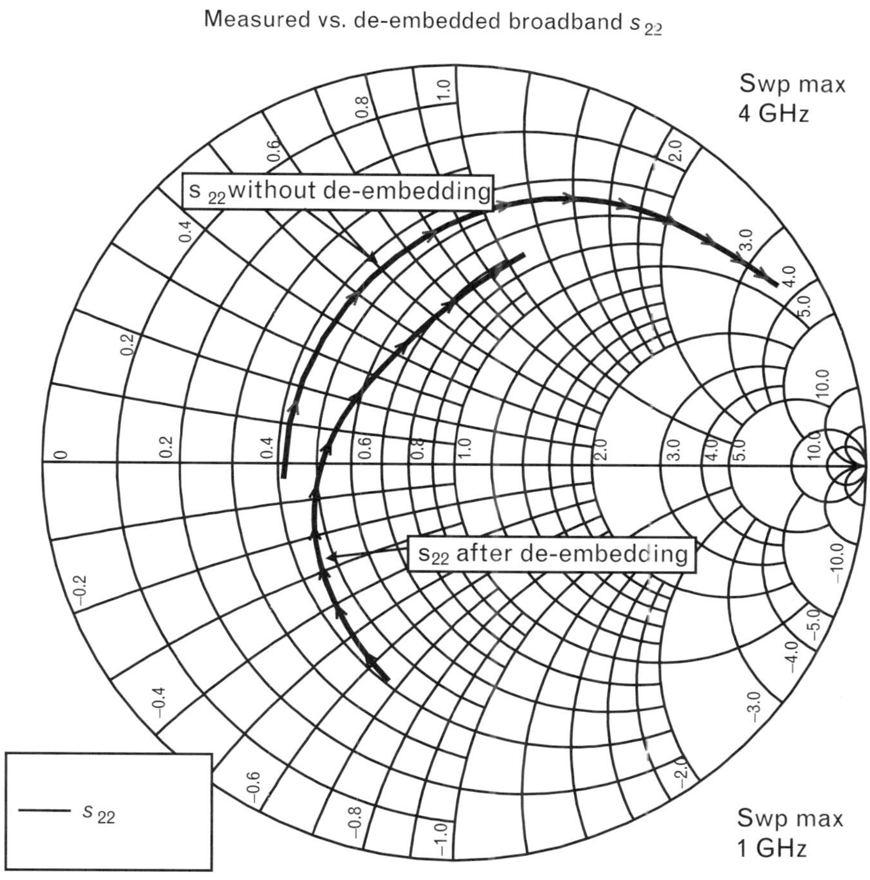

FIGURE 4.9 *Output reflection coefficient of an RF power transistor with and without 0.5-nH bond-wire inductance. The Smith chart is normalized to 5Ω to improve the resolution of the low-impedance data. Arrows indicate increase of frequency from 1 GHz to 4 GHz.*

Using an RF circuit simulator program, we added a −0.5-nH series inductor to the one-port that contains the measured reflection coefficient data. Viewing Figure 4.9, we can clearly see the significant effect of the small inductance in the gigahertz frequency range.

4.8 Negative resistance and the extended Smith chart

Impedances with negative real parts are located on the left side of the rectangular coordinate system. Converting such an impedance to a reflection coefficient gives us greater than unity magnitude: Γ lies outside of the unit-radius chart. In such cases the extended (or sometime called compressed) Smith chart is employed (Plate 20), wherein impedances with negative real parts also can be viewed. We can also plot $-Z$ as $1/\Gamma$.

All positive resistors and both negative and positive reactances (passive elements) are displayed inside the conventional Smith chart. When the reflection coefficient magnitude exceeds unity ($\rho > 1$), the circuit contains one or more active elements (tunnel diodes, transistors, and so forth), which produce negative resistances. Although these elements are only covered in Volume II, we want to point out the existence of the compressed chart.

Only negative resistance and negative conductance can cause Γ to cross the boundary of the unit-radius Smith chart. Similarly, if $\rho > 1$ only the addition of a positive resistance or conductance can transform the impedance locus back to the inside of the unit-radius chart.

Note that, while we have drawn the extended Smith chart in Plate 20 for values of ρ up to 3.0, there is no limit to how large the chart might actually be. Accordingly, extended Smith charts of any radius might be drawn and used. Of course, real circuits having very large reflection coefficients would be difficult to stabilize and use as amplifiers; hence, such charts may be of only academic value.

By duality everything true for impedances has a counterpart for admittances. An extended admittance Smith chart can also be created using the same developmental procedure as was presented for the admittance Smith chart earlier.

4.9 Transmission line manipulations on the Smith chart

The impedance Smith chart conveniently handles series *R-L-C* additions, while for parallel lumped elements the admittance Smith chart is more convenient. As we will see shortly, the latter is also suitable for parallel open-circuited and short-circuited transmission line stubs. Before covering the parallel stubs however, let's look at the behavior of cascade transmission lines on the Smith chart.

4.9.1 Cascade transmission lines

Figure 4.10 shows a generalized form of cascade transmission line with characteristic impedance of Z_{TL}, placed between a signal source of Z_S impedance and an arbitrary load termination of Z_L. Equation (2.37) already defined the input impedance, Z_{IN}, as a function of the transmission line parameters and the load termination.

$$Z_{IN} = Z_{TL} \frac{Z_L + jZ_{TL} \tan\theta}{Z_{TL} + jZ_L \tan\theta}$$

We can simplify (2.37) by setting the source with impedance equal to Z_{TL} and normalizing the circuit of Figure 4.10 with respect to Z_{TL}, obtaining $z_L = Z_L/Z_{TL}$ and $z_{TL} = Z_{TL}/Z_{TL} = 1$. If the transmission line is lossless, the

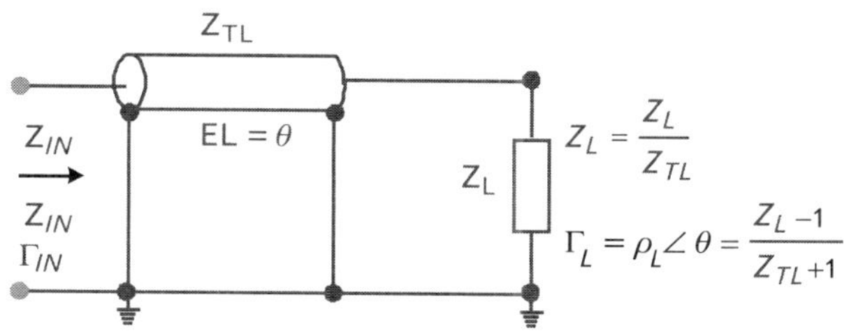

FIGURE 4.10 *In the general case, the input impedance of a terminated transmission is a complex value, depending on the transmission line parameters and the termination. In the properly terminated case, when $Z_L = Z_{TL}$, Z_{IN} is also equal to Z_{TL} regardless of the transmission line's electrical length.*

amount of reflected power is uniform throughout the length of the line, and only the phase of the reflected wave changes. To find the input reflection coefficient at any arbitrary phase angle from the load, we first write the normalized input impedance, z_{IN},

$$z_{IN} = \frac{z_L + j\tan\theta}{1 + jz_L \tan\theta} \tag{4.4}$$

Substituting (4.4) into (4.1), the input reflection coefficient of the normalized circuit can be written as,

$$\Gamma_{IN} = \frac{(z_L - 1) + j(1 - z_L)\tan\theta}{(z_L + 1) + j(1 + z_L)\tan\theta} \tag{4.5}$$

Rationalizing and rearranging (4.5), we have a simple expression for the input reflection coefficient,

$$\Gamma_{IN} = \Gamma_L e^{-j2\theta} = \rho_L \angle(\phi - 2\theta) \tag{4.6}$$

where ϕ is the phase angle of the load reflection coefficient [see Figure 4.11(a)].

Note that the reflection coefficient at the input has the same magnitude as the load, $\rho_L = |\Gamma_L|$, but the phase angle is *rotated from ϕ clockwise, through twice the electrical length of the line, 2θ*. The reason for this rotation is that a wave applied at the input travels through twice the length of the line by the time it is reflected back to the input.

Our expression also applies to any transmission line impedance, since we are normalizing the load termination to whatever the characteristic impedance of the line may be. For example, if our load termination is 100Ω and we are using a 50-Ω characteristic impedance transmission line,

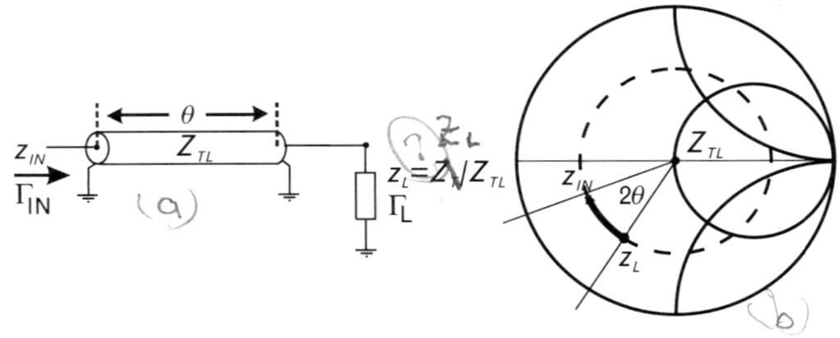

FIGURE 4.11 *The Smith chart may also be used to find the input reflection coefficient of a transmission line terminated with an arbitrary impedance. (a) First we need to normalize the load to the characteristic impedance of the line, and (b) mark the load on a normalized Smith chart. Rotating from the normalized load through an angle that is twice the electrical length of the line moves us to the input reflection coefficient of the circuit.*

the normalized load impedance will be $100\Omega/50\Omega = 2.0$ units. On the other hand, if the characteristic impedance of a transmission line is 33.3Ω, the normalized load becomes 3.0 units. As long as we normalize our termination to the characteristic impedance of the line, (4.6) is applicable.

Equation (4.6) is directly applicable to the Smith chart [see Figure 4.11(b)]. To find the input impedance or input reflection coefficient of a terminated cascade transmission line with the Smith chart we need to use the following procedure:

1. Normalize the termination to the characteristic impedance of the transmission to be used, $z_L = Z_L/Z_{TL}$.
2. Mark z_L on a normalized Smith chart.
3. Draw a concentric circle on the Smith chart going through z_L.
4. Rotate on this concentric circle *clockwise* two times the electrical length (2θ) of the transmission line.
5. The end result of our rotation is the new input impedance, z_{IN}, and input reflection coefficient, Γ_{IN}, of the transmission line. Remember that you are working on a chart normalized to Z_{TL}, which is not necessarily equal to 50Ω!

Doubling the transmission line length is necessary because any wave applied to the input of the line travels through *twice* the length of the line by the time it arrives back to the input. The clockwise rotation is derived from (4.6). This procedure works for all cases where a transmission line is

3. Even the commercial Smith chart, shown in Plate 10, has the two different directions. Perhaps it would be more clear to state, "moving away from the load" and "moving toward the load," the latter meaning removing transmission line length.

terminated with any arbitrary impedance. Many courses and textbooks describe two different ways[3] of finding the impedance along the transmission line, depending on whether we travel *toward the load or the generator* in the circuit. This dual rule is often misinterpreted, so in our text we talk about only one direction, which is a clockwise rotation, not differentiating the actual function of the termination. Our rule is: *When we terminate a transmission line with any impedance, whether of a source or a load, the input impedance of that transmission line is always computed by the five-step procedure outlined above.* Moving away from a termination on a transmission line always results in a clockwise rotation on the Smith chart.

Here is a suggestion to help you remember the clockwise direction. Enter the Smith chart at the short circuit impedance point and note that a clockwise rotation results in movement into the top half of the chart, which is the inductive region—consistent with the reactance to be expected looking into a short length of short circuited transmission line. You may also remember we said earlier that real physical circuits describe traces on the Smith chart that follow clockwise direction as frequency increases. If you link these two rules together, you should not have problems while using Smith charts.

4.9.2 Parallel transmission line stubs

We mentioned in Chapter 2 that parallel open-circuited stubs behave like capacitors, and parallel short-circuited stubs act like parallel inductors, as long as their electrical lengths are less than 90°. We also mentioned that the behavior of these stubs change cyclically through every 90° interval. A parallel open-circuited stub acts like a capacitor until 90° length, like an inductor between 90° and 180°, again a capacitor form 180° to 270°, and so on.

We have closed-form expressions for the susceptances of parallel open- and short-circuited stubs in Section 2.2. To show the effect of a parallel stub on the Smith chart, we compute first the equivalent susceptance at the input of the stub, and apply it to the Smith chart as a parallel susceptance (i.e., inductive or capacitive).

An alternative way to find the input susceptance of an open or short-circuited stub is to use the admittance portion of the immittance chart instead of the equations mentioned above. For example, to find the normalized input susceptance of an open-circuited stub having 75-Ω characteristic impedance and 45° electrical length, we refer back to our five-step rule stated earlier [see Plate 21(a) for details].

1. Normalize the termination to $Z_{TL} = 75\Omega$:

$$z_L = \frac{Z_L}{Z_{TL}} = \frac{\infty}{75\Omega} = \infty \Rightarrow y_L = \frac{1}{z_L} = \frac{1}{\infty} = 0$$

2. Mark the y_L on a normalized admittance chart: y_L equals zero admittance (open circuit).
3. Draw a concentric circle through the normalized termination: in our case the concentric circle is the circumference of the Smith chart.
4. Rotate clockwise on the circumference $2\theta = 2(45°) = 90°$, which moves us to $-90°$ on the admittance Smith chart.
5. Read off the susceptance from the admittance chart: on the circumference at $-90°$, as $b_{OS} = 1$. The input admittance of the open-circuited stub, $y_{OS} = (g_{OS} + jb_{OS}) = (0 + j1)$ unit.

If instead of an open-circuited stub we have a short-circuited stub, the same procedure is followed, but instead of the open termination we use a short circuit at infinite admittance (zero impedance), as shown in Plate 21(b). The rotation then starts at short circuit. Otherwise the procedure is exactly the same.

4.9.3 Important points to remember about transmission lines

Let us summarize some key issues related to transmission line elements:

- Transmission lines with electrical lengths less than 90° behave inductively for short-circuited loads and capacitively for open-circuited loads.
- An RF short can be produced at any point in a circuit by using a short-circuited transmission line with half-wavelength electrical length. The short circuit reappears for every multiple of half wavelength.
- A transmission line's input impedance is always equal to the termination at adjacent ends of the line if the characteristic impedance of the line is the same as the termination.
- Transmission lines have extremely large transformation capabilities. We saw that short circuit can be transformed to an open circuit by using a 90° long transmission line. The opposite is also true; an open circuit can be transformed to a short with a 90° line.
- Cascaded transmission lines follow concentric circles on a normalized Smith chart if the load termination is normalized to the characteristic impedance of the transmission line.
- Parallel open- and short-circuited stubs behave inductively and capacitively as long as their electrical lengths are less than 90°.

- Parallel stubs always move on the constant conductance circles.
- We stated in Section 4.5.3 that a properly selected *L-C* combination can always transform from any point of the Smith chart to any other point. With slight modification the statement also applies to transmission lines. A properly selected combination of a parallel stub and cascade transmission line can transform from any point to any other one on the Smith chart. Selection of topology depends on the relationship of the two points with respect to each other. Sometimes the cascade line is used first, while at other times we have to start with the parallel stub.

4.9.4 Illustrative example: impedance transformation with transmission line and lumped elements

The equivalent input impedance of an RF power transistor at 900 MHz is represented by an $R_S = 2.5\Omega$ resistor in series with $L_S = 0.88$ nH inductance. Use the immitance Smith chart, normalized to 50Ω, to find the new input impedance of the device with the impedance transforming circuitry added (Plate 22).

Solution

Calculate the normalized inductive reactance of the 0.88-nH series inductor at 1 GHz, using (2.16).

$$x_{LS} = 0.1257 f_{\text{GHz}} L_{\text{SnH}} = 0.1257(0.9)0.88 = 0.1$$

The normalized series input resistance, r_S

$$r_S = \frac{R_S}{50\Omega} = \frac{2.5\Omega}{50\Omega} = 0.05$$

Starting from the normalized input impedance of the device, $z_{IN} = (0.05 + j0.1)$, that is, label *A* on the chart. [On a high-resolution chart we could also find the equivalent parallel admittance as $y_{IN} = (4 - j8)$.] Looking at the admittance circles, we add $b_{C1P} = 8$ unit of parallel susceptance that is equivalent to a 28.3-pF capacitor, computed from Table 2.2,

$$C_{1P} = \frac{3.813 b_{C1P}}{f_{\text{GHz}}} = \frac{3.813(8)}{0.9} = 28.3 \text{ pF}$$

Moving on the constant conductance circle of $g = 4$, the added capacitive susceptance of $b_{C1P} = 8$ takes us down to the real axis (label B), where y

= (4 +j0). Measuring the reflection coefficient at point B gives us $\Gamma_B = 0.6 \angle 180°$. Next, we move to intersect the $g = 1$ constant conductance circle with a cascade transmission line as oulined below.

Since our Smith chart is normalized to 50Ω, the cascade transmission line follows a clockwise rotation on a concentric circle with radius of 0.6, through an angle of 2(26.5°) = 53°. Starting at 180°, the rotation ends at 127° (label C). Read the admittance at location C as $y = (1 - j1.5)$.

Computing the value of the second parallel capacitor, C_{2P}

$$C_{2P} = \frac{3.183 b_{C2P}}{f_{\text{GHz}}} = \frac{3.813(1.5)}{0.9} = 5.3 \text{ pF}$$

The added susceptance of the 5.3-pF capacitor moves from point C, following the $g = 1$ constant conductance circle through $b_{C2P} = 1.5$ unit, taking us to the center of the chart (label D).

The added three-element circuitry transforms the input impedance of the device to 50Ω at 900 MHz.

4.9.5 Illustrative example: computing transmission line lengths with the Smith chart

Find the electrical length of transmission lines needed to transform the of the one-port impedance, Z_L, described in Figure 4.12 to 75Ω. Use two lossless transmission lines with characteristic impedances of 75Ω. Since there are multiple possible solutions with the topology shown, find the one that gives us the shortest electrical length for the transmission lines. Impedance of the one-port termination is $Z_L = (30 - j15)\Omega$.

Solution

Since we work with 75-Ω characteristic impedance transmission lines, we also normalize our Smith chart to 75Ω to maintain the concentric circle rotation of the cascade line. Accordingly, we need to normalize our one-port impedance to 75Ω as well.

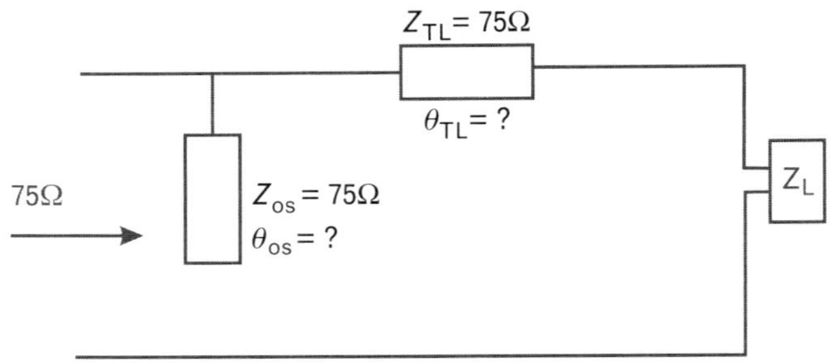

FIGURE 4.12 *Two-element transmission line added to the one-port for impedance transformation.*

$$z_L = \frac{Z_L}{75\Omega} = \frac{(30 - j15)}{75\Omega} = (0.4 - j0.2)$$

Placing the reflection coefficient of the one-port on the immitance Smith chart (label A, reflection coefficient, $\Gamma_A = 0.45\angle -153.4°$) gives us the starting point for the rotation caused by the 75-Ω cascade line. Drawing a concentric circle with radius of 0.45 indicates the path of rotation.

Before we can determine the electrical length of the cascade line, we need to show the effect of the parallel open-circuited stub, which acts as a parallel capacitor. In order to end the transformation at the center of the chart, the parallel stub needs to follow the $g = 1$ constant conductance circle. Let's mark that circle as the target for the cascade transmission line.

The concentric circle of $|\Gamma| = 0.45$ intersects the $g = 1$ circle at $\Gamma = 0.45 \angle 116.6°$ (label B). Therefore the cascade line rotates through $2\theta_{TL} = (360 - 153.4)° - 116.6° = 90°$. Since the electrical transmission line length is half of the rotation angle, the cascade line's length is $\theta_{TL} = 45°$.

At label B, we can read the intermediate admittance as $y_B = (1 - j1)$. Our goal is at $y_C = (1 + j0)$; therefore, we need to add $j1$ unit capacitive susceptance with the parallel open stub. Taking (2.44) normalizing and solving it for the phase length of the stub,

$$\theta_{OS} = \tan^{-1}\left(\frac{Z_{OS}\, b_{OS}}{Z_0}\right) = \tan^{-1}\left(\frac{75(1)}{75}\right) = 45°$$

The network, therefore, has a cascade line and a parallel open-circuited stub. Both lines have characteristic impedances of 75Ω and electrical lengths of 45°.

4.9.6 Illustrative example: troubleshooting a matching network with the Smith chart

A two-element circuit was designed to transform the high output impedance of a small-signal transistor to 50Ω over the 830- to 860-MHz frequency range. Measuring the first prototype shows s_{22} of the matching network not being at the center of the Smith chart (Plate 23), caused by an incorrect element value for one of the two components. Determine: (1) which component has the problem, and (2) should the element value be increased or decreased?

Solution

Let us reason out the path of proper impedance transformation. Starting from the output impedance of the two-port, the parallel inductor moves upward on a constant-conductance circle to intersect the constant-resistance circle of 1.0. The second element, series capacitor, moves on the unit-resistance circle to the center of the chart.

Since the series capacitor can only move on the unit resistance circle, it must not be the one causing the problem. It appears that the parallel inductor's susceptance is too high (inductance is too low); therefore, its movement goes beyond the unit-resistance circle to Γ_{INT} (actual) instead of Γ_{INT} (desired). To fix the problem we need to increase the inductance.

If the goal is to intercept the unit-resistance circle with the parallel inductors, we need less inductive susceptance to move Γ_{OUT} to the intersection of the $g = 0.2$ and $r = 1.0$ circles, labeled as Γ_{INT} (desired). At that intersection, $b = -0.4$. Then, the necessary inductive susceptance is

$$b_{LP} = 0.19 - (-0.4) = 0.59$$

From Table 2.2,

$$L_{\text{nH}} = \frac{7.96}{f_{\text{GHz}} b_{LP}} = \frac{7.96}{0.845(0.59)} = 16 \text{ nH}$$

We could also determine the actual inductance mistakenly placed into the circuit. The existing inductor transformed from Γ_{OUT} to the intersection Γ_{INT} (actual), corresponding to inductive susceptance, $b_{LP} = 0.79$. Computing the equivalent inductance, $L_P = 11.9$ nH. After replacing the inductor with a 16-nH value, the new output reflection coefficient moves to the center of the Smith chart.

4.10 Matrix descriptions of networks

The description of a linear bilateral network having n ports can be described using n equations. Similar to one-port characterization, two-port networks are customarily described by their impedance or admittance parameters. The resulting frequency-dependent two-port parameters define the terminal voltage/current relationships of the two-port under specific operating conditions, such as dc bias, input signal level, and temperature. A matrix version of Ohm's law with impedance parameters can be written as

$$\begin{bmatrix} v_1 \\ v_2 \end{bmatrix} = \begin{bmatrix} z_{11} & z_{12} \\ z_{21} & z_{22} \end{bmatrix} \begin{bmatrix} i_1 \\ i_2 \end{bmatrix} \tag{4.7}$$

where v_1, v_2, i_1, and i_2 are the voltages and currents shown in Figure 4.13:

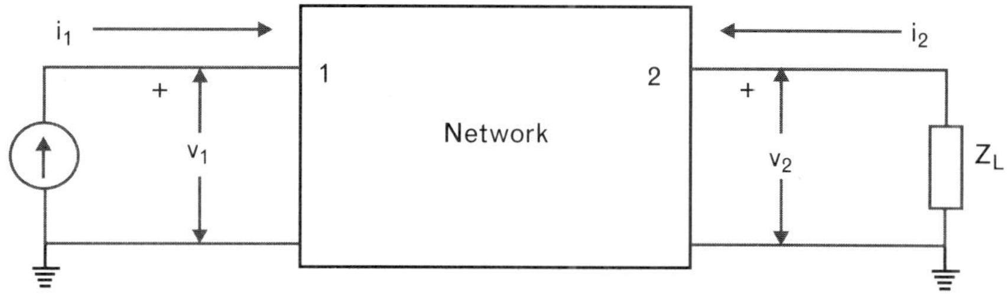

FIGURE 4.13 *Two-port network voltage and current relationships allow characterization of the network by frequency-dependant impedance or admittance parameters.*

- z_{11} is the input impedance of the two-port measured with $i_2 = 0$.
- z_{22} is the output impedance of the two-port measured with $i_1 = 0$.
- z_{21} is the transimpedance of the two-port, measured with $i_2 = 0$.
- z_{12} is the transimpedance of the two-port, measured with $i_1 = 0$.

Notation of the parameter subscripts is in the backward order. The second index shows where the signal is applied, and the first one is where it is detected, as shown below for the generalized case.

$$x_{(\text{Port where signal is detected})(\text{Port where signal is applied})}$$

where x is the parameter type, z, y, h, and so forth.

For example, z_{21} means signal applied at Port 1 and detected at Port 2; therefore, z_{21} is a transfer parameter. On the other hand, z_{11} means signal is applied and detected at Port 1, so z_{11} is the impedance of Port 1.

Breaking (4.7) into two single equations for the input and output voltage, we get

$$v_1 = z_{11} i_1 + z_{12} i_2 \tag{4.8}$$

$$v_2 = z_{21} i_1 + z_{22} i_2 \tag{4.9}$$

The next task is to determine the four z-parameters of the two-port, by applying the measured voltages and currents. Viewing (4.8) and (4.9), we find that both equations contain two unknowns. Since we cannot solve an equation with two unknowns, we need to eliminate one of the unknowns.

If we leave the output port open-circuited, the output current i_2 is set to zero. Substituting zero values into (4.8), we can now compute z_{11}, the input impedance with open circuited output,

$$z_{11} = \left. \frac{v_1}{i_1} \right|_{i_2=0}$$

The same can be applied to (4.9), computing the forward trans-impedance, z_{21}, as

$$z_{11} = \left. \frac{v_2}{i_1} \right|_{i_2=0}$$

The two z-parameters represent half of the 2×2 z-matrix. Switching the current source to the output port and leaving the input side open-circuited, the current i_1 is set to zero. Applying zero i_1 to (4.8) and (4.9), we can solve the remaining two z-parameters, z_{12} and z_{22}.

An alternative to the impedance parameter is the admittance. In matrix form, the current vector is the product the admittance matrix and the voltage vector,

$$\begin{bmatrix} i_1 \\ i_2 \end{bmatrix} = \begin{bmatrix} y_{11} & y_{12} \\ y_{21} & y_{22} \end{bmatrix} \begin{bmatrix} v_1 \\ v_2 \end{bmatrix} \tag{4.10}$$

We can also split the y-matrix equation into an individual one for the input and the output currents. Once again we face a problem similar to the one we had earlier with the z-matrixes. Both equations have two unknowns.

$$i_1 = y_{11} v_1 + y_{12} v_2 \tag{4.11}$$

$$i_2 = y_{21} v_1 + y_{22} v_2 \tag{4.12}$$

This time, however, we will apply short-circuit load termination[4] instead of open circuits, setting v_2 to zero. After finding y_{11} and y_{21} from (4.11) and (4.12), we excite the output port and short-circuit the input. From the second set of measurements, we can compute y_{12} and y_{22}.

For the z- and y-parameters to be valid, the input signal must be within the linear operating range of the two-port. Their limitations come from the difficulties of measuring voltages and currents at RF, as well as the open or short-circuit terminations on which they are based.

We already mentioned that at high frequencies voltage and current measurements are not practical. Applying open- or short-circuit terminations during the measurements are even more difficult. Although

4. We refer to an RF short that still allows dc biasing, if needed.

Y-parameter test equipment was used in the past to characterize components up to the hundreds of megahertz frequency range, it was a very tedious form of measurement. Short circuits were applied to the specific ports by reflecting through a half-wavelength long transmission line, which was frequency- dependent. Every time the frequency was changed, the transmission line length needed to be adjusted. That was both slow and prone to errors. Since active circuits do not like to be terminated with short circuits, unwanted oscillation was a frequent partner of broadband parameter measurements.

4.11 The scattering (S) matrix

Although the impedance and admittance parameters are familiar to most engineers, their measurement is impractical at higher frequencies. This was the motivation to introduce a new set of parameters [6], based on the traveling waves that enter and leave an n-port network. Incident waves are denoted as *a-waves,* and exiting waves from the network are denoted as *b-waves* (Figure 4.14). These normalized traveling voltage waves, a and b, were defined in (2.46) and (2.47).

A matrix equation is formed to relate the incident and reflected wave of Figure 4.14 by the S-parameter matrix.

$$\begin{bmatrix} b_1 \\ b_2 \end{bmatrix} = \begin{bmatrix} s_{11} & s_{12} \\ s_{21} & s_{22} \end{bmatrix} \begin{bmatrix} a_1 \\ a_2 \end{bmatrix} \tag{4.13}$$

As before for the z- and y-parameters, (4.13) can also be written in two parts, expressing the reflected waves at the input and output ports.

$$b_1 = s_{11}a_1 + s_{12}a_2 \tag{4.14}$$

$$b_2 = s_{21}a_1 + s_{22}a_2 \tag{4.15}$$

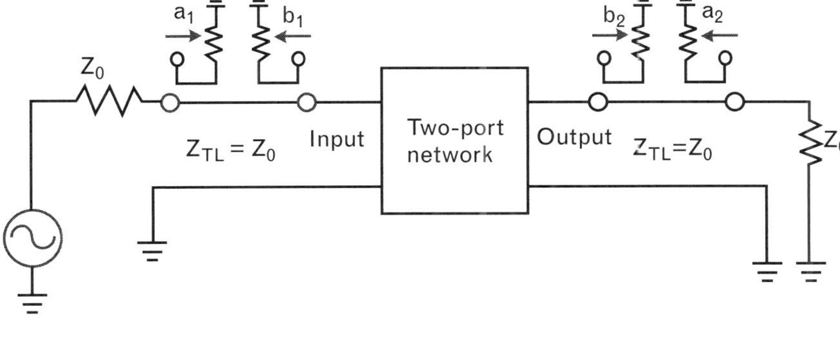

FIGURE 4.14
The scattering parameters as applied to a two-port network. Normalized incident and reflected traveling voltage waves are separated by dual-directional couplers at the input and output port.

Once again, we have four unknown parameters but only two equations. This time, however, if the load termination at Port 2 is Z_0, there is no reflection, a_2, between the load and the transmission line with Z_0 characteristic impedance. If $a_2 = 0$, from (4.14) and (4.15) we can compute s_{11} and s_{21}. Next, feeding the a_2 signal into the output port and terminating the input port with Z_0, a_1 is eliminated, enabling us to calculate s_{12} and s_{22}.

Conceptually, scattering parameters for an *n*-port network are defined similarly to *z*- and *y*-parameters in that they are also based on *n* independent equations. However, there are three important distinctions that apply to *s*-parameters.

1. Instead of voltages and currents, incident and reflected waves (*a* and *b* waves, respectively) are used to define the independent and dependent variables, respectively. These waves are defined such that they are proportional to the incident and reflected voltages at the ports divided by the square root of the characteristic impedance, Z_0, of the lines attached to these ports. When the magnitudes of these waves are squared, $|a|^2$ and $|b|^2$, they represent power flow incident on and reflected from the network, respectively.
2. By definition, *S*-parameters are ratios of normalized voltage waves; therefore, they are dimensionless.
3. *S*-parameter values depend not only on the *n*-port network parameters but also on the source and load impedances (or reflection coefficients) terminating the ports. Although most frequently *S*-parameters are measured in a 50-Ω system, the reference impedance may vary. We must specify along with the *S*-parameters the characteristic impedance of the system in which they were evaluated.

The latter property of *s*-parameters is often overlooked, namely that the values of *S*-parameters depend upon the characteristic impedance of the measurement system. If we obtain the *S*-parameters of a transistor measured in a 50-Ω system, the parameters should not be used without the appropriate conversion in a design for which the characteristic impedance is 75Ω. A new set of *S*-parameters can always be expressed in terms of the actual source and load reflection coefficients, Γ_S and Γ_L. We shall cover this later.

The complex 2 × 2 *S*-matrix relates the incident and reflected waves to characterize a two-port at a given frequency and operating conditions. Perhaps because the image of waves impending upon a network with partial transmission and partial reflection is suggestive of a scattering phenomenon, the *S*-parameters are called the scattering parameters.

Each of the two-port *S*-parameters has physical significance useful for describing small-signal, linear, steady-state operation.

$$s_{11} = \left.\frac{b_1}{a_1}\right|_{a_2=0}$$ Input reflection coefficient while output is terminated with Z_0

$$s_{21} = \left.\frac{b_2}{a_1}\right|_{a_2=0}$$ Forward transmission coefficient between Z_0 terminations

$$s_{12} = \left.\frac{b_1}{a_2}\right|_{a_1=0}$$ Reverse transmission coefficient between Z_0 terminations

$$s_{22} = \left.\frac{b_2}{a_2}\right|_{a_1=0}$$ Output reflection coefficient while input is terminated with Z_0

S-parameters are ratios of normalized voltage waves. Squaring their magnitudes provides normalized power. If we have a lossless passive two-port, the power applied to the input port is either reflected or transmitted. Accordingly,

$$|s_{11}|^2 + |s_{21}|^2 = 1.0 \tag{4.16}$$

The *S*-matrix always exists for all physical circuit terminations, including open and short circuits. Network analyzer measurements are performed with resistive (50Ω for most cases) source and load, providing stable and physically realizable broadband terminations.

The advantage of *S*-parameters is that they can be evaluated by attaching directional couplers to all of the networks ports. The directional couplers separate incident and reflected power waves directly, simplifying the measurement procedure. Furthermore, the conditions required for determining the individual *S*-parameter values always use properly terminated transmission lines at the various ports. *S*-parameters can easily be measured and *S*-parameter-based techniques quickly became the standards in high-frequency circuit design [7].

4.12 The network analyzer

The basis of modern RF and microwave network measurement is the vector network analyzer, which consists of a computer controlled automated measurement setup. The network analyzer contains the necessary directional couplers along with switching to direct and measure incident waves alternately applied to one or two-port networks.

The full S-parameter characterization of a two-port requires two different test setups. The basic network analyzer has a calibrated RF source with Z_0 output impedance, and three RF input ports.[5] Of the three, Port R (reference) is used to measure the incident voltage (i.e., a_1) of the applied RF signal, and the other two to measure incident and reflected waves.

We can determine the forward parameters, s_{11} and s_{21}, with the arrangement shown in Figure 4.15. The directional couplers have Z_0 characteristic impedances; therefore, they present Z_0 terminations to the two-port that includes the *device under test* (DUT). Reflection at Port 1 is computed from the complex ratio of b_1/a_1. Forward transmission from the source to the load is sampled by b_2/a_1. To measure reflection at Port 2, and the reverse transmission from Port 2 to Port 1, we need to change to a different test setup.

The second half of the 2 × 2 S-matrix is determined by applying the signal to the output port and terminating the input port of the DUT with Z_0 through Coupler 1, as shown in Figure 4.16. Now we can measure the reflection of the output port by the ratio of b_2/a_2. Reverse transmission is computed from b_1/a_2.

The computerized storage and use of calibration data permits the realization of measurement accuracy that would be impractical to obtain using manually collected and processed data. The network analyzer was a profound addition to the RF and microwave measurement art in the 1960s and is today an indispensable device in the engineering laboratory. Cost and size of the equipment have been nearly proportionally reduced (Figure 4.17), making it manageable and affordable even to small businesses these days.

Computer-controlled test equipment providing accurate and repeatable measurements in the gigahertz frequency range was unheard of until the introduction of the automated network analyzer system. I was traveling with the first HP demo unit and while giving a talk at one of the major East

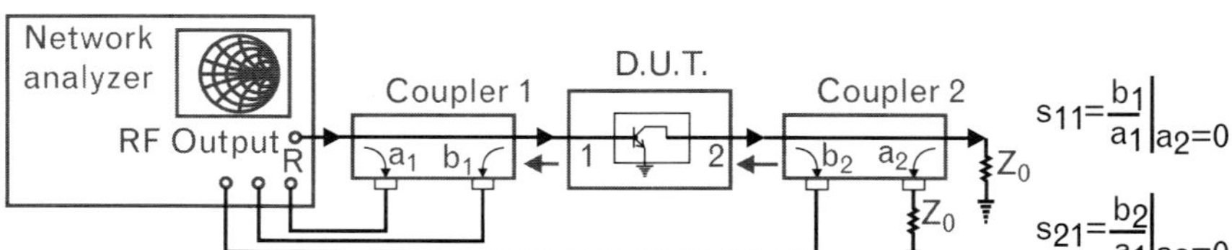

FIGURE 4.15 *The method of measuring two-port S-parameters, using a transmission line system terminated with Z_0 at both ports. The two forward S-parameters (s_{11} and s_{21}) are measured by applying the signal at the input port with a Z_0 termination at the output port. All connecting lines shown, including the two couplers, are transmission lines with Z_0 characteristic impedance.*

5. On the newer analyzers the port switching is done internally, and only the input and output connectors are on the front panel.

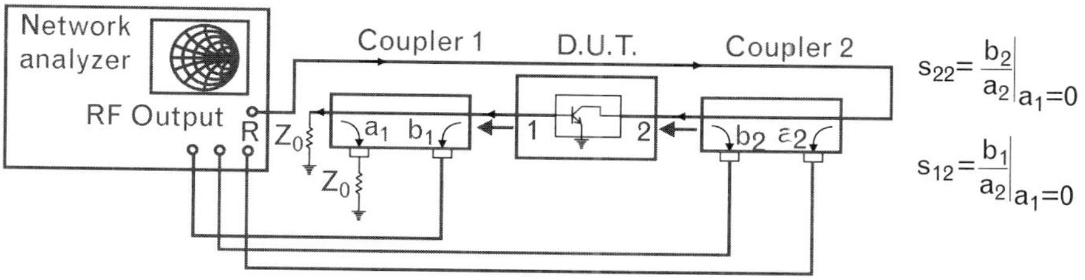

FIGURE 4.16 *To measure s_{22} and s_{12}, the signal is applied at the output port, with Z_0 at the input, setting a_1 to zero. Coaxial switches enable us to change between the setups of Figures 4.15 and 4.16.*

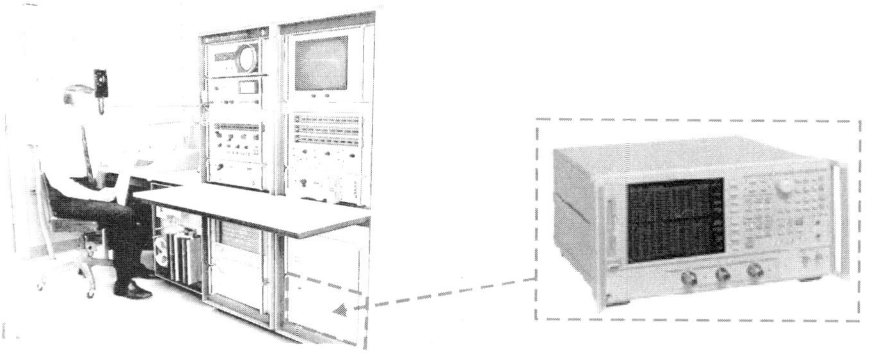

FIGURE 4.17 *Comparison of the first generation automated network analyzer system, HP 8410 (which cost well over \$200,000 in 1969) and the Agilent 8753E (which cost about \$40,000 in 2003). Also in the photo is the 1969 version of one of the authors. (Courtesy of Agilent.)*

Coast–based defense contractors, I noticed their senior engineer sitting through our presentation, expressing obvious disbelief. At the end of our talk he asked if "this big monster could really provide accurate measurements of transistors at microwave frequencies, without the risk of oscillation?" Next, he wanted to know if we "repeated the measurement the next day, would we still get the same results?" After hearing our positive replies, he told us, "You are just too young to know better—those things cannot be done at microwave frequencies," and he stormed out the room.

Actually, the gentleman was partially right. We were young enough not to have the long and frustrating experience of trying to measure components on the General Radio *Y*-parameter test equipment that relied on short-circuited terminations. High-frequency transistors did not like to work into short circuits, and they usually expressed their unhappiness by going into oscillation. The broadband resistive terminations of the network analyzer eliminated that problem. Computer-controlled error correction provided repeatable and accurate results.

4.13 S-parameter measurements

S-parameters of one- and two-port networks are conveniently and accurately measured using a network analyzer (Figure 4.18). The instrument contains electrically controlled coaxial switches, which apply the measuring signal alternately to the input and output ports of a two-port network. Incident and reflected waves are sampled by broadband directional couplers.

The key to accurate measurements lies in the calibration of the system at every test frequency. This is accomplished by applying precision loads, short and open circuits to the test ports during a calibration measurement sequence. The magnitude and phase of the incident and reflected waves are recorded by an internal computer, and this data used to correct actual measurements made with the device under test. Commonly a measurement accuracy of less than ±0.1 dB and ±2° is obtainable for transmitted and reflected wave amplitudes from a few megahertz to 20 GHz, or more.

Calibration is vitally important to remove the imperfections of the internal system components (couplers with finite directivity and match, losses, phase shifts, and reflections caused by internal bias networks, coaxial switches, attenuators, connectors, and transmission lines), as explained in Section 4.13.1. Calibration procedure is a simple semi-automated process that is always required to obtain accurate *S*-parameter measurements.

4.13.1 Measurement errors

Network analyzer *s*-parameter measurements have three types of errors that affect measurement accuracy: *systematic*, *random*, and *drift errors*. *Systematic errors* are those that are stable and repeatable, and can be identified, characterized, and removed (mathematically) from the measurement system. This process, called calibration of the network analyzer, results in the mathematical removal of the systematic errors. For a full two-port set of

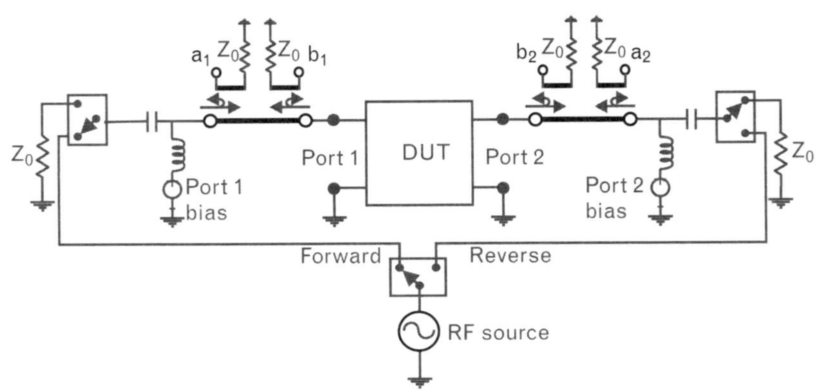

FIGURE 4.18 *Simplified block diagram of the RF portion of a network analyzer.*

s-parameters, such errors include coupler directivity, source and load mismatch, transmission and reflection tracking, and finite port isolation. To quantify and remove the systematic errors, a set of measurements are made using calibration standards or known entities. For a one-port case the systematic error coefficients are reduced to three kinds: coupler directivity, load mismatch, and finite port isolation. The calibration process establishes reference plane(s) for the measurement of device *s*-parameters. The reduction of systematic errors through calibration is ultimately limited by the quality of the standards, and the magnitude of random and drift errors.

Random errors are those that are random in nature and cannot be characterized and removed. Examples of random errors are noise and connection repeatability. There are two different types of noise encountered in any measurement system: low level noise (system noise floor) and the phase noise of the source (or high level noise).

The broadband system noise floor of the receiver can be varied through averaging, or reducing the IF bandwidth. The low level noise limits the dynamic range and absolute accuracy of the measurements as the signal approaches the noise floor level. Any instability in the source results in high level noise or jitter of the trace data. This noise is reduced to 0.004 dB typically for a synthesized source used in place of a sweeper.

Connector repeatability is the random variation encountered when connecting a pair of RF connectors to each other. Variations in both reflection and transmission can be observed. Connector repeatability limits the measurement accuracy for very low reflection coefficients or resolution of a low-loss transmission device. To minimize measurement problems, keep connectors clean and free of defects. Avoid touching the contacts of the cables, connectors, and calibration standards. When making connections, minimize the center conductor wear by rotating only the outer ground sleeve. While random errors cannot be eliminated, they can be reduced by averaging over a number of measurements. Modern vector network analyzers facilitate this process.

Drift errors fall into two classes: instrumentation and source drift. Instrumentation drift is largely due to temperature variations in the measurement environment. The temperature drift is composed of two parts: magnitude (typically 0.0015/°C) and frequency-dependent phase (typically 0.1°/°C+0.05°/GHz/°C). Although the network analyzer can store calibration coefficients, after the environmental conditions change, recalling the stored coefficients does not help. Drift errors can only be reduced by recalibrating the network analyzer before important measurements are taken.

One-port or two-port components are generally measured in test fixtures, which add complexity to the measurement system. A test fixture moves the component to be measured from the calibrated network analyzer's error reference plane(s) to a different position(s). The fixture must be characterized to distinguish the component *s*-parameters from the fixture/component *s*-parameter measurements. The process of mathematically removing the fixture, leaving the net component *s*-parameters, is known as deembedding.

4.13.2 One-port calibration

Network analyzer measurements are based on the ratio of reflected and incident waves inside a Z_0 characteristic impedance system. Between the swept signal source and the port of measurements we find several components, such as transmission lines, switches, connectors, adapters, and directional couplers. These system components represent losses and discontinuities; they form an internal two-port subnetwork with unknown transmission loss and input/output reflection coefficients (Figure 4.19). As a result, a calibration procedure is needed to remove the error produced by the component imperfections. A simplified description of the error correction is given below. For a more detailed procedure we refer to specialized publications [8–10].

The internal subnetwork in Figure 4.19 may be represented by three frequency-dependent two-port parameters: input reflection coefficient s_{11}, output reflection coefficient s_{22}, and a transmission coefficient $s_T = s_{12} = s_{21}$, as shown in Figure 4.20. These three parameters may be determined by three independent measurements using three known terminations—generally open-circuit, short-circuit, and a Z_0 load.

The input reflection coefficient of the passive two-port representing the internal subnetwork is given by

$$\Gamma_{IN} = s_{11} + \frac{s_{21} s_{12} \Gamma_L}{1 - s_{22} \Gamma_L} = s_{11} + \frac{(s_T)^2 \Gamma_L}{1 - s_{22} \Gamma_L} \tag{4.17}$$

During calibration, with three different known Γ_L terminations we get three measured values for Γ_{IN}. If the measurements are error-free, writing

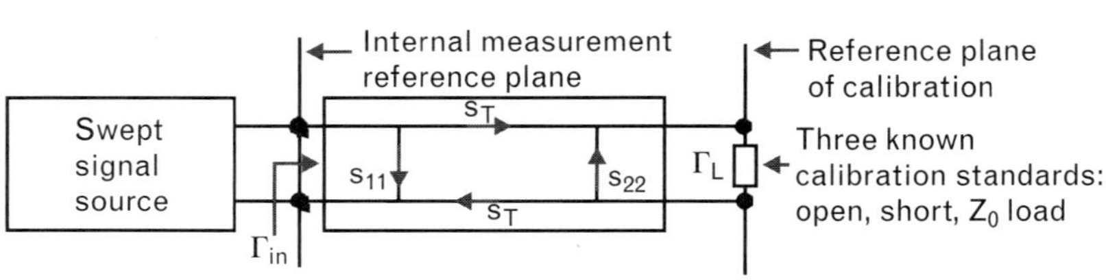

FIGURE 4.19 *At one of the two-ports, we can combine all applicable internal elements of the network analyzer into a two-port for calibration purposes. The losses and mismatches of this internal subnetwork introduce systematic errors to measurements. By characterizing the subnetwork and creating suitable calibration coefficients, we can remove the related measurement errors.*

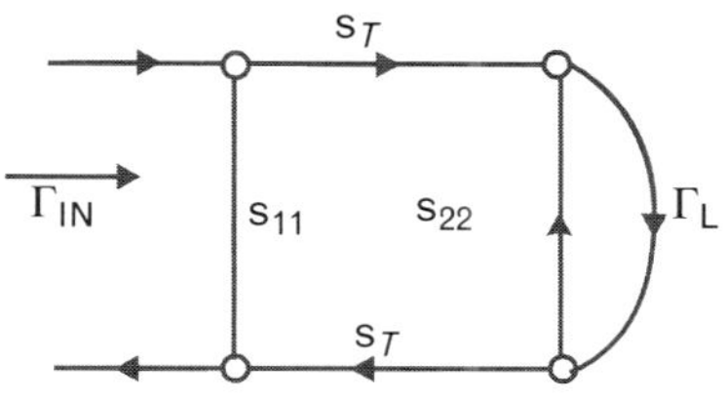

FIGURE 4.20 *Flow-graph representation of the internal passive subnetwork and external termination. Using known standard terminations for Γ_L, the three unknown parameters of the two-port can be found mathematically. These two-port error parameters are later used to mathematically correct any data measured at the external reference plane.*

and solving a system of linear equations using (4.17), the three error parameters s_{11}, s_{22}, and s_T are found. In practical measurements however, due to unavoidable small errors, the three equations do not converge and the coefficients are found by approximation techniques. Once the internal subnetwork is characterized, an error-corrected expression is established to find an arbitrary termination Γ_L, as a function of the measured data and error parameters.

4.13.2.1 One-port "home-made" test fixture for chip components

Ideally, for high-level accuracy, the network analyzer should be calibrated at the plane of measurement by using open, short, and load (Z_0) standards provided by the manufacturer. When precision fixtures and standards are not available, low-cost panel-mount type connectors may be used for chip components with reasonable results up to 1 GHz to 2 GHz. Shaving off the protruding center pin of the connector, calibration standards can be fabricated, by using a brass plate for short, and two 100-Ω chip resistors connected in parallel for a 50-Ω load.[6] Components to be measured are soldered between the center pin and the body of the connector.

4.13.3 Two-port calibrations

Although there are many calibration methods [11], including some variations, the three dominant procedures supported by the network analyzer are *short-open-load-thru* (SOLT), *thru-reflection-line* (TRL), and *line-reflection-match* (LRM). These techniques have relative advantages and disadvantages; therefore, a careful comparison is recommended to find the optimum for each specific application. For example, in measuring low

6. Up to several gigahertz, chip resistors between 80-Ω and 100-Ω values are mainly resistive. Lower value parts are too inductive and higher ones are controlled by self-capacitance (see Chapter 7 and Volume II, Chapter 1 for more details).

reflection coefficients of well-matched products, such as attenuators and terminations, use the calibration that includes a load.

Of these three methods, SOLT is the most commonly used that includes (1) one-port short, open, and load terminations at both ports, similar to that covered in Section 4.13.2, and (2) thru connection and isolation test between the two-ports. Each of the four measurements leads to three error terms, a total of 12 calibration coefficients. Since three of the 12 terms are computed redundantly, there may be questions about which results should be used. (Remember the old saying: when you have a watch, you know what time it is. When you have two watches, you can never be sure.) Accuracy obtained with the SOLT calibration primarily depends on the quality of standards, so be very careful about handling those standards. A SOLT variation, *short-open-load-reciprocal* (SOLR), is used when the thru connection is difficult to obtain [10].

TRL calibration techniques are based on transmission line standards of various lengths. One advantage of this approach is that the reflection standard does not have to be an ideal open or short since the phase length is the more important parameter. Another advantage is that only eight error terms need to be calculated. On the other hand, the accuracy of the line standard's characteristic impedance is critical to the quality of calibration, because it determines the test system's reference impedance. A commonly used variation of TRL is called *line-reflect-line* (LRL) that replaces the thru with another precision line standard.

LRM is similar to TRL and it also requires only eight error terms. The precision line standard is replaced by a match (load) element, giving wider frequency range coverage.

4.13.4 Time-domain reflectometry

An alternative to one-port frequency-domain reflection analysis is *time-domain reflectometry* (TDR) [8]. In this method a voltage step is sent into the network under investigation, and the incident and reflected voltage waves are monitored by a time-domain oscilloscope. The advantage of TDR analysis is the ability to isolate and identify the type and location of impedance variations along a transmission-line circuit. For a network with multiple element-induced impedance variations, each element reflects a portion of the incident signal with a different time offset. This provides a means of identification of each element in the network flow on the oscilloscope trace and its impact on the incident signal. Reflection coefficient measurements do not single out the individual components or element impedance variations along the network; they only indicate their aggregate effect.

The waveform of the TDR analysis can be used to determine element types (capacitive, inductive, resistive, or distributive) and configuration (series or parallel) in a system. Portions of the TDR waveform can also be adjusted to remove or deembed connector repeatability errors. The time-domain analysis can then be processed back into the frequency-domain for improved measurement accuracy. TDR analysis offers an insight into the network under investigation not possible with frequency-domain analysis alone. Special options of some network analyzers are available to convert frequency-domain measurements to time-domain via the inverse Fourier transform.

Since active circuits with sensitive devices may not be able to withstand the large voltage step of the TDR, the technique is primarily used in passive circuits.

4.14 Two-port gain expressions in terms of S-parameters

The forward transmission coefficient of a two-port, s_{21}, is the ratio of two normalized voltage waves, measured between Z_0 terminations. Using a network analyzer, the $|s_{21}|$ reading may be called gain if it is greater than unity (+dB), and loss when it is less than unity (−dB). Squaring $|s_{21}|$ amounts to the power gain or loss from a Z_0 source into a Z_0 load. In the actual applications, however, our two-port devices often face complex source and load terminations other than Z_0.

We need to be careful about our definitions because there are many different output/input power ratios. Do we reference to the power delivered into the two-port, or the power available from the source? At the output side, is it the power emerging from the two-port, or what is truly delivered to load that counts? It appears that there are different forms to express gain or loss, depending on our definitions. It is important that we are clear about their use.

The four commonly used expressions, based on the power definitions and terminations, are:

- *Transducer power gain,* G_T, in the generalized form is:

$$
\begin{aligned}
G_T &= \frac{\text{Power delivered to the load}}{\text{Power available from the source}} \\
&= \frac{\left(1-|\Gamma_S|^2\right)|s_{21}|^2\left(1-|\Gamma_L|^2\right)}{\left|(1-s_{11}\Gamma_S)(1-s_{22}\Gamma_L)-s_{12}s_{21}\Gamma_S\Gamma_L\right|^2}
\end{aligned}
\tag{4.18}
$$

where "power available" refers to matched condition, and

$$\Gamma_S = \frac{Z_S - Z_0}{Z_S + Z_0}$$

$$\Gamma_L = \frac{Z_L - Z_0}{Z_L + Z_0}$$

Using a network analyzer, $Z_S = Z_L = Z_0$, reduces $\Gamma_S = \Gamma_L = 0$; therefore, G_T is simplified to

$$G_T = G_0 = |s_{21}|^2 \tag{4.19}$$

Or, converted to decibels,

$$G_T = 10 \log|s_{21}|^2 = 20 \log|s_{21}| \tag{4.20}$$

- *Available power gain, G_A:*

$$\begin{aligned} G_A &= \frac{\text{Power available from the two-port}}{\text{Power available from the source}} \\ &= \frac{|s_{21}|^2 \left(1 - |\Gamma_S|^2\right)}{\left(1 - \left|\dfrac{s_{22} - (\Delta)\Gamma_S}{1 - s_{11}\Gamma_S}\right|^2\right) |1 - s_{11}\Gamma_S|^2} \end{aligned} \tag{4.21}$$

where Δ is the determinant of the S-matrix,

$$\Delta = s_{11}s_{22} - s_{21}s_{12} \tag{4.22}$$

- *Operating power gain, G_P:* (also called *power gain* in some textbooks):

$$\begin{aligned} G_P &= \frac{\text{Power delivered to the load}}{\text{Power applied to the input of the two-port}} \\ &= \frac{|s_{21}|^2 \left(1 - |\Gamma_L|^2\right)}{\left(1 - \left|\dfrac{s_{11} - (\Delta)\Gamma_L}{1 - s_{22}\Gamma_L}\right|^2\right) |1 - s_{22}\Gamma_L|^2} \end{aligned} \tag{4.23}$$

- *Insertion loss, IL:*

$$IL = \frac{\text{Power delivered to the load without the two-port}}{\text{Power delivered to the load with the two-port inserted}}$$
$$= \frac{\left|(1 - s_{11}\Gamma_S)(1 - s_{22}\Gamma_L) - s_{12}s_{21}\Gamma_S\Gamma_L\right|^2}{|s_{21}|^2\left(1 - |\Gamma_S\Gamma_L|^2\right)} \tag{4.24}$$

Of the four listed expressions, three refer to gain—(4.18), (4.21), and (4.23)—while the fourth one, (4.24), is defined for loss. Although the term insertion loss is frequently used in the RF industry, its mathematical definition is different from the reciprocal of transducer gain and they should not be used interchangeably except when the source and load terminations are real and equal to each other. Otherwise, you can actually get insertion gain (instead of insertion loss) while matching to unequal terminations with a passive network.

4.14.1 Illustrative exercise: transducer gain versus insertion loss

A lossless impedance transformer network with measured *S*-parameter (referenced to $Z_0 = 50\Omega$) is connected between 10-Ω source and 50-Ω load, that corresponds to 5:1 *VSWR* (recalling that *VSWR* between two resistive terminators is just the ratio of larger and smaller resistances). The measured 50-Ω *S*-parameters are:

$$S_N = \begin{bmatrix} 0.667\angle 180° & 0.745\angle 180° \\ 0.745\angle 180° & 0.667\angle 0° \end{bmatrix}$$

Find the transducer gain and insertion loss of the lossless network using the source and load termination specified.

Solution

The transducer gain expression of (4.18) may be applied here by using *S*-parameters of the two-port between these two terminations. The source and load reflection coefficients, using $Z_0 = 50\Omega$, are

$$\Gamma_S = \frac{Z_S - Z_0}{Z_S + Z_0} = \frac{10\Omega - 50\Omega}{10\Omega + 50\Omega} = 0.667\angle 180°$$
$$\Gamma_L = \frac{Z_L - Z_0}{Z_L + Z_0} = \frac{50\Omega - 50\Omega}{50\Omega + 50\Omega} = 0.0$$

$$G_T = \frac{\left(1-|\Gamma_S|^2\right)|s_{21}|^2\left(1-|\Gamma_L|^2\right)}{\left|(1-s_{11}\Gamma_S)(1-s_{22}\Gamma_L)-s_{12}s_{21}\Gamma_S\Gamma_L\right|^2}$$
$$= \frac{\left(1-|0.667\angle 180°|^2\right)|0.745\angle 180°|^2(1-0)}{\left|\left(1-|0.667\angle 180°|^2\right)(1-0)-0\right|^2} = 1.0$$

or, in decibels,

$$G_{T\text{dB}} = 10\log(G_T) = 10\log(1) = 0 \text{ dB}$$

If we just connect the source directly to the load, the mismatch loss computed from (2.54) is

$$ML_{\text{dB}} = -10\log\left(1-|\Gamma_S|^2\right) = -10\log\left(1-|0.667|^2\right) = 2.55 \text{ dB}$$

It looks like the matching network fully recovers the mismatch loss. Now, let's compute *insertion loss* of the lossless network from (4.24),

$$IL = \frac{\left|(1-s_{11}\Gamma_S)(1-s_{22}\Gamma_L)-s_{12}s_{22}\Gamma_S\Gamma_L\right|^2}{|s_{21}|^2\left(1-|\Gamma_S\Gamma_L|^2\right)}$$
$$= \frac{\left|\left(1-|0.667|^2\right)(1-0)-0\right|^2}{|0.745|^2(1-0)} = 0.555$$

Converting the insertion loss to decibels,

$$IL_{dB} = 10\log(IL) = 10\log(0.555) = -2.55 \text{ dB}$$

The result is −2.55-dB loss = 2.55 dB gain. However, do not rush out to apply for a patent, just because a passive circuit seems to provide gain! It is not true. The result shows the difference between the two definitions.

4.14.2 Illustrative exercise: transistor gain calculations—1

The measured small-signal *S*-parameters of the BFP405 transistor at 1,900 MHz are shown in Table 4.6. For simplicity, we shall neglect the effects of the reverse transmission coefficient, by setting $s_{12} = 0$.

TABLE 4.6 MEASURED SMALL-SIGNAL S-PARAMETERS OF THE BFP405 TRANSISTOR AT 1,900 MHZ

s_{11}		s_{21}		s_{12}		s_{22}	
Mag.	Ang.	Mag.	Ang.	Mag.	Ang.	Mag.	Ang.
0.707	–67°	5.45	119°	0	0	0.84	–32°

Compute the decibel gain of this device under four different sets of conditions:

1. The basic 50-Ω gain, without performing any impedance matching;
2. If we add a lossless matching network to the input of the device to eliminate ML at the input;
3. Instead of working on the input side, we add a lossless matching network to the output of the device to eliminate ML at the output;
4. When both ports are matched, *assuming* the matching networks have no effect on each other.

Solution

One of the advantages of using S-parameters to characterize transistors is that important aspects of their operation can be surmised directly by inspection of the individual S-parameter values at each frequency.

When mismatch is eliminated at one of the ports, the resulting gain of the device (expressed in decibels) increases by the amount of the reflection loss that that mismatch would have caused.

1. In a 50-Ω system the basic decibel gain of the device is

$$G_{T\text{dB}} = G_{0\text{dB}} = 10\log|s_{21}|^2 = 20\log|s_{21}| = 20\log(5.45) = 14.7 \text{ dB}$$

2. Impedance matching the input port to 50Ω eliminates the input mismatch loss and increases the decibel gain by $G_{1\text{dB}}$:

$$G_{1\text{dB}} = -10\log\left(1-|s_{11}|^2\right) = -10\log\left(1-0.707^2\right) = 3.0 \text{ dB}$$

3. Impedance matching the output port to 50Ω eliminates the output mismatch loss and increases the decibel gain by $G_{2\text{dB}}$:

$$G_{2\,\mathrm{dB}} = -10\log\left(1-|s_{22}|^2\right) = -10\log\left(1-0.84^2\right) = 5.3 \text{ dB}$$

4. Adding the transforming networks simultaneously, and assuming that they do not interact with each other, eliminates mismatch losses at both sides and increases the decibel gain by ($G_{1\mathrm{dB}} + G_{2\mathrm{dB}}$) = 8.3 dB. This is called the *unilateral gain* assumption and its validity is dependent upon the condition that $|s_{12}| = 0$. In practice $|s_{12}|$, while often small, is not insignificant and the total gain so predicted will be in error by an amount that is a function of all four *S*-parameters, as we shall see later in Volume II, Chapter 1. Nevertheless, the unilateral gain may provide a quick first estimate of a transistor's performance.

Filling the results of our computations into Table 4.7 allows quick comparison of the gain improvement achieved by eliminating mismatch losses. Of the two-ports, the output has greater mismatch and we can recover 5.3 dB. At the input, the mismatch loss is 3.0 dB, so between the two-ports there is a significant amount of potential gain that we do not achieve by using 50-Ω terminations. In the next chapter we cover impedance matching in detail.

Let us emphasize again that the gain improvements were computed without any allowance for matching component losses. In case (d) we also assumed that there is no feedback in the transistor ($s_{12} = 0$).

4.14.3 Illustrative exercise: transistor gain calculations—2

The 50-Ω *S*-parameters of an RFIC gain block (Table 4.8) has internal feedback to maintain flat gain through a broad frequency range. (Note that the tabulated s_{21} magnitudes are in decibels.) In fact, the amplifier has only 1 dB gain roll-off between 100 and 600 MHz.

TABLE 4.7 THE 50-Ω SYSTEM, IN WHICH THE TRANSISTOR PROVIDES-14.7 dB GAIN AT 1 GHz

CASE	G_1(dB)	G_0(dB)	G_2(dB)	TOTAL GAIN (dB)
a	—	14.7		$G_T = G_0 = 14.7$
b	3.0	14.7		$G_T = G_1 + G_0 = 17.7$
c	—	14.7	5.3	$G_T = G_0 + G_2 = 20.0$
d	3.0	14.7	5.3	$G_T = G_1 + G_0 + G_2 = 23.0$

Note: Looking at the large reflection coefficient magnitudes, we can expect noticeable gain improvement when the mismatches are eliminated.

TABLE 4.8 MEASURED BROADBAND S-PARAMETERS OF AN RFIC GAIN MODULE SHOW EXCELLENT IMPEDANCE MATCH INTO THE GIGAHERTZ REGION

FREQUENCY (GHz)	s_{11}		s_{21}		s_{12}		s_{22}	
	MAG.	ANG.	MAG.	ANG.	MAG.	ANG.	MAG.	ANG.
0.1	0.08	171	19.0	173	0.073	2	0.10	−13
0.2	0.07	161	18.9	167	0.075	6	0.11	−27
0.3	0.07	152	18.7	160	0.077	8	0.10	−39
0.4	0.06	143	18.5	143	0.76	11	0.11	−49
0.5	0.05	133	18.3	147	0.79	13	0.11	−59
0.6	0.04	115	18.0	141	0.81	17	0.12	−67
0.8	0.03	79	17.5	130	0.087	20	0.12	−83
1.0	0.04	−14	16.8	119	0.098	23	0.12	−96
1.5	0.08	−52	15.0	96	0.112	26	0.10	−116
2.0	0.12	−87	13.2	78	0.131	24	0.08	−134
2.5	0.15	−112	11.7	67	0.147	25	0.07	−135
3.0	0.19	−132	10.3	54	0.156	22	0.07	−129
3.5	0.24	−148	8.9	41	0.170	18	0.09	−117
4.0	0.26	−159	7.7	29	0.177	13	0.13	−106
4.5	0.27	−170	6.6	18	0.184	8	0.17	−105
5.0	0.27	−175	5.7	8	0.192	5	0.20	−106

At 1 GHz the device has 16.8-dB basic gain in the 50-Ω system. Using the S-parameters, compute the mismatch losses at both ports to find out the amount of gain improvement we could get my eliminating the mismatch.

Solution

Finding the input and output mismatch losses from $|s_{11}|$ and $|s_{22}|$ at 1 GHz,

$$ML_{1\,\text{dB}} = -10\log\left(1-|s_{11}|^2\right) = -10\log\left(1-|0.04|^2\right) = 0.007\ \text{dB}$$

$$ML_{2\,\text{dB}} = -10\log\left(1-|s_{22}|^2\right) = -10\log\left(1-|0.12|^2\right) = 0.063\ \text{dB}$$

The total mismatch of both ports is less than 0.1 dB. Clearly, for such a well-matched amplifier the minute amount of gain improvement does not justify the effort.

When the input and output reflection coefficients of a two-port have magnitudes less than 0.1, very little improvement is obtained by matching the circuit. A 0.2 magnitude reflection coefficient still only converts to 0.18-dB mismatch loss and in many cases the added components have comparable amount of dissipative losses. Unless there are other compelling reasons, it is still not justified to add matching components.

As the reflection coefficient magnitude increases beyond 0.25 to 0.30, the mismatch loss becomes more important and justifies impedance matching.

4.15 Cascading two-ports with S-parameters

When the interstage port reflection coefficient products of two cascaded two-ports, S_A and S_B, are equal to zero ($|s_{A_{22}} s_{B_{11}}| = 0$), the overall gain is just the product of the two individual gain factors

$$s_{AB_{21}} = s_{A_{21}} s_{B_{21}} \tag{4.25}$$

Or, if the gain of the two-ports are in decibels, then

$$s_{AB_{21\,\mathrm{dB}}} = s_{A_{21\,\mathrm{dB}}} + s_{B_{21\,\mathrm{dB}}} \tag{4.26}$$

Generally, however, this is not the case, and the interstage ports are mismatched to some extent. In these cases, the overall S-matrix of the cascade shown in Figure 4.21 becomes

$$S_{AB} = \begin{bmatrix} s_{A_{11}} + \dfrac{s_{A_{12}} s_{A_{21}} s_{B_{11}}}{1 - s_{A_{22}} s_{B_{11}}} & \dfrac{s_{A_{12}} s_{B_{12}}}{1 - s_{A_{22}} s_{B_{11}}} \\ \dfrac{s_{A_{21}} s_{B_{21}}}{1 - s_{A_{22}} s_{B_{11}}} & s_{B_{22}} + \dfrac{s_{B_{12}} s_{B_{21}} s_{A_{22}}}{1 - s_{A_{22}} s_{B_{11}}} \end{bmatrix} \tag{4.27}$$

Looking at the overall S-parameters in (4.27) tells us that the individual terms are very significantly affected by the interstage terms. Keep in mind

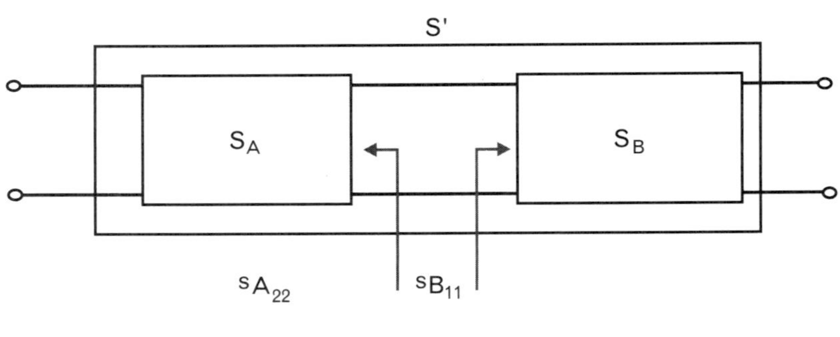

FIGURE 4.21
The solution for the overall S-parameters of two cascaded two-ports is strongly influenced by the match or mismatch between two-ports S_A and S_B.

that both $s_{A_{21}}$ and $s_{B_{21}}$ are complex numbers; therefore, their product may be positive, negative, or a complex quantity. Accordingly, the values of the four fractional terms in the matrix may vary through large extremes. Visualize calculating the overall gain when $s_{A_{21}} s_{B_{21}} = 1.0$. That is a lot of gain!

4.15.1 Illustrative exercise: performance of two cascaded filters

We need a bandpass filter for the 850-MHz passband with at least 20-dB attenuation at 880 MHz. The eager salesperson of our supplier offers a great deal on filters that easily meet our passband requirements, but only provides 11 dB loss at 880 MHz. Could we cascade two such filters for our needs?

Solution

Before signing an order, it is a good idea to measure, or to simulate the performance of the cascaded filters. Most filters are frequency-dependent signal reflectors. In the passband they are well matched and suffer only small dissipative losses. In their stopbands, they provide isolation by reflecting unwanted power. Reflection means poor match.

When the filters are individually tested on the network analyzer, they see solid Z_0 terminations. Cascading two of them, they are only matched in their passbands. In the stopband they are very poorly matched and the overall performance is hard to predict without accurate characterization. In fact, the output of the first filter may just be matched to the input of the second at some arbitrary impedance.

Figure 4.22 shows the response of single and two cascaded bandpass filters with a short segment of 50-Ω connecting transmission line. Between 840 MHz and 860 MHz each filter has just over 1 dB dissipative loss. Cascading two of them doubles the loss. At 880 MHz, however, a single filter provides 11-dB rejection, but the cascade can only get 6 dB. In fact, if the simulation were performed with ideal filter components, the loss at 880 MHz would be much less. Looking at the stopband attenuation of the cascaded filters at other frequencies, we can see that it is generally not twice the attenuation of a single filter.

Note that the reflection coefficients, facing each other between the filters, have the same magnitudes but opposite type angles—that is a conjugate match. Depending on the length of the interconnecting transmission line, such a match, and multiple other matches, can occur almost anywhere.

There are two ways to help with problems caused by mismatch between cascaded components. Both methods require inserting an additional two-port into the interstage connection:

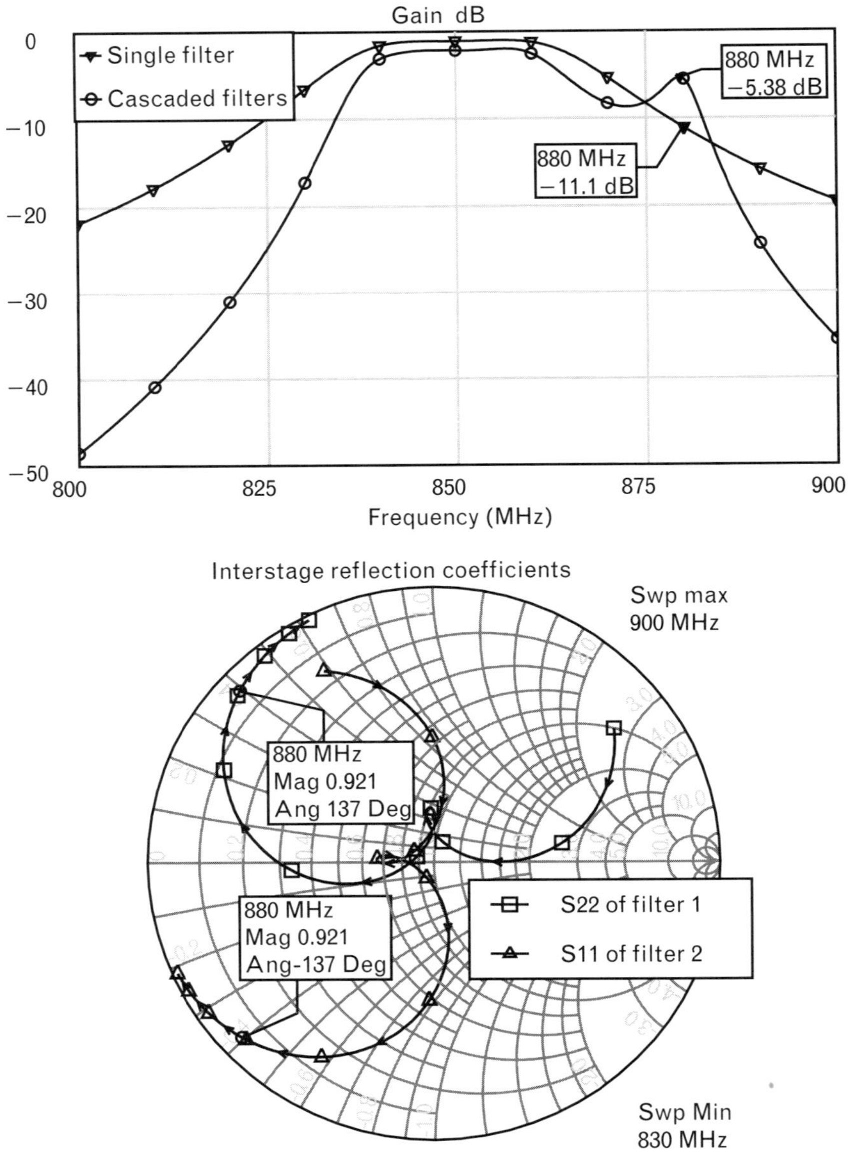

FIGURE 4.22
Stopband mismatch error resulting with the cascade combination of two filters. At some frequency the attenuation of the cascade is greater than twice that of the single filter, while at other frequencies it is less. At 880 MHz the two filters together provide less attenuation than the single one because the interstage impedances are nearly complex conjugates of each other.

1. Adding a matched attenuator reduces the reflected signals by twice the loss of the attenuator. Typically 3- to 6-dB attenuation, which adds 6 dB to 12 dB to the return loss, offers appreciable improvement. The price we pay, in addition to the increased cost and space requirements, is the reduced gain and dynamic range of the system.

2. Instead of the attenuator, we can use a matched isolator to attenuate the interstage reflection. Isolators cause a nominal loss (0.2 to 0.3 dB) in the forward direction, but they cut reverse transmission by

–20 dB to –30 dB. They generally require more space and cost more than attenuators. Their frequency coverage is more limited also. Figure 4.23 shows the performance of the cascaded filters with an isolator added between the two stages. The loss of the isolator slightly increases the forward attenuation of the overall circuit.

In Volume II, Chapters 1 and 2, we look at cascaded active circuits and see that interstage mismatch can cause even greater problems. Combining two 10-dB gain amplifiers, we may get less or even *more* than 20 dB overall gain! Or, if we are unlucky, we may also experience unwanted oscillation.

4.15.2 Mismatch error

The example shown in Section 4.15.1 reveals how two identical filters have actually less attenuation at some frequency than the single one has. At other frequencies the cascaded filters exceeded the expected attenuation. The variation in spacing influences the interaction of their interstage reflection coefficients.

The uncertainty mismatch between two two-ports, having arbitrary interstage reflection coefficients of $s_{A_{22}}$ and $s_{B_{11}}$ was defined in (2.57) as

$$MU_{\text{dB}} = 20\log\left(1 \pm \left|s_{A_{22}}\, s_{B_{11}}\right|\right)$$

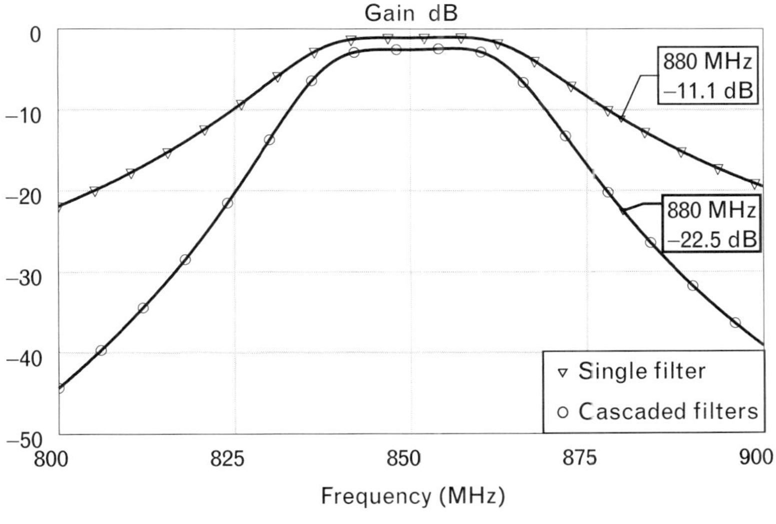

FIGURE 4.23 *An example showing how isolation improves mismatch error. After placing an isolator between the two filters of Figure 4.22, the loss of the cascaded filters is equal to twice that of the single filters through the 800- to 900-MHz frequency range.*

In this example the reflections are large; $s_{A_{22}}$ and $s_{B_{11}}$ each have magnitudes of 0.92. Since either of these reflections could result in a reflection of 83% of the incident power, it is not difficult to understand that their interaction with each other produces a mismatch error, ranging from +16.2 dB to –5.3 dB. In our cascaded filter illustration, each individual filter had 11.1 dB attenuation (–11.1 dB gain) at 880 MHz, when measured between 50-Ω terminations. Cascading the two filters with a segment of lossless 50-Ω transmission line, the *actual attenuation* at 880 MHz may be anywhere between $22.2^{+5.3}_{-16.2}$ dB, depending on the exact phase relationship of the interstage connection.

The mismatch error is the difference between the actual loss (or gain) through a network compared with the sum of the transmission losses/gains predicated on the matched source and load s-parameter measurements of the constituent elements.

4.15.3 The cascadable scattering transfer matrix

We saw that the *S*-matrix is not cascadable in its original form. That is to say, one cannot matrix multiply the individual *S*-matrices of cascaded two-port networks to find the overall resultant *S*-matrix of the cascade. For this reason a scattering transfer matrix, *T*, is defined which is cascadable. Matrix multiplication of the scattering transfer matrices for the two cascaded two-port networks shown in Figure 4.24 results in the *T*-parameters of the overall cascade. The same approach may be extended to any number of cascaded two-ports.

$$\begin{bmatrix} a_{A1} \\ b_{A1} \end{bmatrix} = \begin{bmatrix} T_{A11} & T_{A12} \\ T_{A21} & T_{A22} \end{bmatrix} \begin{bmatrix} b_{A2} \\ a_{A2} \end{bmatrix} \text{ and } \begin{bmatrix} a_{B1} \\ b_{B1} \end{bmatrix} = \begin{bmatrix} T_{B11} & T_{B12} \\ T_{B22} & T_{B21} \end{bmatrix} \begin{bmatrix} b_{B2} \\ a_{B2} \end{bmatrix} \quad (4.28)$$

Looking at Figure 4.24, we see that the traveling wave vector at the output side of two-port *A*, $[b_{A2}, a_{A2}]$, is equal to the vector at the input side of two-port *B*, $[a_{B1}, b_{B1}]$. We can write the equality, and from the second part of (4.28) substitute the matrix product for $[b_{A2}, a_{A2}]$.

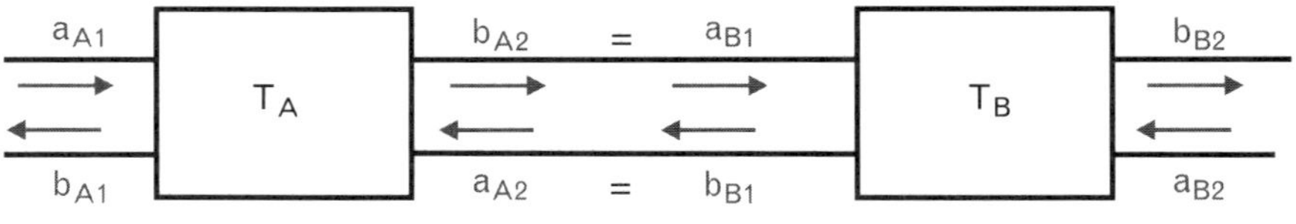

FIGURE 4.24 *Matrix multiplying the scattering transfer matrices of two-ports A and B, T_A and T_B, gives us the T-matrix of the cascaded circuit, T_{AB} since $b_{A2} = a_{B1}$, and $a_{A2} = b_{B1}$.*

$$\begin{bmatrix} b_{A2} \\ a_{A2} \end{bmatrix} = \begin{bmatrix} a_{B1} \\ b_{B1} \end{bmatrix} = \begin{bmatrix} T_{B11} & T_{B12} \\ T_{B22} & T_{B21} \end{bmatrix} \begin{bmatrix} b_{B2} \\ a_{B2} \end{bmatrix} \tag{4.29}$$

Combining (4.28) and (4.29), we get the relationship between the input vector of two-port A and the output vector of two-port B, as the matrix product of the two T-matrices.

$$\begin{bmatrix} a_{A1} \\ b_{A1} \end{bmatrix} = \begin{bmatrix} T_{A11} & T_{A12} \\ T_{A21} & T_{A22} \end{bmatrix} \begin{bmatrix} T_{B11} & T_{B12} \\ T_{B22} & T_{B21} \end{bmatrix} \begin{bmatrix} b_{B2} \\ a_{B2} \end{bmatrix}$$

or in a simpler matrix form,

$$[T_{AB}]_{2\times 2} = [T_A]_{2\times 2}[T_B]_{2\times 2} \tag{4.30}$$

Conversion from S-parameters to T-parameters may be given as

$$\begin{bmatrix} T_{11} & T_{12} \\ T_{21} & T_{22} \end{bmatrix} = \frac{1}{s_{21}} \begin{bmatrix} 1 & -s_{22} \\ s_{11} & -DET[S] \end{bmatrix} \tag{4.31}$$

Converting back from T to S,

$$\begin{bmatrix} s_{11} & s_{12} \\ s_{21} & s_{22} \end{bmatrix} = \frac{1}{T_{11}} \begin{bmatrix} T_{21} & DET[T] \\ 1 & -T_{12} \end{bmatrix} \tag{4.32}$$

where $DET[S]$ and $DET[T]$ are the determinants of the corresponding 2×2 S- and T-matrices.

Note that some textbooks may use a different definition for the $[T]$ matrix for the conversions between the two matrices [12]. Those other definitions are also correct, but those expressions must not be mixed with the ones given here.

4.16 Multiport S-parameters

The $[S]$ matrix is very widely used to describe two-port transistor amplifiers, and it may be expanded to define any n-port network, where n is any positive integer. Some examples for one- to four-port networks are shown

$$s_{1\times1} = \begin{vmatrix} s_{11} \end{vmatrix}$$

$$s_{2\times2} = \begin{vmatrix} s_{11} & s_{12} \\ s_{21} & s_{22} \end{vmatrix}$$

$$s_{3\times3} = \begin{vmatrix} s_{11} & s_{12} & s_{13} \\ s_{21} & s_{22} & s_{23} \\ s_{31} & s_{32} & s_{33} \end{vmatrix}$$

$$s_{4\times4} = \begin{vmatrix} s_{11} & s_{12} & s_{13} & s_{14} \\ s_{21} & s_{22} & s_{23} & s_{24} \\ s_{31} & s_{32} & s_{33} & s_{34} \\ s_{41} & s_{42} & s_{43} & s_{44} \end{vmatrix}$$

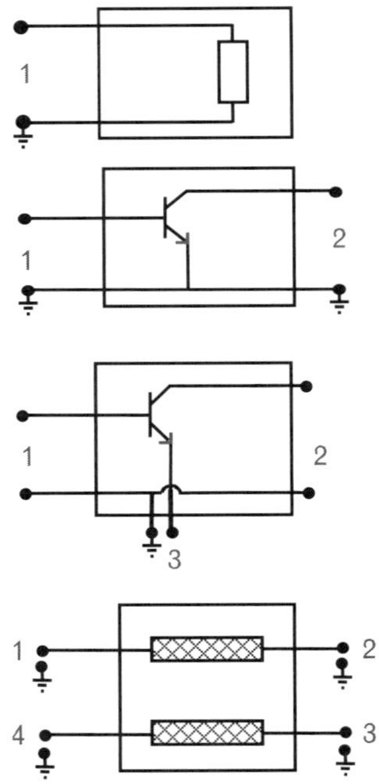

FIGURE 4.25 *Examples of one through four-port components and their S-matrices.*

in Figure 4.25. Note that the order of the matrix is simply equal to the number of independent network ports. Also note that a bipolar transistor not having a common terminal (such as grounded base, grounded emitter, or grounded collector) may be represented as a three-port network and characterized by a 3 × 3 S-matrix.

The *S*-parameter matrix can be determined for any *n*-port circuit, always resulting in finite entries for all terms. Two-port parameters for series and parallel impedances as well as for some of the commonly used ideal circuit components are given in Table 4.9.

Some of the matrix element entries may not be intuitively obvious. For example, ask a colleague the following question: Using $Z_0 = 50\Omega$, what is s_{11} of a two-port containing only an ideal 50-Ω series resistor? The typical answer is "0," forgetting that s_{11} is measured with a Z_0 termination connected to the output port, which gives $s_{11} = 0.33 \angle 0°$.

4.17 Generalized two-port S-parameters

S-parameter measurements are generally made with a network analyzer, which provides well-matched Z_0 (either 50Ω or 75Ω) source and load impedance to the two-port. However, when we use that device it often sees different terminations, particularly when it is a transistor. We may wish

TABLE 4.9 TWO-PORT S-MATRICES OF EIGHT FREQUENTLY USED CIRCUIT ELEMENTS IN THEIR IDEAL FORMS

SERIES IMPEDANCE:

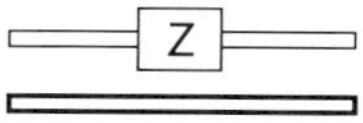

$$S = \frac{1}{Z + 2Z_0}\begin{bmatrix} Z & 2Z_0 \\ 2Z_0 & Z \end{bmatrix}$$

TRANSFORMER:

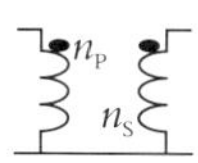

$n = n_S/n_P$

$$S = \frac{1}{1+n^2}\begin{bmatrix} 1-n^2 & 2n \\ 2n & n^2-1 \end{bmatrix}$$

VOLTAGE-DEPENDENT CURRENT SOURCE:

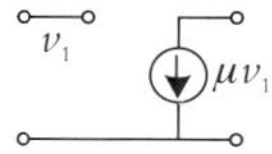

$$S = \begin{bmatrix} 1 & 0 \\ 2\mu & -1 \end{bmatrix}$$

CURRENT-DEPENDENT VOLTAGE SOURCE:

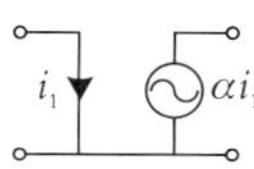

$$S = \begin{bmatrix} -1 & 0 \\ \frac{2\alpha}{Z_0} & 1 \end{bmatrix}$$

SHUNT IMPEDANCE

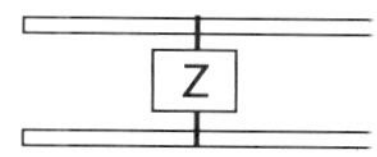

$$S = \frac{1}{Z_0 + 2Z}\begin{bmatrix} -Z & 2Z \\ 2Z & -Z_0 \end{bmatrix}$$

TRANSMISSION LINE:

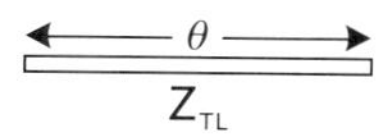

$$\Gamma = \frac{Z_{TL} - Z_0}{Z_{TL} + Z_0}$$

$$S = \frac{1}{1-\Gamma^2 e^{-j2\theta}}\begin{bmatrix} \Gamma(1-e^{-j2\theta}) & (1-\Gamma^2)e^{-j2\theta} \\ (1-\Gamma^2)e^{-j2\theta} & \Gamma(1-e^{-j2\theta}) \end{bmatrix}$$

CURRENT-DEPENDENT VOLTAGE SOURCE:

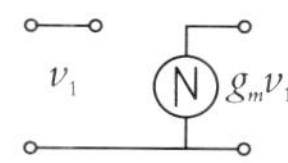

$$S = \begin{bmatrix} 1 & 0 \\ -2g_m Z_0 & 1 \end{bmatrix}$$

CURRENT-DEPENDENT CURRENT SOURCE:

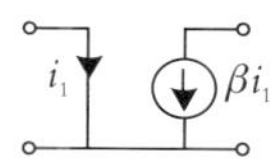

$$S = \begin{bmatrix} -1 & 0 \\ -2\beta & 1 \end{bmatrix}$$

to provide a specific mismatch at the input to obtain (1) lowest noise performance, (2) particular load impedance for maximum absolute power output, (3) complex source and load for maximum gain, or (4) some intermediate source and load combination to provide a specific value of gain at a given frequency.

The problem is that the original Z_0-based S-parameters change when they are referenced to different terminations. For example, the input reflection coefficient of a transistor is s_{11} when the load termination is Z_0. If we transform the load to a new impedance, perhaps to eliminate the mismatch, the input reflection coefficient changes to a new value.

For this reason we need to have generalized S-parameters, which can be applied when any specified values of complex source, Γ_S, and complex load, Γ_L are applied to the two-port, as shown in Figure 4.26. In other words, we need to transform the Z_0-referenced source and load s-parameters to a new set, which we label as the S'-parameters. These parameters are derived from the original S-parameters but include the effects of having assignable values of Γ_S and Γ_L. This formulation is given in (4.33) through (4.36). We see in Volume II, Chapters 1 and 2 that the generalized S-parameters are the key to transistor amplifier design.

$$s'_{11} = \frac{A_1^*}{A_1}\frac{(1-\Gamma_L s_{22})(s_{11} - \Gamma_S^*)\Gamma_L s_{12} s_{21}}{D} \tag{4.33}$$

FIGURE 4.26 *Generalized S-parameters are derived from the Z_0-based parameters for a new set of source and load, Γ_S and Γ_L. Conversion requires complex computations and we generally use an RF circuit simulator for the task.*

$$s'_{21} = \frac{A_1^*}{A_1} \frac{s_{21}\left(1-|\Gamma_L|^2\right)}{D} \tag{4.34}$$

$$s'_{12} = \frac{A_2^*}{A_2} \frac{s_{12}\left(1-|\Gamma_S|^2\right)}{D} \tag{4.35}$$

$$s'_{22} = \frac{A_2^*}{A_2} \frac{(1-\Gamma_S s_{11})(s_{22}-\Gamma_L^*)\Gamma_S s_{12} s_{21}}{D} \tag{4.36}$$

where

$$\Gamma_S = \frac{Z_S - Z_0}{Z_S + Z_0}$$

$$\Gamma_L = \frac{Z_L - Z_0}{Z_L + Z_0}$$

$$A_1 = \frac{1-\Gamma_S^*}{|1-\Gamma_S|}\sqrt{1-|\Gamma_S|^2}$$

$$A_2 = \frac{1-\Gamma_L^*}{|1-\Gamma_L|}\sqrt{1-|\Gamma_L|^2}$$

$$D = \left((1-\Gamma_S s_{11})(1-\Gamma_L s_{22}) - \Gamma_S \Gamma_L s_{21} s_{12}\right)$$

Note that for the special case in which the source and load remain Z_0 but the reference planes of the terminations are shifted by θ_S and θ_L degrees, only the angles of the original S-parameters are shifted in the following forms:

$$\angle s'_{11} = \angle s_{11} - 2\theta_S \qquad \angle s'_{12} = \angle s_{12} - \theta_S - \theta_L$$
$$\angle s'_{21} = \angle s_{21} - \theta_S - \theta_L \qquad \angle s'_{22} = \angle s_{22} - 2\theta_L$$

For example, if we add a 45° length of transmission line to the input side and a 30° line to the output, the phase angle of the Z_0-based s_{11} shifts 90°, the angles of s_{21} and s_{12} shift 75°, and s_{22} moves 60°. Negative signs indicate clockwise phase angle rotations.

4.18 Mixed-mode S-parameters

While RF and microwave engineers working with single-ended circuits have been using *S*-parameters through several decades, balanced circuit designers have not been able to take advantage of the techniques. We mentioned in Section 2.19.2 that the lack of component characterization presented a stumbling block, since there was no convenient way to measure balanced circuits and components directly with a single-ended network analyzer. The introduction of a new line of test equipment [13] now offers designers direct *S*-parameter measurements of balanced circuits and components, and the development of a new design methodology [14, 15] provides ways to effectively use the measured data. Since signal processing in differential and common-mode is distinctly different from signal propagation in standard single-ended circuits, scattering parameter definitions needed to be extended to differential, common, and mixed-mode signal propagation. Simultaneous propagation of both differential and common-modes is referred to as mixed-mode propagation. Specific circuit examples benefiting from the new design methods are differential low-noise amplifiers, 180° and 0° splitters/combiners, four-port couplers, RF transformers, and baluns.[7]

An RF differential circuit (Figure 4.27) has four signal terminals, arranged as a set of two terminal pairs that may be considered as balanced input and output ports. The circuit may be characterized in single-ended mode by creating a 4 × 4 S-matrix. Since the single ended network analyzer has only two-ports, we can only characterize two-ports at a time while terminating the remaining two-ports with Z_0. Switching the ports around, eventually we can fill in all 16 members of the 4 × 4 matrix. Needless to say, every time we change connections, random errors may be introduced to our results.

An alternative approach is to employ two baluns, one at the input and another at the output side, to convert the single-ended ports of the

7. A transformer to convert balanced to unbalanced operation (see Chapter 7).

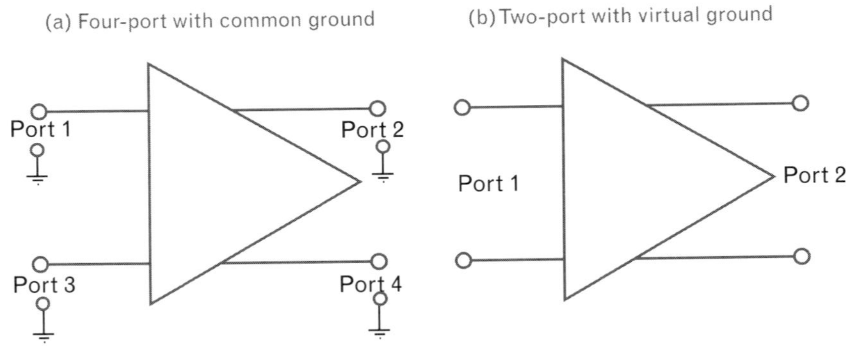

FIGURE 4.27 *A differential RF circuit may be characterized as (a) a single-ended four-port or (b) a differential two-port. Conventional single-ended network analyzers can only apply signal to one unbalanced port at the time and therefore cannot properly characterize a differential circuit.*

network analyzer to balanced ports. Calibration problems associated with baluns can lead to errors [16] and limitations as follows:

- The reference plane of the measurement cannot be defined because there are no calibration standards available for balanced operation. Therefore, it is not a recommended procedure.
- The balun cannot be characterized in the balanced mode. Total losses of two back-to-back connected baluns may be measured with the two-port analyzer, but phase balance and reflection coefficients of the balanced ports cannot be accurately determined.
- Baluns do not pass common-mode signals, and therefore, they cannot be used to evaluate the full mixed-mode operation.

Differential-mode port voltage is defined as the voltage difference across terminals of the port (Figure 4.28), while common-mode port voltage is the voltage sum across the terminal pairs divided by two. In a similar

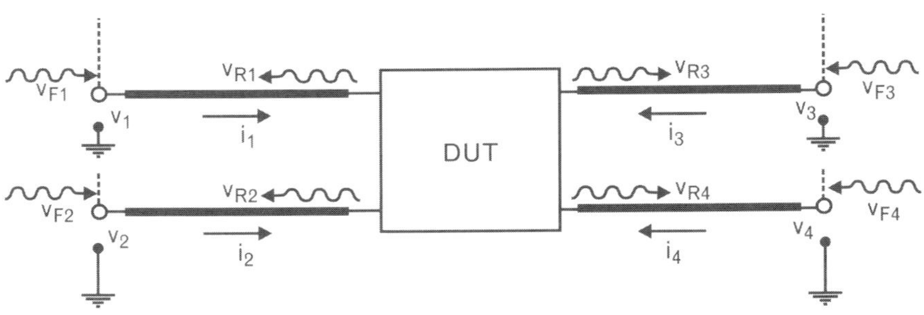

FIGURE 4.28 *Incident and reflected voltages and currents of a four-terminal circuit travel in similar ways to the two-port example covered earlier. The two sets of input and output terminals here may be treated as two balanced ports.*

manner the differential-mode port current is the terminal pair current difference divided by two, while the common-mode port current is the sum of the terminal pair currents. Note that the differential current does not require ground as the return flow conductor, unlike the common mode current, which requires a good quality RF ground.

$$v_d \equiv v_1 - v_2$$
$$v_c \equiv \frac{1}{2}(v_1 + v_2)$$
$$i_d \equiv \frac{1}{2}(i_1 - i_2)$$
$$i_c \equiv (i_1 + i_2)$$

Differential and common-mode impedances at the input side can be expressed in terms of their respective traveling-wave voltages and currents.

$$Z_d \equiv \frac{v_{Fd}}{i_{Fd}}$$
$$Z_c \equiv \frac{v_{Fc}}{i_{Fc}}$$

In Figure 4.28 the subscript F again represents the forward propagation (into the circuit) of voltage and current, and subscript R represents the reverse propagation (away from the circuit). Common-mode signal sources are defined as equal magnitude in-phase voltages and currents applied to each of the input terminal pairs. As we can see, differential and common-mode input signals propagate into different impedances.

Alternatively, if we have access to a mixed-mode network analyzer, another 4 × 4 matrix can be measured having four submatrices, corresponding to four different modes of operation:

1. Differential input drive working into differential load;
2. Single-ended drive working into single-ended load;
3. Single-ended drive working into differential load;
4. Differential input drive working into single-ended load.

The last two combinations are the mixed-mode forms. Figure 4.29 compares the two different 4 × 4 matrices obtained by single-ended and mixed mode S-parameter measurements.

The mixed-mode S-parameter index has four characters, two letters followed by two numbers. The letters refer to specific modes at the input

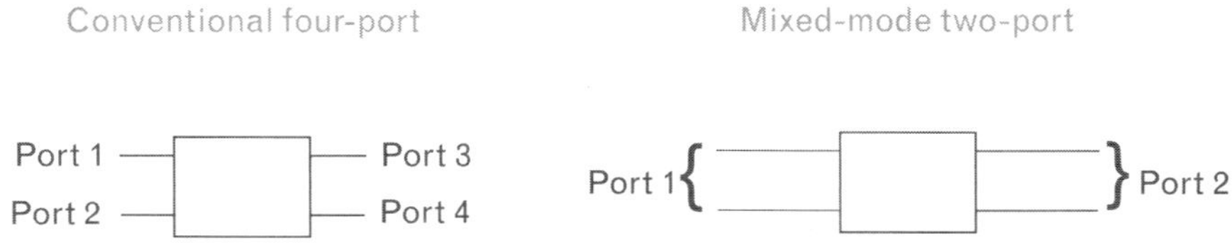

$$\begin{bmatrix} b_1 \\ b_2 \\ b_3 \\ b_4 \end{bmatrix} = \begin{bmatrix} s_{11} & s_{12} & s_{13} & s_{14} \\ s_{21} & s_{22} & s_{23} & s_{24} \\ s_{31} & s_{32} & s_{33} & s_{34} \\ s_{41} & s_{42} & s_{43} & s_{44} \end{bmatrix} \bullet \begin{bmatrix} a_1 \\ a_2 \\ a_3 \\ a_4 \end{bmatrix}$$

$$\begin{bmatrix} b_{d1} \\ b_{d2} \\ b_{c1} \\ b_{c2} \end{bmatrix} = \begin{bmatrix} s_{dd11} & s_{dd12} & s_{dc11} & s_{dc12} \\ s_{dd21} & s_{dd22} & s_{dc21} & s_{dc22} \\ s_{cd11} & s_{cd12} & s_{cc11} & s_{cc12} \\ s_{cd21} & s_{cd22} & s_{cc21} & s_{cc22} \end{bmatrix} \bullet \begin{bmatrix} a_{d1} \\ a_{d2} \\ a_1 \\ a_2 \end{bmatrix}$$

FIGURE 4.29 *Conventional single-ended and mixed-mode characterizations of a balanced amplifier provide different forms of 4 × 4 S-parameter matrices.*

and output, while the numbers have the same functions as in the conventional *S*-parameters, as shown in Table 4.10.

For example, s_{dd21} is the forward transmission coefficient with differential source and differential load, while s_{cc11} is the input reflection coefficient with single-ended source and single-ended load.

Differential drive and termination may be applied two ways, either with a real RF ground or with a floating (virtual) ground, as shown in Figure 4.30. Since the ground currents of the two differential channels cancel each other, there is no need for RF ground—a major advantage of balanced operation.

Single-ended forward wave terms v_{F1} and v_{F2} are used to solve for the differential forward wave, while the reverse single-ended wave terms v_{R1} through v_{R4} are used to solve for differential reverse waves.

Partitioning the 4 × 4 mixed-mode S-matrix into four 2 × 2 submatrices allows us to separate the parameters of different modes from each other. The 2 × 2 S_{dd} portion can be used to design differential circuits, using the same methodology we use for single-ended operation and the conventional single-ended 2 × 2 *S*-matrix. If the differential channels are in perfect balance, the upper right and lower left 2 × 2 submatrices (S_{cd} and S_{dc}) are zero. In a differential amplifier all of the differential input signal would be processed at the differential output. There would be no loss due to mode conversion of the differential input signal to a common-mode output signal from either mixed-mode port. In addition, any common-mode signals coupled to the amplifier's input are rejected in processing at the output as a differential mode signal, ($s_{dc21} = 0$). This is also known as *common-mode rejection,* which is a specification parameter on low-frequency differential amplifiers. With the use of mixed-mode parameters this low-frequency

TABLE 4.10 COMPARISON OF CONVENTIONAL AND MIXED-MODE *S*-PARAMETER NOTATIONS

CONVENTIONAL *S*-PARAMETERS	MIXED-MODE *S*-PARAMETERS
$s_{\text{Port out, Port in}}$	$s_{\text{Mode out, Mode in, Port out, Port in}}$

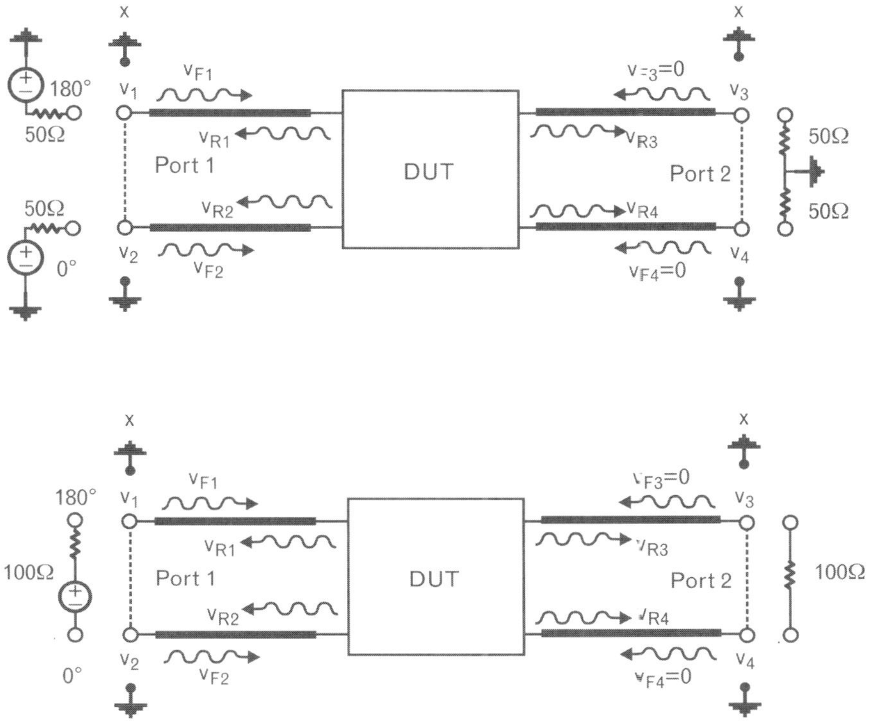

FIGURE 4.30
The two circuits shown operate exactly the same way, even though the lower one does not require RF grounding.

specification parameter can be measured and applied to RF and microwave frequency applications. Figure 4.31 illustrates the functions of the four partitions.

Obtaining ideal zero-value cross-mode submatrices (S_{cd} and S_{dc}) in reality is impractical. However, with the use of mixed-mode *S*-parameters, a design can be evaluated to determine what the cross-mode values are and their impact on the desired performance. If the cross-mode terms are very small compared to S_{dd} and S_{cc} terms, the cross-mode terms may be neglected when using the mixed-mode *S*-parameters. This means that for a differential amplifier, gain, stability, and impedance matching can be evaluated using the four S_{dd} parameters only. These principals can be extended to more complex applications, such as a balun where the desired signal processing is one of the cross-mode propagations.

4.18.1 Standard S-parameter to mixed-mode S-parameter transformations

As we explore this new technology of mixed-mode *S*-parameters the question of a transformation between these new parameters and standard *S*-parameters comes up. Does it exist in both directions and what are the limits and issues? The answer is, yes, there is a very simple transformation

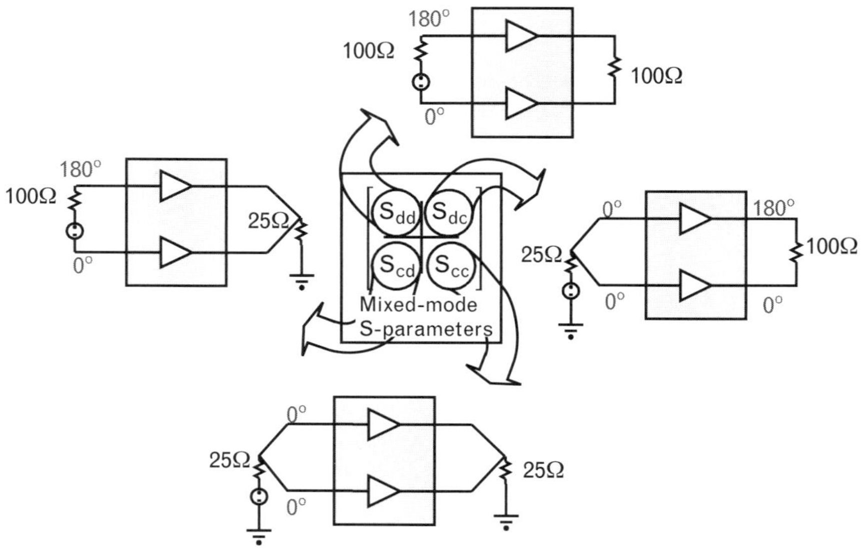

FIGURE 4.31
The four 2 × 2 submatrices of the 4 × 4 mixed-mode S-matrix refer to four different modes.

that can be applied in both directions. What this means is a standard four-port *S*-parameter measurement can be made on a differential circuit and transformed to relevant mixed-mode parameters. It also means a mixed-mode measurement can be applied on a standard *S*-parameter application and transformed into relevant single-ended four-port *S*-parameters. Is there an advantage or significant limitation of one set of *S*-parameters over the other? We look at that issue after showing the transformations.

The matrix relationship between mixed-mode *S*-parameters (S^{mm}) and standard *S*-parameters (S^{std}) is shown in (4.37) through (4.39).

$$\begin{bmatrix} S_{dd} & S_{dc} \\ S_{cd} & S_{cc} \end{bmatrix}^{mm} \Leftrightarrow \begin{bmatrix} s_{11} & s_{12} & s_{13} & s_{14} \\ s_{21} & s_{22} & s_{23} & s_{23} \\ s_{31} & s_{32} & s_{33} & s_{34} \\ s_{41} & s_{42} & s_{43} & s_{44} \end{bmatrix}^{std} \tag{4.37}$$

$$S^{mm} = MS^{std}M^{-1}$$

where M is defined as

$$M = \frac{1}{\sqrt{2}} \begin{bmatrix} 1 & -1 & 0 & 0 \\ 0 & 0 & 1 & -1 \\ 1 & 1 & 0 & 0 \\ 0 & 0 & 1 & 1 \end{bmatrix} \tag{4.38}$$

that implies

$$M\left(M^{*}\right)^{T} = 1 \tag{4.39}$$

An expression for S^{mm} in terms of the standard four-port S-parameters can be used to identify limitations of the transformation.

$$S^{mm} = \frac{1}{2}\begin{bmatrix} \begin{pmatrix} s_{11} - s_{12} \\ -s_{21} + s_{22} \end{pmatrix} & \begin{pmatrix} s_{13} - s_{14} \\ -s_{23} + s_{24} \end{pmatrix} & \begin{pmatrix} s_{11} + s_{12} \\ -s_{21} - s_{22} \end{pmatrix} & \begin{pmatrix} s_{13} + s_{14} \\ -s_{23} - s_{24} \end{pmatrix} \\ \begin{pmatrix} s_{31} - s_{32} \\ -s_{41} + s_{42} \end{pmatrix} & \begin{pmatrix} s_{33} - s_{34} \\ -s_{43} + s_{44} \end{pmatrix} & \begin{pmatrix} s_{31} + s_{32} \\ -s_{41} - s_{42} \end{pmatrix} & \begin{pmatrix} s_{33} + s_{34} \\ -s_{43} - s_{44} \end{pmatrix} \\ \begin{pmatrix} s_{11} - s_{12} \\ +s_{21} - s_{22} \end{pmatrix} & \begin{pmatrix} s_{13} - s_{14} \\ +s_{23} - s_{24} \end{pmatrix} & \begin{pmatrix} s_{11} + s_{12} \\ +s_{21} + s_{22} \end{pmatrix} & \begin{pmatrix} s_{13} + s_{14} \\ +s_{23} + s_{24} \end{pmatrix} \\ \begin{pmatrix} s_{31} - s_{32} \\ +s_{41} - s_{42} \end{pmatrix} & \begin{pmatrix} s_{33} - s_{34} \\ +s_{43} - s_{44} \end{pmatrix} & \begin{pmatrix} s_{31} + s_{32} \\ +s_{41} + s_{42} \end{pmatrix} & \begin{pmatrix} s_{33} + s_{34} \\ +s_{43} + s_{44} \end{pmatrix} \end{bmatrix} \tag{4.40}$$

Each of the mixed-mode S-parameter terms is a summation of four standard S-parameters. If the magnitudes of the various four-port parameters making up a mixed-mode parameter are all similar, the tolerance effects are not significant. However, if the summation consists of two large approximately equal values and two smaller values of opposite signs, the accuracy can be significantly compromised [17]. This is a realistic limitation of the transformation. The same issue exists for the reverse transformation—conversion of mixed-mode to standard S-parameters.

4.18.2 Illustrative example: characterization of a SAW filter

To illustrate the value of mixed-mode S-parameter measurements, we look at the performance of a SAW filter designed for the 110-MHz passband. Differential input and output impedances of the filter are 700Ω.

There are two ways to characterize this filter with S-parameter measurements: one using a single-ended network analyzer and the other is with a mixed mode s-parameter test set. The two sets of port numbering of four-terminal components shown in Figure 4.27 also apply to the filter measurements.

Since the filter has differential input and output impedances of 700Ω, the single-ended measurements were taken with 350-Ω source and load terminations. Plotted results of the 16 four-port S-parameter measurements are shown in Figure 4.32.

The reflection coefficients at all four-ports show that the real part of the input impedance is lower than the reference impedance (the center of the Smith chart) and the reactive part is quite capacitive. The various transmission coefficients are measured between Ports 1-2, 1-4, 3-2, and 3-4,

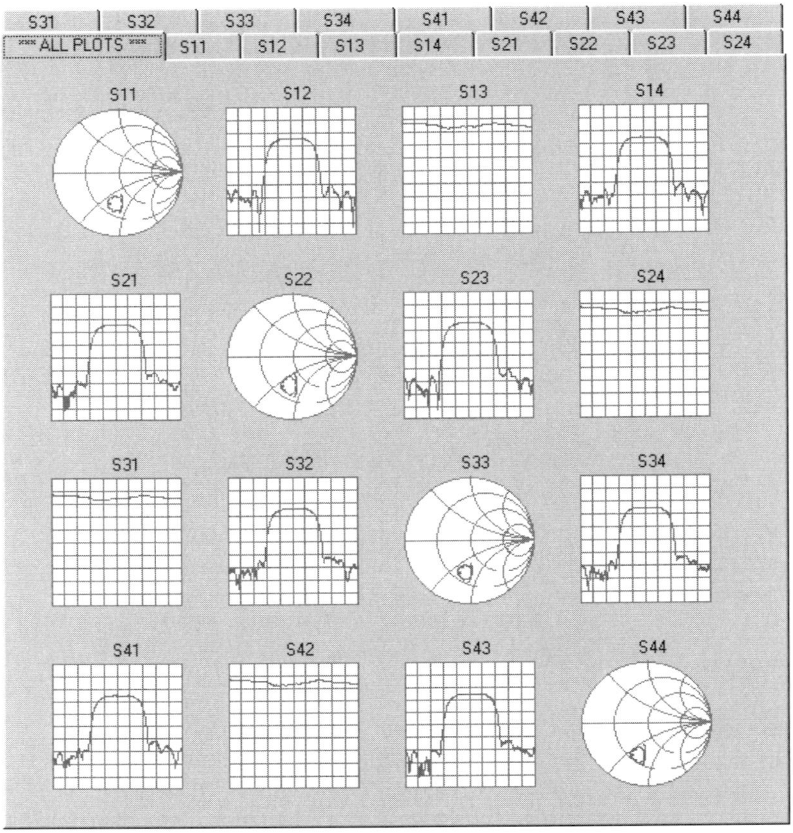

FIGURE 4.32 *Single-ended four-port S-parameter data of the SAW filter referenced to a 350-Ω characteristic impedance level. Reflection coefficients show excess capacitance at all four terminals (s_{11}, s_{22}, s_{33}, s_{44}). Frequency range is 107.8 MHz to 111.8 MHz. Vertical scale of the rectangular plots is 5 dB/division. (Courtesy of Agilent Technologies.)*

averaging around 14-dB loss at the center of the passband. It is clearly hard to tell what the actual loss and port impedances are when the filter is used in differential mode.

Measuring the filter with the mixed mode *S*-parameter type of network analyzer now shows the true performance of the filter in a differential mode. Figure 4.33 shows the four 2×2 submatrices of the 4×4 mixed-mode *S*-parameter set. The 2×2 matrix in the upper left corner tells us how the filter works with differential drive and differential termination. At the band center, the loss of the filter is now 9 dB and the input and output impedances are very near to the center of the Smith chart, which is referenced to 700Ω.

The mixed-mode transmission (upper right and lower left 2×2 submatrices) parameters are low values indicating the balance between the two sides of the filter is quite good. As a result, this filter has very good mixed-mode signal rejection. If the two channels of the filter were not symmetrical, the mixed-mode parameters would be considerably higher. As it stands now for the well-balanced circuit, the ratio of the differential and common

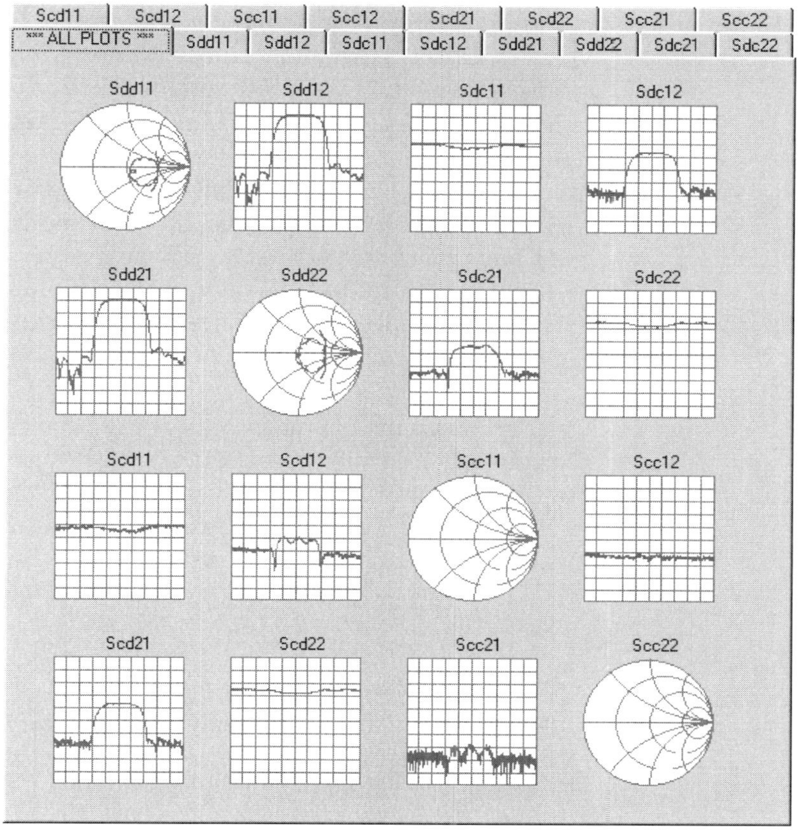

FIGURE 4.33 *Two-port mixed-mode S-parameter data of the SAW filter referenced to 700-Ω differential-mode and 175Ω common-mode characteristic impedance. Frequency range is 107.8 MHz to 111.8 MHz. Vertical scale of the rectangular plots is 10 dB/division. (Courtesy of Agilent Technologies.)*

mode transmissions are well over 50 dB, which indicates excellent *common mode rejection ratio* (CMRR).[8]

4.19 Summary

In spite of the pace of rapidly changing technology, where engineering half-life is estimated to be 3 to 5 years, a graphical design tool developed in the 1930s—the Smith chart—still serves as a useful design aid. Since all of the RF circuit and system simulators as well as the network analyzers can display results on the Smith chart, we can conveniently compare simulated and measured performance. Understanding how the Smith chart works is similar to leaning to ride a bicycle—neither of them can be forgotten. A solid understanding of this valuable tool is a good investment and will be helpful for many years.

8. A parameter that shows how much of an unbalanced signal applied to the differential input, shows up at the differential output.

Scattering parameters have also been with us for some time, and today virtually no design laboratory exists without having at least one network analyzer. Design aids, until recently restricted to single-ended RF/MW circuits, are now also available for mixed-mode differential circuit applications where simultaneous modes of propagation frequently exist. The growing demand for high-density integrated implementations also requires improved isolation along with other specifications. Mixed-mode *S*-parameter technology is important in such an implementation, where each design pass leads to increased product cost. It is expected that the extension of *S*-parameter design techniques to mixed-mode circuits will fundamentally change the way RF multimode circuits are designed and analyzed.

The mixed-mode principles can also be applied within circuit analysis tools such as Agilent's *Advanced Design System* (ADS). Traditional time-domain tools such as MCspice can also utilize mixed-mode tools where frequency-domain transfer function can be used to determine standard or mixed-mode *S*-parameters directly.

References

[1] Smith, P. H., "Transmission Line Calculator," *Electronics,* January 1939.

[2] Smith, P. H., "An Improved Transmission Line Calculator," *Electronics,* January 1944.

[3] Miller, K. S., *Advanced Complex Calculus,* New York: Harper & Brothers, 1960.

[4] Gonzalez, G., *Microwave Transistor Amplifiers: Analysis and Design,* 2nd ed., Englewood Cliffs, NJ: Prentice Hall, 1997.

[5] Analog Instruments, Smith Chart Form ZY-01-N, New Providence, NJ.

[6] Kurokawa, K., "Power Waves and the Scattering Matrix," *IEEE Trans. on Microwave Theory and Techniques,* March 1965.

[7] Hewlett Packard, "S-Parameter Design," Application Note 154, May 1972.

[8] Adam, S. F., *Microwave Theory and Applications,* Los Altos, CA: AMCI, 1992.

[9] Laverghetta, T. S., *Modern Microwave Measurements and Techniques,* Norwood, MA: Artech House, 1988.

[10] Wartenberg, S. A., *RF Measurements of Die and Packages,* Norwood, MA: Artech House, 2002.

[11] Agilent Network Analyzer, Application Note.

[12] Rizzi, P. I., *Microwave Engineering, Passive Circuits,* Englewood Cliffs, NJ: Prentice Hall, 1988.

[13] Cole, B., "Fully Characterize Balanced Devices," *Microwaves and RF,* January 1999.

[14] Bockelman, D. E., and W. R. Eisenstadt, "Combined Differential and Common-Mode Scattering Parameters: Theory and Simulation," *IEEE Trans. on Microwave Theory and Techniques,* July 1995.

[15] Bockelman, D. E., and W. R. Eisenstadt, "Combined Differential and Common-Mode Analysis of Power Splitters and Combiners," *IEEE Trans. on Microwave Theory and Techniques,* November 1995.

[16] Betts, L., "Comparing Differential Measurement Techniques," *Applied Microwaves and Wireless,* November 2001, and also as "Theory and Analysis of Differential Measurement Techniques," in January 2002.

[17] Bockelman, D. E., W. R. Eisenstadt, and R. Stengel, "Accuracy Estimation of Mixed-Mode Scattering Parameter Measurements," *IEEE Trans. on Microwave Theory and Techniques*, January 1999.

Selected Bibliography

Free Smith chart software for MS Windows operating systems: http://www.hta-be.bfh.ch/~dellsper

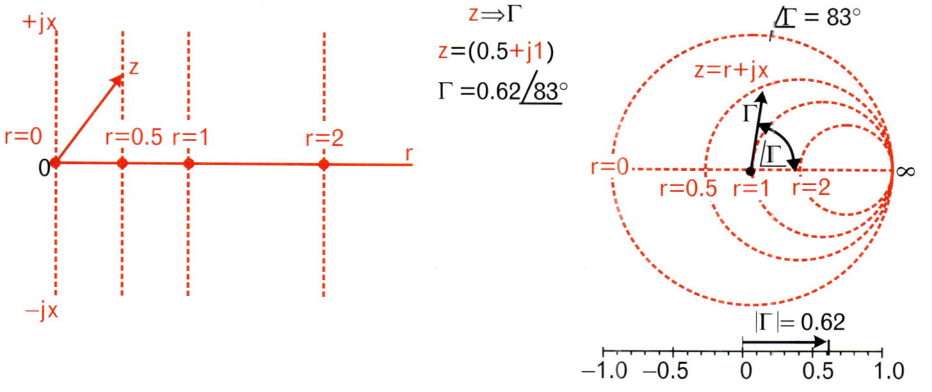

PLATE 1 *Vertical constant-resistance lines from the normalized z-plane are mapped into constant-resistance circles in the Γ-plane. These circles represent half of the impedance Smith chart.*

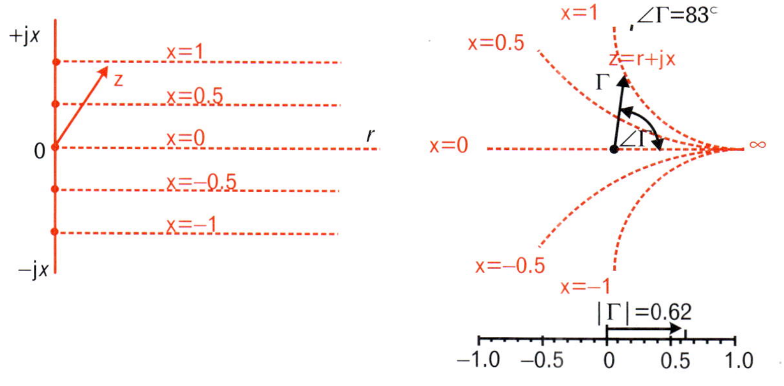

PLATE 2 *Horizontal constant-reactance lines from the normalized z-plane are mapped into constant-reactance circles in the Γ-plane. These circles represent the second half of the impedance Smith chart.*

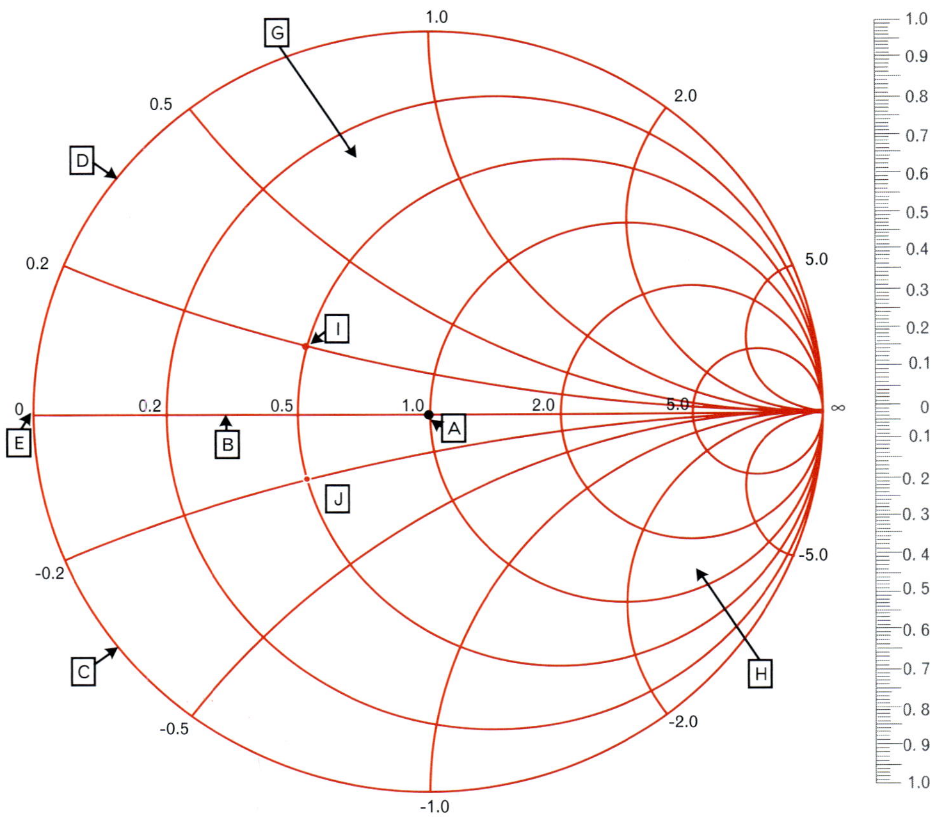

PLATE 3 *The normalized unit-radius impedance Smith chart contains a set of constant-resistance and constant-reactance circular arcs. The circumference represents lossless reactive terminations, leading to 100% reflection. At the center of the chart* $|\Gamma| = 0.0$ *and* $z = 1.0$. *To obtain unnormalized impedance, we need to multiply* z *by* Z_0.

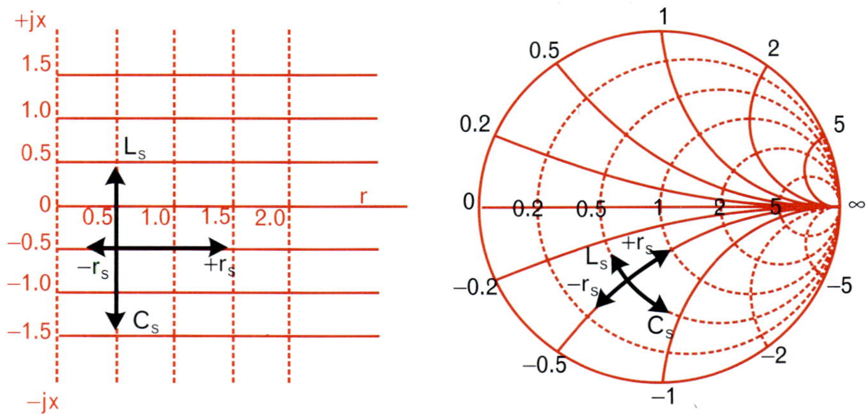

PLATE 4 *Movements of three different lumped elements in the rectangular impedance plane compared to the reflection coefficient plane. Adding a series element in the impedance system always moves toward the direction of infinite impedance (open-circuit) on the chart. The crossing of the Smith chart circles are still at 90° with repect to each other.*

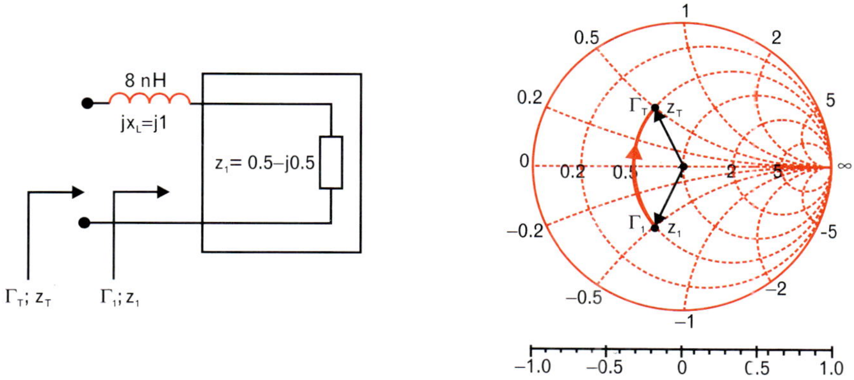

PLATE 5 *For series element addition, the impedance Smith chart conveniently converts the measured reflection coefficient to normalized impedance. Adding the reactance of the series inductor takes us to the sum of the two impedances, z_T, corresponding to the reflection coefficient Γ_T.*

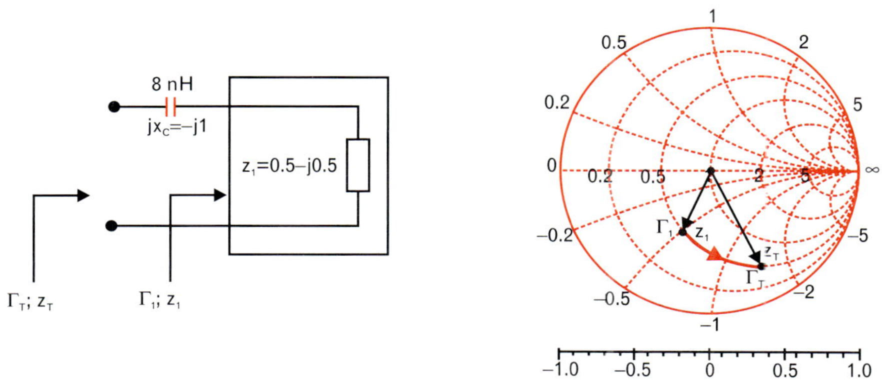

PLATE 6 *Series capacitors also move on the appropriate constant-resistance circle since they do not add resistance to the circuit. Adding $-j1$ normalized capacitive reactance to z_1, takes us to z_T, or Γ_T.*

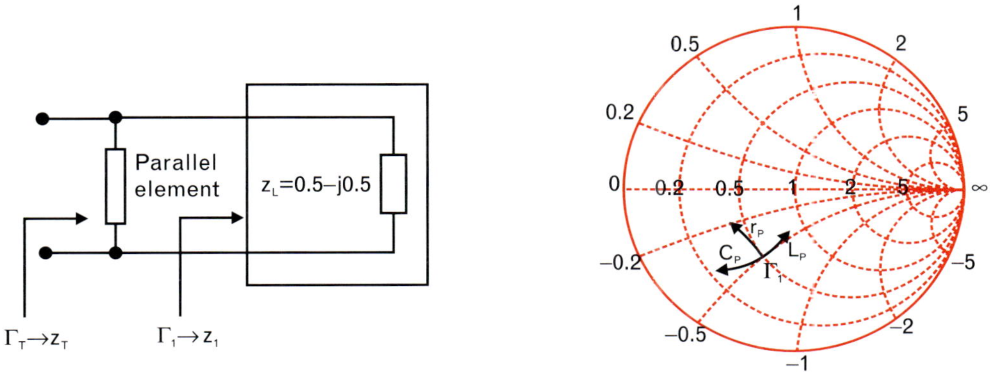

PLATE 7 *On the impedance Smith chart parallel elements added to z_1 do not follow predictable paths. We need to develop a different chart for parallel component manipulations.*

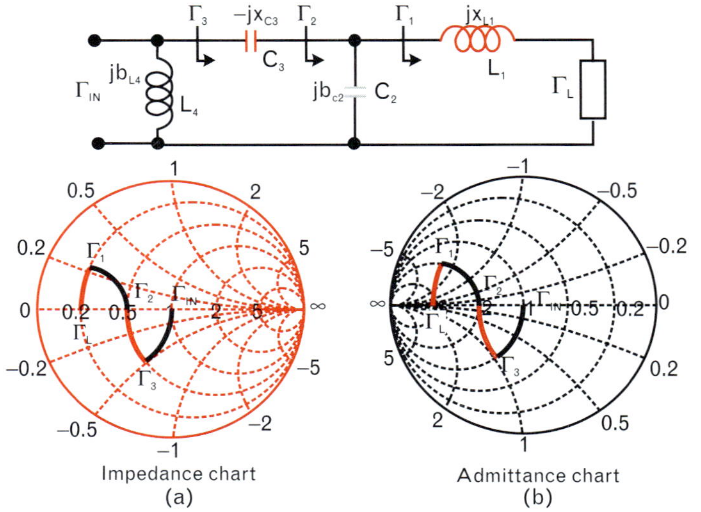

PLATE 8 *Finding the input reflection coefficient of a four-element ladder network graphically requires two separate Smith charts. (a) The impedance chart is used for series elements, while parallel elements are handled on (b) the admittance chart.*

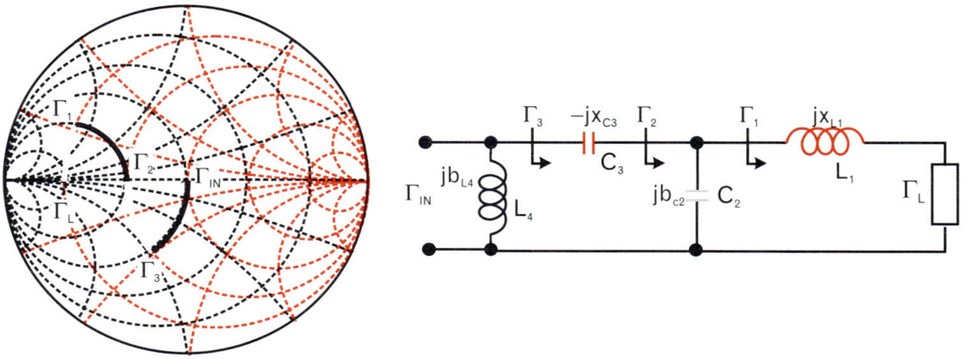

PLATE 9 *Series elements are represented by movements on appropriate impedance circles (constant resistance or constant reactance), while parallel elements are handled on the admittance circles of the immitance Smith chart.*

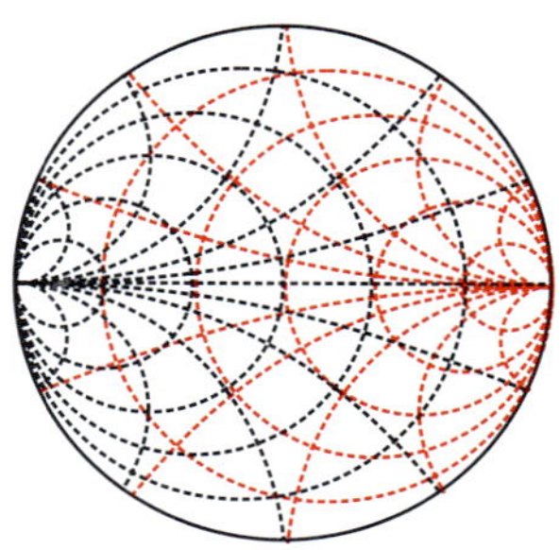

PLATE 10 *Commercial form of the Z-Y Smith chart with normalized impedance and admittance coordinates.*

Element type	Series or parallel	Constant circle	Movement toward
Resistor	Series	x	$r = \infty$
	Parallel	b	$g = \infty$
Inductor	Series	r	$jx = +j\infty$
	Parallel	g	$jb = -j\infty$
Capacitor	Series	r	$jx = -j\infty$
	Parallel	g	$jb = +j\infty$

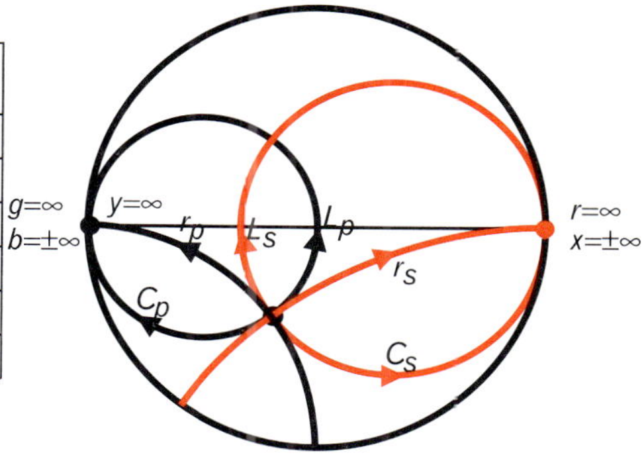

PLATE 11 *Impedance transformation of ideal lumped elements always follows the constant-reactance or constant-resistance circle that intersects the starting impedance. Admittance transformations always occur on the constant-susceptance or constant conductance-circle that intersects the starting point.*

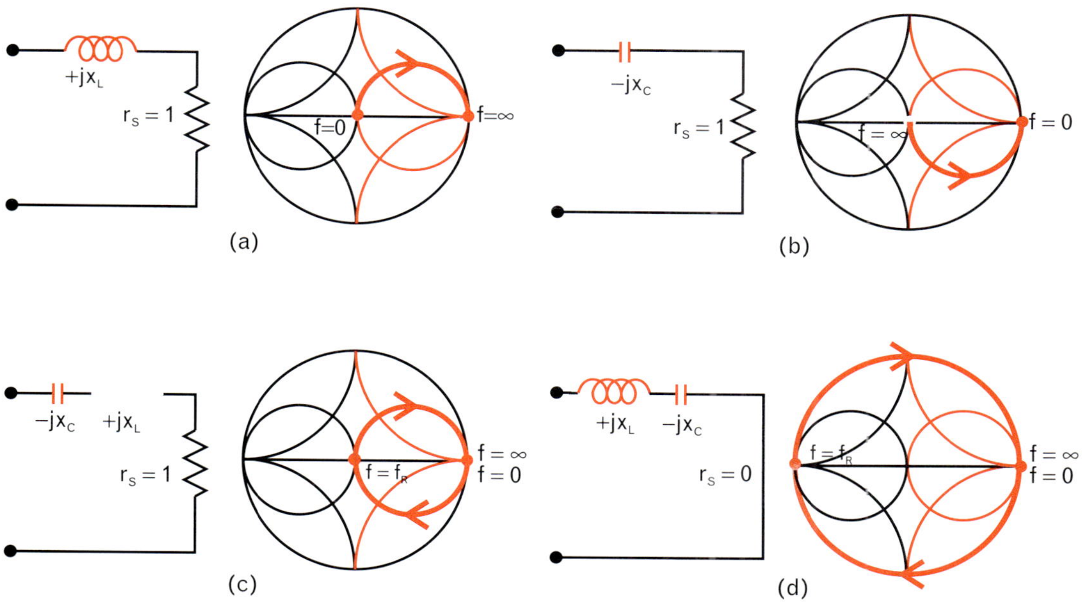

PLATE 12 *Various combinations of series R-L-C circuits and their Smith chart contours as frequency is varied from 0 to ∞. At finite frequencies, circuit (a) is always inductive and (b) is capacitive, and they both stay on the constant resistance circle defined by the series resistance, r_S. Circuits (c) and (d) are capacitive below f_R and become inductive above.*

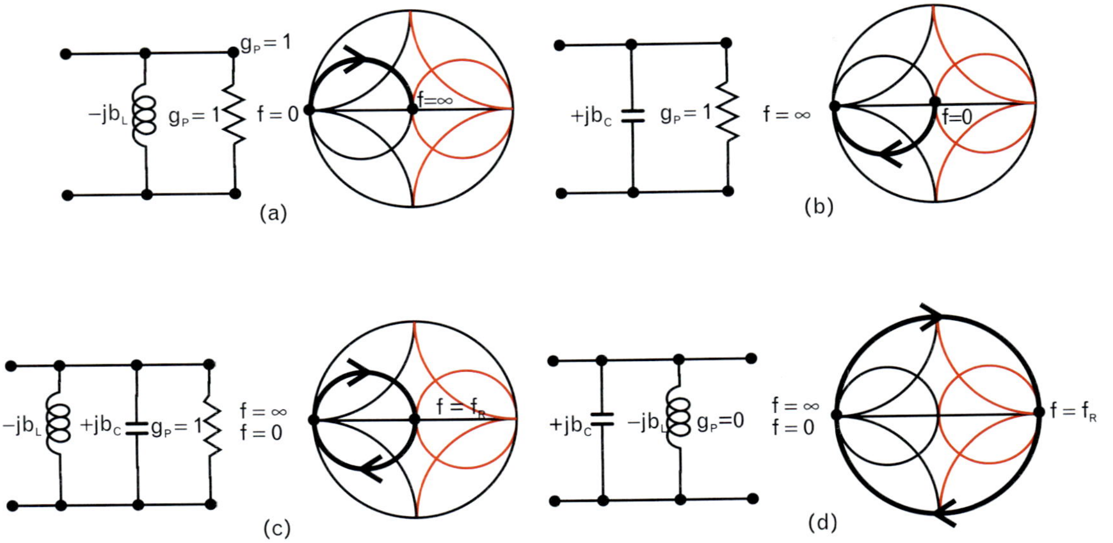

PLATE 13 *Various combinations of parallel R-L-C circuits and their Smith chart contours as frequency is varied from 0 to ∞. Here, circuits (a–c) follow the constant conductance determined by the value of g_P. The two resonant circuits, (c) and (d), are inductive below f_R and capacitive above.*

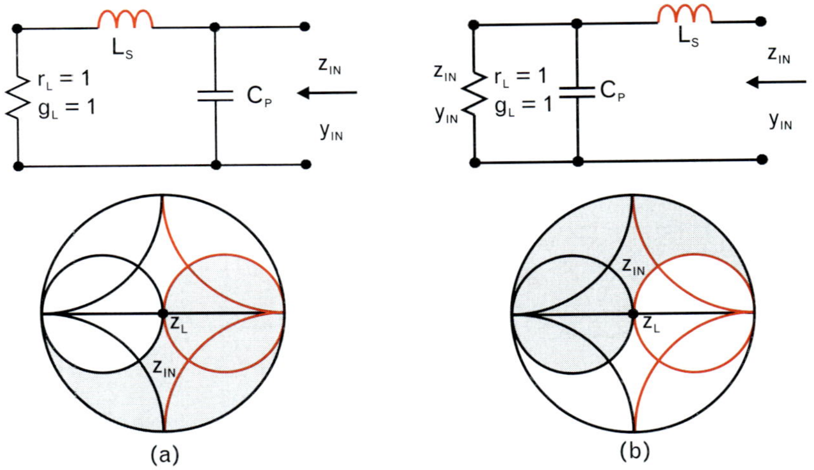

PLATE 14 *Two lossless L-C impedance transformer configurations: (a) if the first reactive element next to z_L is in series, the L-C section transforms to lower conductance [$Re(y_{IN}) < Re(y_L)$]; and (b) When the first reactive element is in parallel with z_L, the transformation is to a lower resistance [$Re(z_{IN}) < Re(z_L)$]. Gray shaded regions show possible input impedances with $r_L = 1$.*

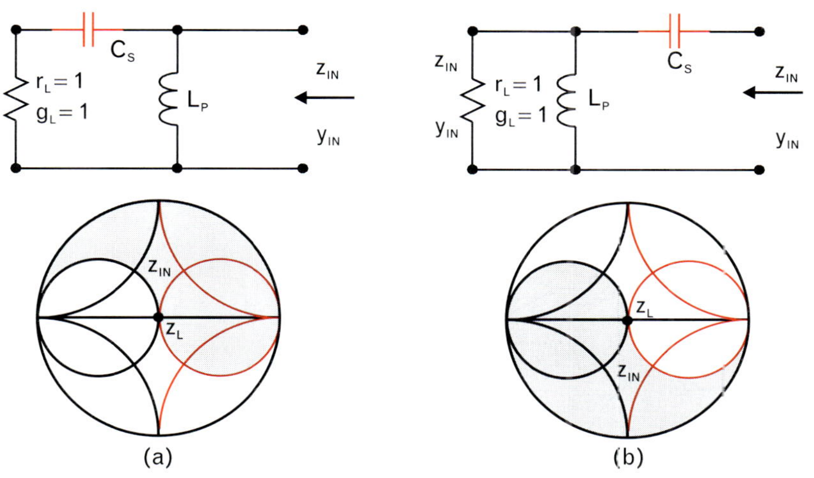

PLATE 15 *Highpass transformer section configurations. (a) If the first element next to z_L is in series, the transformation is to a lower conductance [$Re(y_{IN}) < Re(y_L)$]. (b) If the first element is in parallel with z_L, the transformation is to a lower resistance [$Re(z_{IN}) < Re(z_L)$].*

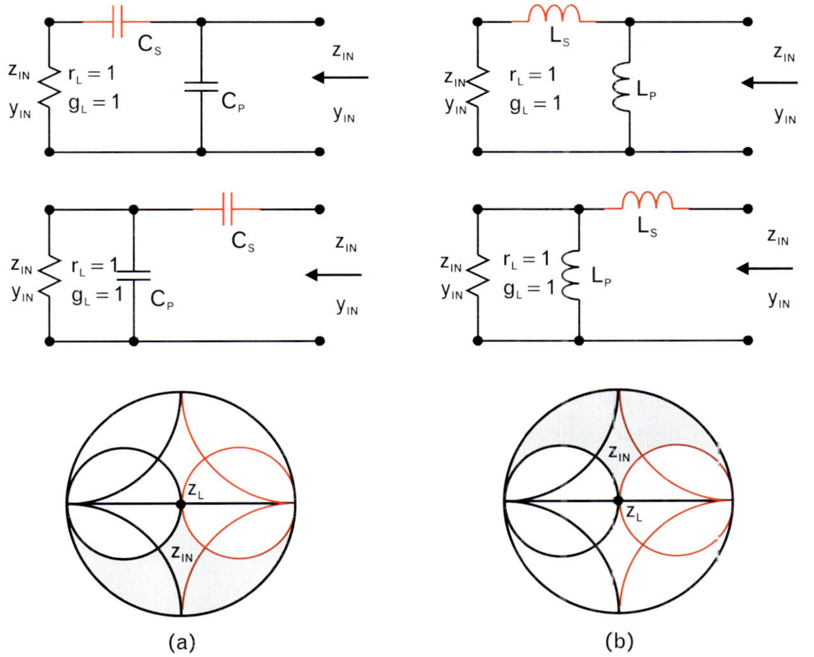

PLATE 16 *Bandpass networks lower the real parts of both z_L and y_L [$Re(z_{IN}) < Re(z_L)$ and $Re(y_{IN}) < Re(y_L)$]. (a) Capacitive networks transform to the lower half of the chart, while (b) inductive sections transform to the upper half.*

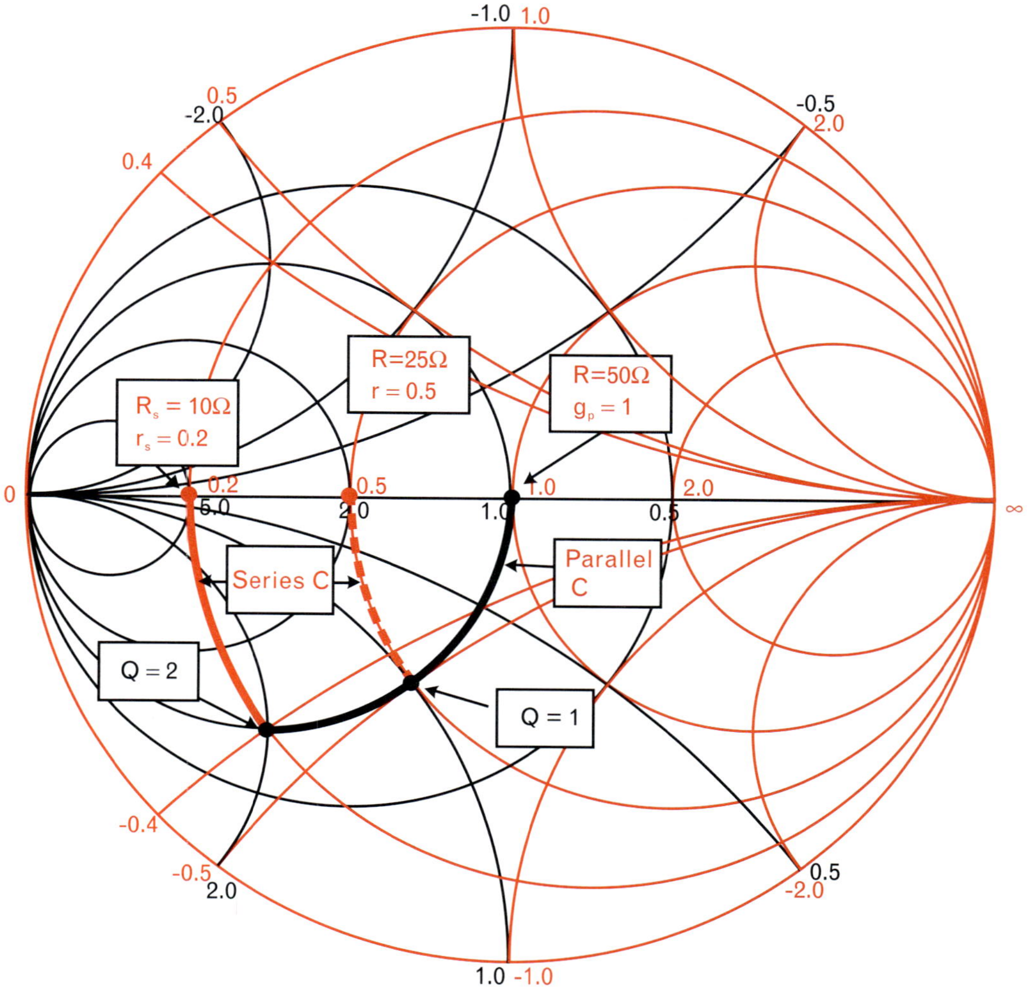

PLATE 17 *Comparing the reflection coefficients of series R-C and parallel R-C circuits at 1 GHz. The same chart may also be used to derive series-to-parallel equivalence with any other series resistor, assuming that $r_S < r_P$. If the required part of the series R-C circuit is 25Ω ($r_s = 0.5$), the required normalized series reactance (shown with dashed arc) is $X_{cs} = -0.5$. The corresponding capacitive susceptance of the parallel R-C circuit, $b_{cp} = 1.0$.*

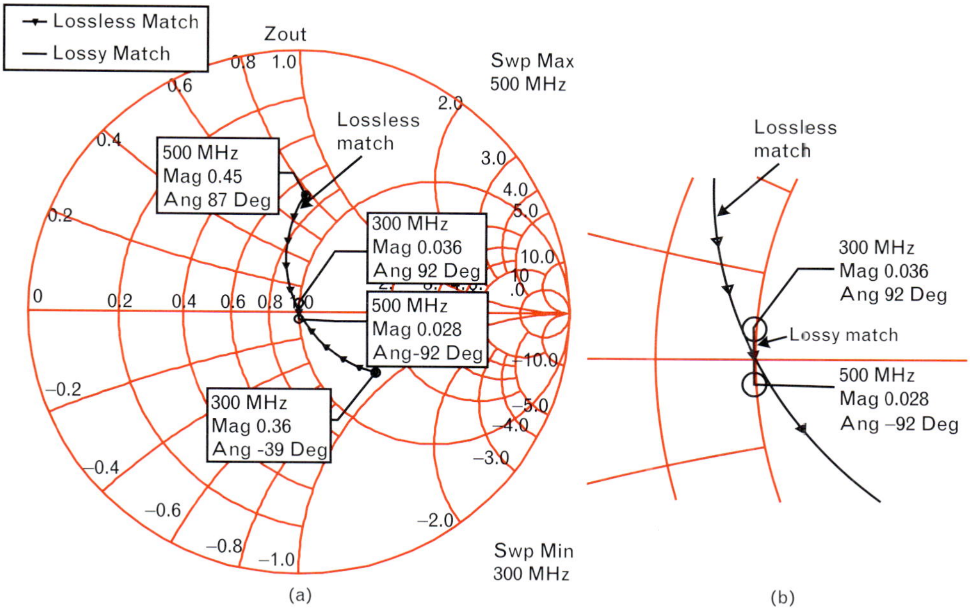

PLATE 18 *Comparing the output reflection coefficients of the two impedance transformer circuits of Figure 4.6 shows better performance of the lossy parallel R-L network. Maximum value of* $|s_{22}|$ *for the R-L circuit for the is only 0.04 through the 300- to 500-MHz frequency range, while the L-C circuit's* $|s_{22}|$ *reaches 0.45 at 500 MHz. Plots shown on (a) the unit-radius chart, and on (b) the zoomed center portion of the chart.*

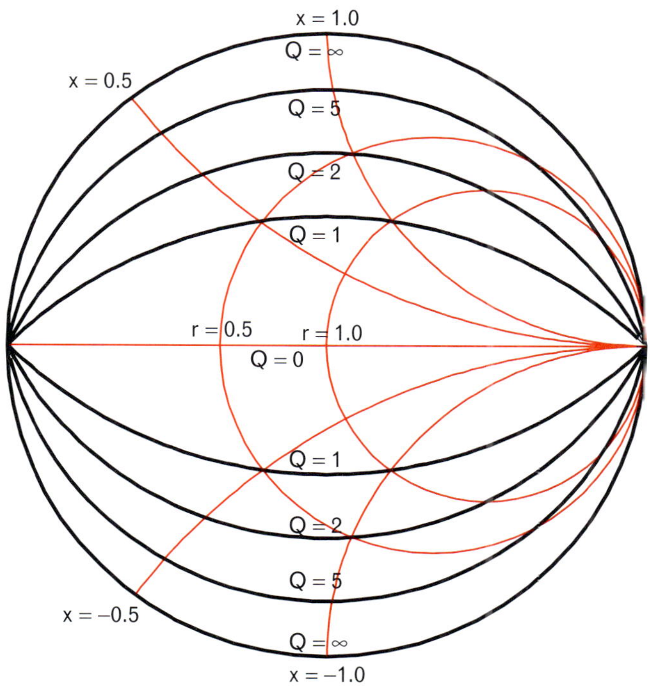

PLATE 19 *Constant Q contours on the Smith chart connecting together identical x/r (and b/g) values.*

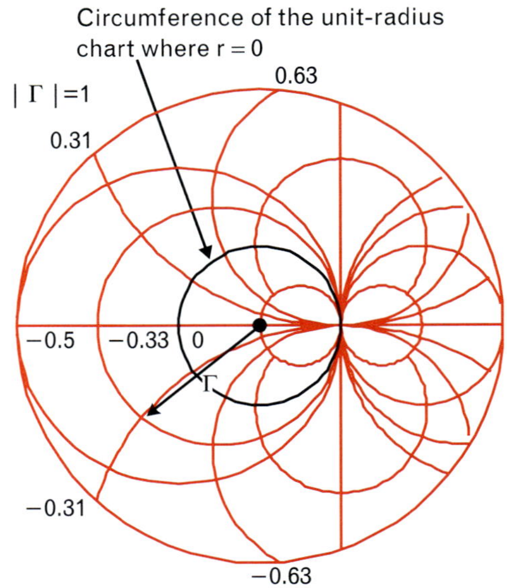

PLATE 20 *The extended impedance Smith chart with radius of 3.0 to be used with circuits having $\rho > 1$. Reflection coefficient vector shown is $\Gamma = 1.8 \angle -150^{\circ}$. The corresponding $z = (-0.33 - j0.31)$ that translates to $Z = (16.5 - j15.5)\Omega$, using $Z_0 = 50\Omega$ reference impedance.*

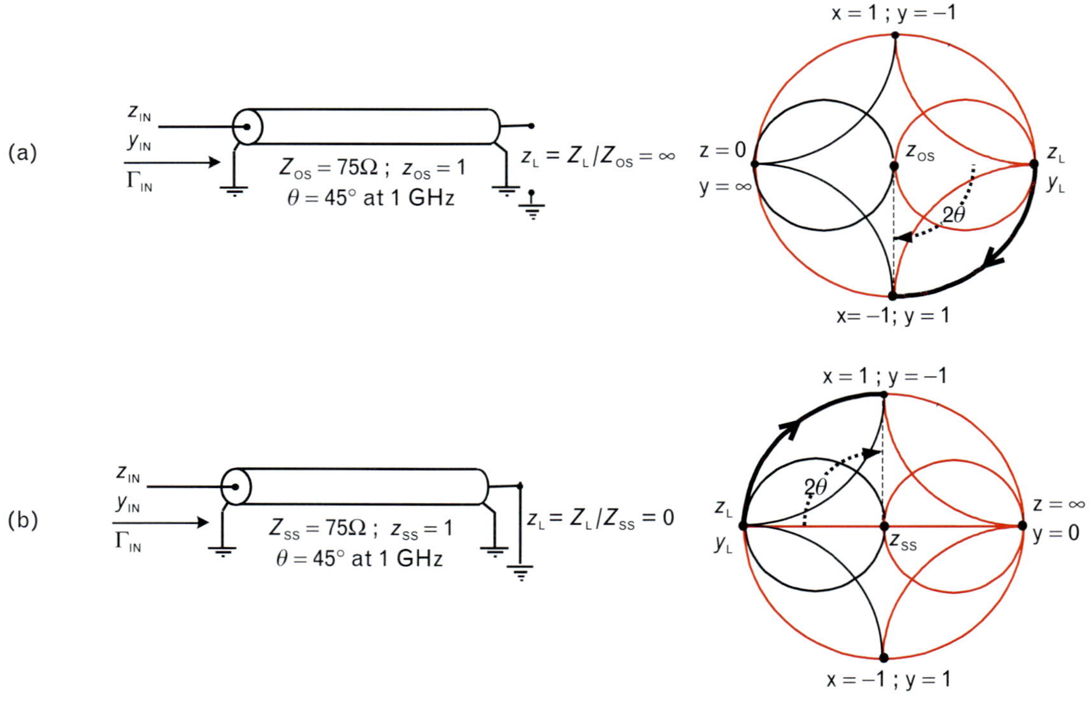

PLATE 21 *(a) Finding the input susceptance of an open-circuited parallel stub requires a clockwise rotation of 2θ from the normalized termination. (b) A short-circuited stub is treated similarly, but the starting point is at short circuit $Z_L = 0$.*

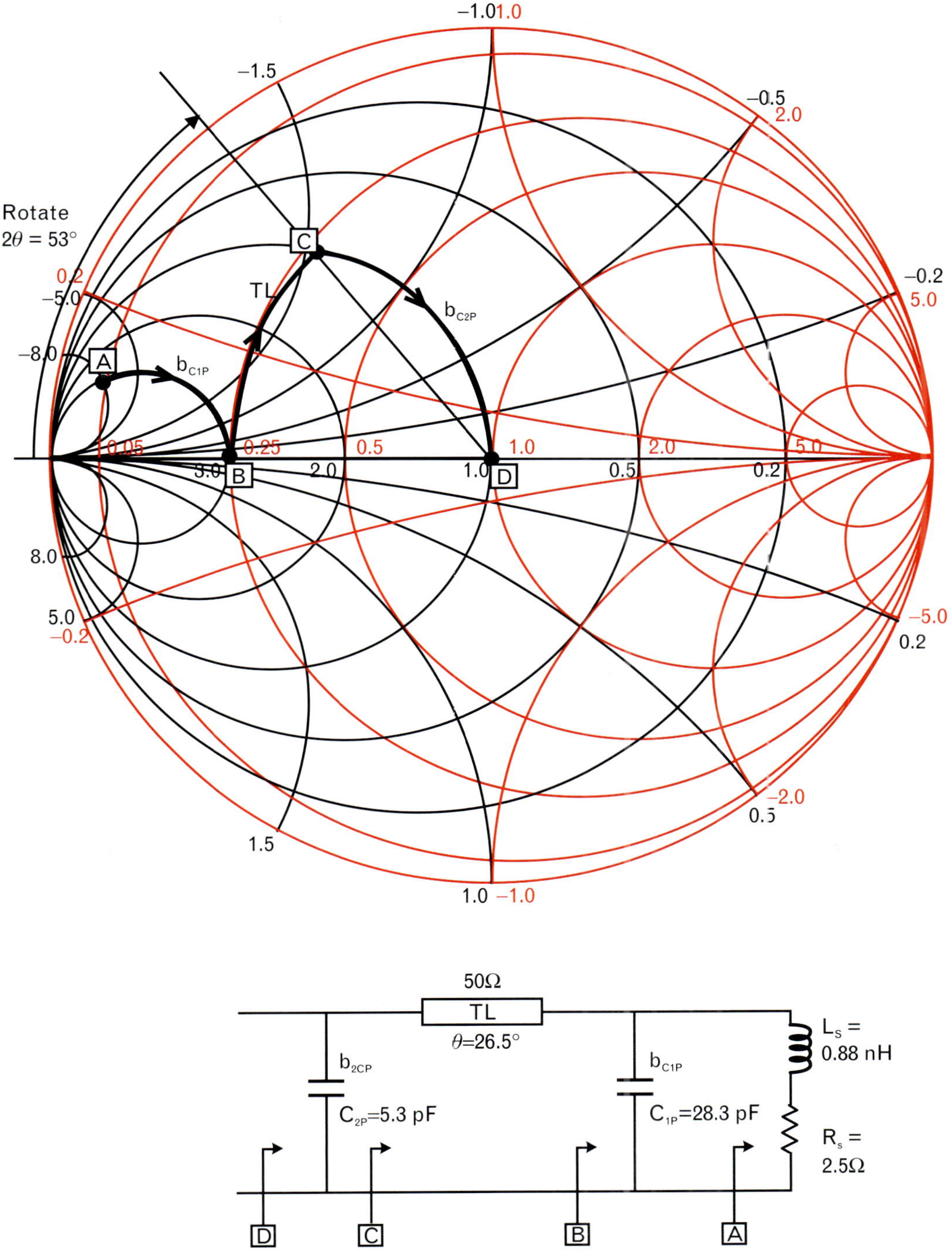

Plate 22 *Impedance transformation with two parallel capacitors and a 50-Ω transmission line. All components are assumed to be lossless. Label "A" indicates the input impedance of the transistor at 900 MHz. Labels "B" and "C" are the intermediate points of the transformation, after adding the 28.3-pF capacitor and the transmission line, respectively. The 5.3-pF capacitor transforms from label "C" to "D."*

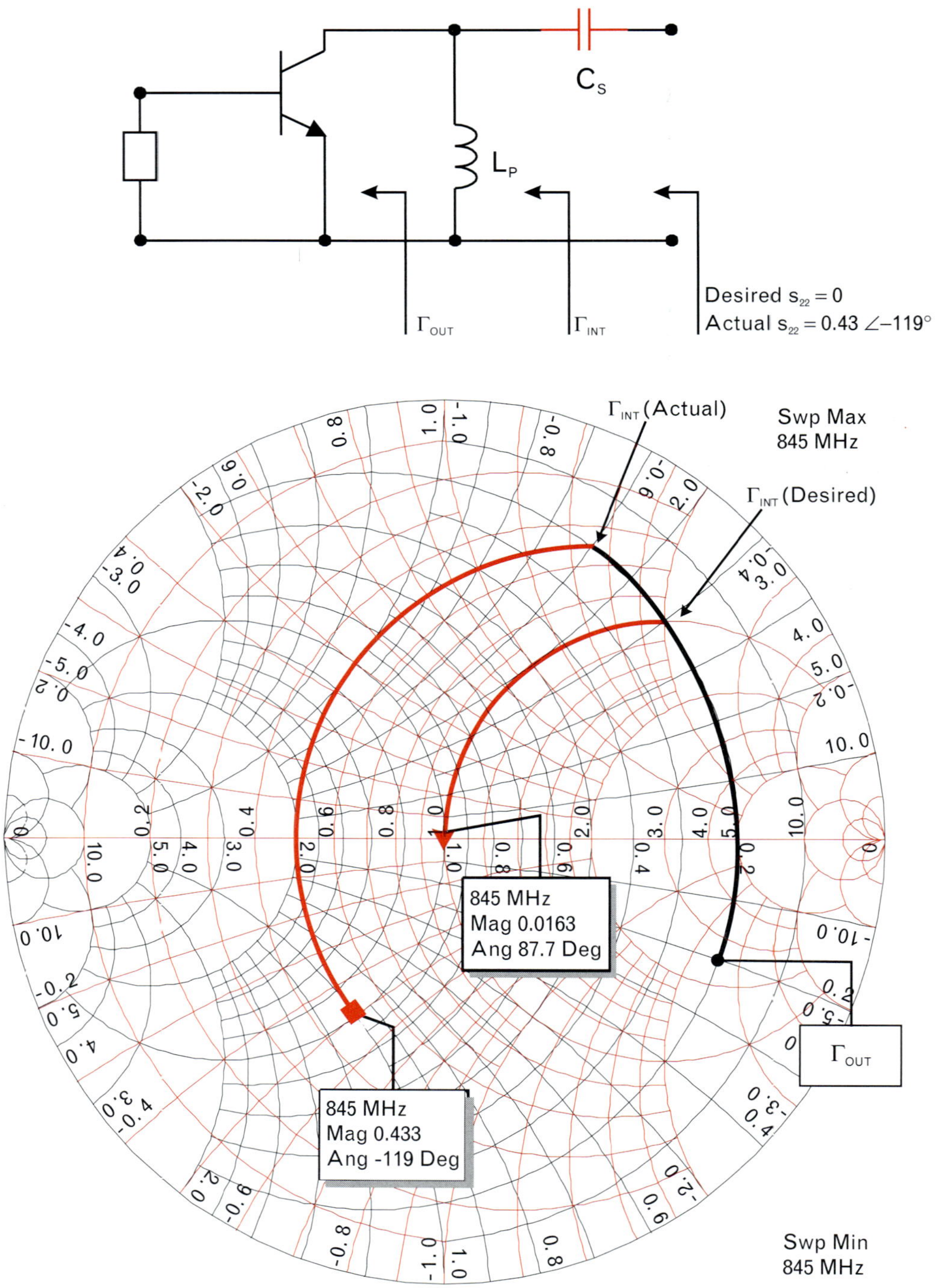

PLATE 23 *One of the two circuit elements' value is incorrect. As a result, Γ_{OUT} of the transistor (marked with solid circle) is not transformed to the center of the chart. The actual s_{22} at 845 MHz is $0.43 \angle -119°$ instead of the center of the chart, where it is supposed to be. The marker at the center is the simulated result with proper element values. Output reflection coefficient of the transistor without the two-element circuit is at $\Gamma_{OUT} = 0.82 \angle -22°$.*

CHAPTER 5

Impedance matching techniques

In this chapter we look at the analytical and graphical techniques to maximize RF power transfer from one system block to another. If all of our components came with port impedances of Z_0, cascading any number of them would not present any problem. While at the system block level components may be well matched, it is certainly not the case for components. For example, the real part of a transistor's port impedances may change from several hundred ohms to tens of ohms through the RF range. To make things worse, the reactive portion may be capacitive at first and by the time we get to the higher frequencies it might be completely dominated by the inductive parasitics. To maintain the highest possible dynamic range, we want to minimize reflection and maximize transmission. Therefore, impedance matching is necessary.

We mentioned in Chapter 2 that optimum power transfer is one of the most important design considerations in RF systems. Although there are exceptions, such as the input circuitry of a low-noise amplifier, in most cases we want to transfer all signal power available from a specific source into the next system block. The signal source may be the receiver antenna, the output port of a mixer, or an amplifier, just to name a few. The load may be represented by the input port of the following stage in the system.

In Chapter 4 we saw that for a lossless reciprocal two-port network, the sum of normalized transmitted power and reflected power must be equal to unity. In dissipative (lossy) networks, a third term enters into the equation, which is the power dissipated in the networks. Our goal in this chapter is to maximize transmission and minimize reflection, and working with lossy components makes the computations much more complex. Since the losses of reasonably high component-Q ($Q > 20$) do not affect the computed impedance matching element values appreciably, we will initially use lossless matching elements. After the matching circuits are determined, we can add the component losses to see their effects on transmitted power. More detailed component modeling is done in Chapter 7.

5.1 The impedance match

A generator has an equivalent circuit consisting of a voltage source in series with an impedance. Alternatively, it can be represented by a current source in parallel with an admittance. Either way, the conditions for the transfer of maximum power, from the generator to a load, follow similar reasoning. The result is the same-for maximum power transfer: The load impedance must be the conjugate impedance of the generator.

In Chapter 2 we compared RF power transfer in a system to water flow between two circular water pipes. For maximum efficiency, the two adjoining pipes with perpendicular endings must have the same diameters. That is equivalent to two resistive terminations facing each other, having equal resistances. If one water pipe is cut at an oblique angle, then the joining pipe must have the same diameter and complementary oblique cut. The electrical equivalent of this second case is where complex terminations are connected together. For maximum power transfer, the real parts of the impedances must still be equal and the two imaginary parts must be complementary types. We would obviously not try to connect two water pipes with *identical oblique cuts* (Figure 5.1) without rotating one first 180° around the center axis. Yet a common mistake of impedance matching is to connect a $Z_S = R + jX$ source to a $Z_L = R + jX$. The net result is the real part of the source, R, sees an effective load of $R + j2X$, leading to power reflection back to the source.

Now that we appreciate mismatch, let us verify next that terminations must have equal real parts, by terminating a voltage source, having internal series resistance R_S, with a resistive load, R_L (see Figure 5.2). The voltage drop across the load is

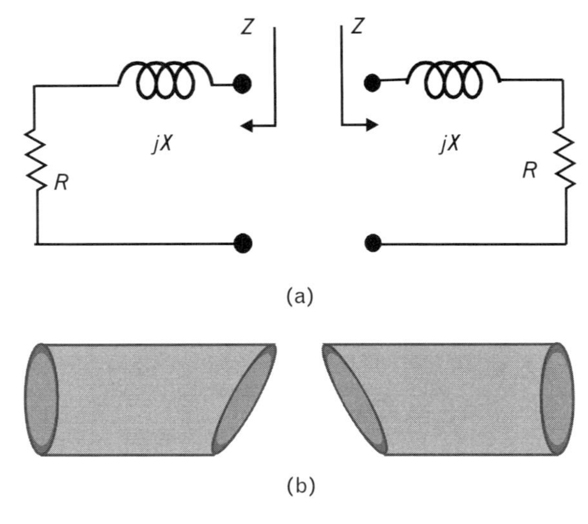

FIGURE 5.1
(a) Interconnecting two identical complex impedances does not lead to maximum power transfer. (b) The equivalent water pipe connection helps reinforce the concept.

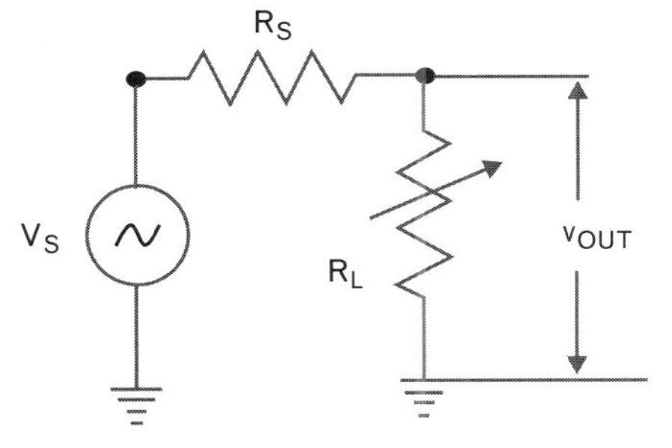

FIGURE 5.2
The connection of a generator to a load to determine the condition for maximum power transfer.

$$v_{OUT} = \frac{R_L}{R_S + R_L}(v_S) \tag{5.1}$$

where v_{OUT} and of v_s refer to rms value. The power delivered into the resistive load is

$$\begin{aligned} P_{OUT} &= \frac{v_{OUT}^{2}}{R_L} = \frac{\left(\frac{R_L}{R_S + R_L}\right)^2}{R_L}(v_S^2) \\ &= \frac{R_L}{(R_S + R_L)^2}(v_S^2) \end{aligned} \tag{5.2}$$

With (5.2) let us calculate and then plot (Figure 5.3) the power delivered to the load for various load resistances. When the load resistor is zero, no power can be delivered to it since it cannot dissipate power. As the resistance R_L increases above zero, the voltage across it increases and with it the power dissipated in it, P_{OUT}, increases. As the resistance R_L further increases, however, the current through it decreases, so eventually the value of P_{OUT} reaches a peak and thereafter decreases. Differentiating the output power with respect to R_L and setting it equal to zero, gives us the optimum load

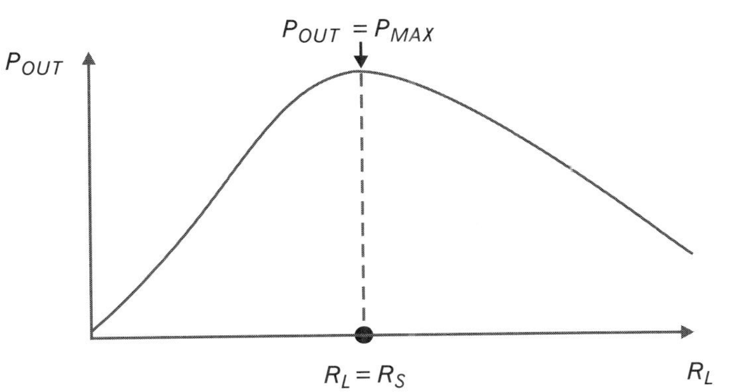

FIGURE 5.3
The derivative of P_{OUT} with respect to R_L goes to zero at the peak where P_{OUT} is maximum and $R_L = R_S$.

resistance for maximum power delivery. This occurs when $R_L = R_S$, as shown below:

$$\frac{\partial P_{OUT}}{\partial R_L} = \frac{(R_S + R_L)^2 - 2R_L(R_S + R_L)}{(R_S + R_L)^2}$$

$$1 - \frac{2R_L}{R_S + R_L} = 0 \tag{5.3}$$

$$R_L = R_S$$

The result of (5.3) confirms that when the load resistance is equal to the source resistance, power dissipation in the load is maximized.

A series reactive element between two equal resistive terminations creates a voltage drop that reduces the voltage across the load. Impedance matching can eliminate or minimize the unwanted reactance through a range of frequencies. The matching process becomes more difficult when the real parts of the terminations are unequal, or the matching network is also required to perform filtering through some undesired frequencies.

It is often not possible to tell the difference between impedance matching and filter networks just by looking at their schematics. Although functional similarities exist between the two categories, there are distinct differences, as follows.

The primarily purpose of a filter is to *reject unwanted signals in stopband(s) without disturbing the passband* performance. Although there are exceptions, most filters are frequency-dependent signal reflectors: they transmit most of the signal in the passband and reject most of it in the stopband. We already saw this earlier that in lossless two-ports the rejection is achieved by having reflection coefficient magnitudes near unity in the stopband. Filters are usually designed to work with resistive terminations and ideal components (i.e., without any losses or parasitics). Some of the more sophisticated filter synthesis programs also apply component Q values (predistortion) to the synthesis, but not component parasitics. Filter design is covered in detail in Chapter 8.

The main purpose of impedance matching is to *match two different terminations through a specified passband*, without having control over the stopband frequencies. Initially, in most cases we may assume that component losses are negligible. However, parasitic effects need to be considered because most of the terminations at RF will have significant reactive portions. While filter design generally follows rigid predefined circuit topologies, impedance matching is more flexible and allows a creative designer to specify the topology. For example, if we need to match a 10-Ω source to a 50-Ω load through a *narrow* frequency range, we can virtually get the same performance by using a highpass or lowpass circuit configuration. While the passband performances are very close for both circuits, the out-of-band performances are significantly different. A lowpass circuit rejects signals at the high end, freely passing them at low frequencies. The highpass network does just the opposite. How we select our matching topologies has a profound effect on the performance outside of the passband.

Before covering the details of impedance matching, let us examine the broadband behavior of reactive components by introducing a new concept, called transmission zero [1]. Although most filter designers work with poles and zeros, we prefer to use the *transmission zeros* because those give us more specific information about the broadband circuit performance. For example, if we have a six-pole bandpass filter, we cannot tell from that information how the frequency response rolls off at the low and high frequencies. If we specify a sixth-order filter with three transmission zeros at dc and three at infinite frequency, we can immediately visualize a symmetrical frequency response in the logarithmic frequency plane. Since we generally do not have access to transfer functions of a proposed impedance matching circuit, an understanding of transmission zero concepts helps us to shape the broadband response of the network.

5.2 Transmission zero definitions

In Chapter 4 we stated that for a lossless, doubly terminated network, the sum of normalized, reflected, and transmitted powers must equal to unity. Recalling (4.16) in terms of *S*-parameters,

$$|s_{11}|^2 + |s_{21}|^2 = 1.0$$

or, in the generalized form of the polynomial functions of reflection, $\rho(s)$, and transmission coefficients, $\tau(s)$, with Z_0 terminations,

$$|\rho(s)|^2 + |\tau(s)|^2 = 1.0$$

Where s is the now complex frequency operator, not to be confused with scattering parameters. The characteristic function, $K(s)$, is the ratio of the reflection and transmission coefficient functions,

$$K(s) = \frac{\rho(s)}{\tau(s)}$$

Zeros of the numerator are called reflection zeros, and zeros of the denominator are transmission zeros. The order of the lossless two-port network is determined by the sum of transmission zeros.

In two-port ladder[1] networks, series and parallel ideal reactive elements stop transmission ($|s_{21}| = |s_{12}| = 0$) at certain frequencies by presenting either a series susceptance or shunt reactance of zero value. A series inductor (or a parallel capacitor) placed into a two-port between two resistive terminations, has no effect at dc, but it blocks transmission at infinite frequency. Mathematically, it adds a *transmission zero at* $f = \infty$ to the characteristic function of the two-port. Inserting either one of these elements in *nonredundant* form increases the order of the network by one. In addition, the presence of the reactive element also introduces a negative monotonic slope of 6 dB/octave (equivalent to 20 dB/decade) to the $|s_{21}|^2$ response, above the 3-dB corner frequency. The term −6 dB/negative slope octave specifies 6-dB additional attenuation every time we double the frequency. If we increase the frequency by a decade, the attenuation increases by 20 dB.

The term *nonredundant* is important because two inductors in series, or two capacitors in parallel, can always be merged into a single component. In such a case, one of the components is *redundant* and does not increase the order of the network.

A nonredundant series capacitor or parallel inductor blocks transmission at zero frequency by adding a transmission zero at $f = 0$ to the transfer function. These components create a 6-dB/octave slope below the 3-dB corner frequency. Each of these elements also adds one to the order of the network.

An L-C resonant circuit in *certain configurations* blocks the transmission at a *finite frequency*, f_R, and causes *a notch in the frequency response at* $f = f_R$. Since a resonant circuit has two elements, it creates a complex pair of transmission zeros, adding *two* to the order of the network.

The *sum* of transmission zeros defines the order of the network, while their *distribution* helps us to have a quick estimate of the circuit's response through a broad range of frequencies. The sum N is defined as

$$N = N_{f=0} + N_{f=\infty} + N_{f=f_R} \tag{5.4}$$

For circuits built with lumped elements, the order can always be determined from the schematics. A good understanding of network order is essential to using a circuit synthesis program. Table 5.1 provides a summary of six L-C component configurations with their associated transmission zeros. Transmission line circuits are more difficult to handle and generally require additional circuit manipulation before the order can be found. We discuss matching networks with distributed components in Section 5.9.

1. Network with alternate series and parallel branches.

Table 5.1 Lumped L-C Element Configurations that Create Single or Double Transmission Zeros, Increasing Network Order by One or Two, Respectively

Frequency Where the Transmission Is Reduced to Zero	Transmission Zero Type and Resulting Frequency Response Between Matched Terminations**		Possible Single-Section Two-Port L-C Circuit Topologies
dc $f = 0$	Single (real)	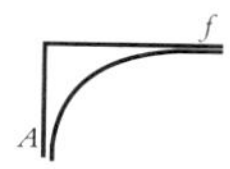	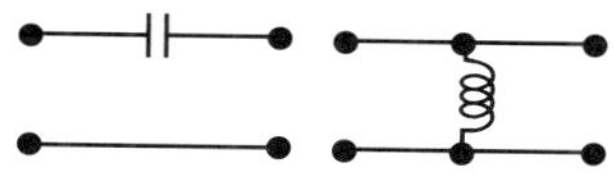
Infinite $f = \infty$	Single (real)	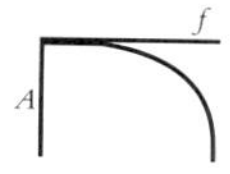	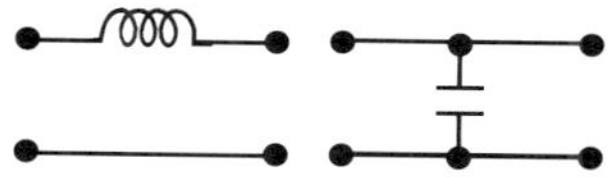
Finite frequency $0 < f_R < \infty$	Conjugate pair (complex)	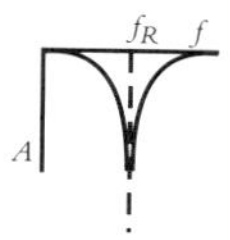	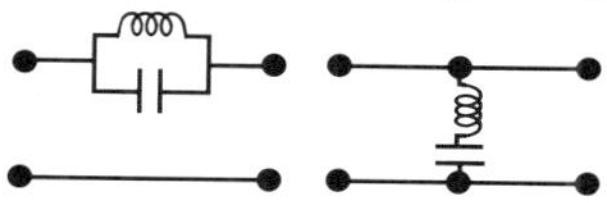

Note: Each of the two resonant circuits, shown at the bottom, blocks transmission at a finite resonant frequency, f_R, and adds two to the network order. On the frequency plots, "A" refers to attenuation.

Table 5.2 summarizes the three types of transmission zeros. Each reactive single nonredundant element creates a transmission zero either at dc or infinite frequency. The two types of resonant L-C configurations shown in Table 5.1 generate transmission zero pairs at finite frequencies.

The order[2] of the L-C network can also be found from its transfer function, generally expressed by the ratio of two polynomials. A large number of textbooks are devoted to the derivations and use of those functions [1–4]. In our books, we rely on commercially available circuit analysis and synthesis programs and focus on their uses rather than how they work. We do, however, categorize and discuss the capabilities of various CAD tools both for linear and nonlinear circuit applications in Chapter 6 and in

Table 5.2 Each Reactive Nonredundant Lumped L or C Component Increases the Network's Order by One

For transmission zero(s) located at frequency	Increase the networks order by
$f = 0$	One for each nonredundant element
$f = \infty$	One for each nonredundant element
$0 < f_R < \infty$	Two for each resonant L-C section

Note: When a resonant circuit blocks transmission at a finite frequency, we increase the order by two.

2. The order of a network is equal to the largest exponent used in the denominator of polynomial describing the transfer function.

Volume II, Chapter 4. Our goal is to provide you with the necessary technical background to use some of the software programs effectively.

5.2.1 Illustrative exercise: determine the order of L-C networks

Let us perform the specified tasks for the four circuits listed next. Then, *assuming lossless elements* for all circuits, verify the results with a circuit simulator by plotting the wideband frequency responses between 1 GHz and 10 GHz.

Circuit 1

Find the order and transmission zero distribution of the two-element circuit shown in Figure 5.4. Using a 10-Ω source and a 50-Ω load, what is the loss of the circuit at low frequencies, and what is the gain slope at the high frequencies.

Solution

This is a lowpass circuit topology without any transmission zero at dc. If the source and load resistances were equal, the low-frequency losses are zero. Since the source and load resistances are different, we do have a mismatch loss, *ML*, at low frequencies.

$$ML = -10\log\left(1 - \left|\frac{Z_S - Z_L}{Z_S + Z_L}\right|^2\right) = -10\log\left(1 - \left|\frac{10-50}{10+50}\right|^2\right) = 2.55 \text{ dB}$$

Elements L_1 and C_2 block transmission at $f = \infty$ and provide −12 dB/octave, or −40 dB/decade, gain roll-off moving toward high frequencies (Figure 5.5). There are no resonant sections in this circuit; therefore, we do not get finite frequency transmission zeros. The order of the circuit, N, from (5.4) is

$$N = N_{f=0} + N_{f=\infty} + N_{f=f_R} = 0 + 2 + 0 = 2$$

When the number of components is the same as the order of the network, the circuit is in *minimum form*, also called *canonic form*.

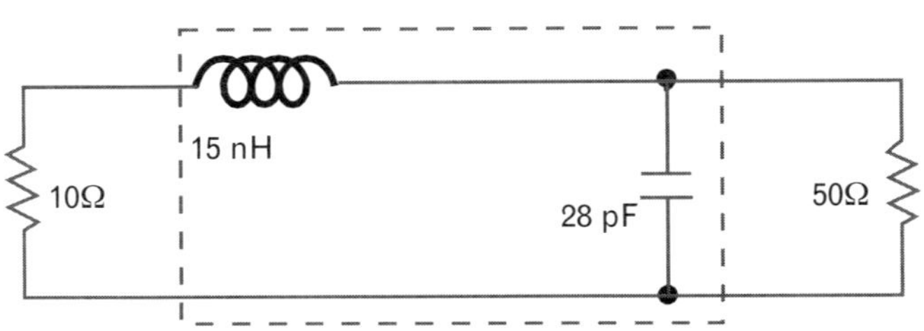

FIGURE 5.4 *Two-element L-C matching network, L_1 and C_2, placed between a 10-Ω source and a 50-Ω load.*

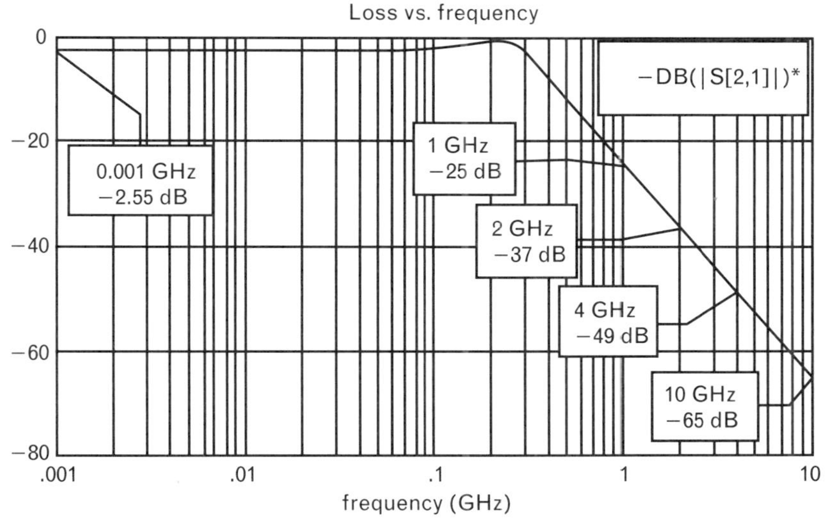

FIGURE 5.5 *Broadband frequency response shows a flat 2.55-dB mismatch loss at low frequencies and the expected 12-dB/octave slope at the high end. If the resistive source and load terminations were equal, the response would be without loss at low frequencies.*

Let us bring up again the 2.55-dB low-frequency loss that is caused by mismatch between the terminations. Such loss troubles many who are new to the transducer gain definition. (*If you are in that category, visualize again our water-pipe illustrations.*) When the source impedance is 10Ω, maximum available power is delivered into a 10-Ω load only. In our case we have a 50-Ω load that reflects nearly half of the available signal power. Since a lowpass matching network is totally ineffective at the low frequencies, the computed 2.55-dB mismatch loss exists. As the matching network gradually begins to function, the reflected loss is reduced. At 220 MHz the circuit matches the two terminations and the mismatch loss disappears. Above 220 MHz the matching circuit slowly loses its effectiveness and begins to reject the higher frequency signals.

Circuit 2

Our second exercise has a five-element circuit shown in Figure 5.6. Find the network order and transmission zero distribution. Plot the frequency response also.

At first look this may appear to be a fifth-order network, but let us take it step by step. At dc, L_1 and L_4 act as short-circuits, leaving a three-element highpass circuit of C_2, L_3, and C_5, giving us three transmission zeros at dc. The predicted slope at low frequencies is 18 dB/octave.

At infinite frequency we seemingly have two transmission zeros caused by L_1 and L_4. Before we conclude, note that C_2 acts as a short circuit, L_3 becomes an open-circuit and C_5 is another short-circuit. These three elements do not affect transmission at infinite frequency and we can remove them from the circuit. At that point we are left with two series inductors L_1 and L_4 that can be combined at infinity into a *single series inductor giving us only one transmission zero at infinite frequency*. The slope at the high frequency

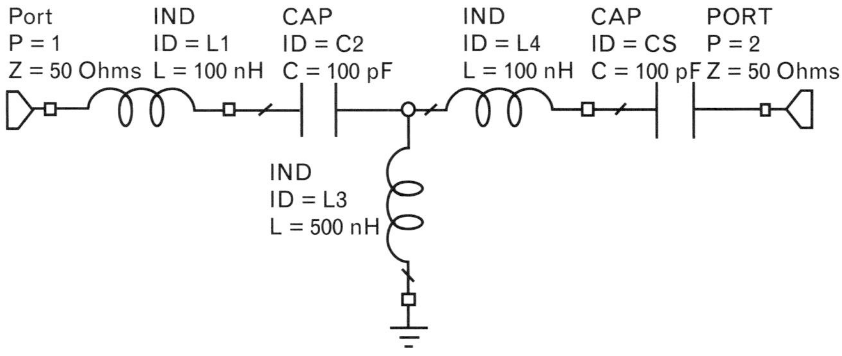

FIGURE 5.6 *The five-element Tee-network has series resonant L-C circuits in the series arms. There is dc isolation between the input and output terminals.*

end is only 6 dB/octave. This is a bandpass network because it has transmission zeros both at dc and infinity.

Finite frequency transmission zeros are always caused by resonant circuits and our topology has two resonant circuits formed by L_1–C_2 and L_4-C_5. However, when these circuits resonate, the signal will simply pass through without any blockage; they do not cause any finite frequency transmission zeros.

The order of the circuit, N, from (5.4) is

$$N = N_{f=0} + N_{f=\infty} + N_{f=f_R} = 3 + 1 + 0 = 4$$

Gain roll-off at the low-frequency side is 18 dB/octave, but only 6 dB/octave at the high-frequency end, giving us the skewed frequency response shown in Figure 5.7.

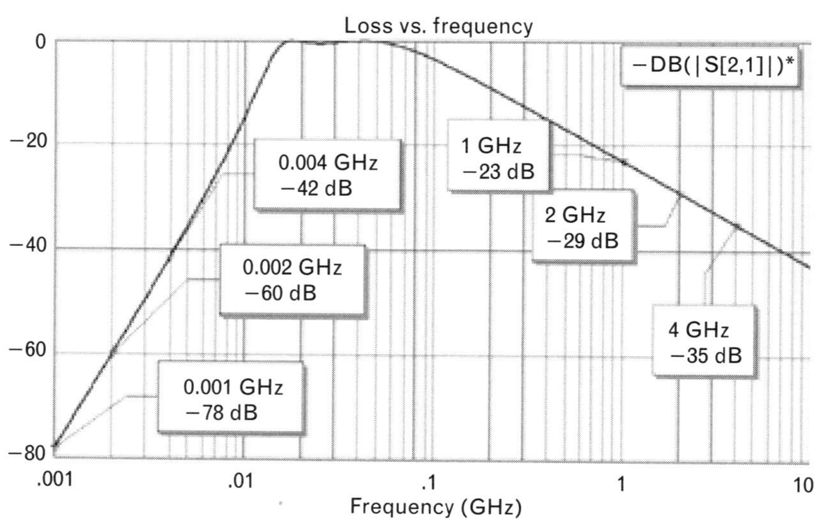

FIGURE 5.7 *The fourth-order network of Figure 5.6 has steeper gain roll-off at low frequencies since it has three transmission zeros at dc. This circuit is a filter, working between two 50-Ω terminations. Since the order is less than the number of components, the circuit is not in its minimum canonic form.*

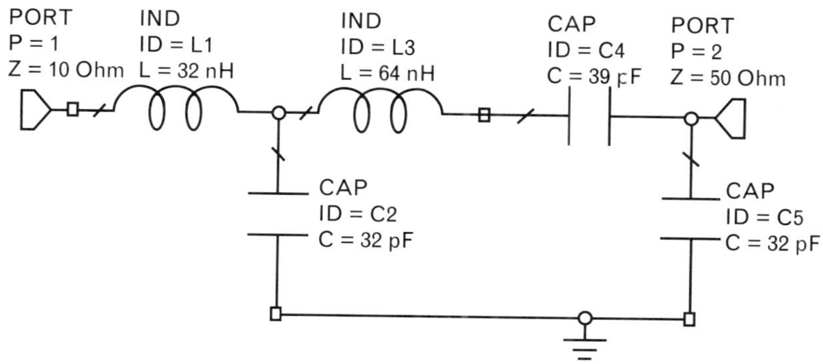

FIGURE 5.8 *Schematic of a five-element network to match two resistive terminations at two separate frequencies*

Circuit 3

Our third example (Figure 5.8) also has five components, including a series resonator. It was designed to match 10Ω to 50Ω at 100 MHz and 200 MHz. Find the network order and determine the gain slope at low and high frequencies. Plot the frequency response to confirm results.

At dc, L_1, C_2, L_3, and C_5, vanish and C_4 generates a transmission zero, corresponding to a 6-dB/octave slope at the low end. At infinite frequency, all but C_4 block transmission, giving us a –24-dB/octave slope at high frequencies.

When inductor L_3 resonates with C_4, they form a series short-circuit, without any blockage of the signal path. The order of this circuit is

$$N = N_{f=0} + N_{f=\infty} + N_{FR} = 1 + 4 + 0 = 5$$

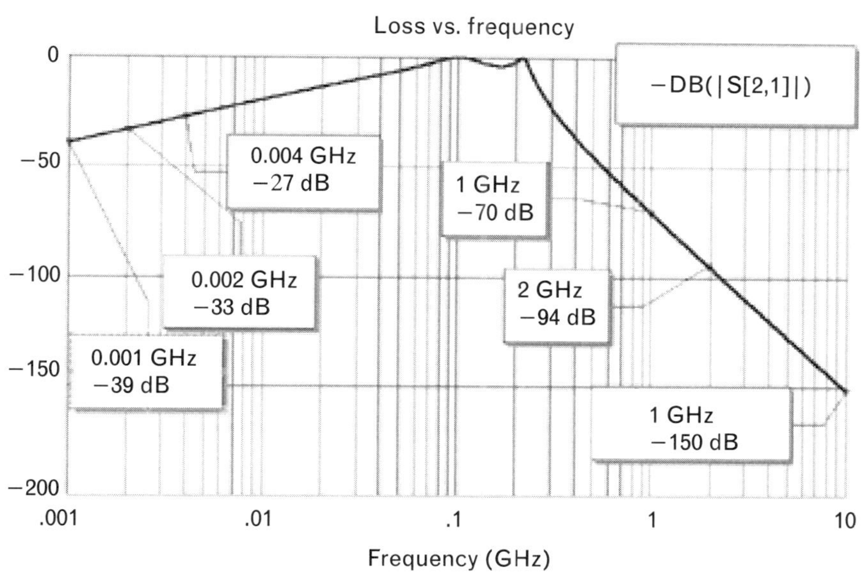

FIGURE 5.9 *The five-element impedance matching network of Figure 5.8 creates a skewed response with −24-dB/+octave slope (− 80 dB/decade) at the high frequency side compared to 6 dB/ octave at the low side. The series resonant circuit does not generate finite frequency transmission zeros.*

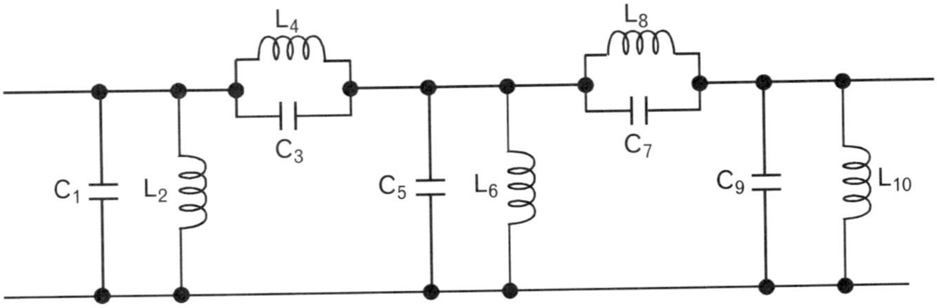

FIGURE 5.10 *Bandpass circuit with five resonant L-C sections.*

Since the order is the same as the number of components, this circuit is in its canonic form. Frequency response is shown in Figure 5.9.

Circuit 4

The fourth example tests your understanding of the transmission zero concepts with a more complex 10-component network, shown in Figure 5.10. The circuit has five inductors and five capacitors forming five resonant sections. Find the order of the network and the distribution of the transmission zeros for the following options:

- Case 1: As the circuit is shown with all 10 components;
- Case 2: After removing inductors L_4 and L_8, leaving eight components;
- Case 3: After removing C_1, L_4, L_8, and C_9, leaving six components.

Which, if any, of the three options are in canonic form?

Solution

This time we let you find the answers by yourself. When finished, compare your results with the solutions provided in Section 5.12.

5.3 Impedance matching into complex termination

Next, consider what happens when the real parts of the source and load are the same, but one of them has a complex part, such as a series reactance (Figure 5.11). Clearly, the presence of any net reactance between R_S and R_L reduces the current in R_L and with it the power dissipated in R_L. Therefore, to restore the dissipation to the maximum that occurs when $R_S = R_L$, the net reactance of the loop must be zero. This occurs when the load and source are made to be complex conjugates of one another—they have the same real parts and opposite type reactive parts.

Location	Impedance to the left	Impedance to the right
A	R	$R + jX - jX = R$
B	$R + jX$	$R - jX$

Note that once the proper type and amount of reactance added to set the net total reactance to zero, the opposite impedances at any point are complex conjugates of each other. For example, in the circuit of Figure 5.11 where the source and load have equal real parts and the load is complex, $Z_S = (R + j0)$, and $Z_L = (R - jX)$, the impedances looking into both directions at locations "A" and "B" are the following:

Although the above may seem to be obvious, it is a very important fundament concerning impedance matching with lossless components. When a source termination is matched to a load with a *passive lossless* two-port network,

- The source is conjugate matched to the input of the two-port.
- The load is conjugate matched to the output of the two-port.

5.3.1 Illustrative example: matching a 50-Ω source to a complex load

Match a 50-Ω resistive source at 100 MHz to a load that has an equivalent circuit of a 50-Ω resistor in series with a 1.59-pF capacitance (Figure 5.12).

Solution

Since the real parts of the terminations are equal, the required *matching circuit* can be a series inductor to negate the reactance of the series capacitance. At 100 MHz the unnormalized capacitive reactance from (2.17) is

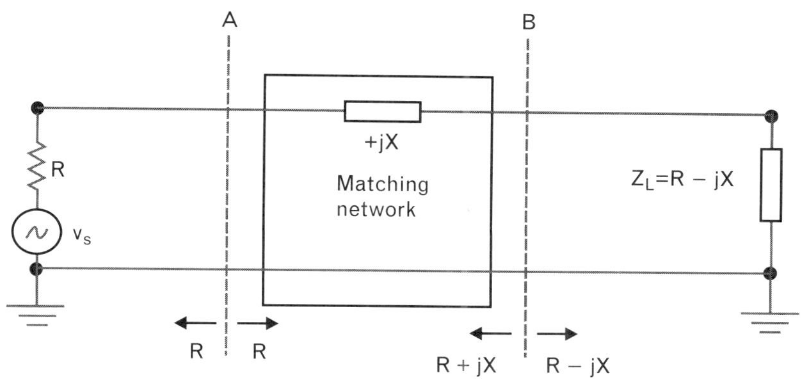

FIGURE 5.11 *Impedance matching a resistive source and complex load for maximum power transfer. When the matching circuit is lossless, conjugate match exists at every interface.*

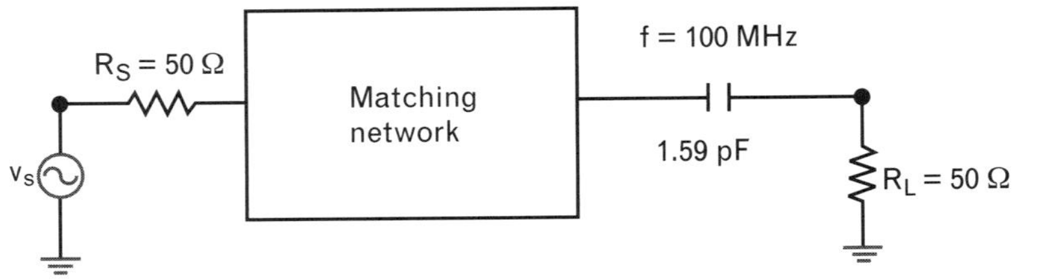

FIGURE 5.12 *Match this complex load, representing the input impedance of an active device, to 50Ω at 100 MHz with the simplest lumped circuitry.*

$$X_C = \frac{159}{(0.1)1.59} = 1{,}000\Omega$$

representing $-j1{,}000\Omega$. From Table 2.2 we can find the necessary inductance to negate this capacitive reactance by adding $+j1{,}000\Omega$ in series,

$$L = \frac{0.159(1{,}000)}{(0.1)} = 1{,}590 \text{ nH} = 1.59\ \mu\text{H}$$

Adding a 1.59-μH series inductor between the two terminations (Figure 5.13) provides match at 100 MHz. At other frequencies some series reactance remains and the power delivered to the resistive part of the load decreases. We will find the 3-dB bandwidth of the circuit in the next section.

Conjugate match is obtained at only one frequency, f_R=100 MHz, where the two reactive elements resonate. As the frequency is increased or decreased from this value, the transmitted power rolls off at a rate determined by the loaded Q of the circuit (Figure 5.14). The 3-dB frequency bandwidth of a resonant matching network is found by the reciprocal of the loaded Q of the network.

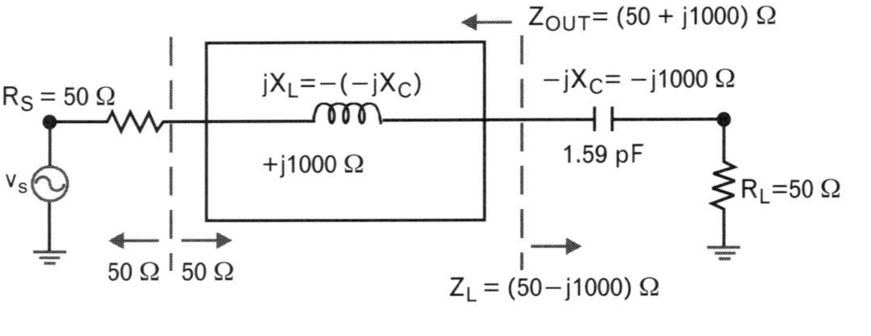

FIGURE 5.13 *Matching a 50-Ω resistive source to a complex load only requires a series reactive element when $R_S = R_L$. Exact match applies only at the frequency where the net reactance of the circuit is zero.*

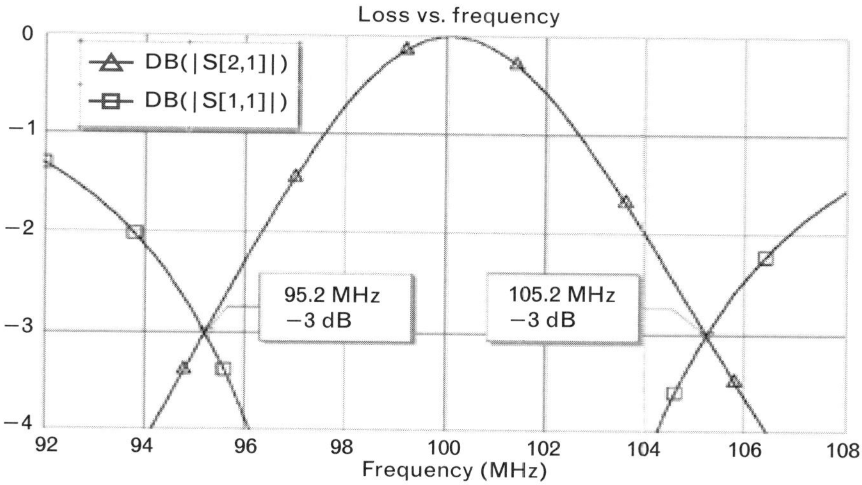

FIGURE 5.14 *Matching network frequency response shows 3-dB bandwidth of 10 MHz. Observe that the transmission and reflection coefficient magnitudes,* $|s_{21}|$ *and* $|s_{11}|$*, are both at −3 dB at the same frequencies.*

$$BW_{3\,\mathrm{dB}} = \frac{f_R}{Q_{SRES}} \tag{5.5}$$

where Q_{SRES} is defined as

$$Q_{SRES} = \frac{\text{Reactance of one resonant element at } f_R}{\text{Total series resistive loading}} \tag{5.6}$$

Using 1,000-Ω reactance and (50 + 50)Ω total resistive loading gives us $Q_{SRES} = 10$. Substituting Q_{SRES} into (5.5) with a resonant frequency of f_R = 100 MHz finds the 3-dB bandwidth, BW_{3dB} = 10 MHz.

At the 3-dB corner frequencies only one-half of the signal power is transmitted. The second half is reflected. Converting the 3-dB return loss gives a reflection coefficient magnitude of 0.707, which indicates a very poor impedance match. For that reason, the 3-dB bandwidth is not a very practical specification to provide, although it is used frequently for comparative purposes.

Note that matching high-Q terminations leads to narrow bandwidths. For a given resistance, high Q results when the associated parasitic reactance is one of the following:

- Large series or small parallel inductance;
- Small series or large parallel capacitance.

When the resistive portion is very small or very large (i.e., input resistance of a high-power transistor, or output impedance of a low-current device), even a small amount of parasitic inductance or capacitance can lead to high Q, making impedance matching a challenging task.

We should point out that if in Figure 5.13 the load capacitance was in parallel with R_L, then a parallel inductor would be used to resonate the load parasitic capacitance. In that case the Q for parallel resonance, Q_{PRES}, is computed as

$$Q_{PRES} = \frac{\text{Total parallel resistive loading}}{\text{Reactance of one resonant element at } f_R} \qquad (5.7)$$

Keep in mind that in parallel resonant circuits, high-value loading resistors lead to high Q, while in series circuits the opposite is true.

5.4 Impedance matching with uneven resistive terminations

In the previous example we had two terminations with equal real parts. Impedance matching in such cases is relatively easy. All we need to do is eliminate the effect of the reactive portion of terminations. In most practical applications, however, the resistive parts are not equal and the terminations are complex. To handle those cases, we first develop a technique to take care of the unequal resistance problem. Then, we expand the technique to also include reactive elements.

Figure 5.15 shows a setup with 50-Ω resistive source and 3-Ω resistive load. The goal here is to match the source to the load at a specific frequency.

Common reaction to this question, often raised in the classroom, is to use a 47-Ω series resistor. Examining the new input impedance tells us we could not do better—the 50-Ω source working into a 50-Ω input impedance gives us a perfect match. However, when we examine the output side of the two-port, there we have a 97-Ω output impedance working into the 3-Ω load. Our initial *VSWR* between the two terminations, without the

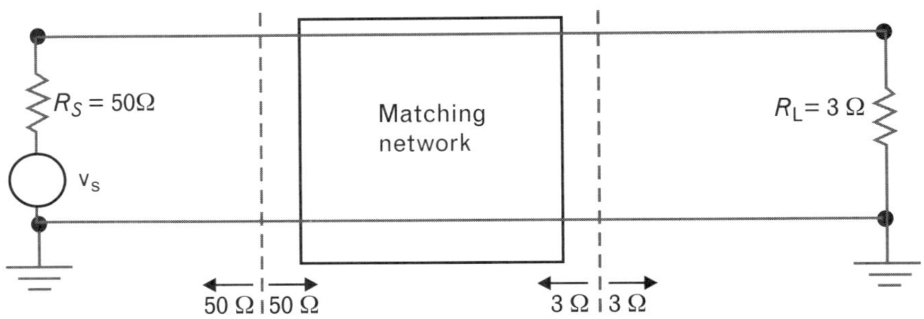

FIGURE 5.15 *Matching a 3-Ω load to a 50-Ω source requires more complex circuitry than the series inductor used in the previous example of Section 5.3.1. At a single frequency, an appropriate pair of L-C elements can always provide perfect match to both terminations.*

series resistor, was 50/3 = 16.7. The new *VSWR* at the output of the circuit is 32.3. Although we improved the input match, the output match is now much worse. An even more significant concern is the transmission loss (more than 12 dB) since most of the signal power dissipates in the 47-Ω "matching" resistor instead of the 3-Ω load.

Using a more complex resistive network can improve the output match but creates more loss. For example, we can find a series-parallel resistor combination to give us 50-Ω input and 3-Ω output impedances simultaneously. Transmission loss, however, further increases (more than 18 dB). The lesson is that using resistors for impedance matching is not a good practice and reactive elements should always be considered first.

Another possible option is to use an ideal transformer. Adding a transformer with a turns ratio of 4.08:1 provides an impedance transformation of 16.7:1, giving simultaneous match from both sides of the two-port. Transformers, however, are difficult to produce and reproduce at RF frequencies, although they can provide much broader bandwidths than traditional *L-C* matching circuits. Since inductors and capacitors are less expensive and more convenient to handle, in this chapter we focus first on

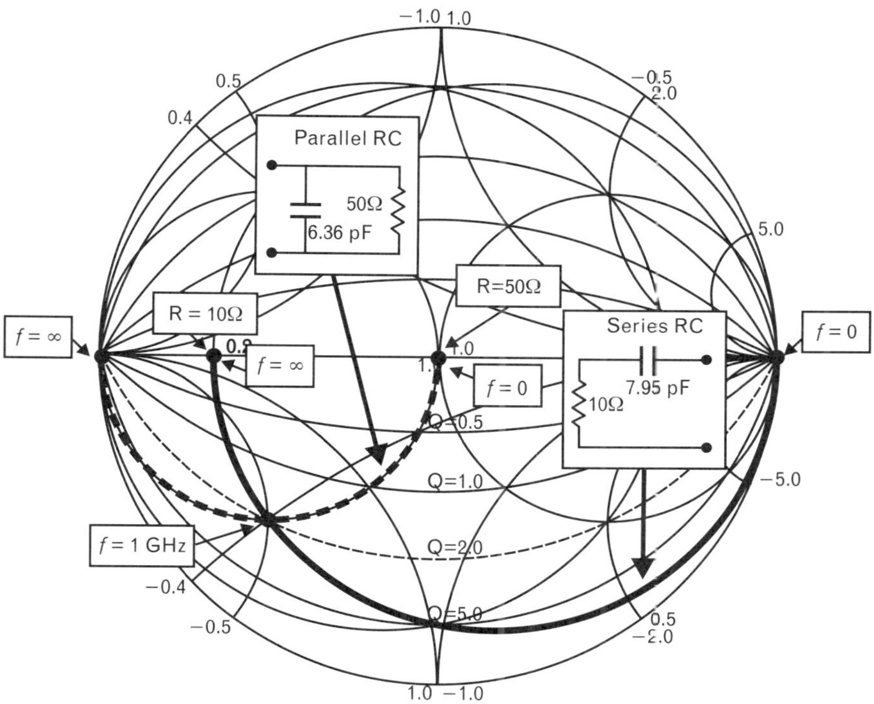

FIGURE 5.16 *Impedence plots of the series and parallel R-C networks from f = 0 to f = ∞. At the intersection of the r = 0.2 and g = 1.0 circles, the two circuits have equal terminal impedances and Q-factors ($Q_S = Q_P$ = 2.0 at 1 GHz). Even though physically the two circuits are different, electrically they are identical at 1 GHz.*

L-C matching sections and later on transmission line sections. Transformer applications are covered in Chapter 7.

Before going further, recall that every series impedance has a parallel equivalent at each particular frequency. In Section 4.5.6 we illustrated the conversion of a series *R-C* network to its parallel equivalent at 1 GHz and noted that the two circuits have the same *Q*s at the frequency of conversion. For easy visualization, we redraw Plate 17 here with slight modification. Our new normalized chart (Figure 5.16) shows the frequency responses of series and parallel *R-C* circuits with the component values given, from dc to infinite frequency. The intersection of the two traces is at 1 GHz for both circuits We can see that the two circuit traces are equivalent when their *Q*s are equal to 2.0.

In Section 2.11 we developed relationships between the series and parallel forms of *R-L* and *R-C* circuits. Table 2.3 showed that the resistive elements of series and parallel equivalent circuits are related by the expressions,

$$R_P = \left(1+Q^2\right)R_S \quad \Leftrightarrow \quad r_P = \left(1+Q^2\right)r_S \tag{5.8}$$

For all *Q* values, $R_P \geq R_S$. Solving (5.8) for *Q* in terms of the two resistors,

$$Q = \sqrt{\frac{R_P}{R_S} - 1} \quad \Leftrightarrow \quad Q = \sqrt{\frac{r_P}{r_S} - 1} \tag{5.9}$$

Equation (5.9) shows the *Q*-factor of both circuits at the frequency where they are electrically equivalent. A quick check of the two *R-C* networks shown in Figure 5.16: $X_P = 25\Omega$, $Q_P = 2$, $R_P = 50\Omega$, and $R_S = 10\Omega$, $X_S = 20\Omega$, $Q_S = 2$, verifies $Q_S = Q_P = Q = 2.0$.

Next, we replace the capacitor of the series circuit with an inductor that has inductive reactance, $X_S = 0.4$ at 1 GHz and compare the frequency responses of the new series *R-L* and previously used parallel *R-C* circuit (Figure 5.17).

Looking at Figure 5.17, we can conclude that we can match a 10-Ω termination to 50Ω at 1 GHz, by adding a series inductor to the 10-Ω side and a parallel capacitor to the 50-Ω side. The values of the two reactive elements, which set $Q_S = Q_P = 2.0$, are determined from (2.19) and (2.24). Although the match is only exact at a single frequency, 1 GHz in our example, depending on the computed *Q*, it may provide reasonable performance through a specified passband, as we show in Section 5.5.2.

We already saw that adding series or parallel resistance to increase or decrease the impedance is not an effective matching technique because it

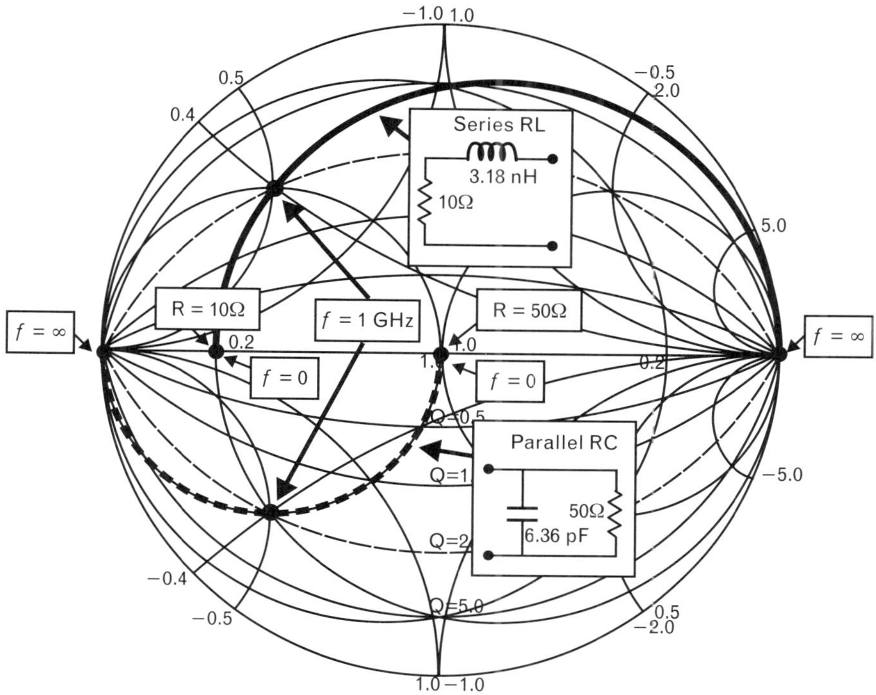

FIGURE 5.17 *Broadband impedence plots of the series R-L and parallel R-C show that they circuits are complex conjugates of each other at 1 GHz. Any two arbitrary resistive terminations may be matched by the same procedure.*

leads to dissipative losses. For that reason we use lossless matching circuits in this book, unless special circumstances justify adding resistors.

5.5 The Q matching technique with L-C sections

Any two resistive terminations can be simultaneously matched by adding two reactive elements between them if we follow four simple rules described below. For better visualization, we name the termination with lower resistance, $R_{SMALLER}$, and the other one R_{LARGER}.

1. Add a series reactive element next to $R_{SMALLER}$, and a parallel one to R_{LARGER}. We have the freedom of choosing whether the inductor or the capacitor is the series element, but the parallel element must be the opposite type, as shown in Figure 5.18. If the series element is an inductor, adding a parallel capacitor creates a lowpass topology. A series capacitor with parallel inductor forms a highpass section. A series-parallel configuration is sometimes called an *inverted-L*, or just *L-network*.

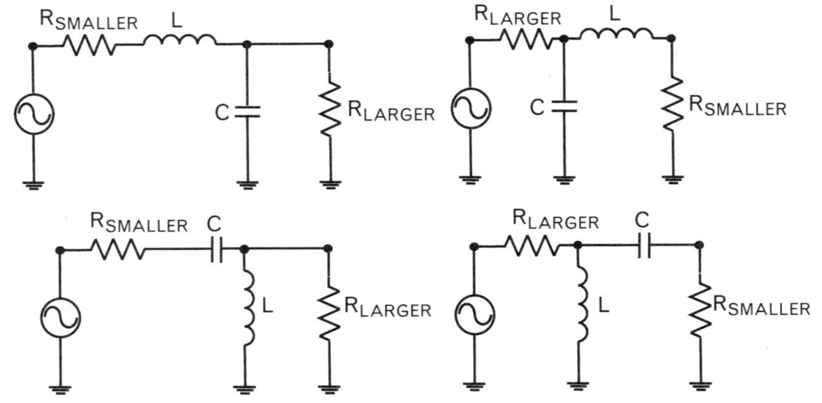

FIGURE 5.18
A single-section L-C circuit may be used to match a pair of arbitrary resistive terminations. The series matching element must be placed next to the termination with lower resistance to increase the impedance level.

2. Adding a series reactive elements to $R_{SMALLER}$ and a parallel one to R_{LARGER} forms two *subnetworks*, one series and one parallel (Figure 5.19). One of the subnetworks is inductive, the other is capacitive, and at the design frequency they must represent complex conjugates impedances to each other, as it was previously in Figure 5.17. Therefore, the *Q*-factors of these two subnetworks must be equal at the frequency where the match is computed. The values of the *Qs* are related to two resistive terminations and the two added reactive components. The *Q*-factors of the subnetworks, Q_S and Q_P are

$$Q_S = \frac{X_S}{R_{SMALLER}} \tag{5.10}$$

$$Q_P = \frac{R_{LARGER}}{X_P} \tag{5.11}$$

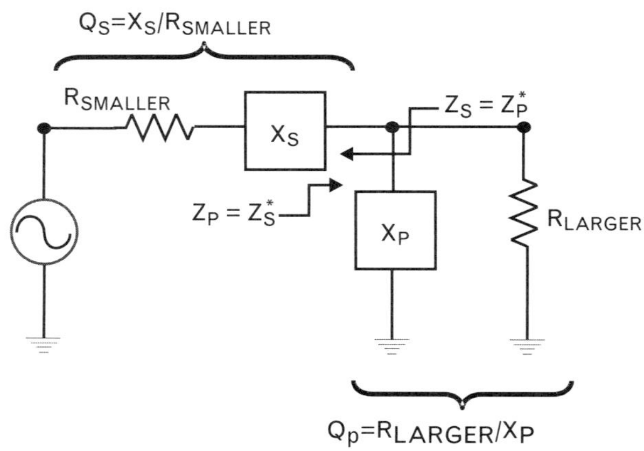

FIGURE 5.19
An L-C section placed between the two resistive terminations creates a series and a parallel subnetwork. When the two subnetworks are conjugate matched to each other, their Qs are equal.

At the frequency where they are matched to each other, the two Q-factors are equal,

$$Q = Q_S = Q_P \tag{5.12}$$

3. Equation (5.3) provides the *Q-factors of the two subnetworks*, with either the unnormalized or normalized resistances of the terminations.

$$Q_S = Q_P = \sqrt{\frac{R_{LARGER}}{R_{SMALLER}} - 1} \Leftrightarrow Q_S = Q_P = \sqrt{\frac{r_{LARGER}}{r_{SMALLER}} - 1} \tag{5.13}$$

4. Knowing the Q values, we can obtain the series and parallel elements, reactances, and compute the inductor and capacitor values by the formulas given in Table 2.2. We can also save the intermediate steps by combining the expressions of Table 2.2 with (5.10) and (5.11).

$$L_{\text{nH}} = \frac{0.159X_L}{f_{\text{GHz}}} = \frac{0.159Q_S R_{\text{SMALLER}}}{f_{\text{GHz}}} = \frac{0.159R_{\text{LARGER}}}{f_{\text{GHz}} Q_P} \tag{5.14}$$

$$C_{\text{pF}} = \frac{159}{f_{\text{GHz}} X_C} = \frac{159}{f_{\text{GHz}} Q_S R_{\text{SMALLER}}} = \frac{159Q_P}{f_{\text{GHz}} R_{\text{LARGER}}} \tag{5.15}$$

The first parts of (5.14) and (5.15) are used when the reactances are known. The second (middle) parts apply to elements in series configurations and the last parts to parallel topologies.

5.5.1 Illustrative example: impedance matching of two resistive terminations

Using the Q-technique outlined, match a 5-Ω source to a 50-Ω resistive load at 850 MHz. Maintain dc connections between the two terminations.

Solution

Adding a series inductor to the 5-Ω side and a parallel capacitor to the 50-Ω side keeps the dc connection and creates the lowpass matching configuration shown in Figure 5.20.

5.5.1.1 Analytical approach

The required Q-factors for the newly created subnetworks are calculated from (5.13).

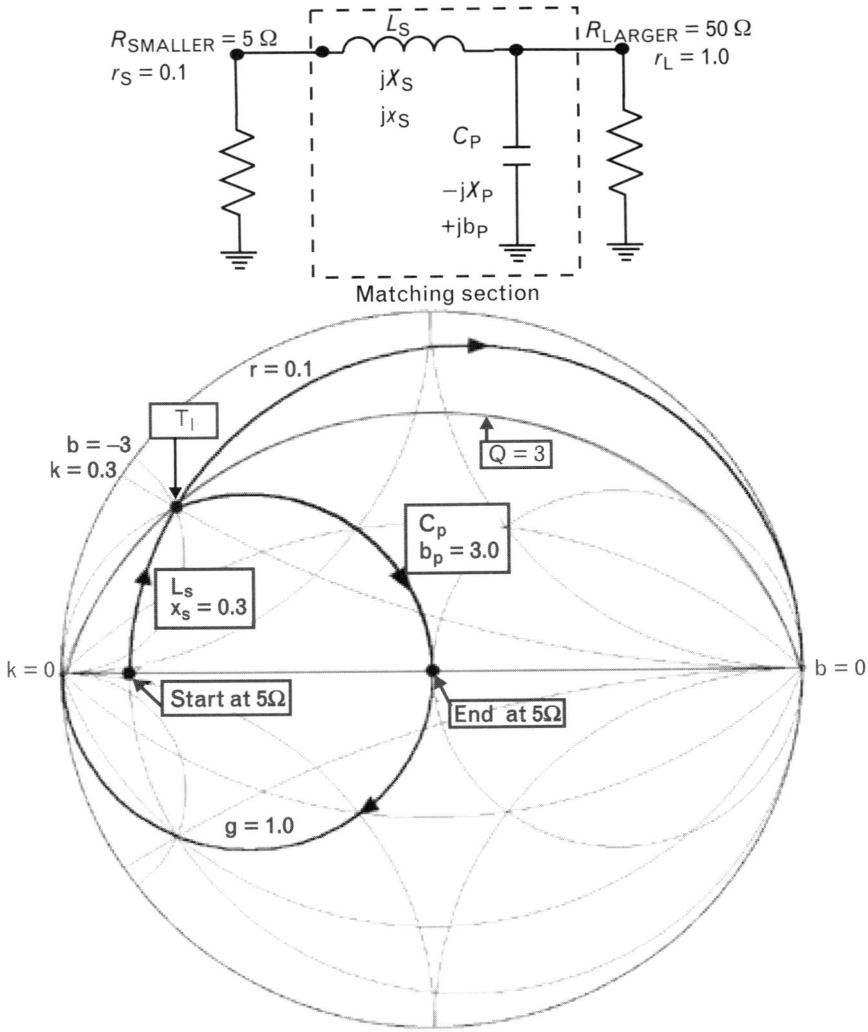

FIGURE 5.20 *Impedance matching a 5-Ω resistive source to a 50-Ω load with a lowpass L-C section. In the graphical approach from the 5-Ω termination move on the constant-resistance circle, r = 0.1 to intersect the constant-conductance circle of g = 1.0 at marker T_1. From that point a parallel capacitor moves to the center of the chart. Frequency response is shown in Figure 5.21.*

$$Q_S = Q_P = \sqrt{\frac{R_{\text{LARGER}}}{R_{\text{SMALLER}}} - 1} = \sqrt{\frac{50}{5} - 1} = 3 \tag{5.16}$$

Finding the inductor and capacitor values at 850 MHz from (5.14) and (5.15),

$$L_S = \frac{0.159 Q_S R_{\text{SMALLER}}}{f_{\text{GHz}}} = \frac{0.159(3)5}{0.85} = 2.8 \text{ nH}$$

$$C_P = \frac{159Q_P}{f_{\mathrm{GHz}} R_{\mathrm{LARGER}}} = \frac{159(3)}{0.85(50)} = 11.2 \text{ pF}$$

5.5.1.2 Graphical solution

Our goal is to transform the normalized source, $r_S = 0.1$ to $r_L = 1.0$. Choosing the lowpass topology, we mark two paths on the upper half of the immitance chart (Figure 5.20):

1. From the source on the $r = 0.1$ constant-resistance circle, where a series inductor can transform to values on the indicated path;
2. Possible transformations to the load with a parallel capacitor, on the $g = 1.0$ constant-conductance circle.

The intersection of the two marked circles is at $z = (0.1 + j0.3)$, equal to $y = (1.0 - j3.0)$. Let us mark the three important points of transformation on the chart as:

1. Starting point at r_S, marked with "Start";
2. Transfer point from constant resistance to constant conductance, at "T_1";
3. Our target at r_L, at "End."

From the impedance coordinates of the Smith chart we read the amount of normalized inductive reactance between locations "*Start*" and T_1, $x_S = 0.3$.

From the admittance coordinates of the Smith chart we read the amount of normalized capacitive susceptance between locations T_1 and End, $b_P = 3.0$

Finally, we convert the normalized reactance and susceptance to inductance and capacitance, using the expressions of Table 2.2.

$$L_S = \frac{7.96 x_L}{f_{\mathrm{GHz}}} = \frac{7.96(0.3)}{0.85} = 2.8 \text{ nH}$$

$$C_P = \frac{3.183 b_C}{f_{\mathrm{GHz}}} = \frac{3.183(3.0)}{0.85} = 11.2 \text{ pF}$$

Results agree with the values obtained from the analytical approach.

After inserting the two-element matching section between the 5-Ω source and 50-Ω load (Figure 5.21), the circuit is ready for simulation. Sweeping through a broad frequency range shows a near symmetrical

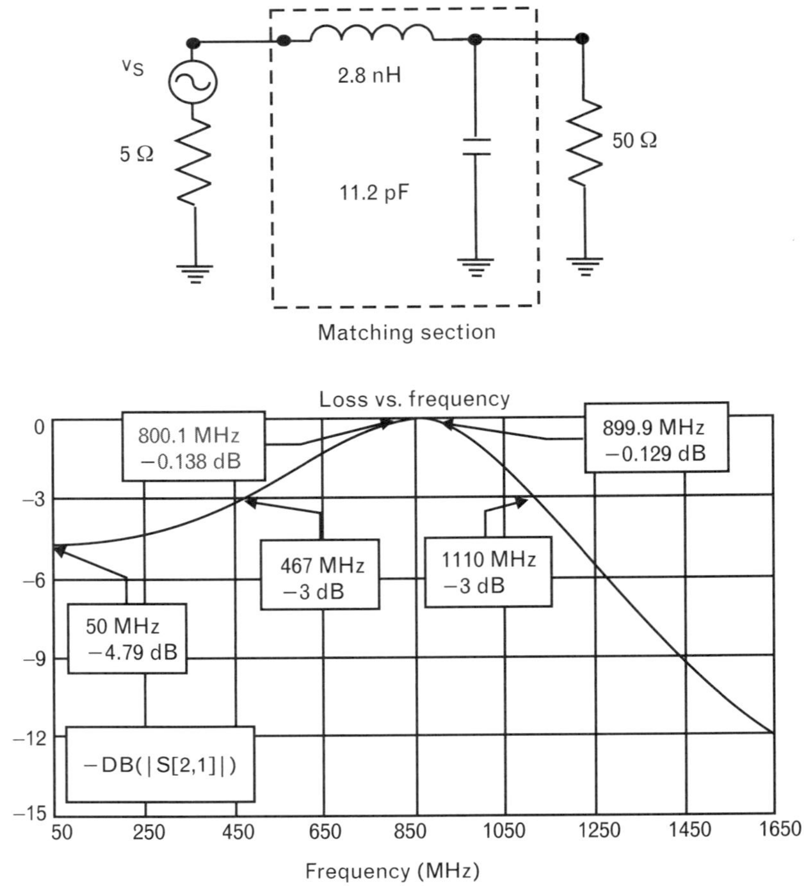

FIGURE 5.21 *A lowpass L-C section with ideal elements to match two resistive terminations shows an asymmetric broadband frequency response. Losses are negligible through the 800- to 900-MHz frequency range.*

response around 850 MHz (transmission loss is less than 0.15 dB through a 100-MHz bandwidth) but distinct differences between the low-end and high-end frequencies. Since we use a lowpass topology, loss at very low frequencies, 4.8 dB, is caused by the mismatch between the two terminations. If we swept to higher frequencies, we would see a −12-dB/octave slope due to the two transmission zeros at infinity.

5.5.2 Bandwidth of L-C matching sections

Equations (5.5) and (5.7) defined the 3-dB bandwidth for single series and parallel resonant type matching networks as the ratio of the center frequency to the loaded Q of the circuit at resonance. Finding a single Q for ladder networks or even for a series-parallel L-C matching section is a more

3. The ratio of x/r (or b/g) computed from the complex impedance (or admittance) of the network at that specific node.

difficult task. Instead of an overall-Q, we refer to a *nodal-Q* [5] at each node[3] of a ladder network, as we will show it later.

Since single-section L-C networks are configured either in lowpass or highpass form, the "3-dB bandwidth" may not even exist because the mismatch between the source and load can be less than 3 dB. For example, if we match 10Ω to 50Ω, the mismatch loss is 2.55 dB. Using a lowpass matching network, the maximum loss at low frequencies is 2.55 dB, and we never reach the 3-dB loss.

Looking at the frequency response in Figure 5.21, we can see *symmetry near 850 MHz*, because for a narrow bandwidth the L-C circuit can be treated as a *resistively loaded resonator.* Since it is not obvious at first glance, let us convert the series R-L part of the circuit to parallel form, as shown in Figure 5.22. After the conversion we see a parallel resonant circuit loaded by two 50-Ω resistors, which amounts to 25-Ω effective loading.

To find the 3-dB bandwidth of the circuit shown in Figure 5.22(b), we first calculate the loaded Q, at resonance with (5.7). The effective load on the resonant circuit is the parallel combination of the two terminations. We use the capacitive reactance in our calculations, although we could also use the inductor, since at resonance the two reactances are equal.

$$Q_{RES} = \frac{R_P}{X_P} = \frac{\dfrac{50(50)}{50+50}}{\dfrac{159}{0.85(11.2)}} = 1.5 \tag{5.17}$$

The bandwidth is then computed from (5.5),

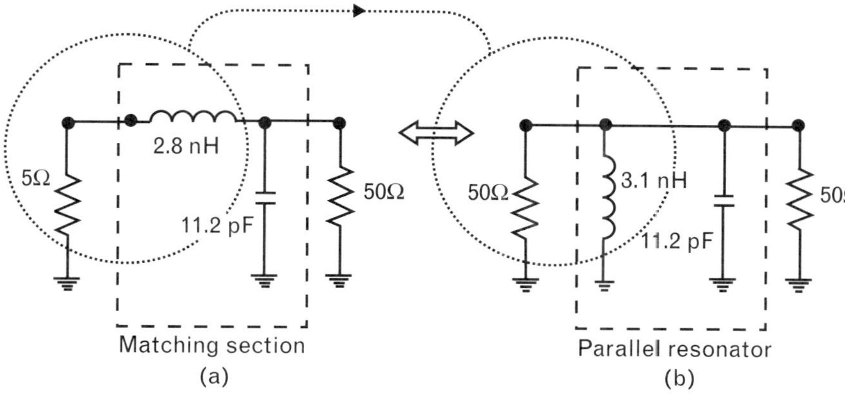

FIGURE 5.22 *Converting (a) the series R-L subnetwork portion of the matching section to (b) parallel form at 850 MHz helps us to understand why the frequency response is symmetrical near that frequency. Moving further away from 850 MHz the equivalence does not hold and the response becomes asymmetrical.*

$$BW_{3\text{dB}} = \frac{f_R}{Q_{RES}} = \frac{850 \text{ MHz}}{1.5} = 567 \text{ MHz}$$

Looking at Figure 5.21 we see a 3-dB bandwidth of (1,110 − 467) MHz = 643 MHz that is more than the computed 567 MHz. Remember, however, that our computation is based on a resistively loaded parallel resonant network that only represents the *L-C* matching section in the vicinity of 850 MHz. Still, the approach we used is useful to approximate the actual bandwidth for narrowband (10–20%) applications, as shown in Figure 5.23.

With this background now we can modify (5.5) to *approximate* the 3-dB bandwidth of an *L-C* matching section as

$$BW_{3\text{dB}} = \frac{f_0}{\frac{Q_N}{2}} = \frac{2f_0}{Q_N} \tag{5.18}$$

where Q_N is the nodal-Q mentioned above. In our illustrative example 5.5.1 Q_N is equal to $Q_S = Q_P = 3.0$ computed from (5.16). In the graphical solution Q_N is read from the Smith chart of Figure 5.20 at location T_1.

We emphasize that (5.18) shows *what the 3-dB bandwidth would be if the circuit had a symmetrical bandpass response*. An *L-C* matching section's response shows symmetry only near the frequency where the match is computed, and the 3-dB bandwidth is meaningless if the mismatch loss between the two terminations is less than 3 dB. Still, for narrowband applications (5.18) can be useful.

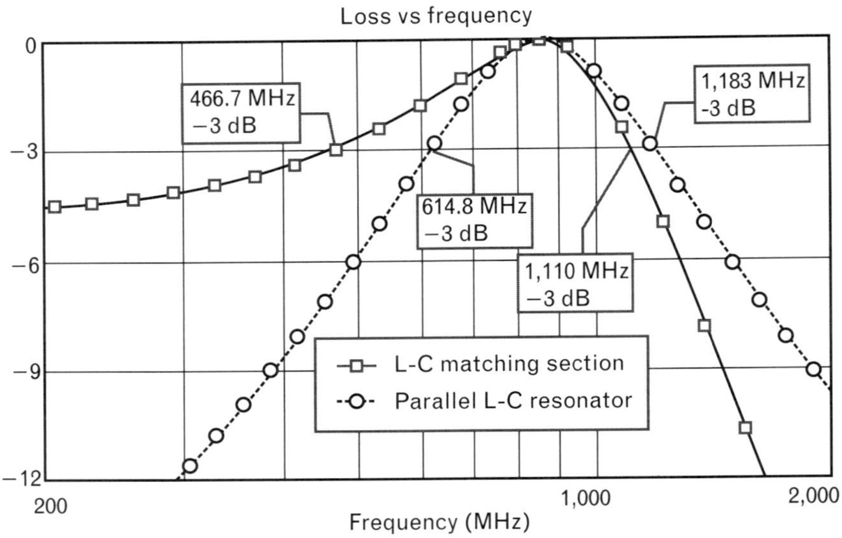

FIGURE 5.23 *Frequency response comparison of the L-C matching section and parallel resonant circuits of Figure 5.22 shows a remarkable closeness for about 15% bandwidth and a reasonable approximation up to nearly 50% bandwidth.*

5.6 Impedance matching of complex terminations

Impedance matching of complex terminations with lossless components has a simple rule: to match two complex terminations, transform one termination to the *complex conjugate* of the second termination. Figure 5.24 shows two possible ways to begin the matching process.

A properly designed passive lossless matching network automatically matches both complex terminations simultaneously. If there is no reflection at the input port, then all signal power is delivered to the load, without any reflection at the output port. Stating it another way, when a network is found which transforms a given load to the complex conjugate of a given source, the reverse is also true. That is, the source is simultaneously transformed by the same network into the complex conjugate of the load

When one or both of the impedances to be matched already has imaginary parts, two possibilities exist for the analytical approach: either absorb the parasitic into the matching network or eliminate it by resonance.

5.6.1 Absorbing the parasitics of the terminations

Input and/or output reactances or susceptances may be absorbed into the required L-C matching section if the Q of the termination is less than the nodal-Q computed from (5.13). For example, the 5-Ω to 50-Ω matching circuit of Figure 5.20 has a 2.8-nH series inductor followed by an 11.2-pF parallel capacitor. The 5-Ω source forms a series subnetwork with the series inductor, having a Q-factor of 3.0 at 1 GHz. If the source, instead of being purely resistive, had parasitic series inductance, up to 2.8 nH, that inductance can become part of the matching network. Series source inductance greater than 2.8 nH creates Q greater than 3.0; therefore, it cannot be absorbed by the matching network. Similarly, at the output side parallel

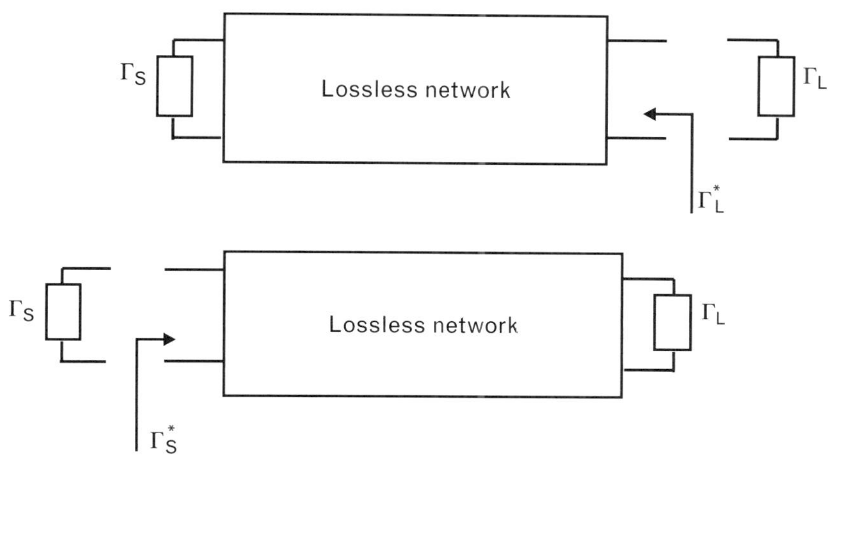

FIGURE 5.24 *Whether we add an impedance matching two-port to the source or load side, the effect must be the same. Connecting the lossless matching network to one of the terminations, the impedance at the adjacent side of the network must be the complex conjugate of the second termination.*

load capacitance can be absorbed up to 11.2 pF, but not above that limit. Figure 5.25 illustrates the absorption of a 0.93-nH source inductance and 3.74-pF load capacitance separately, but they can also be absorbed simultaneously at both sides of the network. When the parasitic capacitance or inductance is used as part of the matching network, the frequency response is the same as it would be with resistive terminations.

5.6.2 Resonating excessive parasitic inductance or capacitance

When the parasitic of a termination exceeds the maximum value that the matching network can absorb, we either (1) fully resonate out the parasitics and proceed matching to the leftover resistive part of the termination, or (2) resonate only the excessive part of the parasitic and use the remainder as part of the matching circuit.

Both of these resonance-based techniques reduce the bandwidth-the more reactance or susceptance we have to resonate, the narrower the frequency response becomes. Figure 5.26 illustrates the full resonant matching approaches, using the circuit of Figure 5.20 where the maximum allowed parallel capacitance at the load side was 11.2 pF.

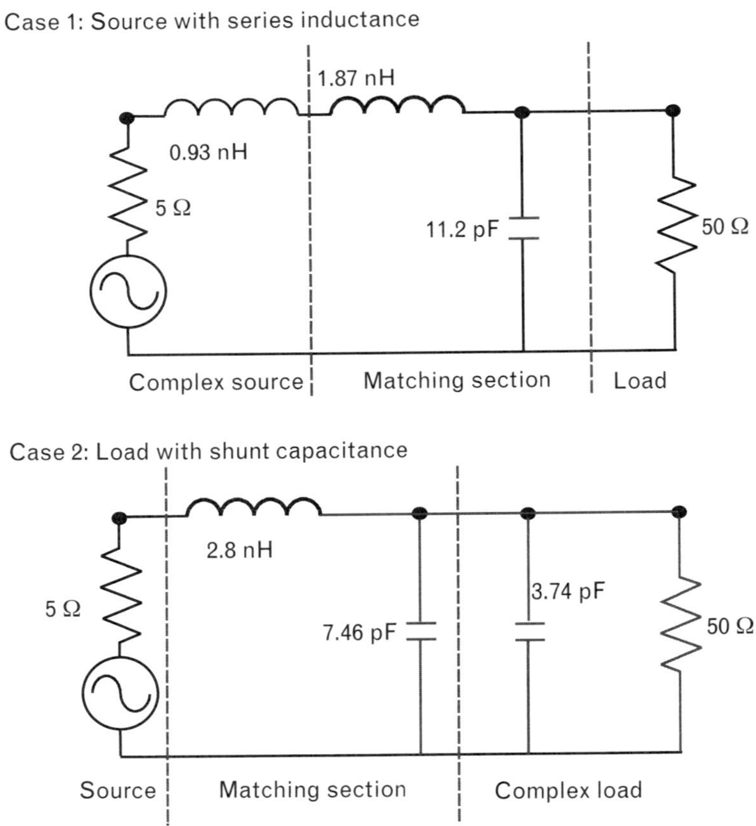

FIGURE 5.25 *Source and load inductance or capacitance may be incorporated into the matching network as long as the Q-factor of the termination does not exceed the computed nodal-Q of the L-C network. Parasitics of the terminations in this example must not exceed 2.8 nH at the input and 11.2 pF at the output. The limit is set by the resistance ratio of the terminations to be matched.*

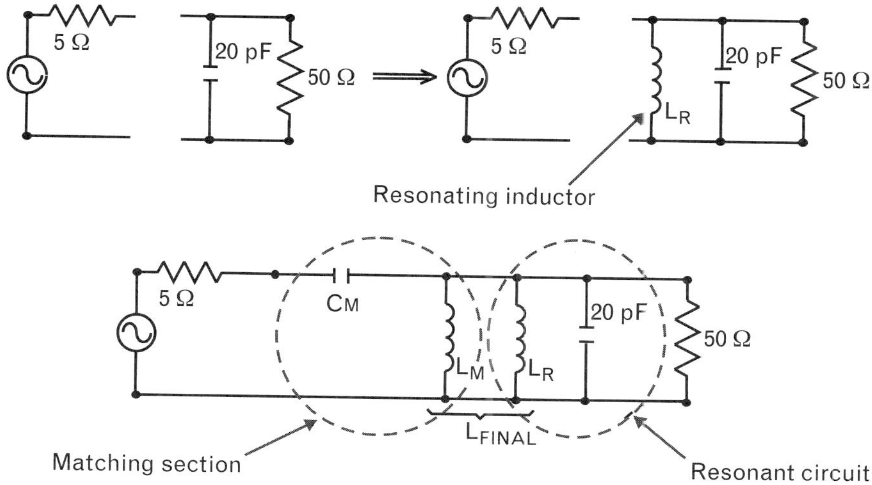

FIGURE 5.26 *When the Q of a termination exceeds the limit set by the nodal-Q calculations, we need to neutralize the excessive amount of parasitic reactance. In this example, the 20-pF parallel load capacitance is 8.8 pF higher than the maximum limit, set by the computed Q of 3.0, and it is fully resonated by the inductor,* L_R

In the top circuit of Figure 5.26 the 20-pF capacitor is fully resonated by a parallel inductor, L_R, leaving the load to be resistive at the resonant frequency. The inductor value to resonate 20 pF at 850 MHz is found from (2.25) as

$$L_R = \frac{25.33}{f_{RGHz}^2 C_{pF}} = \frac{25.33}{0.85^2(20)} = 1.75 \text{ nH}$$

Although it is difficult to create such a small parallel inductor[4], we continue to investigate the resonance effects on bandwidth.

Since the load capacitance is resonated out, the two resistive terminations can then be matched with the highpass configuration *L-C* section C_M and L_M, as shown in the lower schematic of the same figure. This takes us back to impedance matching 5Ω to 50Ω. In Section 5.5.1 we already computed the nodal-Q of this transformation ($Q_S = Q_P = 3.0$), so we can find the highpass matching element values, L_M and C_M from (5.14) and (5.15).

$$L_M = \frac{0.159 R_{\text{LARGER}}}{f_{\text{GHz}} Q_P} = \frac{0.159(50)}{0.85(3)} = 3.12 \text{ nH}$$

$$C_M = \frac{159}{f_{\text{GHz}} Q_S R_{\text{SMALLER}}} = \frac{159}{0.85(3)5} = 12.47 \text{ pF}$$

4. In Chapter 8 we discuss techniques that may help to overcome such problems by transforming parts of a circuit to higher impedance levels.

Finally, to save a component we can combine L_M and L_R into a single parallel inductor, L_{MR},

$$L_{MR} = \frac{L_R(L_M)}{L_R + L_M} = \frac{1.75(3.12)}{1.75 + 3.12} = 1.12 \text{ nH} \tag{5.19}$$

Remember that the parallel resonant *L-C* section only appears to be an open-circuit to the RF signals at one (resonant) frequency and its presence *reduces the 3-dB bandwidth* of the frequency response. We will compare the frequency response of this circuit with the previous example after looking at the partial resonance approach.

We do not have to fully resonate out the 20-pF parallel load capacitance because an 11.2-pF portion of it can be used in the impedance matching circuit, as shown in Figure 5.27. Adding a 2.8-nH series inductor to the input side forms a matching *L-C* section with the 11.2-pF leftover part of the load capacitance. Since only the excess 8.8-pF load capacitance is resonated out, the reduction of 3-dB bandwidth is less severe than it would be if the full 20 pF were resonated.

The inductor value to resonate 8.8 pF at 850 MHz is given by (2.25)

$$L_R = \frac{25.33}{f_{\text{RGHz}}^2 C_{\text{pF}}} = \frac{25.33}{0.85^2(8.8)} = 3.98 \text{ nH}$$

The effect of load capacitance is clearly seen by comparing the frequency responses (Figure 5.28) of the following three circuits, all of them having resistive 5-Ω source terminations:

1. Complex parallel *R-C* load, $R_L = 50\Omega$, $C_P = 3.74$ pF. Parasitic absorption is used, since the load capacitance can be fully merged into the matching capacitance (Figure 5.25).

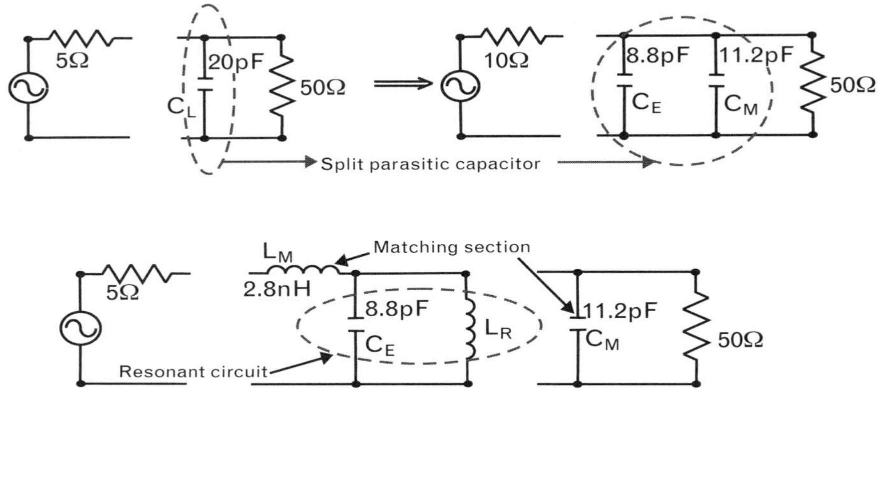

FIGURE 5.27 *If inductor L_R is selected to resonate 8.8 pF of the 20-pF load capacitance, the remaining 11.2 pF can be used with inductor L_M to match the two terminations. We get more bandwidth with this approach compared to the circuit shown in Figure 5.26.*

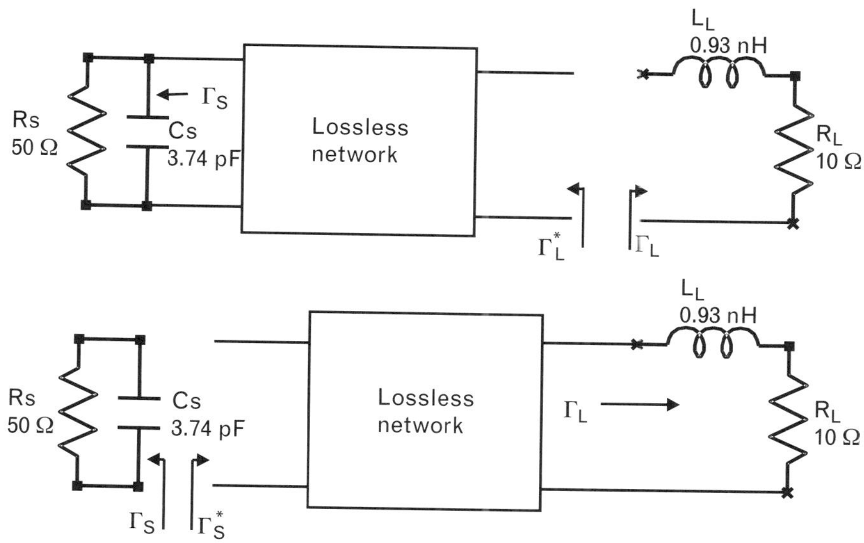

FIGURE 5.28 *Comparison of three matching examples (Figures 5.25 through 5.27) shows that resonating load capacitance reduces bandwidth. When the parasitic capacitance is fully absorbed, the bandwidth is the same as it was for the resistive terminations.*

2. Complex parallel *R-C* load, $R_L = 50\Omega$, $C_P = 20$ pF. Load capacitance is fully resonated and the two resistive terminations are matched with a highpass *L-C* section (Figure 5.26).

3. Complex parallel *R-C* load, $R_L = 50\Omega$, $C_P = 20$ pF. Only 8.8-pF parallel capacitance is resonated. The two resistive terminations are matched with an added 2.8-nH series inductor and the leftover 11.2-pF portion of the load capacitance (Figure 5.27).

5.6.3 Illustrative exercise: impedance matching complex terminations with the Smith chart

Let us illustrate complex impedance matching on the Smith chart. An inductive complex source, $\Gamma_S = 0.82 \angle 168.5°$, is to be matched to a capacitive complex load, $\Gamma_L = 0.45 \angle -116.6°$. In order to compare the results of the graphical and analytical techniques, the two reflection coefficients are equivalent to the series *R-L* and parallel *R-C* terminations used in Figure 5.25. Again, we use lossless reactive elements for the specified task.

Solution

We can either begin at the source or the load side. In this example we work from the source toward the load, moving from lower impedance toward higher impedance. We stated earlier that in order to match complex terminations we transform one of them to the complex conjugate of the second one. Starting from the source, our goal then is to transform to the complex conjugate of the load. We label the starting point at Γ_S as *START* and the completion of the transformation at Γ_L^* with *END* (Figure 5.29).

FIGURE 5.29
Two complex terminations are matched using a circuit which transforms by setting the source and load reflection coefficients to become complex conjugates of one another. (a) Starting from Γ_S we transform it to Γ_L^* with a (b) series inductor-parallel capacitor network.

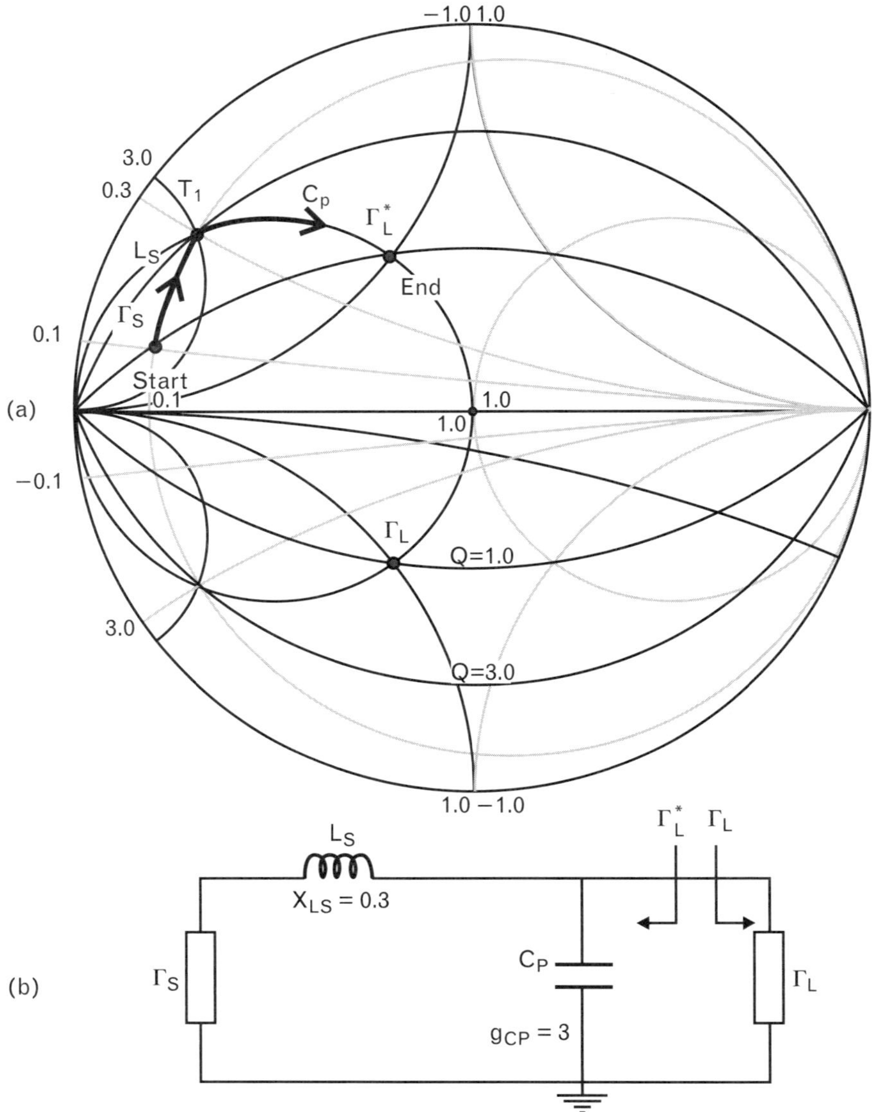

To move from *START* to *END* on the Smith chart requires two elements since there is no common arc between the two circles. Since we are heading toward the higher termination portion of the chart, the logical start is to use a series element first. Using the *Q*-matching approach, there are two circuit choices available, beginning either with a series capacitor (*highpass configuration*) or a series inductor (*lowpass configuration*). Both of them will move us on the constant-resistance circle of $r = 0.1$. If we use a two-element *L-C* section, the second element must be a parallel type that moves on the constant conductance circle. If we mark the $g = 1.0$ circle, depending on the starting point, we can get to the center with either a parallel capacitor or a parallel inductor.

The intersection of the $r = 0.1$ and $g = 1.0$ circles is labeled on the chart as T_1 on the top side and T_2 on the lower side of the chart. These are the possible transfer points for our transformation. We can see that there are two choices from start to end. A series-L-parallel-C lowpass network transfers from constant resistance to constant conductance at T_1. With a series-C-parallel-L highpass topology the transfer point is T_2.

The lowpass circuit shown in Figure 5.29(b) matches Γ_S to Γ_L at 850 MHz. At that frequency the transformation is an exact conjugate match. At other frequencies we experience some mismatch. Frequency response will be shown later.

The second option for the same matching task is to use the highpass circuit mentioned above. The initial and the end points of the transformation are exactly the same as in the previous case. There is a different transfer point, T_2, where we switch from the series capacitor to a parallel inductor. Figure 5.30 shows the movements of the highpass elements on the Smith chart, as well as the resultant highpass matching circuitry.

The lowpass and highpass circuits give us exactly the same match at 850 MHz but have different performance at other frequencies as shown in Figure 5.31. Since the nodal-Q of the highpass transformation is exactly the same as it was for the lowpass circuit, we expect the same bandwidth, but that is not case. Looking at Figure 5.31 shows about one-third less 1-dB bandwidth (~200 MHz versus ~300 MHz) for the highpass circuit because its circuit elements serve dual functions: (1) resonate the parasitics of the terminations, and (2) impedance match 5Ω to 50Ω. Since only parts of the series capacitor and parallel inductor perform effective impedance matching, the highpass circuit cannot cover the same bandwidth as the lowpass circuit, where the circuit elements and the parasitics "team up" for the impedance matching task.

Equivalent circuits of the terminations are often not available and when using an analytic approach it is hard to tell if a matching network absorbs or resonates the parasitics. Comparing the impedance transformation paths of different circuits on the Smith chart can be helpful to visualize the matching process. Also, as a general rule, the shorter the trace of component transformation, the less sensitive is the circuit to frequency changes. Comparing Figures 5.29 and 5.30, we can see that the lowpass circuit transformation path is much shorter than the same of the highpass circuit. Consequently, the lowpass circuits frequency response shows less roll-off as the frequency moves away from the design center.

Matching networks are often expected to perform additional tasks, such as dc biasing, or gain shaping outside of the passband. Accordingly, we may not be able to choose the optimum topology from bandwidth considerations alone, and a single-section matching network may not provide the

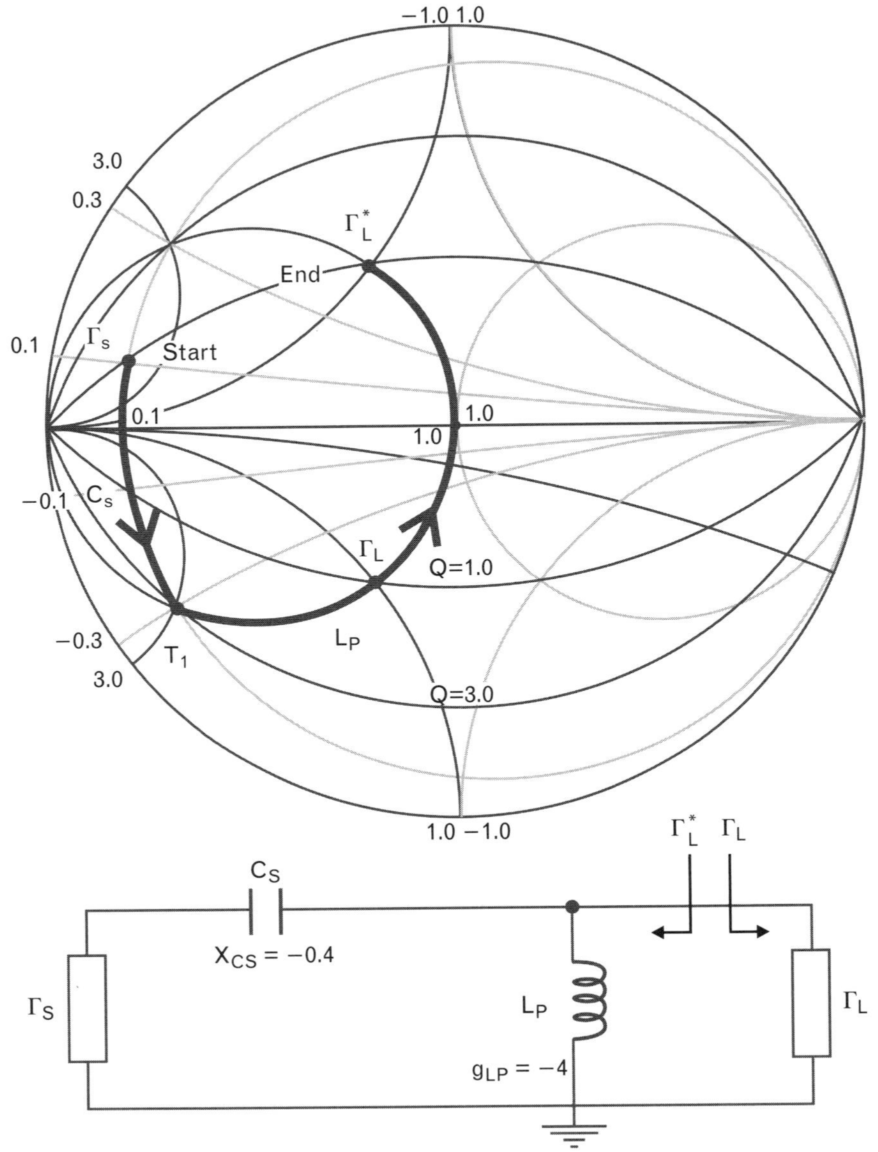

FIGURE 5.30
A two-element highpass section can also transform Γ_S *to* Γ_L^*. *A closer look tells us that both matching components resonate out the parasitics of the terminations, and we can expect less bandwidth compared to the lowpass option.*

bandwidth to meet specifications. In such cases, increasing the order of the matching network may be the only solution.

5.7 Multisection impedance matching to increase bandwidth

If the bandwidth obtained by a single section *L-C* network is not sufficient, we can always increase it by adding another section, as shown in Figure

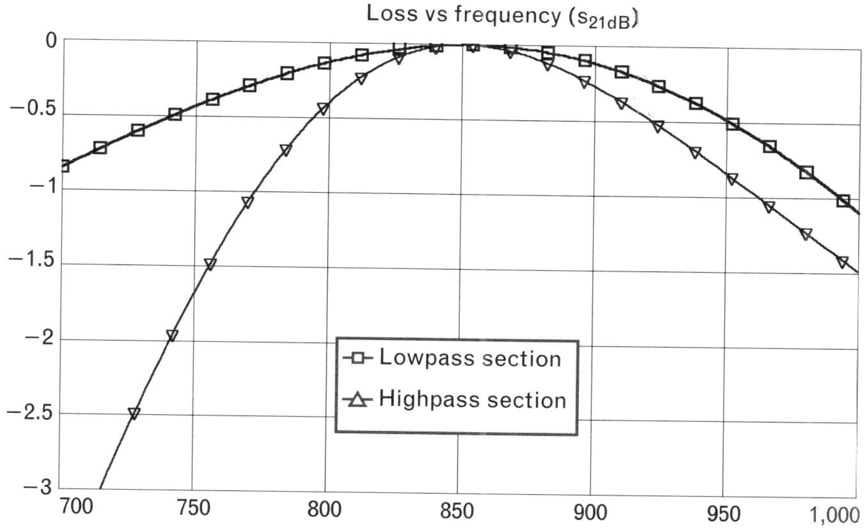

FIGURE 5.31 *Comparing the frequency responses of the lowpass and highpass impedance matching networks shows a considerable difference. The lowpass network absorbs the parasitics on both sides while the highpass circuit resonates them, resulting in narrower bandwidth.*

5.32. If instead of transforming with one L-C section from 5Ω to 50Ω, we first transform to an intermediate impedance level, R_{INT}, and finish the matching job with the second L-C section, the bandwidth increases significantly. If the increase is still not sufficient, adding more sections brings further bandwidth improvement. The improvement is caused by reducing the termination resistance ratio, which in turn reduces the Qs of the matching sections.

For optimum performance this intermediate step should be geometrically spaced between R_S and R_L. Then, the transformation ratio from R_S to R_{INT} is the same as it is from R_{INT} to R_L. Mathematically this means

$$\frac{R_S}{R_{INT}} = \frac{R_{INT}}{R_L} \tag{5.20}$$

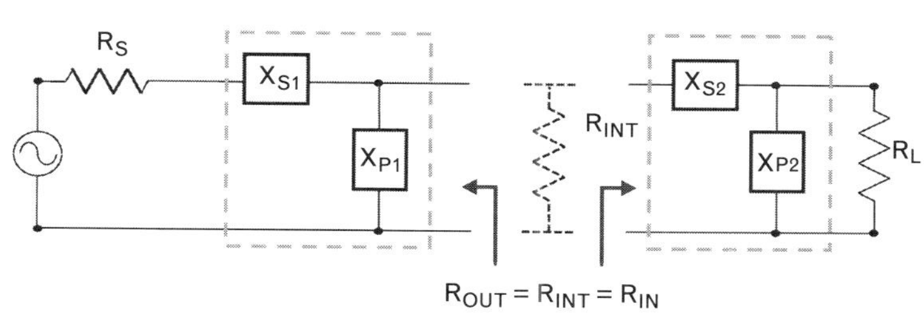

FIGURE 5.32 *Breaking the impedance transformation into two steps increases the bandwidth, as long as the interstage impedance $R_S < R_{INT} < R_L$. For equal component sensitivities, set $R_{INT} = \sqrt{R_S R_L}$. Note that R_{INT} is not an actual component; it only indicates the intermediate impedance level.*

Hence, from (5.20) the intermediate impedance level between two sections is

$$R_{INT} = \sqrt{R_S R_L} \tag{5.21}$$

In our two-section *L-C* network, R_S is transformed to $R_{OUT} = R_{INT}$ at the output of the first section, and R_L is transformed to $R_{IN} = R_{INT}$ at the input of the second section. This way the output of the first section is matched to the input of the second stage, and two networks provide a gradual two-step match between R_S and R_L. Since $R_L/R_{INT}=R_{INT}/R_S < R_L/R_S$, the Qs of the two new matching sections are lower than that of a single matching network, providing wider bandwidth (see Section 5.7.1).

Changing from a single section to a two-section match brings significant bandwidth improvement. Adding a third section does not offer quite the same ratio of improvement. Additional sections bring gradually less and less improvement, and after six to eight sections the improvements are hardly noticeable. With physical circuits, we reach much sooner a point of diminishing return, after which added sections actually *reduce* the effective bandwidth. Depending on the component Q-factors, the practical limit of *L-C* matching sections is somewhere between two to five sections.

The generalized form of the impedance level of a specific interstage, R_m, for n cascaded sections with n - 1 interstages, is

$$R_m = R_S^{(n-m)/n} R_L^{m/n} \tag{5.22}$$

where

n = number of matching sections

$m = 1, 2, 3, \ldots (n - 1)$ is the interstage index counted from the source side.

For example, the interstage resistance levels, R_1 and R_2, of a three-section network, designed to match $R_S = 5\Omega$ to $R_L = 50\Omega$, are computed with $n = 3$, $m = 1, 2$,

$$R_S = 5\Omega$$

$$R_1 = R_S^{(2)/3} R_L^{1/3} = 10.77\Omega$$

$$R_2 = R_S^{(1)/3} R_L^{2/3} = 23.21\Omega$$

$$R_L = 50\Omega$$

We stated earlier that the bandwidth of the transformation is inversely related to the nodal-Q of the *L-C* sections. High nodal-Q leads to narrow bandwidth and vice versa. Since the nodal-Q is a function of the termination ratio, high resistance ratio gives us narrow bandwidth. Exact bandwidth calculation of multisection matching circuits is a complex function of termination ratio, termination and component Qs, and the order of the network [6].

The two *L-C* sections of Figure 5.32 may be realized in four different sets of four-element circuits. Both networks may be lowpass, or highpass, or they can be mixed-one being a highpass and the other one a lowpass. The response of a four-element lowpass network rolls off quickly at the higher frequencies, and the all-highpass network does just the opposite. The remaining two highpass-lowpass combinations give us the best, most symmetrical, bandpass response.

We should emphasize that the intermediate resistance, R_{INT}, is a phantom value, *not a real physical component*. R_{INT} is the resistance that would be seen looking into the matching network if it were split at the middle. When we connect the two matching sections together, the resistor R_{INT} does not exist physically. Instead, the input impedance of the second section represents the load, equivalent to R_{INT}, to the first section. The output impedance of the first section, also equal to R_{INT} acts as the source impedance for the second section. Figure 5.33 may be helpful in finding the applicable reactance-resistance combinations while computing the individual component values.

5.7.1 Illustrative exercise: two-section impedance match for wider bandwidth

Compute a new set of elements to match 5Ω to 50Ω with two matching sections. Use a lowpass section at the 5Ω side and follow with a highpass section for maximum symmetry of the frequency response. Choose the intermediate impedance level for equal amount of impedance transformation by the two sections.

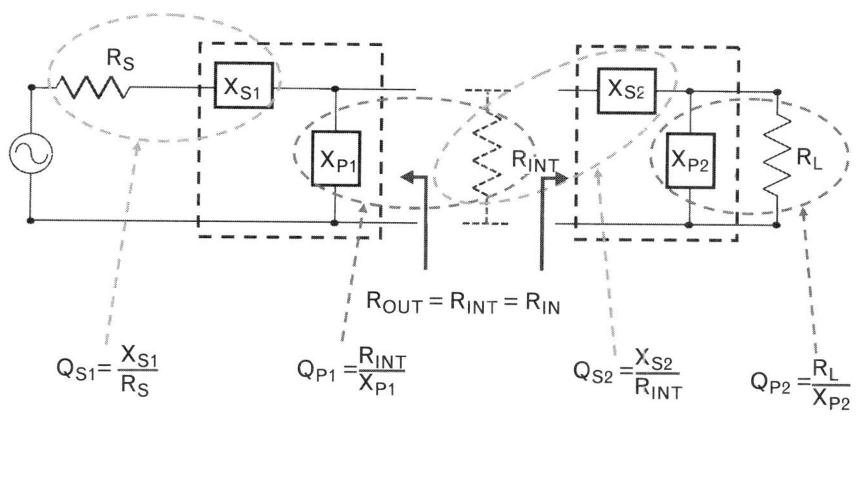

FIGURE 5.33 *The interstage impedance, R_{INT}, may be used to find the values of matching components facing the interstage. Computed Q-factors apply to the series and parallel subnetworks, formed by encircled component pairs.*

Solution

Using the circuit topology of Figure 5.34, we first compute the intermediate impedance, R_{INT}, from (5.21). By selecting the impedance level at R_{INT}, both sections perform the same transformation ratio.

$$R_{INT} = \sqrt{R_S R_L} = \sqrt{5(50)} = 15.81\Omega$$

Computing the nodal-Q values of the lowpass and highpass sections, Q_1 and Q_2,

$$Q_1 = \sqrt{\frac{R_{INT}}{R_S} - 1} = \sqrt{\frac{15.81}{5} - 1} = 1.47$$

$$Q_2 = \sqrt{\frac{R_L}{R_{INT}} - 1} = \sqrt{\frac{50}{15.81} - 1} = 1.47$$

Notice that the two Q values are equal since we selected R_{INT} at the geometric mean of R_S and R_L. Next, we can compute the reactances and component values for both sections, using Figure 5.33 for guidance.

$$X_{S1} = Q_1 R_S = 1.47(5) = 7.35$$

$$X_{P1} = \frac{R_{INT}}{Q_1} = \frac{15.81}{1.47} = 10.75$$

$$L_1 = \frac{0.159 X_{S1}}{f_{\mathrm{GHz}}} = \frac{0.159(7.35)}{0.85} = 1.37 \text{ nH}$$

$$C_1 = \frac{159}{f_{\mathrm{GHz}} X_{P1}} = \frac{159}{0.85(10.75)} = 17.4 \text{ pF}$$

$$X_{S2} = Q_2 R_{INT} = 1.47(15.81) = 23.24$$

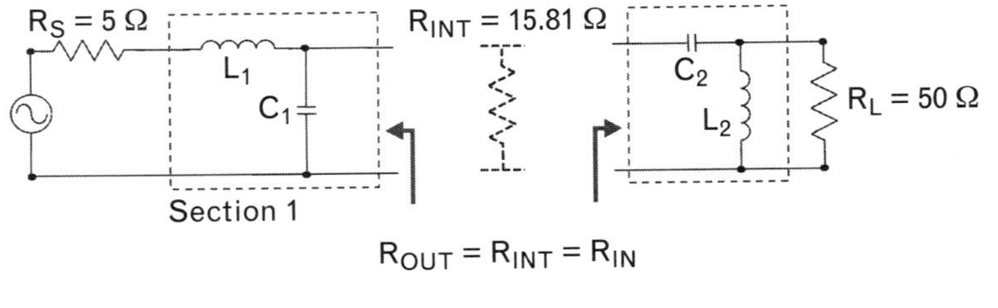

FIGURE 5.34 *Circuit configuration of the two-section, wideband, impedance matching network. Remember that resistor R_{INT} is not part of the circuit. Its purpose is only to show the intermediate impedance level.*

$$X_{P2} = \frac{R_L}{Q_2} = \frac{50}{1.47} = 34.01$$

$$L_2 = \frac{0.159 X_{P2}}{f_{\text{GHz}}} = \frac{0.159(34.01)}{0.85} = 6.36 \text{ nH}$$

$$C_2 = \frac{159}{f_{\text{GHz}} X_{S2}} = \frac{159}{0.85(23.24)} = 8.05 \text{ pF}$$

Simulation of the two-section circuit (Figure 5.35) shows a considerable bandwidth improvement over the single-section performance. The insertion loss is less than 0.1 dB between 805 MHz to 897 MHz (a 92-MHz bandwidth) for the single-section network but covers 706 MHz to 1,028 MHz (322-MHz bandwidth) for the two-section match. Thus, if we define the bandwidth according to the frequencies at which the mismatch loss reaches 0.1 dB, there is more than a 3 to 1 improvement obtained by going from one to two matching sections. Since this required only two additional components, the level of improvement might well be worth the added circuit complexity in many practical applications. Optimizing the circuit for equal-ripple response can further reduce the attenuation at the band-edges if we can accept some loss at the band center (see Chapter 8 for differences between *maximally flat* and *equal-ripple* responses).

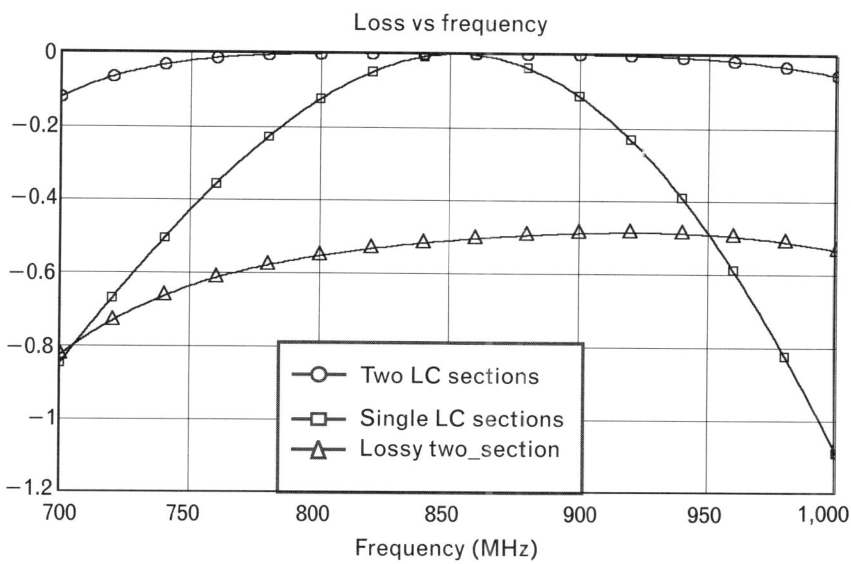

FIGURE 5.35 *Comparison of single-section and two-section lossless networks designed to match 5Ω to 50Ω. Adding losses for all four components, at 850 MHz (i.e., assuming that the components' unloaded Q is 50), introduces 0.5 to 0.7 dB loss through the frequency range.*

We have been using lossless components in our examples, and we will cover real-life components in Chapter 7. If the component Q values are reasonably high, the dissipative losses are generally in the 0.1 to 0.2-dB range per circuit element. Our simulation in Figure 5.35 shows the total loss related to using Q-factors of 50 at 850 MHz is about 0.15 dB per component.

Addition of a second matching section always changes bandwidth. Whether the bandwidth increases or decreases depends on the intermediate impedance level. Our example used an intermediate impedance of 15.81Ω, which is *within* the range of 5Ω and 50Ω. We could actually *reduce* the bandwidth if the intermediate impedance is *outside* the range of the two terminations—less than 5Ω or greater than 50Ω in our previous example.

5.8 Multisection impedance matching to decrease bandwidth

We mentioned earlier that impedance matching circuits may need to perform additional tasks, such as tailoring the frequency response outside of the passband. In the previous example we performed the matching to maximize the bandwidth, but this might not always be the desirable course. Transistor amplifiers, for example, should not have high gain outside the bandwidth of use. Particularly at low frequencies the excessive gain of transistors may lead to undesirable effects, and we may want to limit the passband of the impedance matching circuitry.

To reduce the transmission bandwidth of a matching circuit, we again transform to an intermediate resistance value, but in this case R_{INT} lies outside the range of R_S and R_L. Two possibilities can be employed as shown in Figure 5.36. We can select R_{INT} to be considerably below or above the range of resistance between R_S and R_L. The decision where to place R_{INT} depends on the impedance levels of the terminations and practically realizable component values. For example, matching 5Ω to 50Ω is difficult to realize at impedance levels of a few ohms, therefore, we should select R_{INT} to be on the high side above 50Ω. On the other hand, matching 300Ω to 50Ω requires just the opposite, and R_{INT} of lower than 50Ω impedance is the more practical choice.

In the top circuit of Figure 5.36, the intermediate impedance level is chosen below the smaller termination as shown in Figure 5.18. We can tell that by the topology of the matching sections, noticing that both of the series arms, X_{S1} and X_{S2}, are placed next to R_{INT}. This is a clear indication that R_{INT} is lower than either of the two terminations as we showed in Figure 5.18. In the second circuit schematic the parallel elements of the two matching sections are on the inner side indicating that R_{INT} must be greater than R_S and R_L.

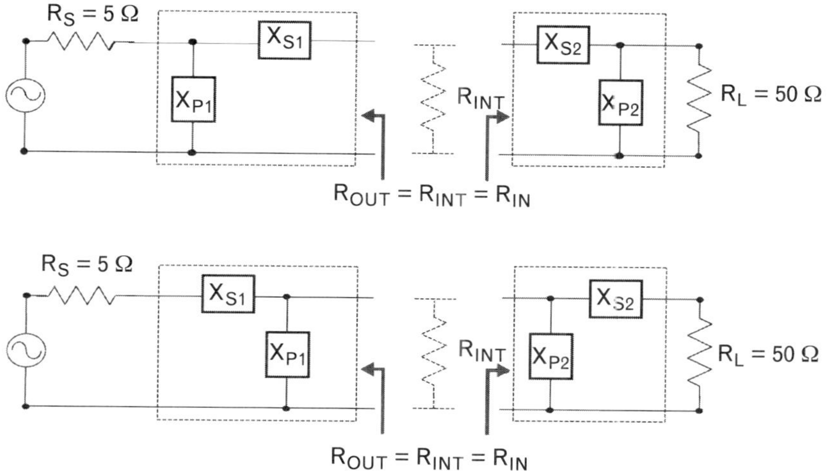

FIGURE 5.36 *Selection of an intermediate R_{INT}, which is either above or below the range of R_S and R_L, provides higher Q and reduced transmission bandwidth. How R_{INT} is selected affects the topology of the two matching sections since the parallel element must always be next to the higher value termination.*

Since R_{INT} is not a real physical element, we can combine the two inner elements of both circuits of Figure 5.36 into single ones. X_{S1} and X_{S2} may be merged into a one series element while X_{P1} and X_{P2} can be reduced into a single parallel component. By merging the two elements together, the top circuit reduces to a Pi-network while the lower one will give us a three-element Tee-network. The reactance of the parallel element, X_{COMB}, is

$$\pm jX_{COMB} = \frac{(\pm jX_{P1})(\pm jX_{P2})}{(\pm jX_{P1}) + (\pm jX_{P2})} \tag{5.23}$$

where X_{P1} and X_{P2} are the reactances of the two elements connected in parallel. Note: Keeping separate highpass and lowpass sections provides the most symmetrical frequency response, but we can save a component by combining the two adjacent parallel components into a single one.

Once again when we build a circuit we have several options. Both subcircuits may be lowpass or highpass or we can use highpass and lowpass

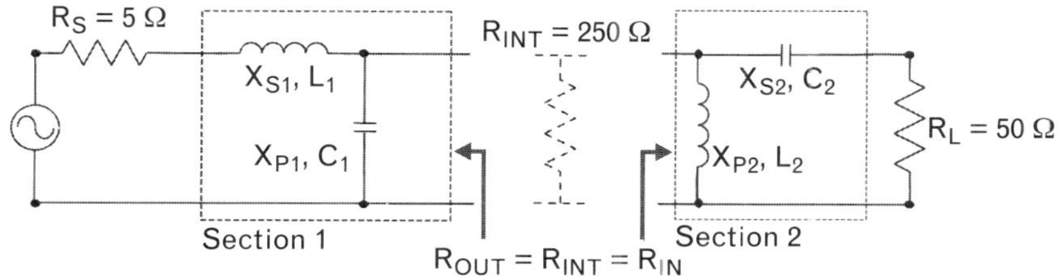

FIGURE 5.37 *Another 5-Ω to 50-Ω matching example using a two-section circuit with a high intermediate resistance, R = 250Ω, to produce a narrow transmission bandwidth. The termination ratio of the lowpass section is 1:50 compared to 5:1 for the highpass section. Overall bandwidth is reduced by the high termination ratio of the lowpass circuit.*

mixed in two different forms. The broadband frequency response of these networks is determined by their transmission zero distributions.

5.8.1 Illustrative exercise: two-section impedance match for narrow bandwidth

As an example of this approach we repeat the 5Ω to 50Ω matching with two sections, but set the intermediate resistance to 250Ω (Figure 5.37). There is no closed-form expression to find the value of R_{INT} for a specific bandwidth, but the farther we move from the $R_S - R_L$ range, the resultant Q gets higher and the bandwidth narrower. Of course the Qs of the two matching sections are not equal, because they face different sets of termination ratios (250/5 and 250/50). The bandwidth of the overall transformation is determined by the higher of the two Q-factors.

Solution

Computing the nodal-Q values of the lowpass and highpass sections, Q_1 and Q_2,

$$Q_1 = \sqrt{\frac{R_{INT}}{R_S} - 1} = \sqrt{\frac{250}{5} - 1} = 7.0$$

$$Q_2 = \sqrt{\frac{R_{INT}}{R_L} - 1} = \sqrt{\frac{250}{50} - 1} = 2.0$$

Notice that the two Q values are not equal since we arbitrarily selected R_{INT} to be 250Ω. The lowpass section with higher Q is the limiting factor of the bandwidth.

Next, we can compute the reactances and component values for both sections, using the Q-values obtained above.

$$X_{S1} = Q_1 R_S = 7(5) = 35\Omega$$

$$X_{P1} = \frac{R_{INT}}{Q_1} = \frac{250}{7} = 35.71\Omega$$

$$L_1 = \frac{0.159 X_{S1}}{f_{\text{GHz}}} = \frac{0.159(35)}{0.85} = 6.55 \text{ nH}$$

$$C_1 = \frac{159}{f_{\text{GHz}} X_{P1}} = \frac{159}{0.85(35.71)} = 5.24 \text{ pF}$$

$$X_{P2} = \frac{R_{INT}}{Q_2} = \frac{250}{2} = 125\Omega$$

$$X_{S2} = Q_2 R_L = 2(50) = 100\Omega$$

$$C_2 = \frac{159}{f_{\text{GHz}} X_{S2}} = \frac{159}{0.85(125)} = 1.5 \text{ pF}$$

$$L_2 = \frac{0.159 X_{P2}}{f_{\text{GHz}}} = \frac{0.159(100)}{0.85} = 18.7 \text{ nH}$$

As we will see, the bandwidth of this matching circuit is considerably narrower than the previous ones, based on lower Qs. Before comparing bandwidths of the different types of circuits, let us examine the differences among the various narrowband matching topologies.

In Figure 5.37 the first section is lowpass while the second one is a highpass configuration. However, as we have seen, each transformation can be implemented using either a lowpass or highpass circuit. For a two-section matching circuit there are four possible combinations (Figure 5.38), each one having a different out-of-passband frequency response.

At the design center frequency, all four options provide an exact match, and from the outside terminals they are exactly equivalent to one another. However, the bandwidth and shape of the frequency response contour varies considerably with each of the four options (Figure 5.39). The all-lowpass circuit (LPLP) converges to 4.8-dB mismatch loss-determined by the mismatch between the two terminations at the low frequencies-and shows an 18-dB/octave roll-off at the high end. The all-highpass network (HPHP) has mirror-image symmetry to the lowpass option and converges to 4.8 dB loss at the high frequencies. The two

FIGURE 5.38
Four possible options exist for the narrowband matching design by mixing lowpass (LP) and highpass (HP) sections. The middle two elements of each circuit can be combined into one with (5.23) to get the three-element Tee-networks shown.

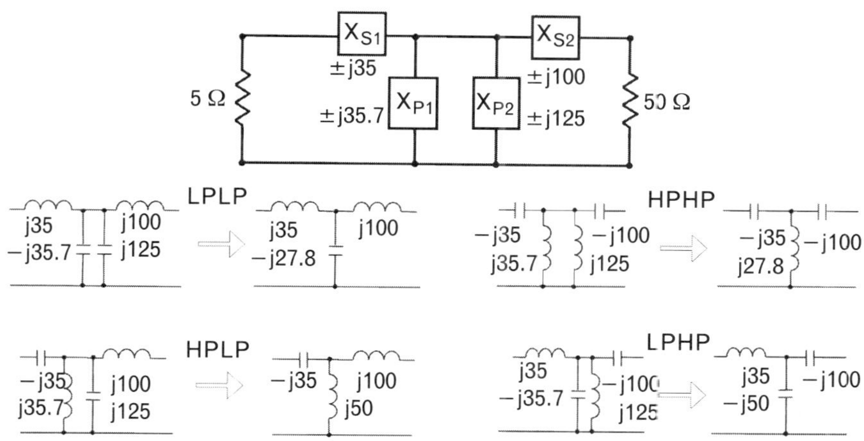

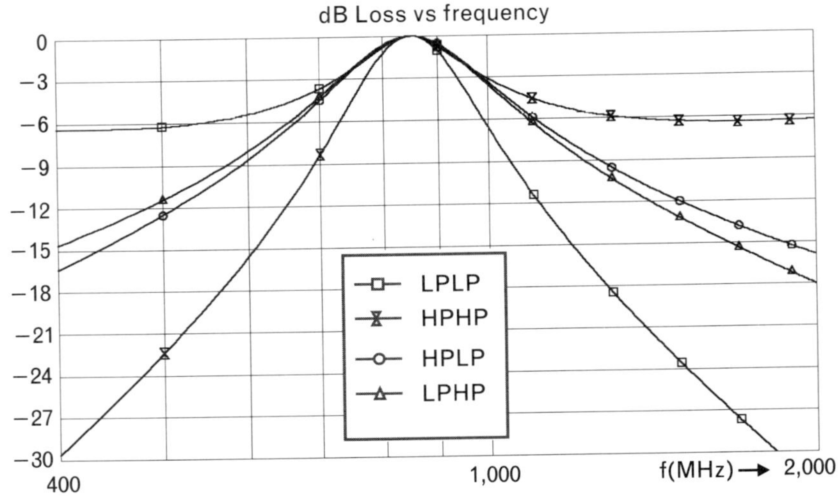

FIGURE 5.39 *Comparison of the four different Tee-circuits used for high-Q matching. For maximum selectivity at high or low frequency, use all lowpass or all highpass topology. Mixing lowpass and highpass elements provides the most symmetrical bandpass frequency response.*

mixed lowpass-highpass combinations (LPHP and HPLP) have the most symmetrical bandpass response near the center frequency.

The effect of relatively small changes in the nodal-Q of matching circuits creates quite a dramatic change in the bandpass characteristics. Figure 5.40 compares the three methods for matching between 5Ω and 50Ω with maximum nodal-Q values of 1.47, 3.0, and 7.0. If we extrapolate the attenuation of the Q = 7.0 case, it yields a 3-dB bandwidth of 180 MHz.

5.9 Impedance matching with transmission line components

We stated earlier that two properly chosen *L-C* elements can always match any point of the Smith chart to any other point. If the two targeted terminations were located on the same constant-conductance circle one single reactive element is sufficient. Since a transmission line has two variables, characteristic impedance and electrical length, we have more flexibility and a wider range of terminations can be matched with a single cascade transmission line. The proper combination of one cascade transmission line and a parallel stub is always capable of matching any two passive terminations.

5.9.1 Impedance matching with a single cascade transmission line

At one frequency, matching two arbitrary complex terminations with a single uniform cascade line (Figure 5.41) requires a unique characteristic impedance, but the length of the line has multiple solutions at additional 180° intervals[5]. Although the physical form of the transmission line does

5. If the solution gives us θ°,there are also repetitive solutions at (θ +180)°, (θ +360)°, and so on.

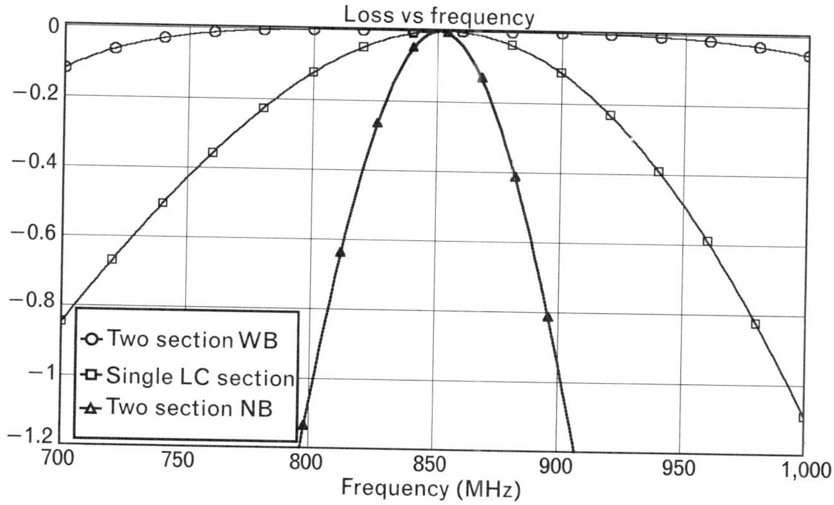

FIGURE 5.40 *Comparison of the transmission bandwidths for matching circuits having nodal-Q values of 1.47 (WB), 3.0 (single section), and 7.0 (NB).*

not affect the impedance match, certain types are more adoptable and present less discontinuities when connected to other components.

The required characteristic impedance, Z_{TL}, and the electrical length, θ, of the transmission line to match two arbitrary complex terminations are

$$Z_{TL} = \sqrt{\frac{\left(R_S^2 + X_S^2\right)R_L - \left(R_L^2 + X_L^2\right)R_S}{R_S - R_L}}, R_S \neq R_L \tag{5.24}$$

$$\theta = \tan^{-1}\left[\frac{Z_{TL}\left(R_L - R_S\right)}{X_S R_L - X_L R_S}\right] \tag{5.25}$$

Note in (5.24) that for cases when real parts of the terminations are equal ($R_S = R_L$), the characteristic impedance goes to infinity, and we do not have a workable solution. However, it is not just that one specific case that gives us physical transmission line realization problem. The char-

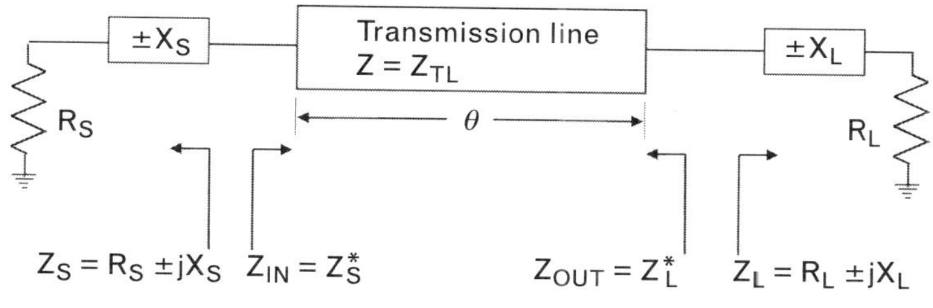

FIGURE 5.41 *A single uniform cascade transmission is capable of matching a wide range of passive terminations, limited by the physical realization of extreme transmission line impedances.*

acteristic impedance may be so low or so high that realization is not practical. Therefore, there is no guarantee that a single transmission line is always a workable solution. Since the tangent function is repetitive, (5.25) leads to multiple solutions.

When both terminations are real, $X_S = X_L = 0$, both (5.24) and (5.25) are simplified to

$$Z_{TL(REAL)} = \sqrt{R_S R_L} \tag{5.26}$$

$$\theta_{(REAL)} = 90°, 270°, 450°, \ldots \tag{5.27}$$

Equations (5.26) and (5.27) are simple enough to memorize and can be very helpful for quick calculations. They also display the wide range of potential impedance transformations that can be achieved with a transmission line. For example, a quarter-wave-long 50-Ω line can transform 25Ω to 100Ω, 10Ω to 250Ω, 5Ω to 500Ω, and, yes, a short circuit to an open circuit. These transformations are bidirectional, also working from high impedances to lows. Although (5.27) indicates multiple choices for minimum loss, the 90° option is the best.

5.9.2 Illustrative exercise: impedance matching with a cascade transmission line

Find the parameters of the single cascade transmission line to match two complex terminations shown in Figure 5.42, at 1 GHz, using $Z_S = (10 - j20)\Omega$ and $Z_L = (50 + j100)\Omega$. Find the physical dimensions of the line on FR-4 dielectric, having relative dielectric constant, $\varepsilon_R = 4.2$, and dielectric thickness, $h = 31$ mil $= 0.79$ mm, copper conductor thickness, $t = 0.7$ mil $=$ 17.8 μm, dielectric loss factor, $\tan\delta = 0.025$.

Solution

Finding the characteristic impedance from (5.24),

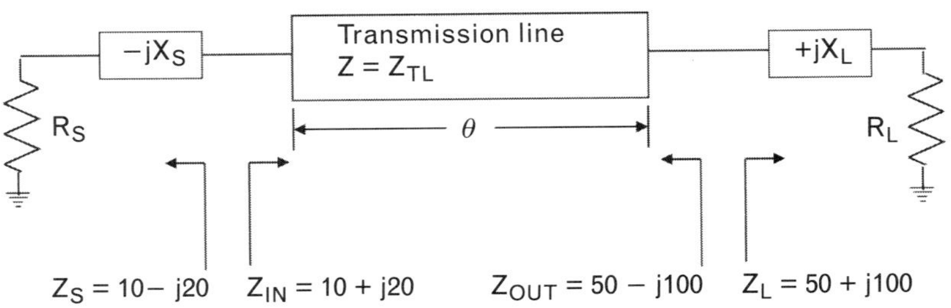

FIGURE 5.42
If the transmission line parameters are reasonable, the two complex terminations may be matched with a single transmission line.

$$Z_{TL} = \sqrt{\frac{\left(R_S^2 + X_S^2\right)R_L - \left(R_L^2 + X_L^2\right)R_S}{R_S - R_L}}$$

$$= \sqrt{\frac{\left(10^2 + 20^2\right)50 - \left(50^2 + 100^2\right)10}{10 - 50}} = 50\Omega$$

The shortest electrical length is computed from (5.25),

$$\theta = \tan^{-1}\left[\frac{Z_{TL}\left(R_L - R_S\right)}{X_S R_L - X_L R_S}\right] = \tan^{-1}\left[\frac{50(50 - 10)}{(-20)50 - (100)10}\right] = 135°$$

Electrical parameters of the transmission line are converted to physical dimensions with one of the free software packages [7, 8] available through the Web. The dimensions are: ℓ = 2,470 mil = 6.27 cm, w = 60.8 mil = 1.54 mm. Losses are computed as 0.26 dB for the line length.

5.10 Impedance matching with transmission lines on the Smith chart

The Smith chart is very convenient for impedance matching problems with transmission line elements. Cascade transmission lines always follow a clockwise rotation on the chart. *If the chart is referenced (normalized) to the characteristic impedance of the transmission line, the rotation takes place on a concentric circle, through twice the electrical length of the line.* The radius of the concentric circle is determined by the normalized termination. A parallel stub is treated as an equivalent parallel inductor or capacitor at specific frequencies, depending on what type of susceptance it represents.

If we use several cascade lines with different characteristic impedances, the Smith chart must always be *renormalized* to the appropriate impedance and the process becomes more tedious for manual work. We highly recommend one of the many standalone commercially available Smith chart software programs to augment specific CAD programs since these smaller programs are generally more flexible and reasonably priced. We will talk more about CAD in Chapter 6.

In the previous chapter, while introducing transmission line manipulations on the Smith chart, we mentioned that many schools teach transmission movements into two different directions on the chart. Even the commercially available immittance chart shows two directions: one labeled "wavelengths toward generator," which is a clockwise movement, and one labeled "wavelength toward load," a counterclockwise direction that may be misinterpreted. We suggest a simpler rule for uniformity: *Moving away from a termination on a transmission line, always follow a clockwise circular rotation on the Smith chart. If the chart is normalized to the characteristic impedance of the transmission line, the rotation is along a concentric circle.*

In our classrooms the question comes up frequently: If you have so much respect for the Smith chart, why don't you follow the old rules written on it? Let me tell you a true story that shows how I got into trouble misinterpreting those rules. Perhaps my experience convinces you also to use our simpler method.

As a fresh-out-of-college engineer, while working in the microwave division of Hewlett Packard, I, Les Besser, needed to match the output impedance of an active circuit, $Z_{OUT} = (50 - j60)\Omega$ at 4 GHz to a load $R_L = 50\Omega$. The port also required dc ground return. If the ground return were not needed, and if the real part of the source termination were not equal to the real part of the load, perhaps a single cascade line could have worked. (The problem with $R_L = R_S$ is that the impedance of the matching cascade line goes to infinity.) Then, the parameters of the line could have been found from (5.24) and (5.25). As an alternative, I chose a two-element topology starting with a cascade transmission line, followed by a short-circuited parallel stub, as shown in Figure 5.43.

I turned to the Smith chart to determine the parameters of the two transmission lines of Figure 5.43. After all, I had three semesters of microwave courses in my undergraduate program where I had learned lots of theory and was ready to apply some of it. Like most other people, not wanting to complicate the task by using different transmission line impedances, I decided on 50Ω for both components and used a chart normalized to 50Ω. After marking the normalized source (generator) impedance, $z_S = (1 - j1.2)$ on the normalized chart, I was ready to find the electrical lengths of the two elements.

Since I was moving away from the generator the active circuits toward the load, I followed a counterclockwise direction on the chart as shown on Figure 5.44(a). After building the circuit, it did not work anywhere near my expectation. Instead of 50Ω, the real part of the output impedance was closer to 10Ω, and highly inductive. Tweaking the circuit did not help. Since this was before the CAD design tools existed, it took me a long time to correct the wrong direction of rotation [Figure 5.44(b)]. Had I known

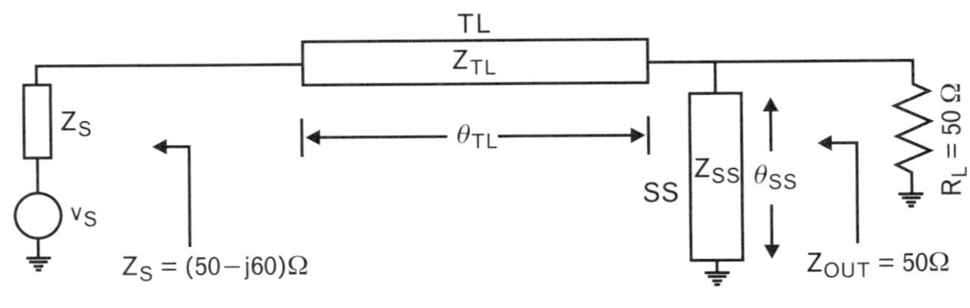

FIGURE 5.43 *The complex source can be matched to the 50-Ω load at 4 GHz with a cascade line and a parallel short-circuited stub.*

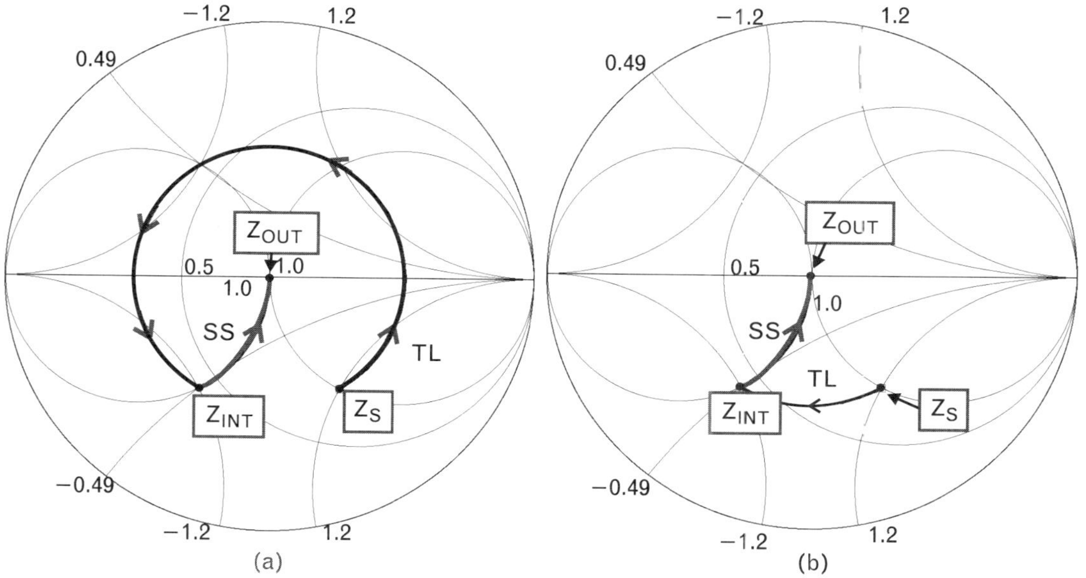

FIGURE 5.44 *Improper and proper solution to the matching problem. The concentric circle movement from Z_S corresponds to the cascade transmission line. From the point of Z_{INT} the parallel stub continues the transformation. (a) Following a counterclockwise rotation on the chart is equivalent to de-embedding, which is incorrect for this application. (b) Moving away from any termination (whether it is called "source" or "load") with a transmission line always leads to clockwise ro-*

the simple rule stated in Section 5.9.2, I would have saved several days of labor and much embarrassment.

Later, reviewing the problem, I finally understood how the two rules need to be interpreted correctly, but from time to time I still had trouble remembering whether to rotate one way or the other. For that reason, during the past (several) decades I have been following the simpler way of rotating clockwise when adding a cascade transmission line to any kind of termination. I have not been wrong since, and neither will you be by doing the same.

5.10.1 Illustrative exercise: impedance matching with a cascade transmission line

The two complex terminations shown in Figure 5.45 represent the output and input impedances of two transistor stages we want to match to each other. Find the parameters of a single cascade line to match the two transistor stages at a single frequency, using the Smith chart.

The reflection coefficient of the complex terminations are given below and are also plotted on the normalized Smith chart in Figure 5.45. The reflection coefficients, as well as the Smith chart, are normalized to 50Ω.

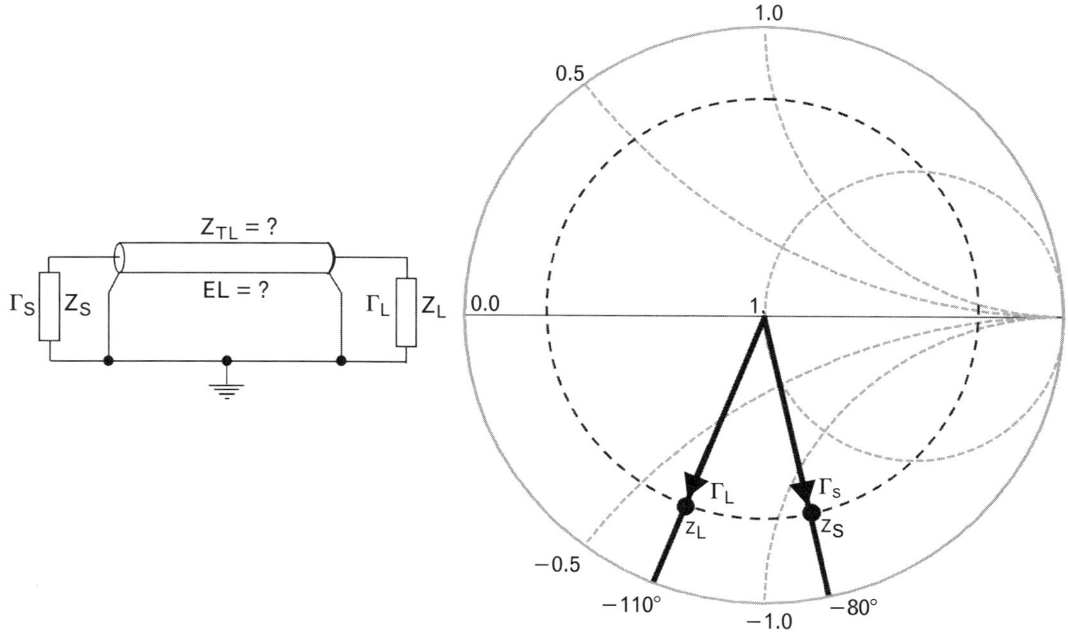

FIGURE 5.45 *Find the characteristic impedance and electrical length of the cascade line to match the two specified terminations at one frequency. The Smith chart is normalized to 50Ω and the magnitudes of* Γ_S *and* Γ_L *are both equal to 0.66.*

$$\Gamma_S = 0.66 \angle -80^\circ$$
$$\Gamma_L = 0.66 \angle -110^\circ$$

Note that the magnitudes of the reflection coefficient are the same for both terminations. Solve the problem first on the Smith chart then verify it by using (5.24) and (5.25). Hint: There is sufficient information on the Smith chart to solve the problem graphically, without resorting to any sophisticated tool or even to a calculator.

Solution

When we assign this problem to our class, the typical answer is 50-Ω characteristic impedance and 30° electrical length. The reasoning behind these answers is that beginning from Γ_S, a 50-Ω characteristic impedance line follows a concentric circle on a 50-Ω normalized chart. A 30° clockwise rotation on a concentric circle moves us from −80° to −110°, and we arrive to Γ_L.

The problem with this answer is that we need to remember that on the Smith chart we always rotate *twice* the electrical length of a transmission line. Thus, the question is whether the 50-Ω impedance and 15° electrical length are the correct answer? If you agree, then you need to go back and reread Section 5.6 where we stated a fundamental rule for impedance matching. Let us state it again here. To match two complex terminations, transform one termination to the complex conjugate of the second termination. By using the 15° long 50-Ω transmission line we are transforming Γ_S to Γ_L rather than matching. We obtain what looks like the two water pipes connected together the wrong way in Figure 5.1. For proper impedance matching, we need to transform Γ_S to the *complex conjugate* of Γ_L, which is located at 0.66 ∠ +110°. Therefore, we need to rotate from −80° to +110°, which corresponds to a 170° movement on the chart. The electrical length of the transmission line is half as much, or 85°. So, the proper answer is 50-Ω characteristic impedance and 85° electrical length. By the way, using the proper matching element, the input side of the lossless cascade transmission line is also automatically matched to the complex source.

5.10.2 Parallel stub manipulations on the Smith chart

Parallel stubs represent inductive or capacitive susceptances, depending on their electrical lengths. Equations (2.42) and (2.44) specify the susceptances of short-circuited and open-circuited parallel stubs with given characteristic impedances (Z_{SS} or Z_{OS}) and electrical lengths (θ_{SS} or θ_{OS}). For convenience of Smith chart transformations, we rewrite here the two equations in normalized forms.

$$b_{SS} = \frac{Z_0}{Z_{SS}\tan\theta_{SS}} \tag{5.28}$$

$$b_{OS} = \frac{Z_0 \tan\theta_{OS}}{Z_{OS}} \tag{5.29}$$

where Z_0 is the reference impedance of the Smith chart, which may be different from the characteristic impedance of a stub. When the susceptance of a stub is known, the electrical length corresponding to a specified characteristic impedance is found from (5.30) and (5.31).

$$\theta_{SS} = \tan^{-1}\left(\frac{Z_0}{b_{SS} Z_{SS}}\right) \tag{5.30}$$

$$\theta_{OS} = \tan^{-1}\left(\frac{b_{OS} Z_{OS}}{Z_0}\right) \tag{5.31}$$

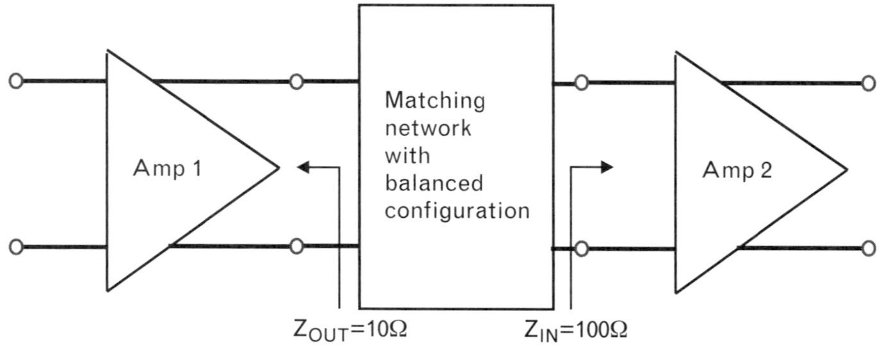

FIGURE 5.46
Matching balanced circuits requires a floating balanced network that seems to be a difficult one to derive.

5.11 Impedance matching of balanced circuits

Impedance matching of balanced circuits can be reduced to that of single-ended circuits by introducing a virtual ground between the terminals of the balanced circuit ports. Then, we can match the individual halves of the balanced circuits the same way, as we did the single-ended circuits. After the networks are derived, the virtual ground may be eliminated to save components, as shown in the following example.

5.11.1 Illustrative exercise: impedance matching of differential amplifiers

Match the 10-Ω output impedance of a differential amplifier to the 100Ω input impedance of the second amplifier at 850 MHz (see Figure 5.46), using the analytical *Q*-matching approach and lowpass *L-C* circuits.

Solution

By creating a virtual ground through the interstage, the 10-Ω output impedance of the first stage is split into two 5-Ω outputs. Applying the same to the input of the second amplifier creates two 50-Ω inputs, as

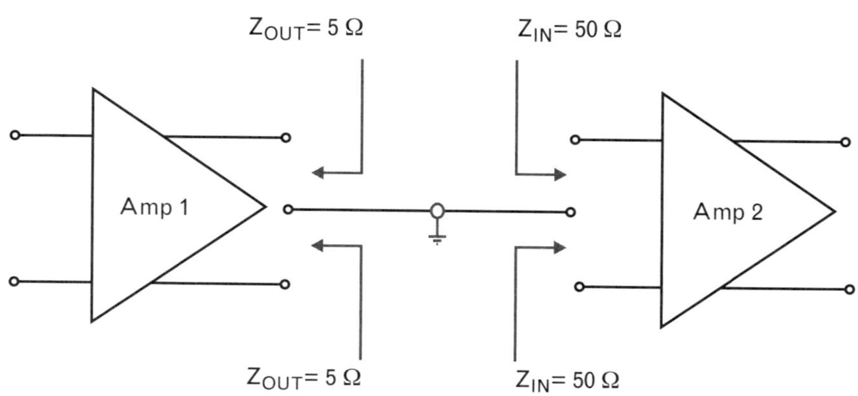

FIGURE 5.47
With the help of a virtual ground the | balanced ports can be split into unbalanced ports. Input and output impedances of the unbalanced ports are half of the balanced ports'.

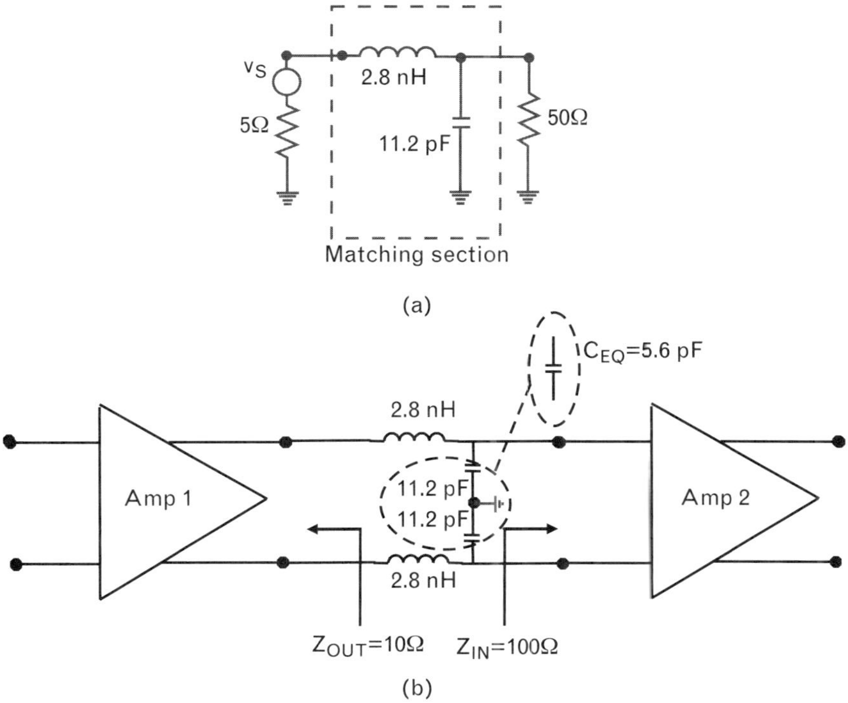

FIGURE 5.48 *(a) The unbalanced L-C circuit to match 5Ω to 50Ω is redrawn from Figure 5.21. (b) After establishing the unbalanced matching circuits the virtual ground can be eliminated. The two 11.2-pF capacitors can be combined into a single floating 5.6-pF capacitor, C_{EQ}. The two balanced ports are now matched to each other at 850 MHz.*

shown in Figure 5.47. As a result, we need to match unbalanced port impedances of 5Ω to 50Ω with two separate circuits.

Using the results of Section 5.5.1, we add the two *series inductor – parallel capacitor* matching sections between the 5-Ω and 50-Ω ports. Lifting the virtual ground combines the two 11.2-pF capacitors into a single 5.6-pF capacitor, leaving a three-element balanced matching network as shown in Figure 5.48.

5.12 Answers to illustrative exercise of Section 5.2.1 (circuit 4)

Case 1

Looking at the number of transmission zeros at dc, our first impression is that we have three parallel inductors L_2, L_6, and L_{10} causing three transmission zeros. After looking more carefully, however, we realize that as we approach dc, inductors L_4 and L_8 become short circuits, connecting L_2, L_6, and L_{10} in parallel with each other. Since the three parallel inductors can be combined into a single one, we get only *one transmission zero at the dc.*

The same effect is true at the high-frequency end. As we approach infinity, capacitors C_3 and C_7 become short circuits, parallel connecting C_1, C_5, and C_9. Once again, even though it appears to have three zeros at infinity, we have only one.

Finally, let us look at finite transmission zeros. Even though we have five resonant circuits, only two of them block transmission at specific finite frequencies. Resonant circuits formed by C_3-L_4 and C_7-L_8 definitely block transmission at finite frequencies. Although the other three parallel networks do resonate at certain finite frequencies, they do not block transmission. The two resonant circuits contribute $2 \times 2 = 4$ zeros at finite frequencies.

We can find the order of the circuit by adding the transmission zeros: one at dc, one at infinity, and four at finite frequencies, for a total of six. This is a sixth-order network, with 6-dB/octave roll-off at the low frequency end, −6 dB/octave at the high frequencies, and two finite frequency resonances that may appear at the low or the high side of the bandpass, depending on the component values of the resonant circuits.

Although we would rarely use such a topology for impedance matching, we wanted to illustrate how to determine the order of a more complex network that may be used later as a filter.

Case 2

We first remove two components, L_4 and L_8, from our initial circuit, reducing the number of components to eight. What is the new order?

The number of transmission zeros at dc now *increases* to five, since we removed the two inductors that previously bypassed capacitors C_3 and C_7. The five transmission zeros are caused by L_2, C_3, L_6, C_7, and L_{10}. The low-frequency response slope of the circuit increases drastically to 30 dB/octave from the previous 6 dB/octave.

At high frequencies the number of transmission zeros has not changed. We can still combine the three parallel capacitors C_1, C_5, and C_9, leaving us only a single transmission zero.

The finite transmission zeros no longer exist since we removed the two inductors, L_4 and L_8, from our resonant circuits.

The order of the network interestingly has not changed, even though we removed two components. In the previous case we had one transmission zero at dc, one at infinity, and four at finite frequencies giving us an order of six. Now we have five transmission zeros at dc, one at infinity and none at finite frequencies. The order is equal to six. The prime difference between the two circuits is that now we have a 30-dB/octave roll-off at the low frequencies with still only −6 dB/octave at the high frequency. This is a highly skewed network response, in favor of the low frequencies.

TABLE 5.3 NETWORK ORDERS FOR ILLUSTRATIVE EXERCISE OF SECTION 5.2.1 (CIRCUIT 4) SHOWING THAT REDUCING THE NUMBER OF COMPONENTS FROM 10 TO 6 DOES NOT CHANGE THE ORDER

	CASE 1 NO REMOVAL	CASE 2 REMOVE L_4, L_8	CASE 3 REMOVE L_4, L_8, C_1, C_9
Number of components	10	8	6
Zeros at $f = \infty$	1	1	1
Finite-frequency zeros	4	0	0
Zeros at $f = 0$	1	5	5
Network order	6	6	6

Note: The slope of the frequency-response at low frequencies changes significantly when moving from case 1 to case 2.

Case 3

Next, we remove two more components C_1 and C_9, cutting the number of elements down to six. Interestingly, the transmission zero distribution has not changed this time. We still have five zeros at dc and one at infinity. Removing the two capacitors *did not change the order* or the shape of the frequency response. The main purpose of those capacitors was impedance transformation.

At this point we have reached what we call the *minimum configuration*, sometimes called the canonic form of the circuit. This is a case were the number of components is equal to the order of the network. If you use a computerized synthesis program, this is probably the initial configuration the program provides, unless impedance transformation is specified. Our exercise showed that it is not a good practice to rely on component count to determine the network's order. Even though the order and the number of components may be equal at times, the safest way to determine the order is to find the sum of nonredundant transmission zeros.

Tabulated results of this exercise are shown in Table 5.3.

5.13 Summary

Impedance matching is one of the most important parts of RF circuit design. We have emphasized the importance of maximum power transfer from one system block to the other and proper impedance matching can help us to achieve that goal.

The performance of lumped *L-C* circuits is often limited by their *Q*-factor and parasitics. In this chapter we did most of our work with ideal components and will look at real-life component models in Chapter 7, using surface mount technology.

Lumped elements are difficult to model beyond the 1- to 2-GHz range and we need to switch to distributed, transmission line components for the higher frequencies. Transmission line models are reliable up into the microwave region but when we use distributed components the printed circuit board becomes part of our circuit. In Chapter 7 we will also look at possible problems related to circuit board materials and losses.

Computer-aided impedance matching synthesis is covered in Chapter 6, using commercially available software.

References

[1] Temes, G. C., and J. W. LaPatra, *Introduction to Circuit Synthesis and Design*, New York: McGraw-Hill, 1977.

[2] Zverev, A. I., *Handbook of Filter Synthesis*, New York: Wiley, 1967.

[3] Matthaei, G. L., L. Young, and E. M. T. Jones, *Microwave Filters, Impedance Matching Networks, and Coupling Structures*, New York: McGraw-Hill, 1964.

[4] Szentirmai, G., (Ed.), *Computer-Aided Filter Design*, New York: IEEE Press, 1973.

[5] Gonzalez, G., *Microwave Transistor Amplifiers Analysis and Design*, 2nd ed., Englewood Cliffs, NJ: Prentice Hall, 1997.

[6] Cuthbert, T. R., Jr., *Circuit Design Using Personal Computers*, New York: Wiley, 1983.

[7] TXLINE, http://www.mwoffice.com/products/txline.html, Applied Wave Research, Los Angeles, CA, 2002.

[8] Ansoft Designer SV, http://www.ansoft.com/ansoftdesignersv.

Selected bibliography

Abrie, P. L. T., *Design of RF and Microwave Amplifiers and Oscillators*, Norwood, MA: Artech House, 2000.

Ludwig, R., and P. Bretchko, *RF Circuit Design, Theory, and Applications*, Englewood Cliffs, NJ: Prentice Hall, 2000.

Pozar, D. M., *Microwave Engineering*, Reading, MA: Addison-Wesley, 1990.

Weber, R. J., *Introduction to Microwave Circuits*, New York: IEEE Press, 2001.

Chapter 6

CAE/CAD of linear RF/MW circuits

6.1 Introduction

Until the 1960s, RF and microwave circuit design was an art rather than a science. The common belief was that component modeling was inaccurate and complex; therefore, *design on the bench* was the common practice. "Design" usually meant a laboratory process of cutting, pasting, shielding, and tweaking circuits that were initially put together by estimates. There was very little correlation between theory and practice. Computers were rather crude, unfriendly, expensive, and design software for technical applications was virtually nonexistent. Learning to use a computer was often more complicated than the task they were intended to perform. To make matters worse, dependable test instruments were not available; most component measurements were the magnitude-only type and did not provide meaningful and complete characterizations.

The push for miniaturization by the space industry provided a strong incentive for accurate simulation of microwave components and circuits. Development of the network analyzer enabled designers to make reliable and accurate measurements (magnitude and phase) of a new parameter set based or traveling waves, instead of voltages and currents. The concept of computer time-sharing, available to the masses, forced the development of simple, understandable operating systems; it also brought powerful computers to the doorsteps of engineers. *Computer-aided engineering* (CAE) and *computer-aided design* (CAD) were born and slowly gained acceptance.

RF/MW CAD initially progressed only in the area of small-signal, linear circuit design—focusing on the analysis and optimization of discrete and hybrid *microwave integrated circuits* (MICs). Linear circuit simulation and optimization software was relatively easy to learn and required little background preparation from the user. Other equally important areas, such as yield-analysis, toleranced design, and statistical data collection, did not progress much since the industry's focus was on technical performance alone, rather than cost and efficiency. Neither had we developed effective computer-aided communication links among the R&D, production, and quality assurance departments. Information on test results and quality were not shared properly with development engineers, and design data was generally not available to production. Without sharing critical data, these groups could not work together as a team.

Introduction of low-cost mini and personal computers in the 1980s brought easily accessible and more user-friendly computing power into the development laboratories. Designers did not have to stand in line and fill out requisitions to have access to computers. The one-shop microwave design "software industry" became competitive, developing new products to handle nonlinear circuits, communication systems, and eventually *electromagnetic* (EM) simulation.

The abrupt end of the Cold War would have been a major setback for microwave CAD had it not been for the timely emergence of personal communications—pagers and cellular telephones. In a short time the focus quickly shifted from defense electronics to more peaceful applications, dropping the frequency range from *microwaves to RF*. Digital technology has moved in to replace some of the analog communication systems and has brought new demands to handle mixed-mode circuits and systems. The cost-conscious consumer market forced large-scale product integration, and CAD tools needed to support a wider range of user base. More and more engineers are becoming generalists instead of being specialized in narrow fields, and they want integrated CAD programs to cover all their needs.

We cannot overemphasize the importance of systematic, computer-aided circuit and system development in today's fierce global competitive environment. Production problems almost always can be traced to poor engineering development practices. Important steps, like finding the optimum circuit solution and performing yield analysis, are frequently overlooked to cut the development-to-market cycle. Design mistakes are expensive to fix in later stages because the cost increases roughly by tenfold at each step of building the product (Figure 6.1). Indirect effects, such as

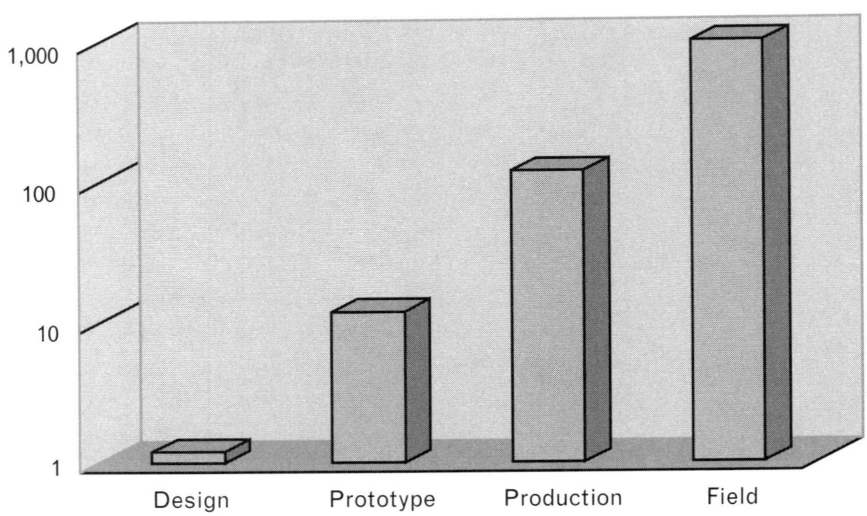

FIGURE 6.1 *The cost of redesign increases by a factor of 10 through each step a product goes through. Getting it right the first time pays off. (After: [1].)*

loss of goodwill when the defect is discovered by the end user, are hard to measure but they can be even more costly.

In this chapter, after a brief historical review, we examine key areas of linear computer-aided design, including circuit, system, and electromagnetic simulation, impedance mapping, tuning, synthesis, optimization, and statistical techniques. We also show illustrative examples and outline the features of the major software products. Our aim is to provide familiarization of the various techniques, their capabilities, and limitations, rather than study the algorithms and related theory. We provide references for those who want to be more deeply involved with any of the techniques.

6.2 Historical review

Computer-aided electronic circuit design was in its infancy in the 1960s. Early mainframe computer programs such ECAP and SCEPTRE were among the first products capable of simulating small-signal frequency response, dc operating point, and transient response with different input signals. Access to mainframe computers was limited to those working at very large companies, and even for them, the unfriendly operating systems posed serious problems. Scientific computer languages such as FORTRAN were also very limited and initially did not handle complex algebra or matrix operations.

Introduction of computer timesharing in the mid-1960s brought computing power to the doorsteps of engineers, with improved user interface at reasonable cost [2]. The various timeshare services offered the first generation circuit simulation packages through their systems, and electronic circuit design was gradually adopted by the engineering community. The first two-port microwave simulator, MICAP, introduced by Tymshare in 1969, was soon followed by SPEEDY [3], which included Fairchild Semiconductors' microwave transistor *S*-parameter database. Neither simulators had editing or graphical output, and they handled cascaded circuit structures only.

In the 1970s the increased demand for integration prompted the University of California, Berkeley to develop a large general-purpose program for integrated circuit design [4]. Their program, called SPICE, initially ran on mainframe computers and later also via timeshare services. Since the main purpose of SPICE was to handle low-frequency analog integrated circuits, it did not have the kind of input/output and component library needed by microwave engineers and was not useful for the microwave community. Other universities and some of the large companies explored microwave circuit simulation [5, 6], but generally there was very little available to the engineering community. (The operation of SPICE is covered in more detail in Volume II, Chapter 4.)

Exactly 30 years prior to publishing this book, the first commercially successful second generation microwave analysis program called COMPACT was introduced through timesharing services [7]. The first release of the program included two-port circuit simulation, multiband, multimode optimization, series and parallel feedback connections, and a transistor *S*-parameter data bank containing microwave transistors of several vendors. Subsequent releases added nodal and noise analysis, printer graphics, and statistical design. By the end of the 1970s the program had about 300 users worldwide and close to 50 on-site installations. A few other microwave programs also appeared through timesharing, like Magic and Match, but they did not survive because they did not offer circuit optimization.

In 1980 COMPACT was completely rewritten and released under the name of SuperCompact [8] for the newly introduced mini-computers such as the DEC PDP-10 system. This third generation version included physical transmission line models and discontinuities, true graphics, matching network synthesis, and an interactive Smith chart subprogram, SmithTool (which, by the way, survived the times and still exists in Ansoft's Designer). Together with the circuit layout program, Autoart, SuperCompact set a new standard for microwave computer-aided design. However, despite its commercial success, SuperCompact only provided help to those who were involved with passive, or linear small-signal active circuits. Engineers dealing with nonlinear problems were still using empirical techniques.

Compact Software's monopoly ended in 1984 when a newly formed company, EEsof, came on the market with the first PC-based program called Touchstone. Although Touchstone initially came with only a fraction of SuperCompact's capability, having microwave design capability on a personal computer became so popular that even the developers of Touchstone could not believe their own success. One of the unique features of Touchstone was its interactive tune-mode that allowed designers to interact with circuits the way they did on the test bench. The fact that the PC-based program cost was about one-seventh that of the mainframe price of SuperCompact also helped, since lower-level management could conceivably purchase the program without waiting for upper management's approval. PC-based microwave circuit design was on its way [9].

Competition between Compact Software and EEsof led to the development of Microwave Harmonica, which included harmonic balance analysis (see Volume II, Chapter 4) for nonlinear circuit simulation. EEsof later countered with their version, called Libra, and also added a system simulator, Omnisys. A few years later, Hewlett Packard entered into the CAD market with their MDS product. At the low end of the price range, new firm Circuit Busters (now Eagleware) entered the market and soon offered a wide range of PC-based programs at low cost.

In the 1990s software products were further polished and by that time were mostly used either on workstations or personal computers. EEsof was acquired by Hewlett Packard (now Agilent Technologies), developing *Advanced Design System* (ADS). Compact Software merged into Ansoft, combining SuperCompact and Microwave Harmonica into Serenade as well as Ansoft Designer. Near the turn of the century a new company AWR entered the market with a highly user-friendly integrated circuit/EM simulation program, MW Office, that gained rapid acceptance among RF designers. Finally there was proof that the operation of a complex program does not have to be complicated. Another large product, from Finland, called APLAC, also made its way into the industry.

EM simulation had a slower start, hampered by the larger computing power requirements. The first commercially successfully product released by Sonnet, called EM [10], was followed by Ansoft's 3D program, HFFS in 1990 [11]. The first integrated EM and circuit simulation product, IE3D, was introduced by in 1992 by Zeland Software. In the same year a Canadian group, OSA, demonstrated optimization with the EM program and later in 1996 with the HFSS. As of this book's publication date there are about a dozen large commercially available EM simulators available.

Most of the RF/MW newer circuit simulators now come with a bundled EM simulator instead of exchanging data with a standalone program. Cosimulation—performing circuit (schematic) level and physical circuit layout design concurrently—is becoming the standard in modern computer-aided design. The old-fashioned form, where circuit engineers did the schematic-level design and turned it over to layout specialists, has been gradually replaced with the new form, where the circuit engineer has complete control of the circuit from inception through prototype realization until verification. As a result, instead of going through several costly and time-consuming physical prototype iterations that used to take months, the cycles now take place in the computer, in a matter of days.

6.3 Analysis versus synthesis and optimization

Much of our circuit engineering work can be categorized into one of three groups, summarized in Figure 6.2. A brief description of the three techniques follows.

1. Analysis, sometimes called *simulation*, is used to find the response of a known circuit for a specified excitation. Circuit theory covers analysis methods like nodal, mesh, state-variable, and harmonic balance analysis that usually lead to a unique solution.
2. Synthesis is a mathematical form of design where we want to find a circuit to provide a desired response for a known excitation. There

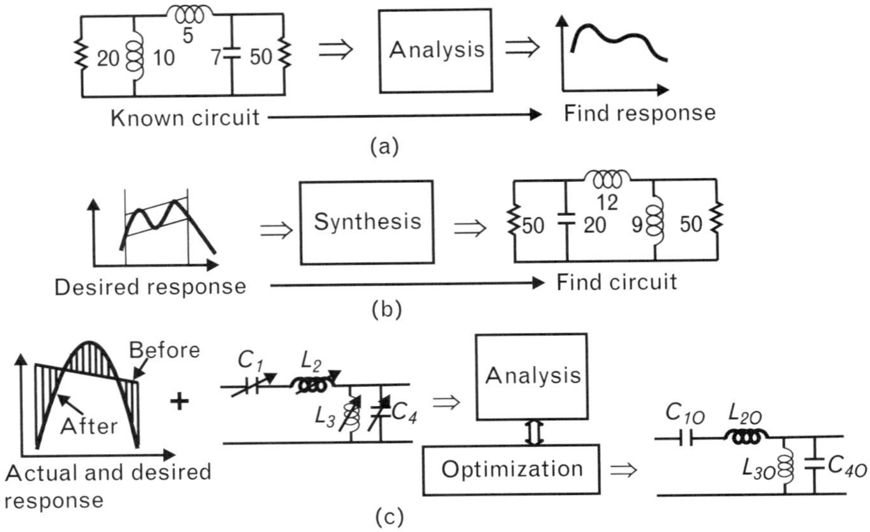

FIGURE 6.2
(a) Circuit analysis and (b) synthesis are inverse procedures. (c) Optimization is an iterative combination of the two to minimize the difference between actual and desired performance.

may be several circuits, or no circuit at all to fulfill the exact requirements. The empirical form of design is based on "cut-and-try" techniques.

3. Optimization is an iterative way of finding a solution when exact synthesis is not available, or when the engineer is not skilled with synthesis. Success of optimization depends on the goals, the quality of the initial circuit, and the numerical technique used.

We will examine computer-aided forms of these techniques and apply them to RF circuit design. Statistical techniques and EM simulation will be covered in Sections 6.8 and 6.10, respectively.

6.4 Circuit simulation techniques

Based on the circuit topology and the component types and element values, a circuit simulator computes the response due to a specific input signal. The simulator formulates circuit equations and finds a numerical solution to them. The program's equations may be relatively simple, such as nodal admittances, or complex nonlinear differential types. If the equations are not solvable directly, one must find an alternative approach.

6.4.1 DC and transient analysis

Both dc and transient analysis are performed in the time-domain, based on a system of algebraic linear and nonlinear differential equations that cannot be solved directly. For dc analysis an iterative approach is used to find a solution, although iterative solutions have no guaranteed convergence. Depending on the initial conditions and the complexity of the circuit, convergence may not be achieved. For transient analysis, direct derivatives are replaced with finite difference approximations to achieve solution. These techniques are explained further in Volume II, Chapter 4.

6.4.2 AC steady-state circuit analysis

Steady-state circuit design may be performed in the time or frequency-domain. Time-domain analysis is mainly used in nonlinear operations and is also covered in Volume II, Chapter 4. Frequency-domain simulation, while being linear, can handle a large percentage of RF circuit designs in a fast and efficient manner. Having speedy analysis also allows us to perform optimization, which may require tens of thousands of iterative analyses to be performed. Most of the early frequency-domain simulators, such as COMPACT, used two-port type analysis by breaking down the circuit into two-port boxes. Next, the two-ports were interconnected in whatever form was most efficient (parallel connections with Y-parameters, series connections with Z-parameters, and so forth). The two-port interconnections initially were not automated and it took a fair amount of interaction from the user to simulate the whole circuit.

Two-port analysis successively reduces the circuit into a single two-port for ease of manipulation and minimum storage requirements, which was an important issue in the early days of computing. As storage capabilities increased, nodal analysis became practical where users specified the node numbers where the elements are connected. Once the nodal admittance matrix is created, it is mathematically manipulated to obtain the desired output. Since the admittance description does not require solving differential equations and has no convergence problem, analysis is fast, allowing circuit optimization in reasonable time.

The admittance matrix does not exist for certain types of components, such as an ideal transformer, and appropriate fixes have to be applied during the matrix calculations. Circuit description was initially done through *netlists*, which have been gradually replaced with schematic entry, although netlist still has advantages, particularly for those working with EM simulation and circuit layout. Active devices are generally characterized by S-parameters that are functions of frequency, temperature, and applied signal level. We have already shown several examples of frequency-domain circuit simulation in the preceding chapters.

6.5 Impedance mapping

Impedance mapping is a very useful graphical technique to visualize the effect of component changes in RF circuits. The response of a linear two-port is a bilinear function of any branch impedance of the two-port [12]. Changes of that branch impedance can be mapped onto one of the plotted responses of the two-port, say, the s_{22} response. Impedance mapping superimposes a second, often distorted, Smith chart on the response plot, which allows us to see how the circuit performance changes as we vary a particular branch or component in the circuit [13]. Having such capability allows us to quickly find a specific type of circuitry needed to achieve the desired performance. It also helps us to find the sensitivity of component changes within our circuit.

To illustrate this concept, let us first look at s_{22} of the Infineon BFP 620 device measured with 2V, 20-mA dc bias at 1,900 MHz (Figure 6.3).

Next, we vary the source termination through all points of the Smith chart and show the corresponding new output reflection coefficients, labeled s'_{22}. The regular Smith chart of Figure 6.4 shows the output impedance of the transistor for all sources, where $|\Gamma_S| \leq 1.0$. The small distorted and super-imposed Smith chart, labeled with various *normalized* z_S values, indicates the source connected to the device and the corresponding s'_{22} for each source[1]. Figure 6.4 shows that the output reflection coefficient could actually *exceed unity magnitude* (a warning sign that the device is capable of oscillating at 1,900 MHz) for low-impedance sources, including a short circuit. For maximum $|s'_{22}|$ we need to use a source with about

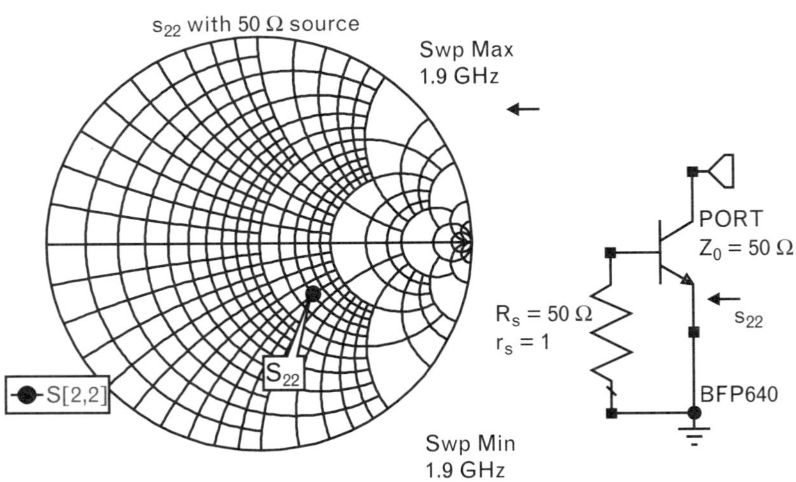

FIGURE 6.3 *Output reflection coefficient of the transistor, s_{22}, measured at 1,900 MHz while the input port is terminated with Z_S = 50Ω. Impedance mapping will help to visualize variations of the output reflection while Z_S is changed.*

1. We use s_{22} for the basic output reflection coefficient measured with Z_S = 50Ω. For all other loads, we use s'_{22}.

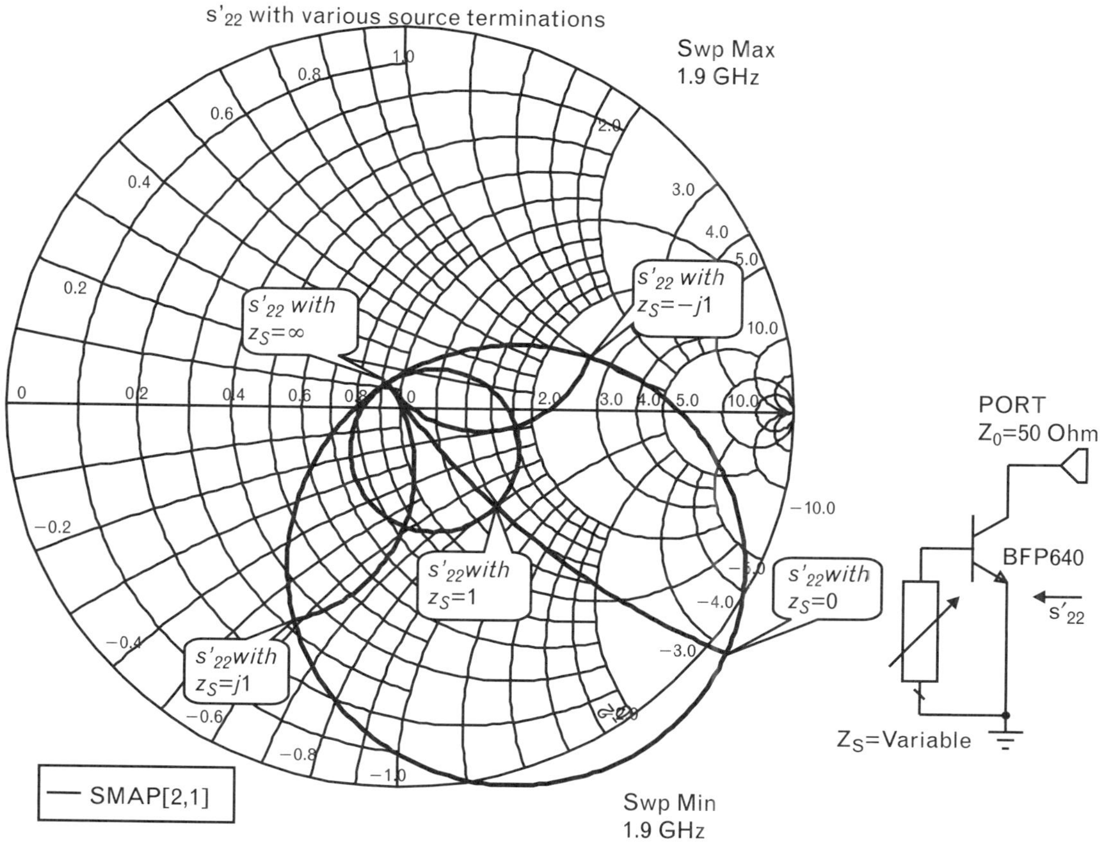

FIGURE 6.4 *Smith chart plots, showing transistor output reflection coefficient changes, caused by using different source terminations. Labels attached to the small transparent "source-plane Smith chart" refer to normalized terminations connected to the input port of the device. The new output reflection coefficients, s'_{22}, caused by different sources are read on the regular larger chart. When the normalized source $z_S = 1.0$, then $s'_{22} = s_{22}$, that is, the basic output reflection coefficient of the device, as displayed in Figure 6.3.*

$j0.2$ normalized inductive reactance. Minimum $|s'_{22}|$ is achieved by using a high-impedance source, which moves s'_{22} close to the center of the Smith chart.

Impedance mapping has many practical applications in amplifier, oscillator, filter, and matching network design. In oscillator design we can find what type of impedance is needed at one of the terminals of a transistor to maximize the reflection coefficient magnitude at a port. Impedance mapping can also help us to determine the feedback element required to neutralize the reverse transmission of a transistor, for buffer amplifier design. A neutralized network has complete isolation between the input and the output. (We cover active circuit design and device neutralization in Volume II, Chapter 1.) Several of the commercially available RF circuit simulators offer impedance matching as one of the standard options.

6.6 Component tuning

A practical form for finding optimum performance is to tune or tweak the circuit. On the workbench, while observing the response on the test equipment, we can tune inductance, capacitance, or other circuit functions and watch how the performance changes. Tuning components in a circuit simulator is extremely handy, particularly when we have several variables and we want to compare the effect of component changes. The main difference between tuning and impedance mapping is that tuning changes a particular element value and shows the effect at various frequencies simultaneously while impedance mapping is performed at a single frequency to visualize the effect of various types of terminations (inductive, resistive, or capacitive) and their values.

I, Les Besser, still remember one of the users of COMPACT telling me how frustrated he was with the amount of time it took to enter into COMPACT's editor, which was a separate subprogram, change a component, and then reenter the program to execute the response. Circuit description at that time was a netlist, which had to be modified every time a change was required. This very same person later became the architect of the Touchstone program and created a tune mode to display immediately the effect of any change typed into the netlist. The tuning concept was later further refined in AWR's Microwave Office program, where by moving a sliding button with the mouse, the changes are interactively shown on the circuit response. This feature was adopted by all the leading simulators, representing a very useful part of the programs.

Figure 6.5 illustrates an application of the tune mode for the troubleshooting exercise described in Section 4.9.6. The two-element network connected to the output of a two-port should match the output to 50Ω. Due to a wrong element value, the measured s_{22} is far from the goal, which is at the center of the chart. Tuning the circuit elements ±30% in the simulator shows that if the inductor had the proper value, the capacitor could only move s_{22} on the $r = 1$ constant-resistance circle. From the measured s_{22} location we can tell that the problem is not caused by the capacitor; it is the incorrect parallel inductor that moves s_{22} to the measured location. To fix the problem, we need to change the parallel inductor to lower susceptance, which means to increase its inductance.

6.7 Circuit optimization

While component tuning can be very effective for a limited number of components, it is not practical when we have a large number of variables, when the changes are *correlated*. For example, in a transmission line circuit

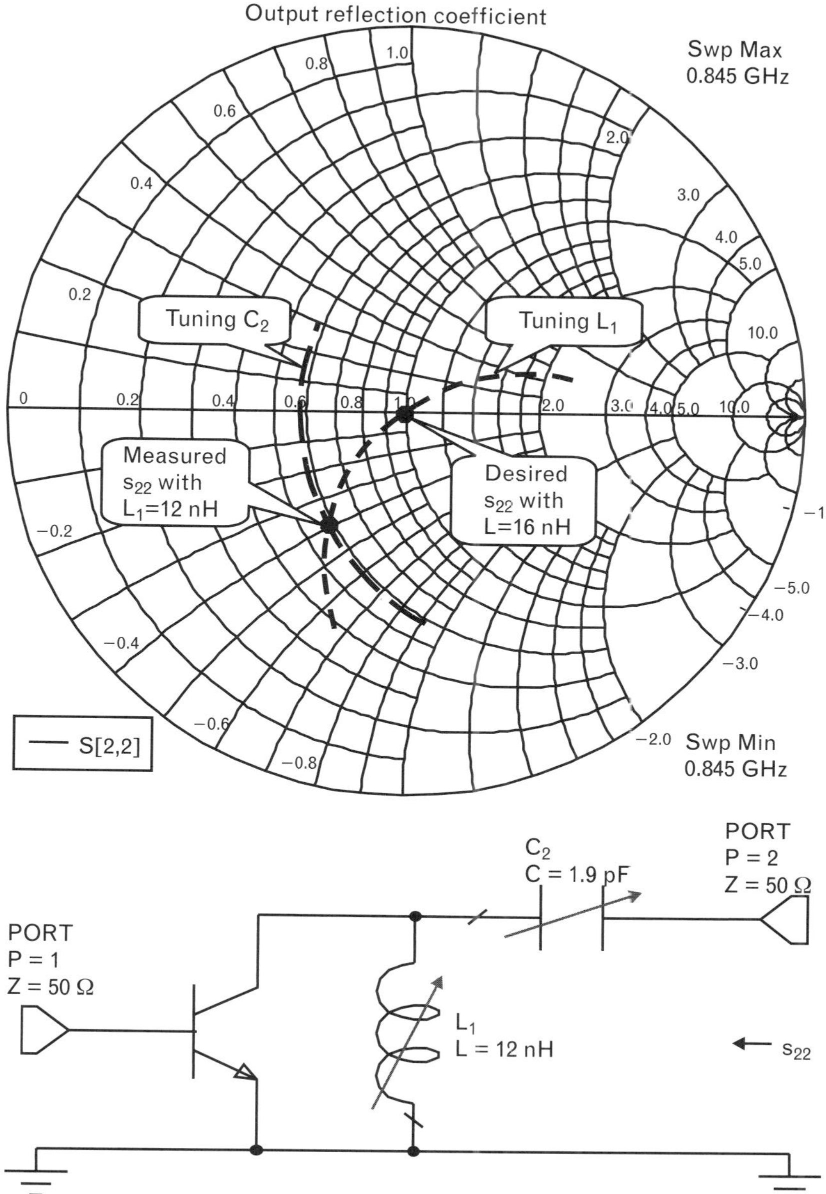

FIGURE 6.5 *Finding the incorrect element of the output matching network in the tune mode is easy. The series capacitor can only transform s_{22} on the constant-resistance circle, and therefore it is not the incorrect element. From the actual measured s_{22} location, only the improper parallel inductor can move us to the desired s_{22}. Note: Since the inductor is electrically behind the capacitor, its change does not track a constant-susceptance circle.*

the line impedances are functions of conductor widths. Over-etching or under-etching the circuit changes the nominal widths of the lines. However, the etching factor *affects all the transmission line widths by the same amount.* Another example is in integrated circuits where capacitance may be controlled by the thickness of a dielectric layer. As the dielectric thickness varies, all the capacitors are affected by the same proportion. Such changes are relatively easy to perform with manual tuning.

Circuit optimization enables us to change any number of variables, correlated or not, to find the optimum performance. Optimization is a successive approximation procedure where after initial circuit simulation, the error between actual and desired performance is computed. Next, the components declared as variables are adjusted simultaneously by the search algorithm. The adjustment is fed back to the initial network and a new analysis is performed, followed by evaluation of the error function. The iterative cycle (Figure 6.6) is repeated until we meet some convergence criteria or, as we mentioned earlier, run into numerical problems that prevent reaching further improvements.

Circuit optimization is now part of all modern RF simulators, which generally include several different types of search techniques. When it was first introduced in COMPACT in the early 1970s, it met with heavy criticism from some members of the academic community, fearing that engineers will not learn exact solutions if they can get the answers iteratively. In a debate at one of the major universities, the most vocal professor summarized his opinion as follows: “I will never allow any of my students to use this feature which is a total cop-out to a scientific approach. A good

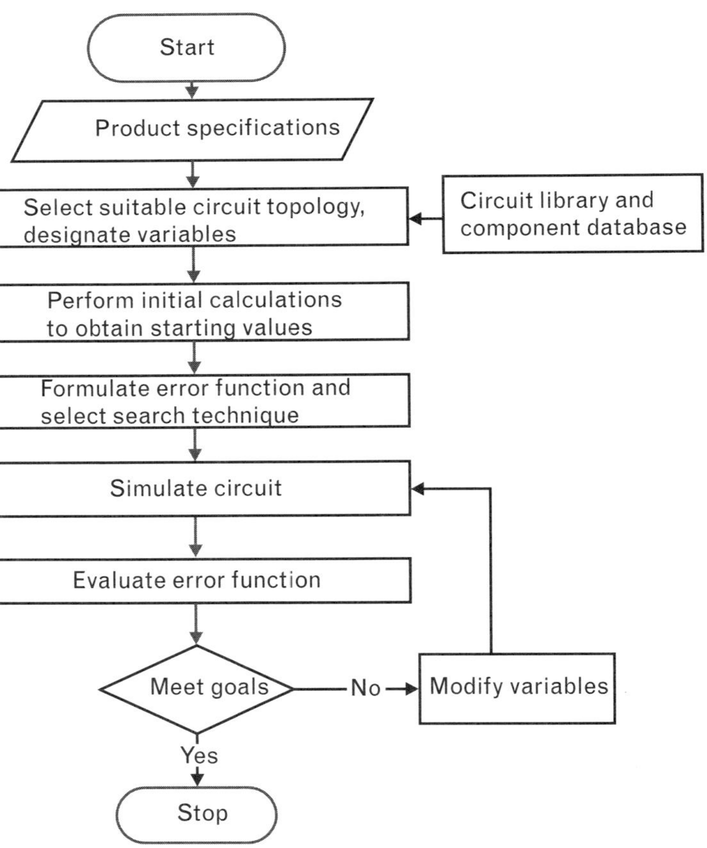

FIGURE 6.6 *Flowchart of computerized circuit optimization.*

engineer should always be able to derive closed-form solutions to microwave circuit problems."

At that time I, Les Besser, totally opposed this view, recalling that I had to spend days of my first engineering project deriving the flow graphs of a two-stage amplifier circuit that included internal overall feedback loops. My effort only gave me two of the four two-port *S*-parameters, and I could still not determine the optimum component values for the performance I wanted. Trying to find the best component combination of more complex circuits, such as a multistage low-noise amplifier, where the design goals may include noise figure, input and output match, flat gain, and RF stability, was practically an impossible task. In contrast to the lengthy time-consuming derivations, COMPACT's optimization routine accomplished the final "tweaking" in seconds.

Thirty years later, however, I am taking a somewhat different stand. After witnessing many "blindfold optimizations" where the designer did not use any effort to find initial component values, I now appreciate the concern of the professor mentioned above. In fact, the strong feelings of both authors are summarized in the opening chapter—we abhor the idea of turning to the optimizer without doing initial engineering. Optimization should always be *one of the last steps* of the design process instead of the first; it should not replace design engineering. Combining good initial design with the power of the optimizers is a much better and safer way to success.

6.7.1 Error function definitions

The *error function*, sometimes called *cost function*, or *penalty function*, can be defined in different ways. One of the possible forms is where we sum the difference between the actual performance and the goal through a specified frequency or time range, depending on whether we operate in the frequency or in a time domain. The general form of a weighed averaged sum of squares in the frequency domain for *k* number of goals can be shown as

$$EF = \frac{1}{n}\sum_{f_1}^{f_n}\sum_{i=1}^{i=k} W_i \left| Actual\ Performance_i - Goal_i \right|^p \tag{6.1}$$

where

f_1 and f_n specify the frequency range of n discrete frequency steps.

W_i is a *weighting factor* of the ith of k circuit parameter goal.

p is an even exponent, frequently set to 2.

A specific form applicable to *S*-parameter calculations in the frequency domain, with goals of $Goal_1$ through $Goal_4$, may be the following:

$$EF = \frac{1}{n}\sum_{f_1}^{f_n} W_1\left|s_{11} - Goal_1\right|^2 + W_2\left|s_{22} - Goal_2\right|^2 \tag{6.2}$$
$$+W_3\left|s_{21\text{dB}} - Goal_3\right|^2 + W_4\left|NF_{\text{dB}} - Goal_4\right|^2$$

where $W_1 - W_4$ are the weighting factors, which emphasize or deemphasize a specific portion of the error function. These weighting factors generally default to a values of 1.0 in all programs and can be very important during practical circuit optimization. NF_{dB} is the noise figure of the circuit in decibels (see Volume II, Chapter 2).

6.7.2 Illustrative exercise: weighting factor determinations

We want to design a low-noise amplifier with the following ideal target specifications:

$|s_{11}|$ = not specified

$|s_{22}| = 0.0$

$Gain_{\text{dB}} = 10$ dB

$NF_{\text{dB}} \leq 0.9$ dB

Find the most suitable set of weighting factors.

Solution

The acceptable performance limits (the worst-case performance we can accept) define the error terms associated with these specifications. We compute the performance limits by taking the difference between the targets and worst acceptable performance, as shown in Table 6.1.

The maximum value of the acceptable noise error term is 100 times smaller than the acceptable gain error, and 10 times smaller than the acceptable output match error. If we run an optimization with default values of the weighting factors ($W_2 = W_3 = W_4 = 1.0$), the search algorithm places equal importance on all three parameters. In reality the noise figure specification is more important than the other two. A 0.1-dB error in noise figure is as important as a 1-dB gain variation; therefore, we need to multiply the squared noise error with a factor of 100. The output reflection coefficient's squared error term is 0.1, requiring a weighting factor of 10. By choosing different weighting factors for the three parts of the error function, we can emphasize the importance of the low noise error and deemphasize the importance of the gain and $|s_{22}|$ errors. By setting the weighting factors as shown in Table 6.1, the product of the weighting factor and the acceptable errors are the same for each term. The error function, using (6.2), is

TABLE 6.1 TABULATED VALUES OF AMPLIFIER SPECIFICATIONS WITH THEIR RELATED WEIGHTING FACTORS

PARAMETER	IDEAL GOAL	ACCEPTABLE PERFORMANCE	(MAX. ACCEPTABLE ERROR)^2	ERROR TERM VALUE	PROPOSED WEIGHTING FACTOR	ERROR × W
$\lvert s_{11}\rvert$	Not specified	N/A	N/A	N/A	N/A	N/A
$\lvert s_{22}\rvert$	0.0	<0.31 dB	$(0.31)^2$	0.1	10.0	1.0
$s_{21\text{dB}}$	10.0	10 ±1 dB	$(1.0)^2$	1.0	1.0	1.0
NF_{dB}	0.9	1.0 dB	$(0.11)^2$	0.01	100.0	1.0

Note: Weighting factors should be selected to equalize numerical differences of the specified parameters.

$$EF = \frac{1}{n}\sum_{f_1}^{f_n} 0 + 10|s_{22}|^2 + 1|s_{21\text{dB}} - 10|^2 + 100|NF_{\text{dB}} - 0.9|^2 \tag{6.3}$$

This way, the search algorithm places equal importance on all three parameters during optimization. Unless the large weighting factor is used, components affecting noise figure are not changed because the contribution of the noise error is very small. Even with the weighting factors set as shown in (6.3), there is no guarantee that we will reach all three goals simultaneously, but this is the best start we can provide.

Using linearly spaced discrete frequency steps, the error function shown in (6.1) forces a *maximally flat, or Butterworth* type of response. When equal-ripple (Chebyshev) response is desired, we need a special case of *minimax* type of error function [14], such as

$$EF = \left(\frac{1}{n}\sum_{f_1}^{f_n}\sum_{i=1}^{i=k} r_i^p\right)^{1/p} \qquad \text{if } r_{MAX} > 0 \tag{6.4}$$

$$EF = \frac{-1}{\left(\frac{1}{n}\sum_{f_1}^{f_n}\sum_{i=1}^{i=k}\left(\frac{-1}{r_i}\right)^p\right)^{1/p}} \qquad \text{if } r_{MAX} < 0 \tag{6.5}$$

$$EF = 0 \qquad \text{if } r_{MAX} = 0 \tag{6.6}$$

where

$r_i = W_i \left| Actual\ \ performance_i - Goal_i \right|$

$r_{MAX} = MAX(r_i)$

Variations of minimax and least pth techniques can be applied to cases to minimize the ripple with respect to a specific performance value, and also where the important parameter is the magnitude of the ripple rather than the absolute value of the performance. For example, we may want to optimize the gain of an amplifier where the mean value of the gain can be anywhere between 12 dB and 15 dB, but the gain ripple should be minimum. Group delay optimization is another case where minimizing the delay ripple throughout a frequency band is more important than the absolute value of the delay.

If the circuit optimization program does not have a minimax type option, we need to increase the number of discrete trial points (frequencies or times) at the band edges to accumulate more error where the performance is falling off. Although it is a "trial and error" procedure, it can bring good results.

An error function with n-variables may be viewed as an $(n + 1)$ dimensional surface where the optimizer's task is to find the extreme low (minimization) or high (maximization) value of the surface[2]. If there is a single unique low point, it is called a *global minimum* and the function is *unimodal*. If there are multiple lows, each of them is referred to as a *local minimum* and the function is *multimodal*. Figure 6.7 compares the two different forms of error functions generated by one variable.

Complex error function specifications can lead to more difficult optimization. This is particularly true if some of the goals conflict with each

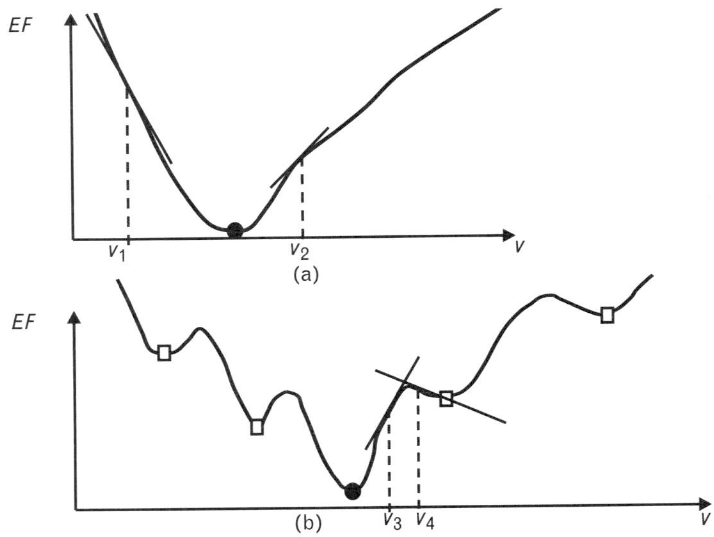

FIGURE 6.7 *With one variable, v, the error function can be plotted in two dimensions: (a) unimodal function with single minimum (●-markers) that is also the global minimum, and (b) multimodal function showing several local minima (□-markers) as well as the global minimum. Tangent lines at v_1 through v_4 indicate the positive and negative slopes of the error function at different variable values. Negative slope at v_1 indicates that the nearest minimum is at a larger v value. Positve slope at v_2 indicates the opposite.*

2. In our examples optimization refers to finding the minimum value of the error function. Maximization is achieved by minimizing the reciprocal of the function.

other. We will see in Volume II, Chapter 2 that in a low-noise amplifier minimum noise is not achieved at perfect input match. Specifying low noise figure and low input reflection coefficient simultaneously can create a sharp ridge on the surface of the error function, leading to a difficult optimization.

The major commercial simulators generally offer several search techniques for optimization. It is a good idea to become familiar with the various forms and learn about their relative advantages and disadvantages. Depending on the level of the initial design, we may have to start with a global search and switch to one that works better in the vicinity of the global minimum.

One weak point of optimization is that we often do not know what the capability of the circuit is. For example, matching two complex terminations for a given bandwidth with a finite number of *physical* components, there is no way to compute the lowest achievable reflection coefficient magnitude. When the optimization converges, it may be the best solution, but the search may also terminate for other reasons, including numerical problems.

6.7.3 Component sensitivities

In circuit optimization we want to know how component value changes affect the error function. Component sensitivities may be defined in various ways. The two kinds we are concerned about here are nominal differential sensitivities and statistical, sometimes called yield sensitivities. In frequency-domain optimization the partial sensitivities may be computed by finite differencing when the exact derivatives are not available. Time-domain partial derivatives are much more complicated and take longer to compute, making time-domain optimization a lengthy task. All gradient search techniques rely on the gradient vector, which is formed by the vector sum of the partial sensitivities.

When computed numerically, nominal sensitivity, S_x^{EF}, perturbs a variable, x, by a small amount, and computes the resultant change of performance while generally having all other components at their nominal values. That is,

$$S_x^{EF} = \frac{\Delta EF}{\Delta x}$$

Checking partial sensitivities before optimization can be useful although most of the commercial programs now do not display these numbers. Since the number of variables inversely affects the probability of success in optimization, it is good to initially eliminate variables with low sensitivities.

One may argue that nominal sensitivities do not provide a realistic view because *none of the products* built in production have all the components exactly at their nominal values. Statistical sensitivities that we will discuss later in Section 6.12 have more meaning because they look at the component sensitivity throughout the whole tolerance range.

6.7.4 Constrained versus unconstrained optimization

Components designated to be variables may be specified with or without constraints, although random search techniques must always have bounds. Placing constraints on a component has advantages and disadvantages. Without constraints component values may become negative or converge to some unrealizable range. We mentioned in Chapter 2 that at RF, elements can only be realized within a limited range of values. A microstrip transmission line may not function properly if the characteristic impedance is below 30Ω to 35Ω or above 120Ω to 150Ω, and it is wise to specify those limits for the optimizer.

Having a limited range to search through can speed up the optimization. On the other hand, there are cases when due to the form of the error function the optimum can only be reached by temporarily being outside the realizable range. If the constraints are placed in such a way, the optimizer path is blocked by the constraints and not allowed to reach the true optimum value. For that reason it might be better to run the first optimization without constraints. If some of the variables tend to go into unrealizable regions, then apply constraints to those specific variables.

The type of limits we mentioned here are explicit variable constraints. There are also implicit error function constraints that are more difficult to describe. For example, we may not want to have a variable reach a value that would decrease the gain of an amplifier below 10 dB. These implicit limits are generally handled by separate statements, specifying the acceptable range of performance instead of a fixed goal.

6.7.5 Search techniques

Search algorithms may be classified into two large categories: direct and indirect techniques [15]. Within these two groups there are a great number of variations. The fundamental difference between the two groups is the way each successive step of the search process is determined.

Direct search techniques are based on some form of gradient directions. The n-dimensional gradient vector may be evaluated by numerical or analytical techniques—the latter one offers definite speed advantages. The power of a gradient approach can be appreciated by looking at the single-variable error function plot of Figure 6.7(a). At variable value v_1 the slope is negative and the minimum is located at a positive direction from v_1. At v_2 it is just the opposite—the slope is positive and the minimum is toward the negative direction. During optimization, the search algorithm changes the variable toward the *negative partial gradient* (the slope in our example) direction, until the sign of the gradient changes. Then, by curve-fitting, the minimum between the last two trials is found. Gradient techniques are very effective when the surface of the error function is unimodal. With multimodal functions, the success of a gradient search depends largely on the quality of the initial approximation. If the starting values are chosen properly, the search already starts in the vicinity of the true *global minimum* and the likelihood of success is high. If initial estimates are poor, a gradient search can only find the nearest *local minimum* of the multimodal function. In Figure 6.7(b), by starting the gradient search at v_4, it moves toward the right, based on the negative slope, and converges to the nearest local minimum instead of finding the global minimum. Beginning at v_3 moves toward the true global minimum.

Another weakness of the gradient techniques is that they do not perform well when the error function has extreme slopes—very low or very high—unless the error function can be adjusted to moderate the slopes.

Random and other indirect techniques do not use gradient evaluations, and their successive steps are based on other criteria. Since a purely random optimizer would be highly inefficient, it is generally updated with the history of past searches. Learning from previous trials, the search is steered away from regions with low probability of success. Indirect search techniques are most effective with multimodal surfaces or when the designer has no way of obtaining reasonable starting values.

Many combinations of the two major categories also exist. Once again, it is a good idea to become familiar with the optimizers offered in your simulator to find the best type of search technique for specific tasks.

As for the future, genetic search algorithms [16] have received considerable attention lately because they require less specialized engineering background and experience. These algorithms are not as sensitive to local minima as direct search techniques, but they require an extensive amount of computations. Genetic optimization uses the mechanics of natural selection to guide search by modifying, creating, and eliminating circuit elements while searching for the best performance. It can actually find an appropriate circuit for the specified goal through crossover and mutation of available components. If the technique lives up to its promise, operators with little engineering skills will be able to "design" circuits.

6.7.6 Illustrative exercise: matching network optimization

To illustrate the power of optimization let us repeat the matching exercise from Chapter 5 where we matched 5Ω to 50Ω at 850 MHz. To make the problem more challenging we specify an octave bandwidth from 600 MHz to 1,200 MHz with a maximum acceptable ripple of 0.2 dB. We start with the single-frequency matching circuit derived analytically in Section 5.7.1, and then we will repeat the exercise with arbitrary component values.

Solution

Schematics of the 5-Ω to 50-Ω impedance matching circuit and frequency response before and after optimization are shown in Figure 6.8. Since we had reasonable initial component values, the gradient search converged in 13 iterations, giving an equal ripple response, just slightly over 0.2 dB. We will solve the same matching problem later using circuit synthesis in Section 6.9.

Next, we repeated the optimization with a new set of component values by arbitrarily setting both capacitors to 20 pF and both inductors to 20 nH. The gradient optimizer was still able to converge to the same results, but it took 246 iterations instead of the 13 required before—a nearly twentyfold increase in computing time. Watching the progress of the search, the struggle to find the optimum was noticeable.

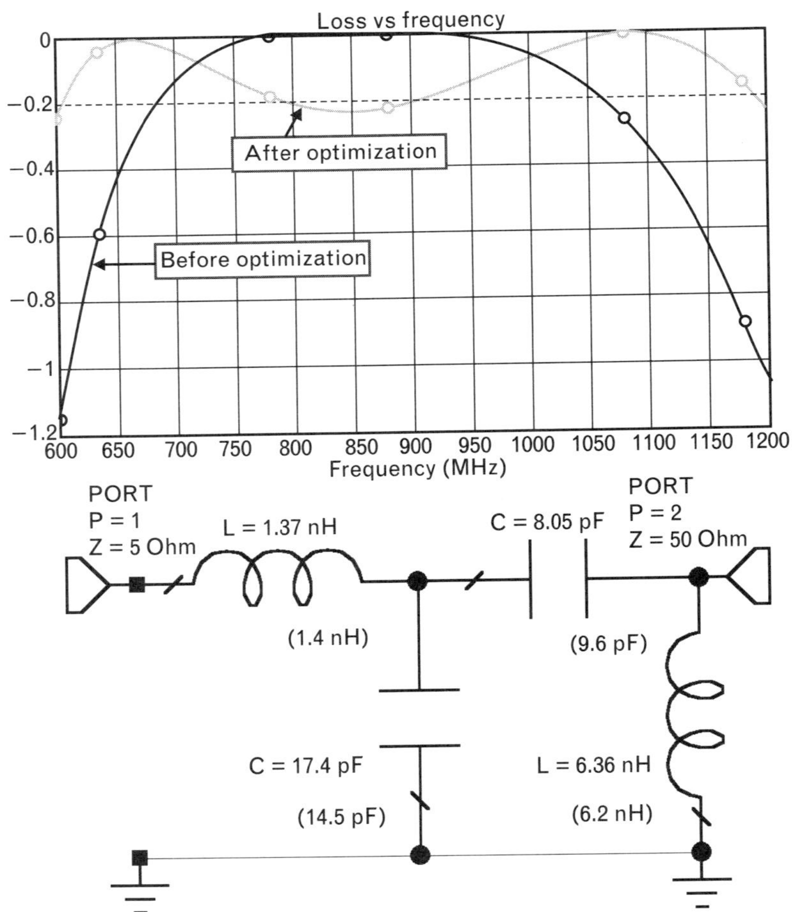

FIGURE 6.8 *Optimizing with good initial values leads to significant improvement over the single-frequency design. When the optimizer is not capable of giving an equal-ripple response, adding more frequency points near the band-edge can be helpful. Optimized values are shown in parentheses.*

Changing the initial values even further, to 50 pF and 50 nH for all components gave a large initial mismatch (not shown) of 34 to 47 dB through the octave frequency band. This time the gradient search converged to a *local minimum* by optimizing effectively only one of the two sections. From there the search could not improve any further. The frequency response was not impressive [Figure 6.9(a)]; it looked like the one we obtained with a single-section matching network since only one of the two sections' element values was changed. Switching to a global random search helped to follow the right direction but was unable to get the equal-ripple solution [dashed plot of Figure 6.9(b)]. Then, an additional gradient search was used to find the optimum in a few iterations [solid plot of Figure 6.9(b)].

We sincerely hope that after reading our books, nobody will try to optimize a matching network with an initial loss of 40 dB. Although for a

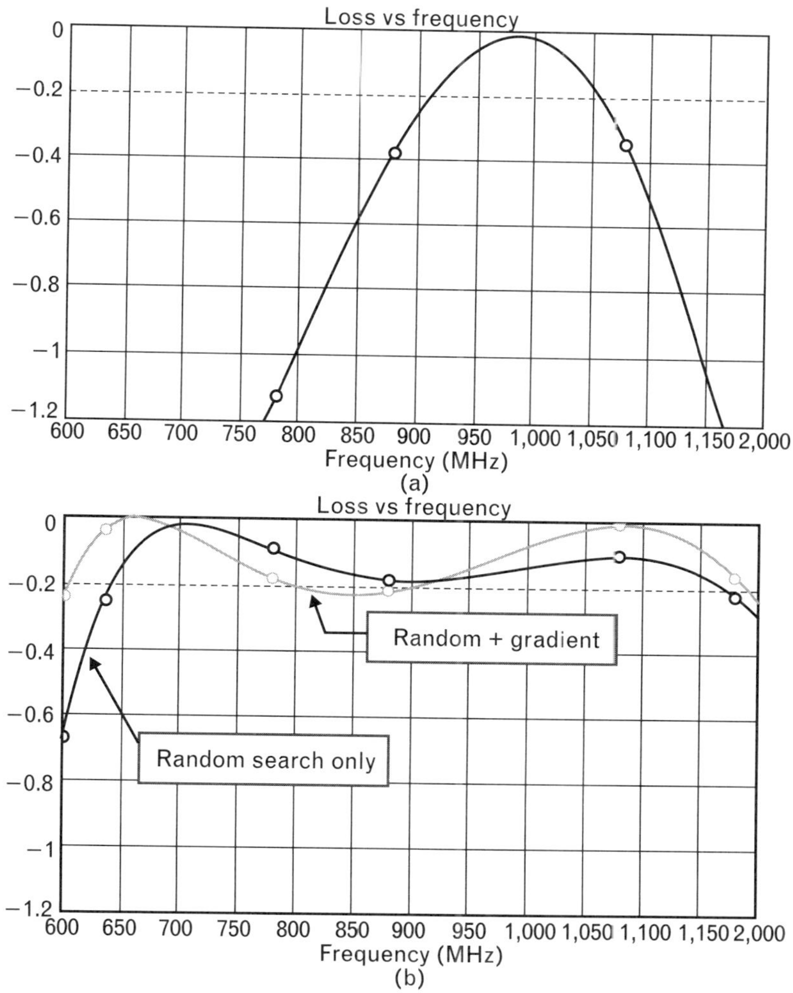

FIGURE 6.9
(a) With the initial components being several hundred percent from the optimum values, gradient search finds a local minimum.
(b) Random optimization brings the performance back to the vicinity of the global minimum and a follow-up gradient search quickly converges to the equal-ripple solution.

simple circuit like the one we showed, the solution can still be found, convergence with more complex networks is very unlikely. Initial design should always precede any optimization. To increase the probability of success:

- Use weighting factors to "emphasize" important specifications and parameters with low residual errors, as we showed in Section 6.7.2.
- Use all available tools to estimate a reasonable set of initial component values. Analytical and graphical techniques are generally available to design portions of a larger circuit, such as matching and gain-shaping networks. If the initial response is poor, try to find what part of the circuit causes the problem.
- If possible, begin with a small section of the circuit. Partition larger circuits and work with each segment individually until their performances are reasonable. For example, a three-stage amplifier may be split into individual stages for initial optimization.
- Initially, keep the number of variables low. If available, look at the nominal sensitivities to identify poorly chosen initial values. Many times one incorrectly selected component can render all other components insensitive to optimization.
- Familiarize yourself with the various search techniques offered in the software package and use them judiciously to get the desired response. Use direct search when the initial response is far from optimum performance and gradient methods for fine-tuning.
- Last, but not least, work *interactively* with the optimizers. One of the most common mistakes is to turn on the optimizer and come back some time later to see what happened. Most likely the final results do not give any hint to what caused the problem if the search failed to find an acceptable solution.

As much as we like optimization, we want to call attention to an important issue here. Optimization is a *process, not the goal*. The objective is to find the best available (these days perhaps the most economical) solution instead of expecting the optimizer to do our design.

6.8 Statistical design techniques

In this section we look at the effect of component tolerances and examine ways to get the highest yield when a product is placed into production. R&D engineers like to blame production groups for not being able to produce their designs, while production staff point fingers at the development groups for designing products that cannot be built. The truth is that often both of these groups are at fault. Although as we mentioned earlier most of the production problems can be traced back to development, design engineers are generally not exposed to the capabilities and limitations of their own production group. (*I actually witnessed a case where the design engineer was not allowed to touch test equipment on the production floor.*) To make matters worse, in many companies design and manufacturing are still two completely separate entities, as shown in Figure 6.10. In such an "R&D-versus-Manufacturing" environment, products are often bounced back and forth several times, both groups blaming each other for not getting things done. Design-related information rarely reaches the production floor and feedback does not flow back to R&D about production problems.

Companies that want to survive need to switch to a cooperative, team oriented arrangement as shown in Figure 6.11. R&D and manufacturing must realize that they both are on the same team. To reduce design-to-manufacturing cycles a high level of cooperation must exist between the two entities; their computer databases need to be linked together to share vital information.

6.8.1 Yield-oriented design

In most of our college-level design-oriented courses we were taught to design for maximum performance. For an amplifier, maximum performance could mean highest gain, lowest noise figure, best input or output match, or a combination of all of the above.

The weakness of the maximum performance–oriented design approach is that component tolerances are not considered until the design is already finished [17]. At that point there is so much time, effort, money, and ego invested in the design that most development engineers do not want to admit that their products will be hard to produce in their own

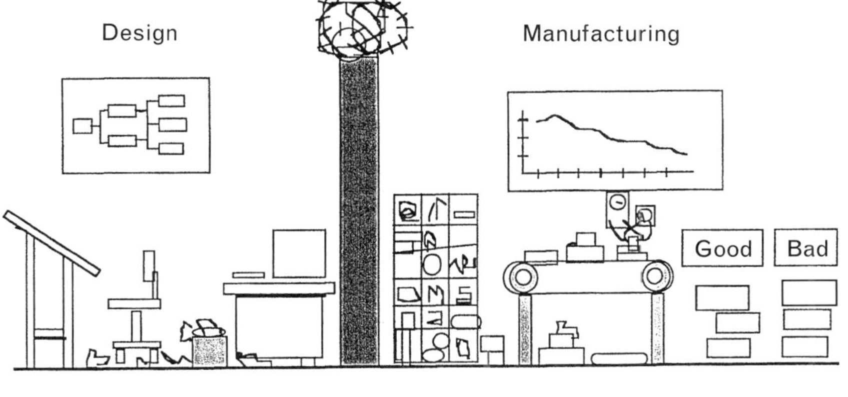

FIGURE 6.10 *Over-the-wall environment, separating manufacturing from design and blocking the flow of vital information between the two groups (From: [17]. © 1993 Artech House, Inc. Reprinted with permission.)*

FIGURE 6.11
Design for manufacture, where manufacturing environment is integrated into the initial design. (From: [17]. © 1993 Artech House, Inc. Reprinted with permission.)

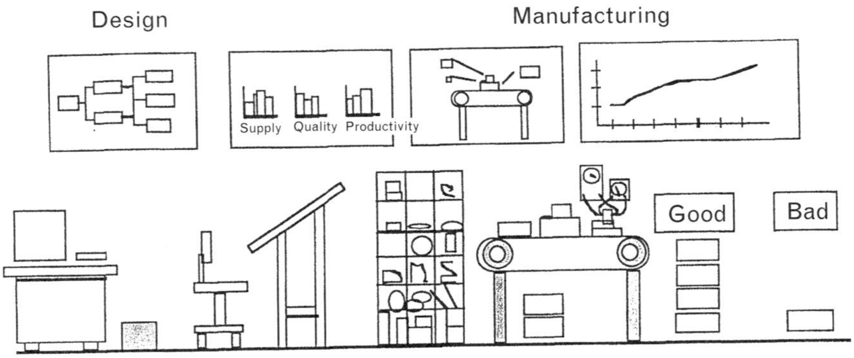

manufacturing environment. To make things worse, in today's global environment the design group may be located in California, manufacturing in Malaysia, and testing in Mexico, making the design-manufacturing team arrangement shown in Figure 6.11 very difficult. However, with proper use of Internet communication and teleconferencing, the effect of geographical separations can be minimized.

The performance plot of Figure 6.12(a) displays one of the performance parameters, such as the gain of an amplifier, at a given frequency, as a function of one of the passive component value, x. Minimum acceptable performance is labeled as P_a, without any upper limit. After seeing the performance plot, our natural reaction is to set the component value at x_p for maximum gain. However, adding the component tolerances to the plot shows that the x_p does not give us best yield [Figure 6.12(b)]. About 30% of

FIGURE 6.12
Instead of maximum performance, component values need to be centered for maximum yield. (a) Designing for maximum performance alone with component value x_P leads to 70% yield. (b) With the component set at x_Y, 100% of the products meet minimum acceptable performance.

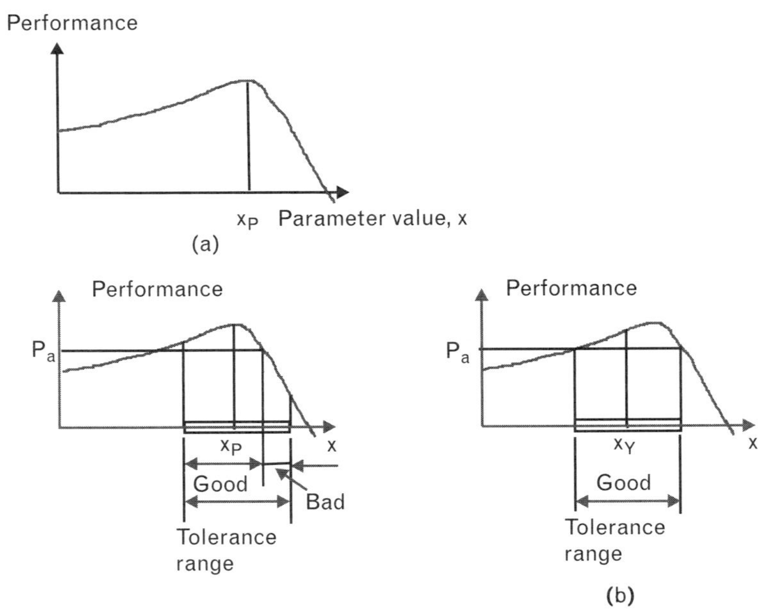

the amplifiers produced would not meet the minimum P_a specification. If we consider component tolerances as well as gain, the component value set at x_Y produces maximum yield [Figure 6.12(b)]. Remember that production does not care how much the performance is above the minimum acceptable level. Their only concern is to meet the minimum acceptable performance.

With a single performance parameter design centering is very easy. As the number of parameters to be tested increases, the task becomes more complex. In amplifiers having minimum gain, G_a, and maximum noise, N_a, specifications simultaneously necessitates component selection that meets both requirements, as shown in Figure 6.13. When the number of parameter specifications exceed three or four, it is virtually impossible to find the optimum without a systematic computerized yield optimization.

A goal of a *robust design*, therefore, is not just performance alone. It includes considerations of reliability, cost, and manufacturability, and is realized in a form least sensitive to uncontrollable parameter variations. Brinkmanship design is to go for highest performance only, where a small parameter or other environmental change can lead to failure.

6.8.2 Component tolerance distributions—probability density functions

Before we can perform yield optimization we need information about the tolerance distribution of our components (Figure 6.14). The distributions may follow a normal, bell-shaped curve centered at the nominal component value with known standard deviation. Assuming that the vendor already screened out parts outside the tolerance range, we still need to know the standard deviation of the distribution before making a meaningful statistical analysis. However, there is no guarantee the mean of the distribution is at the nominal value. Our shipment may have come from a production lot with a higher nominal value, and we receive a sloped distribution within our tolerance range. Alternatively, if someone is willing to pay a higher price for smaller tolerance, we may only get some or no component near the nominal value. Instead, we get a bimodal distribution. When we have no information at all, perhaps the uniform distribution is the one to assume.

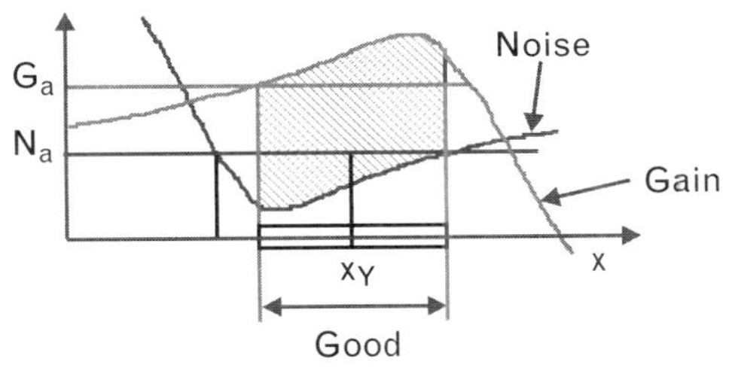

FIGURE 6.13
Only the products with noise performance below N_a, and gain above G_a meet specifications. Setting the component to x_Y provides maximum yield.

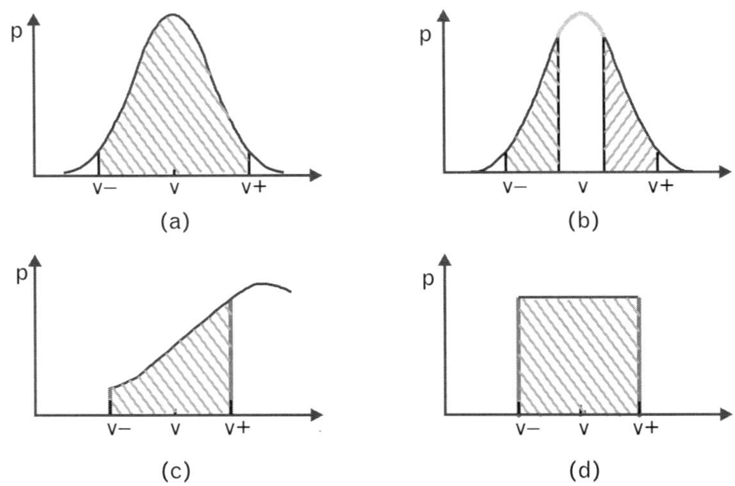

FIGURE 6.14 *Various tolerance distributions: (a) normal (Gaussian), centered at a nominal component value; (b) bimodal; (c) normal, centered at other than nominal value; and (d) uniform. The accuracy of a statistical analysis highly depends on the validity of the tolerance distribution.*

Component vendors seldom provide statistical information voluntarily for each shipment, even though more and more customers are expecting it. Our recommendation to product designers: contact the vendors and obtain statistical data on their products. If the vendor is not open to discuss it, or does not have any information, it may be a warning sign about future problems.

In the absence of data from the vendor, you may have to make arrangements to do samplings within your group or company. We emphasize again that without reliable tolerance information there can not be meaningful statistical analysis.

Once the tolerance distributions are established, we need to set pass/fail limits for each performance parameter specification. After that, we are ready for our Monte Carlo simulation where components are randomly selected and weighted by the appropriate tolerance distribution for each sample trial. The outcome of each trial is tested against the performance specifications.

The number of *Monte Carlo trials*, M, needed is a function of the *actual yield*, Y, the *fractional error* we are willing to accept, ε_F, and the confidence level, C_σ, we attach to the simulation [17]. C_σ is expressed in standard deviations as shown in Table 6.2 for various percentage levels. A confidence level of $C_\sigma = 2$, for example, specifies that the outcome of 95 (95.4%) out of 100 trials is within the range of $(Y \pm \varepsilon_F)$. A closed-form expression for M is

$$M = Y(1 - Y)\left(\frac{C_\sigma}{\varepsilon_F}\right)^2 \tag{6.7}$$

For example, if we expect a yield of 95% and accept a ±1% error with 2σ confidence level, the number of Monte Carlo trials needed are

$$M = Y(1-Y)\left(\frac{C_\sigma}{\varepsilon_F}\right)^2 = 0.95(1-0.95)\left(\frac{2}{0.01}\right)^2 = 1{,}900$$

Although most of us think of statistical analysis and design to predict and compensate for production related variation, they are also useful to deal with other effects, such as:

- Environmental changes (temperature, humidity);
- Component aging (insulation and gain of a device);
- Component model imperfections (discrepancies between assumed and actual structure).

Other considerations, such as the probability of loading the wrong components into the assembly machine, the mood of the operator after a layoff is announced, also affect the performance of a product. These effects are much more difficult to characterize than component tolerance, and we do not attempt to cover them here.

6.8.3 Statistical sensitivities

Statistical optimization, sometimes called design centering, is similar to single point optimization where we try to either minimize or maximize an objective function. Perhaps the most significant difference between the two methods is how the sensitivities are computed, since those have direct effects on how the variables are changed. We mentioned earlier that statistical sensitivities take the full tolerance range into consideration, shown in the form of histograms [18], while nominal differential sensitivities are computed within a very small perturbation of the nominal value.

To illustrate the concept of statistical sensitivity computations, we use the simple example of a voltage divider, published in [17], shown in Figure

Table 6.2 Standard Deviations and Corresponding Confidence Levels

Standard Deviation, σ	Confidence Level, %
1	68.3
2	95.4
3	99.7

6.15. The divider is made up of two resistors, R_1 and R_2 and is connected to a 10-V dc supply. Two specifications for the divider are that the input impedance must be between 50Ω and 70Ω and the output voltage must be between 4V and 5V. For simplicity we use two 35-Ω resistors initially with ±10% tolerances and uniform distribution for both of them.

Tolerance region, T, represents a collection of all possible parameter combinations. Since the two resistors come with uniform tolerance distributions, the probability of any (R_1-R_2) combination is the same within the tolerance region shown in Figure 6.15(b). The acceptable performance space, M_a , is determined by the limits of two different kinds of specifications, input impedance and output voltage. Only the circuits with impedance and voltage measurements within M_a are acceptable.

Mapping M_a into the parameter space can display the acceptable resistor combinations to meet both specifications. Superimposing the tolerance region helps us to estimate the yield. Since M_a shows what resistor pairs meet the specifications, and T shows all possible combinations of the two resistors, the shared space between M_a and T, marked with P_a, represents the resistor combinations that meet both specifications (Figure 6.16). The ratio of areas P_a and T is the yield, Y.

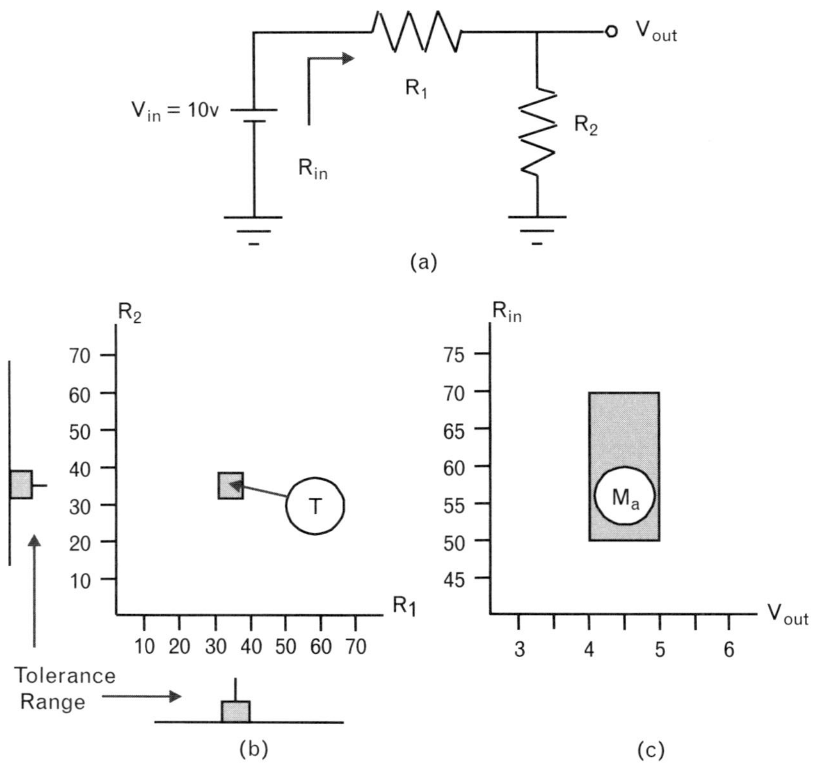

FIGURE 6.15
(a) Voltage divider made by two resistors, R_1 and R_2.
(b) Tolerance region, T, is defined by the combination of individual component tolerances, 35Ω ±10%. (c) Acceptable performance space, M_a, is formed by the intersection of two sets of performance limits: $50\Omega < R_{IN} < 70\Omega$ and $4V < V_{OUT} < 5V$. (After: [17].)

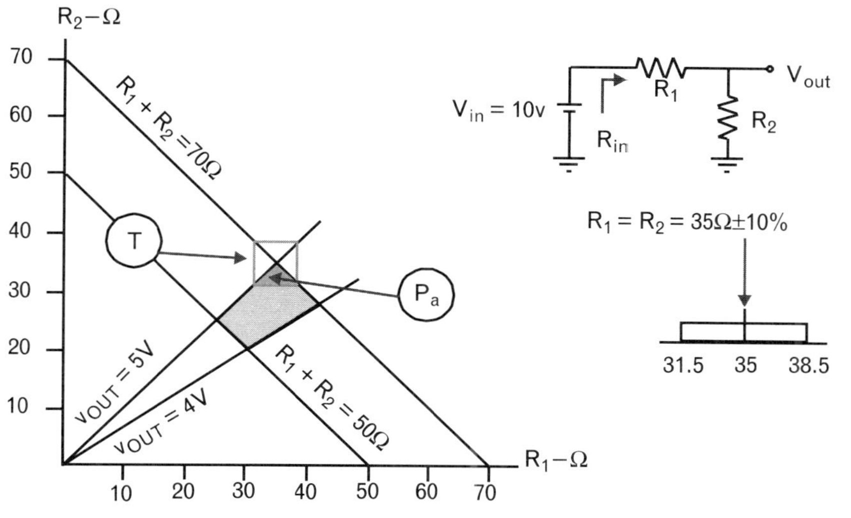

FIGURE 6.16 *The overlap between M_a and T represents the combinations of R_1 and R_2 that meet both specifications. The space of P_a is about one-fourth of the tolerance region, producing a 25% yield. (After: [17].)*

The overlap between the squared tolerance region and the gray acceptable performance space indicates only a 25% yield. Clearly, using smaller tolerances for both resistors does not help because it would only shrink the size of the square. The ratio of acceptable resistor combinations to failures does not change. However, smaller tolerance for R_1 can increase the yield. Before we spend more money on more expensive components, let us find out if different nominal resistor values would work better.

Having two variables only, even a visual inspection tells us what to do. From Figure 6.16 we can tell that moving the tolerance region toward the acceptable performance space increases the overlap between the two spaces. However, if we had more variables, and other than uniform tolerance distributions, the task becomes very difficult and we need some guidance on how to change the components. The statistical sensitivities can provide the help we need.

To construct the yield histograms, or sensitivity histograms related to statistical sensitivities, we plot the tolerance ranges for both components with small steps (Figure 6.17). Next, we vary resistor R through the steps marked and evaluate the yield at each individual step, by varying R_2 through its complete tolerance range. Plotting the resultant yields in form of a histogram, we find that the heights of the histogram bars are proportional to the heights of the shaded bars of the performance triangle. The histogram tells us that the highest yield is achieved when resistor R_1 is at its nominal value of 35Ω. Based on this information, we conclude that R_1 is already at the best value and changing it would only reduce the yield.

Repeating the same procedure for resistor R_2 provides a totally different picture. With a 35-Ω nominal value the yield is virtually zero, but it improves when R_2 is decreased toward its low tolerance limit, as shown in Figure 6.18.

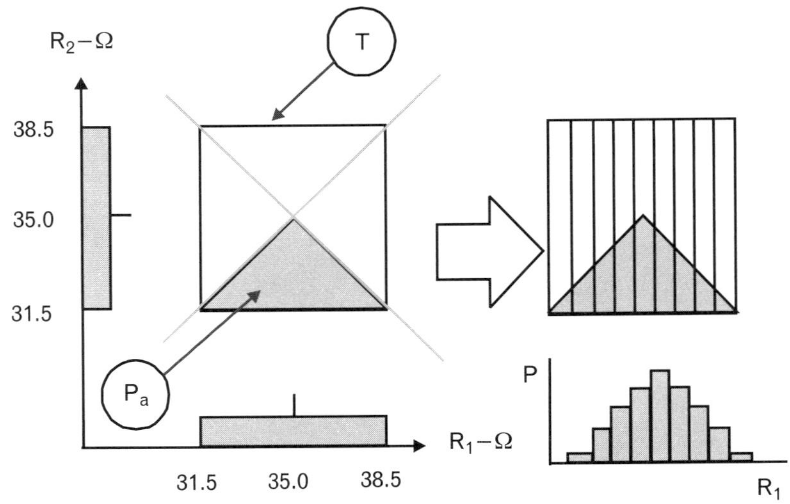

FIGURE 6.17
Varying R_1 *trough its tolerance range and evaluating the yield at each step gives us the sensitivity histogram of the component. The histogram tells us that* $R_1 = 35\Omega$ *is the best choice for maximum yield. (After: [17].)*

After seeing the two sensitivity histograms, we conclude that our first step toward improving yield is to reduce the value of R_2 and keep R_1 at its nominal value. The slope of a line tangent to the histogram bars at any resistance level helps to determine the step size and direction. After lowering the value of R_2, we reevaluate the sensitivities, and keep changing the resistors until the yield histograms show little change throughout the tolerance range. Figure 6.19 shows the end result of the manual yield optimization, with new standard ±10% tolerance element values of R_1=33Ω and R_2 = 27Ω. The new yield is 100% since $T = P_a$.

The important issue here is that we improved the yield from 25% to 100% *without buying more expensive components*. Of course, there is no guarantee that just manipulating the component values always brings such spectacular yield improvement. Generally, after finding the best nominal value combination, we can consider changing the tolerances to further improve

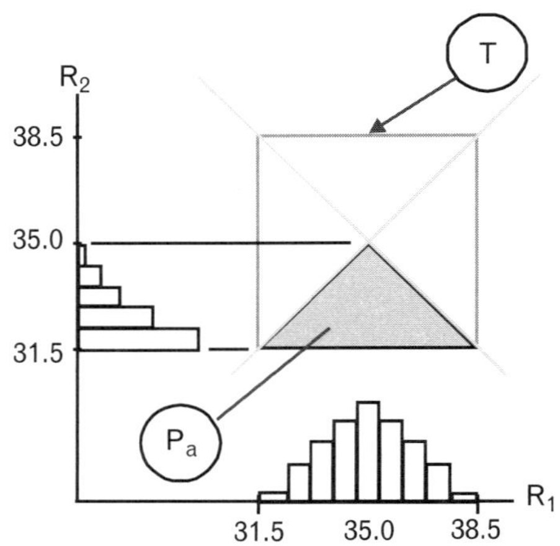

FIGURE 6.18
Yield histogram of resistor R_2 *tells us that lower resistor values increase the yield. (After: [17].)*

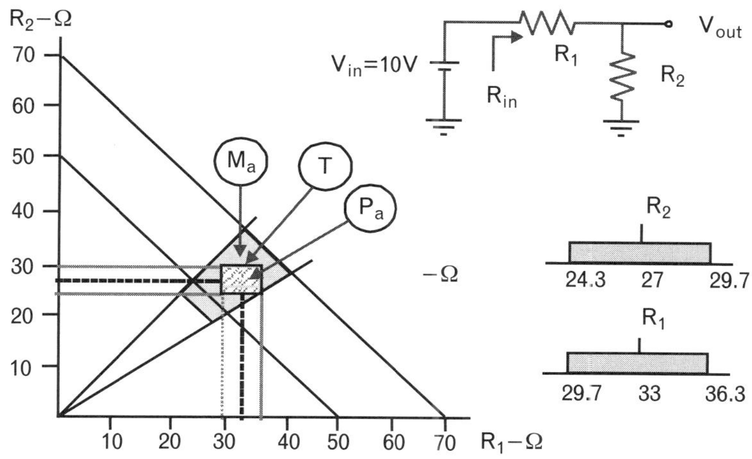

FIGURE 6.19 *With the help of statistical sensitivities, changing the initial resistor values moved P_a fully inside M_a, which is the goal of statistical optimization. (After: [17].)*

the yield. At that point it becomes an economical decision. For example, if our yield in this case had been only 85% and we find that by more expensive components we can increase the yield to 95%, the question becomes would the increased price of the components justify the 10% higher yield? In addition to direct costs, there are other factors as well. If the predicted yield is high enough, we may not have to perform 100% testing, which has additional economic benefits.

6.8.4 Illustrative exercise: design centering of a 500-MHz lowpass filter

The physical equivalent circuit model of a fifth-order lowpass filter, designed by synthesis for a 0- to 500-MHz passband with 0.25-dB ripple, is shown in Figure 6.20. The basic filter components are capacitors C_1, C_3, and C_5, and inductors L_2 and L_4, operating between 50-Ω terminations. The original synthesized capacitor and inductor values have been replaced with the nearest available standard component values (Table 6.3). Parasitic inductances of the capacitors are labeled L_S, and parasitic capacitances of

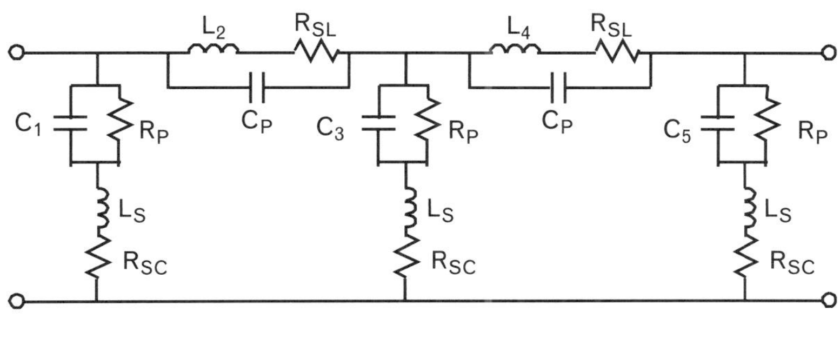

FIGURE 6.20 *A five-element 500-MHz lowpass filter with physical component models. The ideal filter consists of elements C_1, L_2, C_3, L_4, and C_5. The rest of the circuitry represents component parasitics and losses.*

Table 6.3 Synthesized Filter Component Values Initially Changed to Standard Values with ±5% Tolerances

Component	Synthesized Values	Actual Values	Tolerance	Distribution	σ
$C_1 = C_5$	9 pF	9.1 pF	±5%	Normal	0.18
$L_2 = L_4$	21 nH	20 nH	±5%	Uniform	N/A
C_3	14.3 pF	13 pF	±5%	Normal	0.29

Note: Standard deviations were available for the capacitors, but uniform distribution was used for the inductors.

the inductors are C_p. Resistors R_{SC} and R_{SL} represent conductive losses while resistors R_p are the dielectric losses of the capacitors. Since the filter has symmetry, there are only three unique component values: C_1,L_2, and C_3. The first and last capacitors, C_1 and C_5, have the same values, and the two series inductors, L_2 and L_4, are also identical.

Note the unusual way we show the $s_{21\text{dB}}$ specifications in Table 6.4—they are expressed as *negative gain*, instead of positive loss, because that is the way RF simulators handle transmission coefficients. (In Volume II, Chapter 1, we tell you a true story of an unsuccessful filter optimization where the designer inadvertently set goals for 40-dB stopband *gain* instead of loss.)

Adding the parasitics and losses to the inductors and capacitors and replacing the exact design values with nearest available standard components deteriorated the response of the filter as shown on Figure 6.21. The initial 0.25-dB equal-ripple response now gradually rolls off, at 500 MHz it approaches the 1.5-dB maximum loss specified for production. Since the filter just barely meets the production passband specifications at some frequencies, we can suspect poor yield. In the stopband the required 40-dB attenuation was easily satisfied and we are *not showing* that measurement.

With the above specifications, Monte Carlo analysis predicts 52% yield. The filter is built with discrete components so we do not use any correlation among the variables. Our goal is to find out if the yield can be improved by using different component values without paying more for lower tolerances. Instead of performing automated yield optimization, we will change the components manually using only standard ±5% values, based on their statistical yield sensitivities (Figure 6.22).

Table 6.4 Production Specifications for the Filter

Passband, 0–500 MHz	Stopband, 1,000–3,000 MHz
$s_{21\text{dB}} > -1.5$ dB	$s_{21\text{dB}} < -40$ dB
$\lvert s_{11} \rvert < 0.33$	
$\lvert s_{22} \rvert < 0.33$	

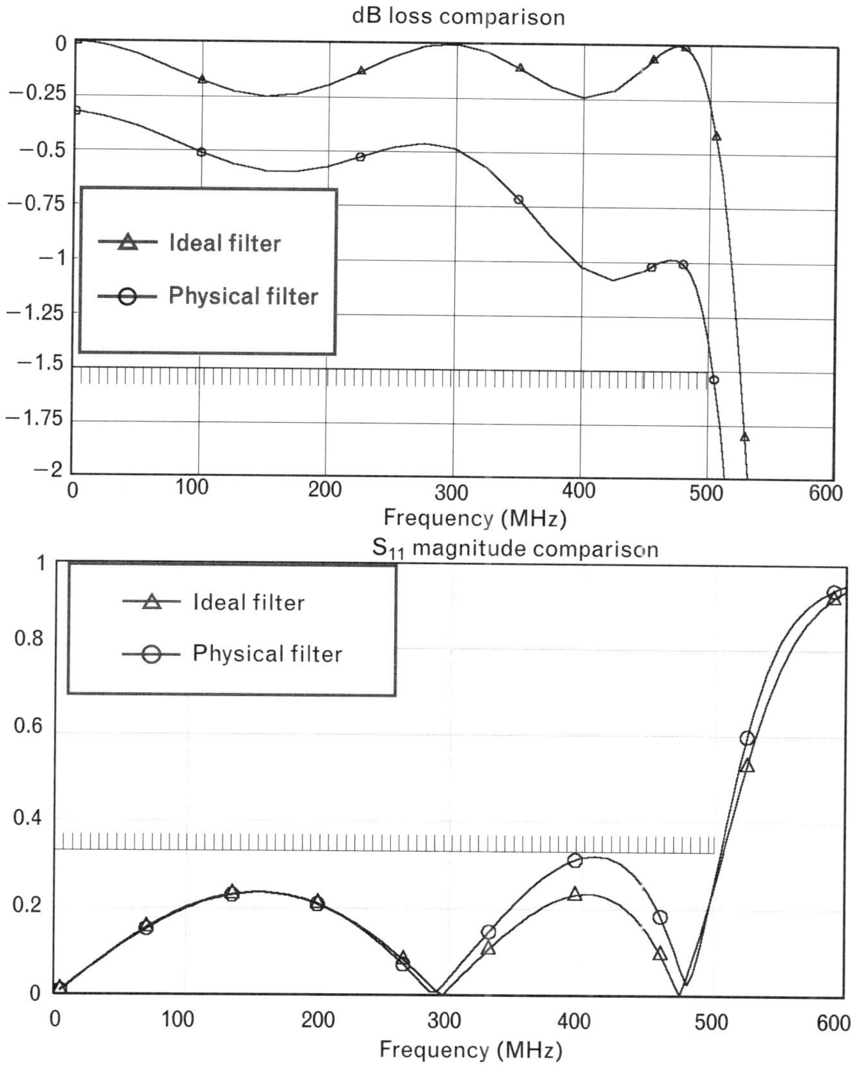

FIGURE 6.21 *Frequency responses of the ideal and physical filters show the effects of losses and component parasitics. Shaded lines indicate production goals for each parameter.*

Since the filter's circuit topology is symmetrical, any change made to C_1 also applies to C_5. The same is true for the two series inductors, L_2 and L_4. Therefore, to save space, we only show the histograms of the first three elements. Looking at the statistical sensitivities of capacitor C_1 (and C_5) shows that we need to reduce the capacitor values. The inductor values could also be lowered, while the center capacitor C_3 is near its optimum value, but it is quite sensitive to tolerance variation.

For best visualization, we should change only one component at the time, but changing the two end capacitors simultaneously maintains circuit symmetry. After lowering C_1 and C_5 to the nearest standard value, of 8.2 pF, we repeat the Monte Carlo simulation. The new yield is up to 88%. Results are shown in Figure 6.23.

FIGURE 6.22 *Sensitivity histograms of the three filter components, C_1, L_2, and C_3. Due to circuit symmetry, we do not need to examine the histograms of L_4 and C_5. The histograms tell us that C_1 and L_2 are too high and C_3 is near its optimum value. Based on these sensitivities, the value of C_1 (and C_5) will be reduced first.*

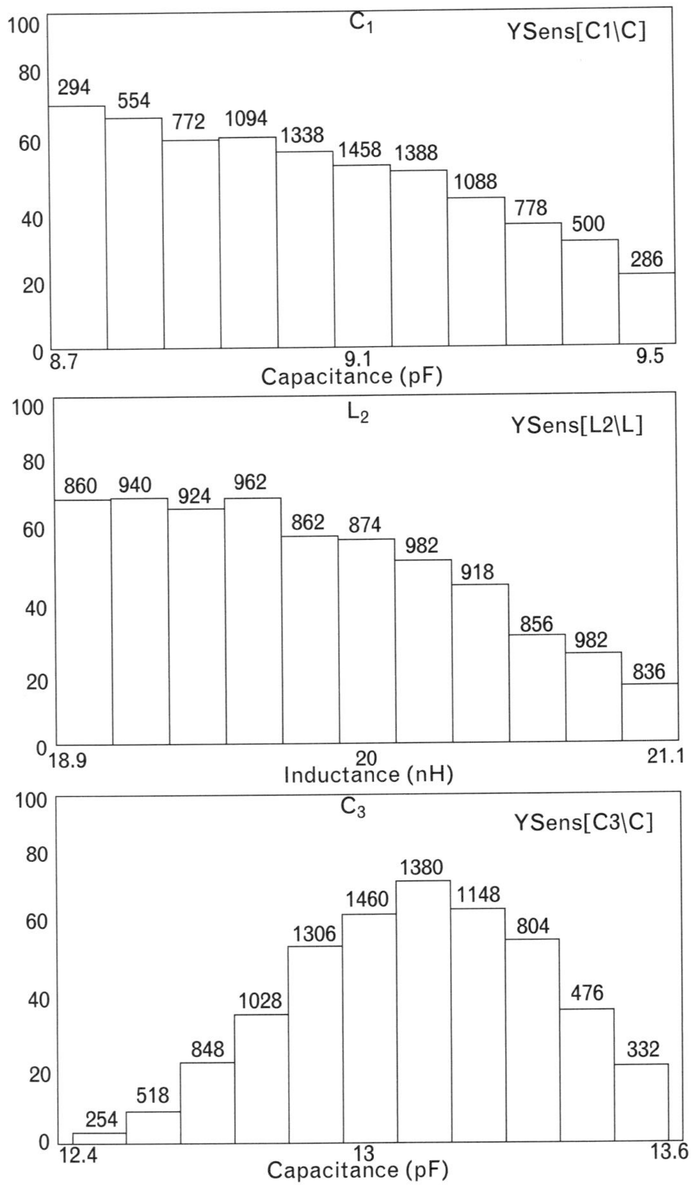

Notice how much the sensitivity histogram of C_3 has changed after switching to the new values of C_1 and C_5. In Figure 6.22 we saw that the nominal 13-pF value for C_3 was just about correct. Now, the 13-pF value seems to be too high, showing the interaction among the components.

Reducing capacitor C_3 to the next standard value of 12 pF increases the yield to 98% (results not shown). However, if instead we reduce the two series inductors, L_2 and L_4, to 18 nH, the yield jumps to 100%! We tested this new component combination through 50,000 Monte Carlo trials and

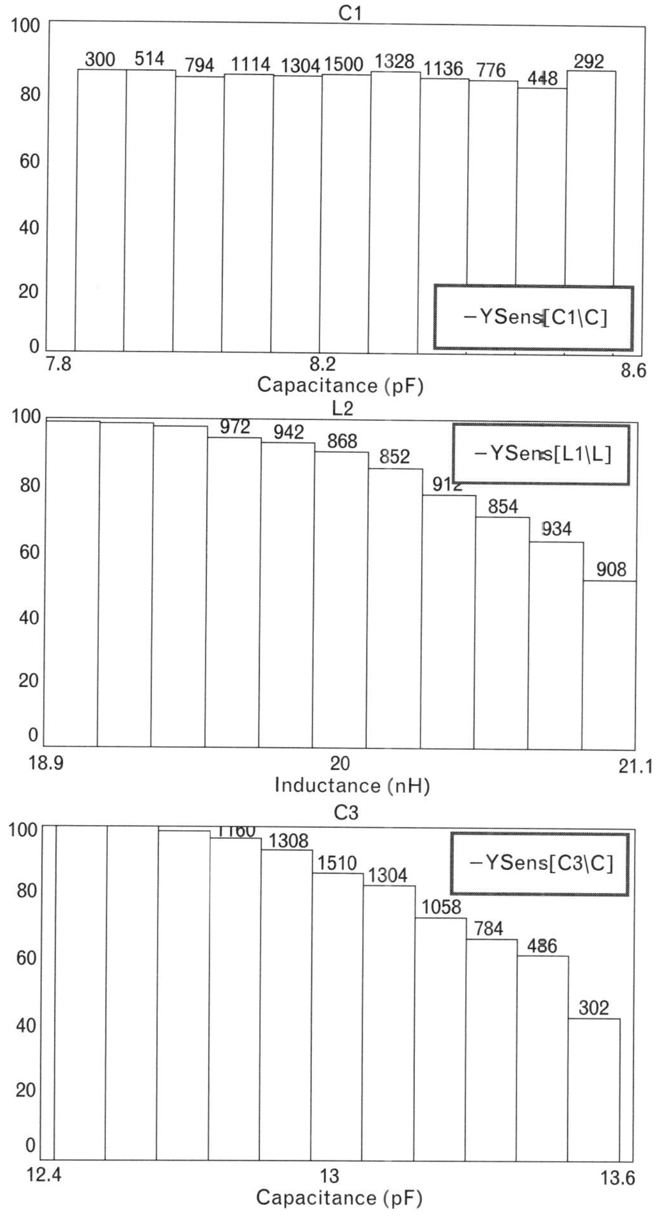

FIGURE 6.23
After the capacitors C_1 and C_5 were changed to 8.2 pF, the statistical sensitivities of all components changed. C_1 shows low sensitivity throughout its tolerance range, but values of L_2 or C_3 need to be lowered to increase yield.

the 100% yield remained. The new statistical histograms now show complete insensitivities of *all components* throughout their respective tolerance ranges (Figure 6.24).

To simulate real-life production of the latest circuit, we display the performance parameters of a large sample of filters with randomly selected component sets. Although from one unit to another, input and output reflection coefficient magnitudes vary, the differences are not noticeable in accumulated plots of the large sample set. Therefore, in addition to the passband and stopband losses, we only show the results of $|s_{11}|$, since $|s_{22}|$ looks identical.

Viewing the cumulated passband and stopband performances (Figure 6.25) shows a comfortable margin for all specifications. Once again, we did not use components with smaller tolerances. If the filter had tighter

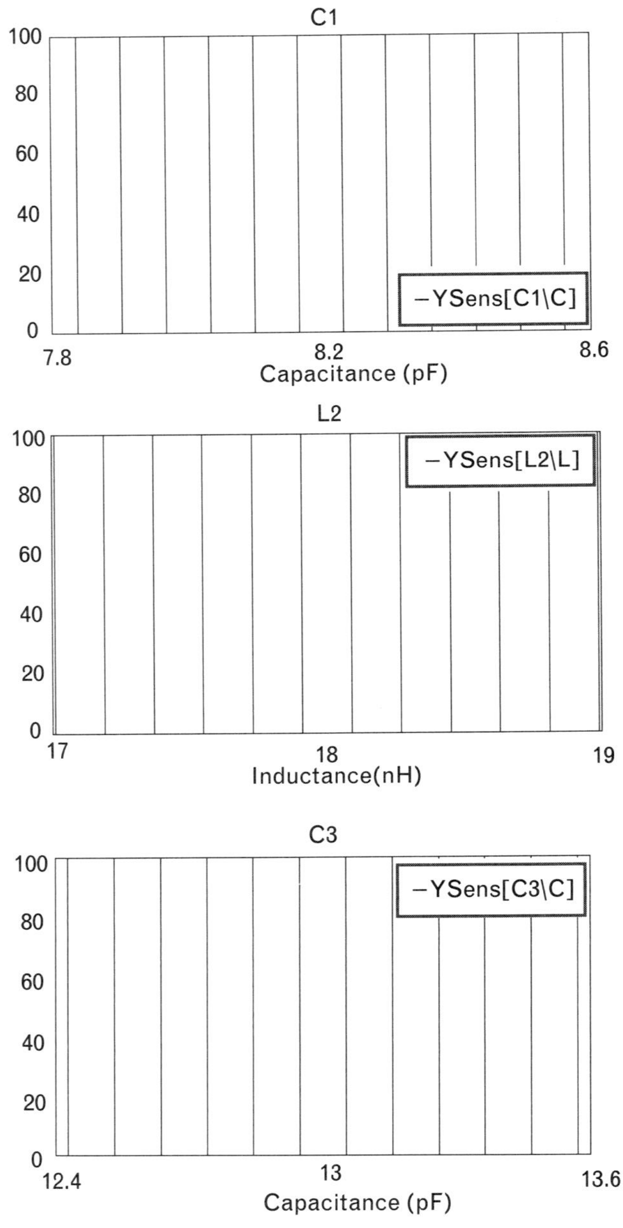

FIGURE 6.24
With 100% predicted yield, the components may be chosen anywhere within the tolerance range. Excluding catastrophic component failure or assembly errors, such circuits would not require testing.

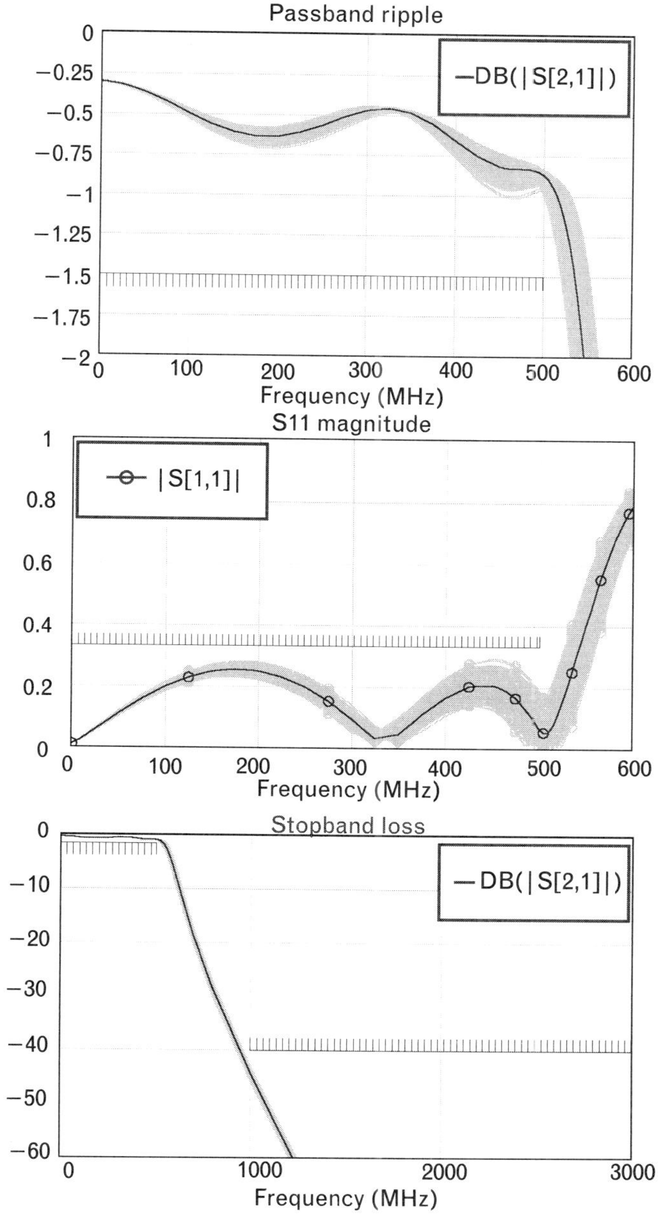

FIGURE 6.25 *After changing four of the five initial component values, the simulated large filter sample meets all production specifications. Only the accumulated measurement of* $|s_{11}|$ *is shown, since that of* $|s_{22}|$ *is virtually the same.*

specifications, it would require better components, and even with those the yield may not be 100%.

6.9 Circuit synthesis

In most RF circuits, two-ports are *doubly terminated*, which means that both sides of the two-port face finite terminations. A *singly terminated* network has an extreme termination at one side that might be an open or a short-circuit. Although the exact synthesis procedures of the two types are somewhat different, they both have two major parts:

1. *Approximation phase to obtain a prototype filter*. Find a rational loss function to approximate the desired response. The loss function is a ratio of polynomials, specified by transmission zeros and reflection zeros. Transmission zeros[3] determine the stopband response, while reflection zeros[4] control the passband performance, including group delay characteristics. If the desired response is a standard type, such as Butterworth or Chebyshev, the functions are already available. For any customized response, the loss function needs to be derived. Inverting the loss function provides a transmission function. For lossless two-ports, from the transmission function we can compute the reflection function, which can be transformed to an impedance function.

2. *Extraction phase to obtain circuit elements*. Starting from one termination, extract circuit elements from the impedance function. Apply algorithms suitable to the topology of the network to determine the circuit element values. In some cases the required algorithm can be as simple as a continued fraction expansion. After the extraction we are left with a residual term that is related to the second termination of the network, but it may *not be equal to the desired value*. Finally, since the prototype computations are done in normalized form, we also need to impedance and frequency scale the components.

The flowchart of Figure 6.26 summarizes the major steps of both phases, as well as the effects of physical component losses and parasitics. After meeting specifications with ideal components, the next step is to find a physical circuit that also meets the requirements. At this point some of the initial specifications may have to be relaxed. If a matching network is synthesized with ideal elements for 0.1-dB equal-ripple loss, the physical circuit will obviously have greater loss. Tolerances must also be considered, as we discussed in the previous section.

3. Related to frequencies where $|s_{21}|^2 = 0$.

4. Related to frequencies where $|s_{11}|^2 = 0$.

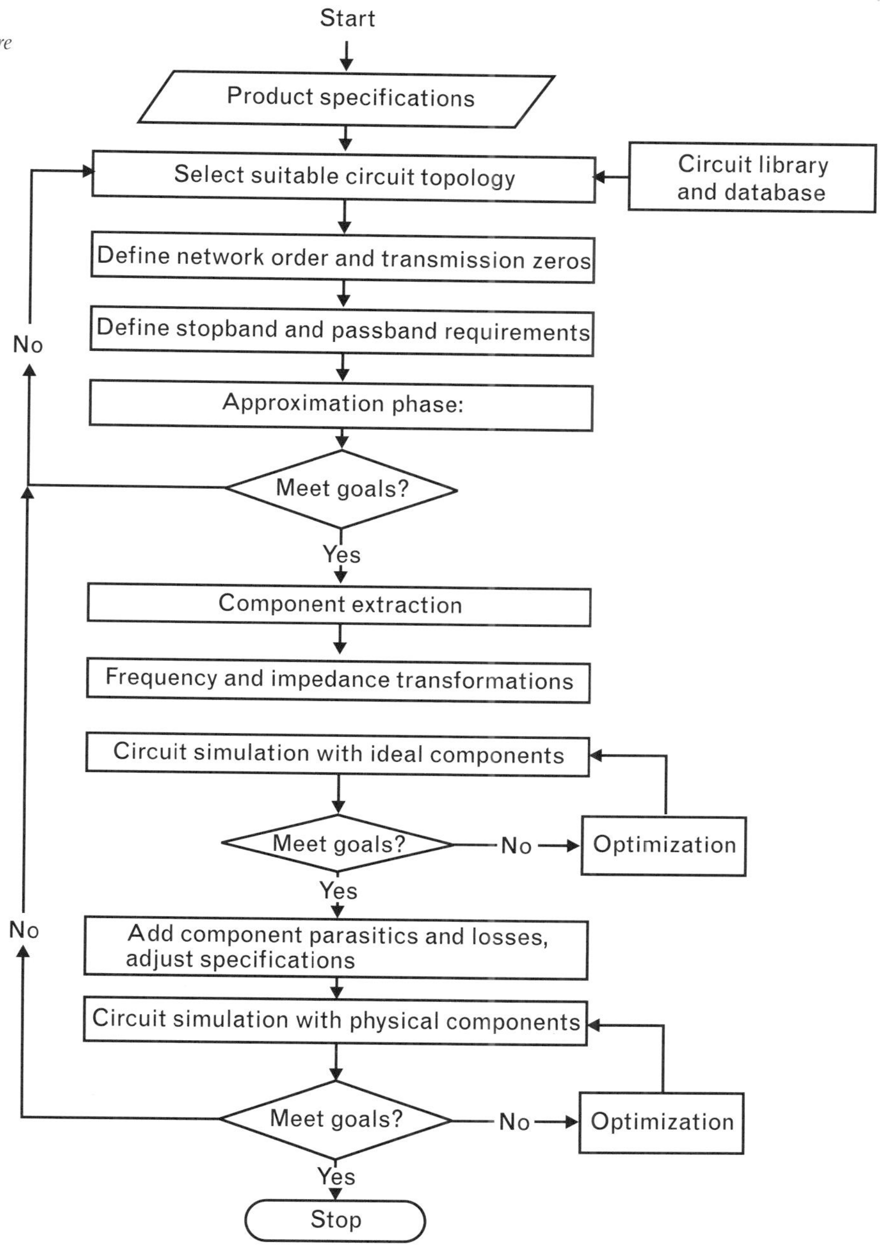

FIGURE 6.26
Exact synthesis procedure from specifications through finding solution with physical circuit components. (After: [19].)

If the circuit is to be realized with lumped elements, the impedances and admittances extracted in the second phase are directly applied to compute the inductance and capacitance values. For distributed circuits, we face a problem because an nth-order transmission line circuit has $2n$ variables of characteristic impedances and electrical lengths, but we only get n coefficients from the extraction. One possible solution is to set the electrical lengths of all transmission lines to a constant. This approach is called the commensurate length design.

Doubly terminated two-port networks are synthesized with resistive terminations and ideal circuit elements. Some of the commercially available matching network synthesis programs allow working into complex terminations, and even into measured one-port impedance or reflection coefficient data. These programs may not be pure synthesis types but also include some form of optimization.

Singly terminated synthesis can be applied to cases where one of the terminations is represented by very high or very low impedance, such as the input of a low-current CMOS device or a high-power transistor, respectively. Another application is for diplexer filter design, discussed in Chapter 8, where two filters are connected in parallel at one of their ports.

A limitation of synthesis is that an exact mathematical solution may not exist for a specific problem. For example, we may want to synthesize a broadband matching network between two terminations through a certain frequency range with a specified ripple. If there is no exact solution for the problem, synthesis generally finds a solution with the proper frequency range and ripple specifications but not for the desired terminations. In such cases applying network transformations, using either Norton or Kuroda's technique, can adjust the initial circuit for the right terminations, without changing the specified frequency response. We will cover these transformations in Chapter 8.

Even if the synthesis does not find an exact solution, the circuit provided may be useful after optimization, as we will show in Section 6.9.3.

6.9.1 Parasitic absorption

Since synthesis is based on real terminations and most RF components have parasitic reactances, we may be forced to incorporate (absorb) the parasitics into the synthesized circuit. This approach places restrictions on the topology selection.

We discussed parasitic absorption in Chapter 5 during impedance matching, so the idea is not new to us. Using the Smith chart, we started impedance matching at one of the complex terminations, which can be interpreted as an equivalent network of a resistor and a reactive element. The reactive part of the termination became part of the matching circuit or we eliminated it by resonance. However, our work was done only at a single frequency, and we later examined the performance at other frequencies also, hoping to meet specifications through the required passband.

Synthesis works through a specified frequency range instead of a single frequency; therefore, we lose the intuitive visualization offered by the Smith chart. Since we cannot resonate a reactive element at more than one frequency, only parasitic absorption is available to us through the following procedure:

1. At the center frequency of the passband, create parallel or series equivalent circuits for the complex terminations. The reactive parts of the equivalent circuits may be lumped or distributed components, depending on the type of network to be synthesized.

2. Select a topology for the matching network that is consistent with the source and load parasitics. For example, if the source parasitic is a series inductor and the load parasitic is a parallel inductor, the matching network must begin with a series inductor and end with a shunt inductor.

3. Temporarily remove the reactive parts from the resistive portions for both terminations and synthesize a matching network between the two resistive terminations.

4. Examine the resultant synthesized components at the input and output side to see if they can absorb the parasitics of the terminations. If yes, split the first and last circuit components into two parts. One part of each split component must be equal to the parasitic of the adjacent termination.

5. If the first or last element of the synthesized network is not suitable to absorb the parasitic, rerun the synthesis with increased ripple and/or *minimum insertion loss* (MIL) specifications. By accepting a finite, nonzero MIL, matching network synthesis programs generally allow flexibility as to which side of the circuit can handle more parasitics. MIL is defined next Section 6.9.2.

6. After we successfully split the input and output elements of the synthesized circuit, the remaining circuitry is the one we have to place between the two complex terminations.

6.9.2 Ripple, slope, and minimum insertion loss specifications

In lossless filter networks the *ripple of transmission coefficient* determines the input and output reflection coefficient magnitude. Changing the ripple also affects skirt selectivity of filters. In matching synthesis, increasing ripple specification allows us to increase the ratio of the resistive terminations to be matched with a given circuit. For example, let us assume that our goal is to match an $R_S = 5\Omega$ source to 50Ω through the frequency range of 600 MHz to 1,200 MHz. If the specified equal-ripple is 0.01 dB, we can only match the 5-Ω source with a synthesized four-element *L*-*C* circuit to a maximum load value of $R_L = 10.3\Omega$. If the ripple is increased to 0.1 dB, the matched load increases to 28.4Ω, while a 0.2 ripple can match to 44.3Ω. To match 5Ω to 50Ω requires 0.238-dB ripple. (The ripple values were determined with the Agilent Esyn program.) Larger ripple also allows us to absorb more parasitics at the terminations.

Minimum insertion loss is a flat loss, evenly applied through the specified passband, that has similar effects on the ratio of matched resistances and parasitics. Introducing relatively small MIL also allows a shift of maximum parasitic absorption capability from one port to the other port. Having a finite MIL, the extraction phase of the synthesis provides more than a single set of coefficients that may be useful to absorb more parasitics. For example, we mentioned above that specifying 0.238-dB ripple, we can synthesize a four-element *L-C* network between 5Ω and 50Ω for the 600- to 1,200-MHz bandwidth. The input element of that circuit is a 1.45-nH series inductor, indicating that the circuit can only absorb a source inductance of up to 1.45 nH. By also adding 0.2-dB minimum insertion loss to the specifications, we obtain several optional solutions with input inductance values ranging from 0.17 nH to 2.13 nH. That is, a nearly 50% increase of the maximum inductive parasitic that could be absorbed without adding MIL.

Slope specification may also be added to the synthesis to compensate the frequency response of other components. For example, in a transistor matching network it can be helpful to design matching networks with positive slope to negate the gain roll-off of the active device.

6.9.3 Illustrative example: matching network synthesis

Synthesize an *L-C* matching network to operate between two complex terminations, as shown in Figure 6.27. The source has an equivalent circuit of 5Ω in series with the $L_S = 0.42$ nH inductance, and the load consists of 50-Ω resistance in parallel with an $L_L = 16.5$ nH inductance. (By providing the equivalent circuits we saved the first step of the recommended procedure since the real and reactive portions of the terminations are already separated.) After the synthesis, compare the results to the solution obtained with the single frequency design and optimization in Section 6.7.6.

Solution

Since our source includes parasitic inductance, our first matching element must be a series inductor. At the output side the last element must be a parallel inductor to represent the parasitic of the load.

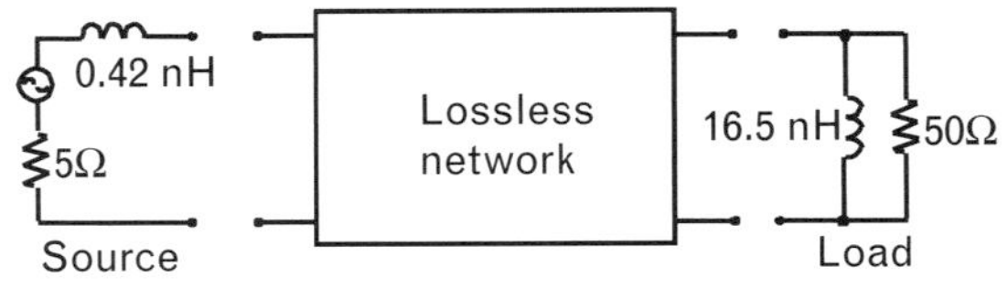

FIGURE 6.27 *Matching complex terminations with synthesis is difficult because most programs accept only resistive terminations. Temporarily removing the parasitic inductances allows the two resistors to be matched, but the circuit must have the proper topology to absorb the parasitics.*

Most of the matching synthesis programs are ready to provide available circuit topologies. If not, by consulting the Smith chart we can come up with a topology that starts with a series inductor and ends with a parallel inductor—going from lower to higher value terminations. A series inductor should be followed by a parallel capacitor, a series capacitor, and finally a parallel inductor as a suitable topology.

Using the Agilent synthesis program, Esyn, we first attempted to synthesize a network between 5Ω and 50Ω with a 0.1-dB ripple specification. (Unlike other filter programs, Esyn asks for "resonators" instead of network order, so we had to enter "2 resonators" for the fourth-order network.) The program did give us the specified matching network; with a 0.1-dB ripple specified, the match is only possible to 28.5Ω, as shown in Figure 6.28.

At this point we can either increase the ripple specification to 0.238 dB, as we have mentioned, introduce a finite MIL, or ask for a higher-order network. A fourth option is optimization, if we are willing to give up the exact equal-ripple response. Using the last option, the convenient built-in optimization routine of the synthesis program quickly reoptimized the circuit to meet the 5-Ω to 50-Ω requirements (Figure 6.29).

Note: Instead of the optimization, we could by trial and error find the necessary ripple requirement for the exact synthesis to match 5Ω to 50Ω. As we stated earlier, it is 0.238 dB. Specifying that ripple would give us synthesized element values within 1% of the optimized values.

Examining the optimized circuit [Figure 6.30(a)], we see that the first inductor on the input side is 1.42 nH, which is greater than our source parasitic of $L_S = 0.42$ nH. That is good news, and we can then split the synthesized inductor into two parts: $L_S = 0.42$ nH is already included in the source and $L_M = 1$ nH will be realized as part of the matching network. At the output side, the synthesized 6.23-nH parallel inductor can be split into

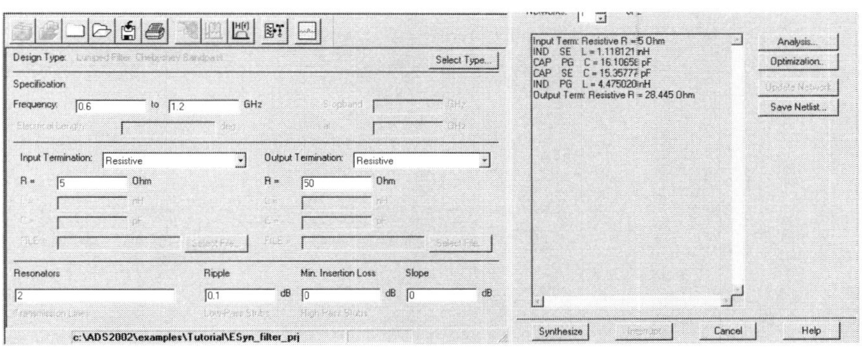

FIGURE 6.28 *With 0.1-dB equal-ripple specification, a fourth-order L-C network can only match 5Ω to a maximum of 28.5Ω for the 600- to 1,200-MHz bandwidth. The desired 50-Ω load could not be reached with the initial specifications.*

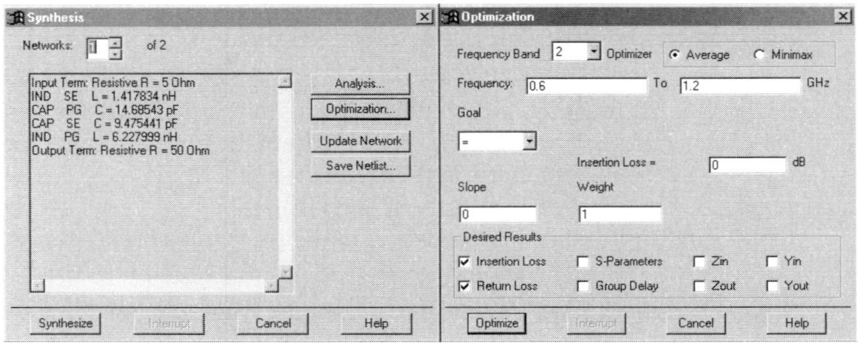

FIGURE 6.29 *Optimization enabled the initially synthesized circuit to match the desired 5Ω to 50Ω, by giving up the exact equal-ripple frequency response.*

two parts, a matching inductor L_M and the load parasitic, L_L. By specifying L_L = 16.5 nH, the L_M is computed as

$$L_M = \frac{L_L(L_T)}{L_L - L_T} == \frac{16.5(6.23)}{16.5 - 6.23} = 10 \text{ nH}$$

After putting the 16.5-nH portion back to the load and leaving the remaining 10 nH as part of our matching network, we have the actual matching circuit, as shown in Figure 6.30(b).

Results obtained here are so close to what we obtained earlier with the optimized single-frequency design in Section 6.7.6 that it is not worth

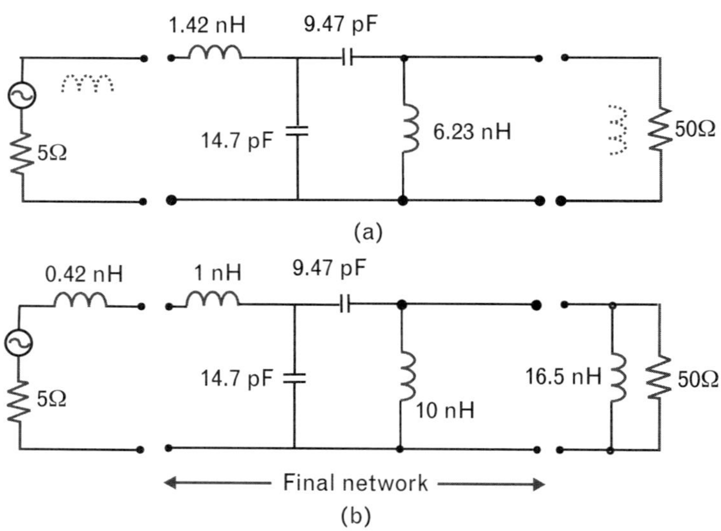

FIGURE 6.30 *(a) Top schematic shows the synthesized-optimized circuit with the reactive parts of the terminations being temporarily removed (shown with dotted lines). (b) In the lower schematic, the first and last elements of the synthesized network are split into two parts, one of which represents the parasitic of the termination. Moving the parasitic portions back to the terminations defines the actual matching network.*

showing the difference. However, it is interesting to see the difference between the initially synthesized circuit of Figure 6.28, matching 5Ω to 28.5Ω, and the optimized circuit shown in Figure 6.30 that matches 5Ω to 50Ω. The comparison is depicted in Figure 6.31.

Between the two options, optimization or synthesis, there is no doubt that the latter is faster and more accurate. As with most design tools, however, synthesis has its limitations. But even when the problem cannot be fully solved by synthesis, the results of the synthesis can provide good initial values for optimization.

Exact synthesis is a rigid mathematical approach and most textbooks are written in that form. One of the few exceptions is a recently published book by Minnis [19] that we highly recommend for those who wish to explore the topic more deeply.

6.10 Electromagnetic field simulation

Electromagnetics (EM), a term that often brought terror to the minds of engineering students, is becoming more appreciated among RF and high-speed digital design engineers. EM simulation allows us to look inside circuits and structures, identifying problems related to interconnections, packaging, multilayer boards, EMI, and EMC. Even when working with the components for which the manufacturer provides measured data, how the part performs often depends greatly on its mounting, nearby components, and the enclosure it is in.

There are several commercially available EM field solvers available and now all the major circuit simulators also have one bundled inside.

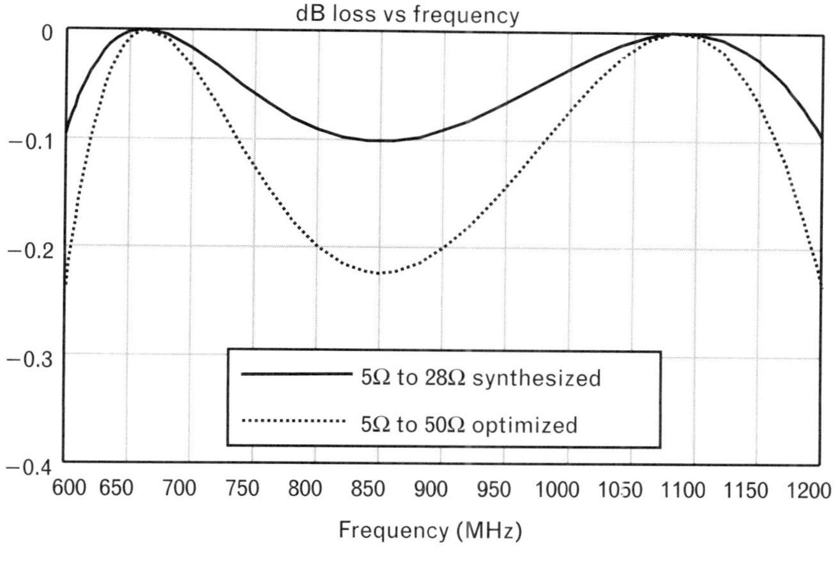

FIGURE 6.31
Matching through a larger resistance ratio leads to larger ripple.

Integrated packages enable engineers to analyze and optimize circuit theory and circuit layout–related performances simultaneously.

Similar to distributed circuit analysis where incremental inductances and capacitances are computed, EM simulation subdivides metal patterns into small cells—a process called discretization. Smaller cell size provides better accuracy, at the cost of increased computational time. Since the EM fields of the cells are described by differential or integral equations, solutions even in linearized form can be time-consuming. Compared to a typical RF circuit with 10 to 20 nodes, where simulation can be obtained in split seconds, EM simulation has thousands of cells, requiring larger matrices to formulate solutions. Consequently, when combined with the more complex algebraic manipulations, EM simulation requires considerably more computing time.

6.10.1 Categories by geometries

EM simulators may be classified either by the type of geometries they solve or the type of numerical technique they use. The former can be split into three main categories, 2D, 2.5D, and 3D geometries, as shown in Figure 6.32.

1. The 2D cross-section solvers handle strips (microstrip, stripline) or slots (coplanar waveguide) with *uniform cross-section*. This group can be split into two categories, closed-box and laterally open, depending on whether there are conductive side walls or not. This is the simplest geometry to handle with an EM simulator; also, it is the easiest to describe. Since it discretizes only the 2D cross-section, it requires the least amount of computing effort of the three types.
2. The 2.5D planar solvers, sometimes called "3D mostly planar types," can handle an arbitrary number of homogeneous dielectric layers with arbitrary shaped metal patterns in several planes, including via holes between the metal layer. (The 2.5D name comes from the fact that vias are allowed.) The 2.5D category may also be

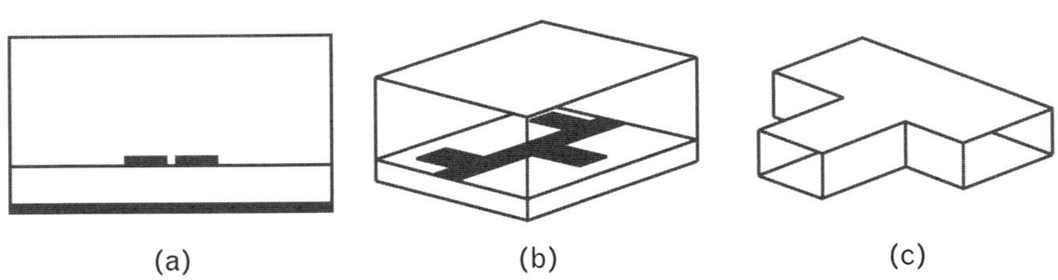

FIGURE 6.32 *EM simulation techniques may be classified by the order of their geometries: (a) 2D with conductors laid out into one direction, (b) 2.5D allowing arbitrary planar metal layer, and (c) 3D with completely arbitrary three-dimensional geometry. (After: [20].)*

split into two groups, laterally open and closed box formulation. Computational time is significantly larger than that of 2D because each of the metal layers is discretized.

3. Arbitrary 3D geometry is the ultimate of field solvers, generally bounded by conductors although the boundaries may also be some absorbent material. The 3D field solvers discretize the entire volume and consequently take much higher amount of computational time. They take also considerably more time to formulate—often requiring some artistic ability from the user.

Table 6.5 summarizes the different geometry groupings, splitting the 2.5D solvers into two subgroups. We recommend getting familiar with all three major categories in order to use the one most appropriate for a given task. There is no justification for using a 3D simulator for a simple problem that can be handled with a 2D type.

Based on the type of numerical technique used, there are four major categories, as follows.

1. The *finite element method* divides a problem into homogeneous tetrahedral elements where the size of the elements depends on the resolution needed. During analysis the technique determines the E field at each corner or node of the elements.
2. The *methods of moments* (MoM) subdivide planar metal layers using rectangles or triangles and solves for the RF current on the conductors, using the appropriate Green's function.

TABLE 6.5 SUMMARY OF FIELD SOLVERS BY GEOMETRY CATEGORIZATION

2.5D PLANAR—LATERALLY OPEN	3D ARBITRARY GEOMETRY (SLOWEST)
No fixed grid	Any specified geometry
Rectangular and triangular elements	Basic formulation is closed box
Numerical Green's function	Absorbing boundaries possible
Symmetry or walls require image theory	Model building time can be significant
Arbitrary spatial resolution	Must discretize the entire volume
2D CROSS-SECTION SOLVERS (FASTEST)	**2.5D PLANAR—CLOSED BOX**
Strips or slots with uniform cross-section	Fixed grid
Easy to find Z_0 and ε_{EFF} for single strips	Rectangular elements
Two subclasses: closed and laterally open	Analytical Green's function
Model building time very low	One plane of symmetry is easy
Discretize only the 2D cross-section	Small features (resolution) can be a problem

Source: [20].

Note: Arrows indicate the increased amount of computational time needed. The order is: 2D→2.5D closed box→2 5D laterally open→3D.

3. The *finite-difference time-domain method* is a direct solution of Maxwell's equations in space and time.
4. The *transmission line matrix method* replaces the circuit with a network of lumped or distributed components and finds voltages and currents that are equated to field quantities.

Textbooks dealing with electromagnetics tend to be highly analytical [21], and sometimes it is difficult to "see the forest through the trees." An exception is a recently published book by Swanson [20], which deals with these complex issues in ways that a typical engineer can understand.

6.10.2 Illustrative example: layout and cosimulation of a 6-GHz Wilkinson power divider

The following describes the conversion of a circuit schematic to an actual transmission line layout. Once the layout is obtained, the physical circuit is submitted for EM simulation and the results are compared to the response of the circuit simulator. Task are formed with the Agilent ADS program.

A detailed circuit schematic of the power divider, including all transmission line discontinuities, is shown in Figure 6.33.

After creating the layout, either manually or automatically using design synchronization, as is done in this example, a Layout Component is created from the EM simulator's menu. The layout is then submitted to the bundled 2.5D EM simulator (Momentum) that splits the metal pattern into small cells for the EM simulation (Figure 6.34). Note that straight lines are partitioned into rectangles and the curved lines into triangles for better approximations—based on Momentum's technology—allowing automatic and transparent integration of planar electromagnetic simulation data at the schematic design level.

After the physical circuit is passed to the EM simulator, the program creates an icon on the tool bar for placement on the circuit schematic. At this point we connect the external terminations to the three ports and place the isolation resistor between the two output ports. If physical model for the resistor is available, it can take place of the ideal resistor. The circuit shown in Figure 6.35 is now ready for cosimulation.

The circuit level schematic either uses previously generated EM simulation results or will cosimulate if no EM results exist or if the layout has been modified. Results of the two simulations are shown in Figure 6.36, displaying input and output return loss, transmission, and isolation. Since the divider is symmetrical, transmission and output return loss are only shown for one side.

The EM simulation shows that the divider's frequency response is centered at about 10% below what was predicted by the circuit simulator, preventing a costly prototype building cycle. Dimensions can now be adjusted

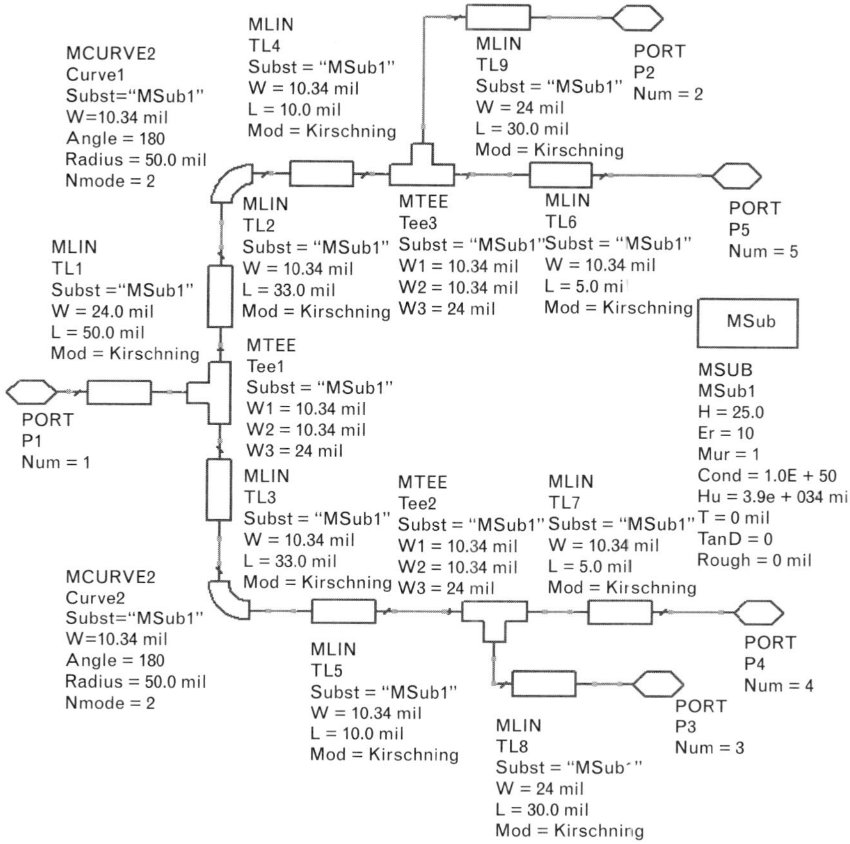

FIGURE 6.33 *Optimized electrical design with transmission line discontinuities. Ports 1, 2, and 3 are the RF input and output terminals. Ports 4 and 5 are the small conductive traces where the 100-Ω isolation resistor is attached. (From: [22]. ©2003 Aglient Technology. Reprinted with permission.)*

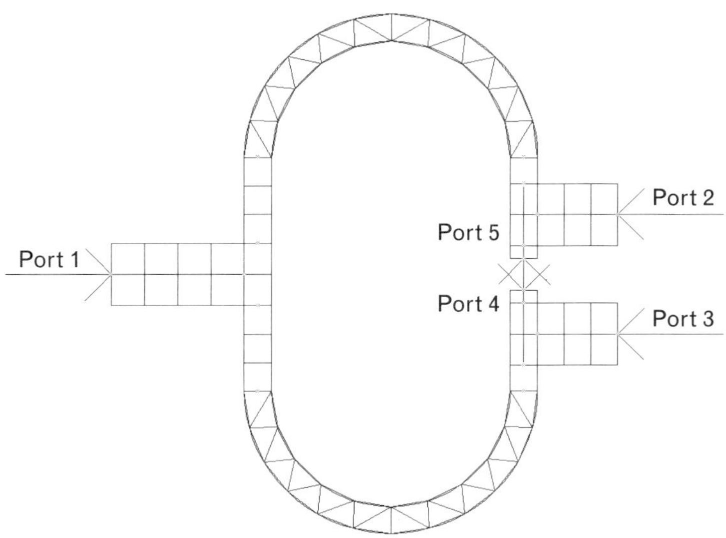

FIGURE 6.34 *Automatic meshing generated by the EM simulator. If needed, the user can override and modify the pattern generated by the program. (From: [22]. ©2003 Aglient Technology. Reprinted with permission.)*

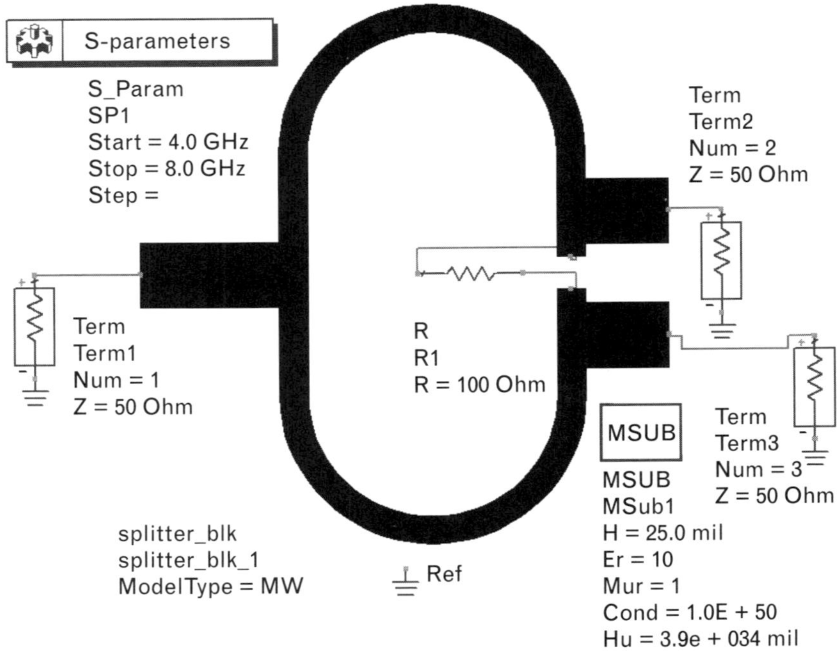

FIGURE 6.35 *The power divider is submitted to the circuit simulator. The five-port transmission line section was already analyzed by the EM simulator and its response is stored in a five-port S-parameter matrix. (From: [22]. ©2003 Aglient Technology. Reprinted with permission.)*

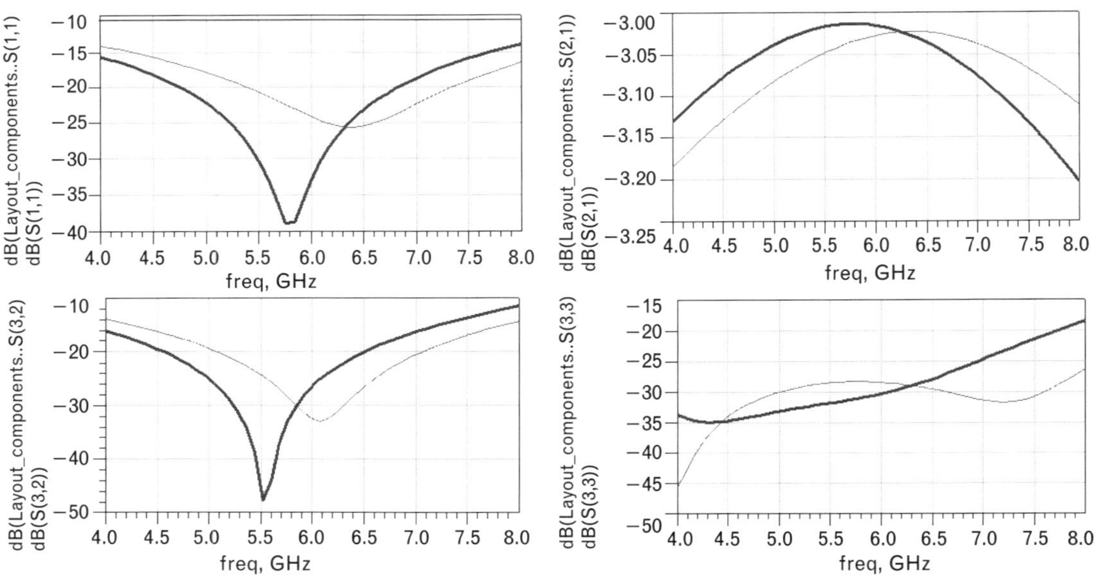

FIGURE 6.36 *Comparison of the circuit and EM simulations for the 6-GHz power divider, resulting from the schematics shown in Figures 6.33 and 6.35. Bold traces come from the circuit that contains the EM analysis. (From: [22]. ©2003 Aglient Technology. Reprinted with permission.)*

until the desired response is reached. Such iterations in the computer can be performed quickly, while the alternative may be a lengthy and expensive process.

6.11 CAD program descriptions

The following four RF circuit/system/EM simulation suite descriptions, extracted from the vendors' Web sites [22–25], provide a cross-section of modern CAD capabilities. The Agilent suite is available for both MS Windows and UNIX, while at this time the Ansoft, AWR, and Eagleware software runs under Windows only. All programs require licensing.

6.11.1 Agilent Advanced Design System

ADS is to RF/MW designers what Photoshop is to professionals of digital photography. The program is capable of designing the entire signal path of wireless or wired communication products—including communications systems, baseband DSP, hybrid and RF PCB, *monolithic microwave integrated circuits* (MMICs), RFIC, electromagnetic simulation, and physical layout, in a single integrated environment. Interfaces are available to other design environments, such as Cadence Design Systems or Mentor Graphics. Functionally, the program can be presented in three layers, as shown in Figure 6.37.

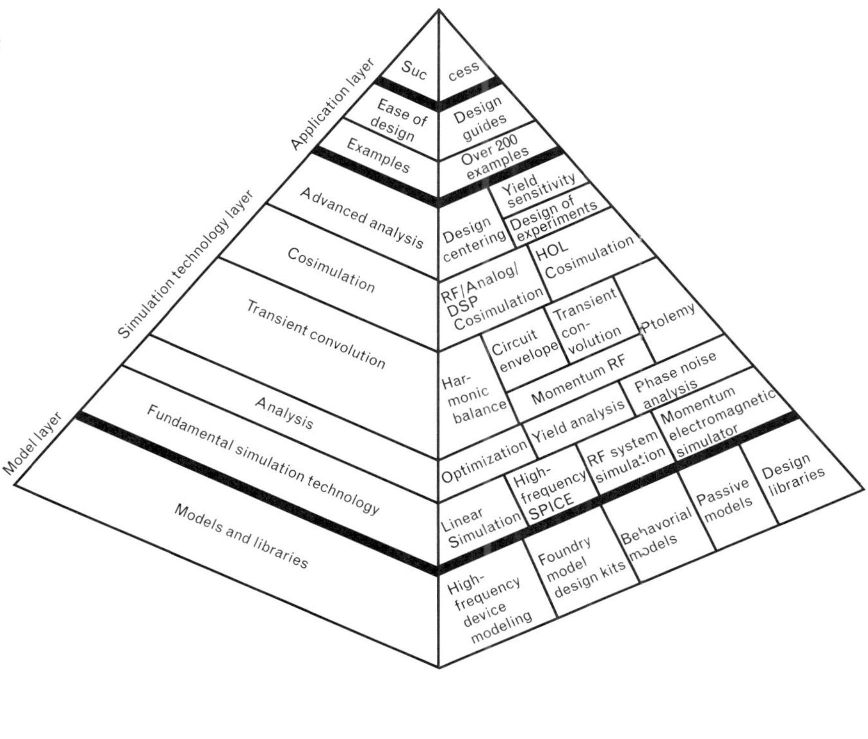

Figure 6.37
Key layers of the Agilent ADS, showing the various capabilities offered. (From: [22]. ©2003 Aglient Technology. Reprinted with permission.)

- The *simulation technology layer,* which includes different simulation technologies;
- The *application layer* with design guides, templates, and circuit libraries for various applications;
- The *models layer* with wide range RF active and passive component models.

The major building blocks of the program are.

- *DC simulation:* common to all RF/Analog simulations. It performs topology check and analysis of the dc operating point.
- *AC simulation:* conventional ac analysis to obtain small-signal parameters in terms of voltages and currents, and linear noise voltage and currents.
- *S-parameter simulation:* provides linear *S*-parameters and noise parameters, in tabulated and graphical forms.
- *Harmonic balance simulation:* a nonlinear harmonic-balance technique to find the steady-state solution in the frequency and time domains.
- *Circuit envelope simulation:* a combination of frequency and time-domain analysis techniques to yield a fast and complete analysis of complex signals such as digitally modulated RF signals.
- *Transient/convolution simulation:* solves a nonlinear circuit entirely in the time domain using simplified models to account for the frequency-dependent behavior of distributed elements.
- *Optimization and statistical design:* for automatic performance optimization based on a variety of optimizers, and statistical design to provide yield analysis and optimization. A manual tune mode is also available.
- *Esyn, filter, and matching network synthesis:* for exact synthesis with lumped and distributed components.
- *LINECALC and transmission line analysis and synthesis:* conversions between physical dimension and electrical parameters of various transmission line forms.
- *Cosimulation and cooptimization:* controlling both circuit component values and layout dimensions to optimize performance.

6.11.2 Ansoft Designer

Ansoft Designer feature highlights

Ansoft Designer suite of design tools fully integrates the power of high-frequency, physics-based electromagnetic solvers (quasi-static, 3D planar, and HFSS 3D) with circuit and system-level simulation. The key to this integration is a unique capability called solver-on-demand that orchestrates the use of multiple solvers, while still giving complete control to the engineer. Ansoft Designer extends this automation and integration even further by enabling work created in other types of design software to be rapidly imported as easily as if it had been created within Ansoft Designer itself.

Three-in-one capabilities

Ansoft Designer's capabilities include modeling, simulation, and automation.

- *Electromagnetic modeling:* Ansoft Designer's integration and advanced multiple electromagnetic solver technology combine to provide the most complete physical design solution available. Ansoft Designer significantly improves accuracy while saving considerable time over traditional empirical characterization methods.
- *Electrical simulation:* Ansoft Designer's time, frequency, and system analysis empower engineers to investigate all electrical performance criteria before committing to fabrication, allowing them to get the design right the first time.
- *Design automation:* Ansoft Designer can address all communication, RFIC/MMIC, and PCB applications with its fully integrated layout editor. It supports Java and Visual Basic scripting, advanced library management, and third-party links.

Desktop

- Advanced data entry:
    - Fully integrated schematic layout editors.
- Dynamic project manager;
- Component library developer and manager;
- Solution manager;
- Advanced results graphing and postprocessing;
- Powerful 3D viewer;
- System, circuit, and component (3D planar EM) analysis:

    - Full multianalyses cosimulation;
    - Optimetrics (optional).
- Full swept parameter analysis (all solvers).

System simulation

- Time, frequency, and mixed-mode analyses;
- Extensive RF and DSP component libraries;
- Compiled and interpretive C and C++ user-defined model cosimulation;
- MATLAB cosimulation;
- Communication standard libraries include: IEEE 802.11 a&b, GSM, EDGE, and so forth;
- WinIQSIM links for WCDMA, TD-SCDMA, CDMA2000, and so forth, accurate waveforms;
- Supports physical design and layout.

Circuit simulation

- Harmonic-balance engine:
    - Fifth generation Krylov solver.
- Nonlinear noise analysis (phase and amplitude);
- Transient analysis;
- Digital modulation analysis (FastACPR);
- Nonlinear stability analysis (Nyquist);
- Load and source pull analysis;
- Design synthesis:
    - Filter, Smith, and transmission lines.
- Extensive model support.

Layout and manufacturing

- Integrated layout/drawing editor;
- Material and stack-up manager;
- Unlimited layers;
- DXF/GDSII import and export;

- Scripted and parameterized footprint (cell);
- JavaScript and Visual Basic scripting;
- Macros ;
- AnsoftLinks for third-party integration (available separately). Links include tools from companies such as:
  - Cadence;
  - Mentor;
  - Synopsys.

3D planar EM simulation

- Fast and discrete frequency sweep: *S*-, *Y*-, and *Z*-parameters;
- Near and far field radiation analyses;
- Static or animated surface currents;
- Single value decomposition (SVD FastSolve);
- Open and closed formulations;
- Thick-metal solver;
- Self-adaptive triangular meshing;
- Advanced port capability:
  - Arbitrary feed angle;
  - Internal and gap ports.
- Infinite array analysis

A student version of the program is available at no charge.

6.11.3 AWR Microwave Office

The Microwave Office Design Suite combines linear and nonlinear circuit simulation with EM analysis and layout to provide a complete design solution for RF and microwave engineering. The program includes an extensive collection of component models that support RF subsystem analysis using harmonic-balance and Volterra simulations, a comprehensive Filter Synthesis Wizard, and a Load Pull Wizard that can incorporate measured and simulated load pull contours. Enhancements to the existing circuit simulator include the addition of oscillator phase noise analysis, and the 3D planar electromagnetic simulation engine has been enhanced to simulate large problems more efficiently by incorporating state-of-the-art solvers.

One of the underlying technologies, *Component Object Model* (COM), provides a foundation for higher-level software integration and uses a single object-oriented database that contains both electrical and physical properties.

The *Visual System Simulator* (VSS) enables circuit and system design engineers to perform top-down analysis of analog and digital communications systems. VSS software is completely integrated with the Microwave Office Design Suite either for "low-to-high level" or "high-to-low level" analysis to be performed. Consequently, the various applied active device signal level effects are incorporated at the system level through system-circuit cosimulation.

The discrete time simulation engine and extensive model libraries of VSS provide a solution for analyzing systems from the channel through the RF and DSP subsystems. This solution is ideal for analyzing wireless communications systems, high-speed wire-line, and electro-optical systems. VSS also enables users to quickly build graphical block diagrams and analyze performance using sophisticated built-in measurements and powerful signal generators that support virtually any modulation scheme, including: AM, FM, OFDM, PSK, MSK, QAM, and others. Users can assess the impact of specifications by utilizing a unique "real-time tuning mode" that immediately displays the impact of parameter changes by updating measurements in real time.

The software includes a comprehensive library of over 230 core elements and mathematical primitives that can be used to build an accurate representation of the most complex communications systems. The library includes encoders/decoders (including Viterbi, Reed-Solomon, convolutional, and others), modulators/demodulators, and filters. Application-specific libraries are optional and support 3G, IS95, GSM, EDGE, 802.11, and other emerging standards. The Design Suite combines a multirate discrete time simulation engine with advanced block processing techniques to efficiently manage the flow of data between blocks. In essence, the product is a data-driven simulation tool with the ability to handle impairments that can be modeled in either the time or frequency-domain.

Block interconnects are modeled using *elastic buffers* that efficiently control the flow of data. These interconnects automatically ensure the proper alignment of data prior to performing bit-error-rate simulations without the need for delay elements for each block. Interconnects support synchronous and asynchronous data transfer, thus implicitly supporting multirate processing. Additional built-in measurements are available, including: power spectral density, *error vector magnitude* (EVM), and *adjacent channel power ratio* (ACPR) measurements. VSS software blocks can also be programmed to run on a sample-per-sample basis, which is an ideal way of handling feedback loops in data-driven simulation tools. VSS solution's block processing techniques provide extremely fast simulation times even for complex system analysis.

Trial versions are available through the Internet.

6.11.4 Eagleware Genesys

Eagleware offers an integrated environment that takes designs from specification through documentation. The GENESYS design flow starts with customer specifications, which are used to develop system architecture. Synthesis tools then convert each block in the system into a circuit representation. The circuit is refined using linear and nonlinear circuit simulators, and then a layout is created. Electromagnetic simulation—including co-simulation with the circuit simulators—verifies the design. Prototype measurements are read into the GENESYS environment for final validation.

As each more detailed design representation is created, more detailed performance characteristics can be included in the overall system simulation. This concept of *Continuous Integration* saves time, facilitates team communication, and catches system errors early.

The GENESYS design system architectures use a wide range of behavioral models such as amplifiers, mixers, splitters, couplers, and filters. Once the architecture is in place, you can simulate the system performance along each signal path. Since SPECTRASYS is a fast full-spectrum simulator, the complete spectrum may be tracked at each node of the design.

Eight synthesis tools take system block specifications and create first pass circuits that are ready for refinement. A range of synthesis modules synthesize matching networks, oscillators, filters (*L*-*C*, distributed, active, and mixed), and group delay equalizers.

Once a circuit is entered, GENESYS simulates *S*-, *Y*-, *H*-, and *Z*-parameters, noise parameters, and stability factors as well as nonlinear dc and spectral (harmonic balance) parameters such as intermodulation distortion. The linear and nonlinear simulation engines are so tightly interwoven that you can perform all simulations and optimizations within a single workspace. A powerful equation and postprocessing engine makes it easy to create custom measurements.

Using highly optimized simulation technology, designs are refined with real-time tuning. The speed of GENESYS is particularly valuable in completing statistical analysis such as Monte Carlo analysis or yield optimization.

The integrated layout module in GENESYS allows engineers to automatically create the layout from the schematic or manually create it through interactive design. Either way, complete layouts are created on up to 128 layers, including metallization, IC processing, solder mask, and assembly layers, and are output using built-in Gerber, GDSII, and DXF writers.

For ultimate accuracy, the metal in the circuit can be analyzed using electromagnetic simulation. Using EM-circuit co-simulation, lumped components such as transistors and capacitors can be simulated together with the metal.

The ultimate step in the design process is to test the design. Integrated into the GENESYS environment, TEST LINK reads data from network analyzers, spectrum analyzers, oscilloscopes, and noise figure meters through GPIB or RS-232 connections into the workspace. This data can be used to compare measured results to simulation, or it can be used directly as a model or as a stimulus to the design.

The last part of any design is completing the documentation. Much of the work is already done since the schematics, layouts, and simulation results are stored in visual form. Comments are easily added to the schematic, graphs and layout, or transferred to office automation tools through simple cut and paste operations.

6.12 Summary

RF and microwave computer-aided design and engineering have matured during the past decade and have proved their value in our industry. Although the cost of software is nearly two orders of magnitude above the price of business programs, companies have to make the investment in RF design tools also. Delayed projects and lost opportunities by losing out to competitors cost far more than the price of the software.

Integration of analog, digital, and electromagnetic simulation is an important requirement. Software vendors have made progress bringing circuit, system, and EM simulation closer to each other. They also need to reach RFIC designers because many of them are still using analog low-frequency design tools.

References

[1] Meehan, M., "Modern Robust Design Techniques," short course, Besser Associates, Mountain View, CA, 1995.

[2] Besser, L, "Combine S-Parameters with Time-Sharing," *Electronic Design*, August 1968.

[3] Besser, L., "A Fast Computer Routine to Design High Frequency Circuits," *IEEE ICC Conference Digest*, San Francisco, CA, June 1970.

[4] Kundert, K. S., *The Designer's Guide to SPICE and Spectre*, Boston, MA: Kluwer Academic, 1995.

[5] Bandler, J. W., and C. Charalambous, "A New Approach to the Computer-Aided Design of Microwave Circuits," *IEEE MTT Int. Symp. Digest*, Arlington Heights, IL, 1972.

[6] Bandler, J. W., et al., "A Microwave Network Optimization Program," *IEEE MTT Int. Symp. Digest*, Boulder, CO, 1973.

[7] Besser, L., S. Ghoul, and C. Hsieh, "Computerized Optimization of Transistor Amplifiers and Oscillators Using COMPACT," *European Microwave Conference*, Brussels, Belgium, September 1973.

[8] Besser, L., et al., "Computer Aided Design for the 1980s," *IEEE MTT Int. Symposium Digest*, Washington, D.C., 1981.

[9] Hamilton, S. E., and D. M. Morton, "User-Defined Elements Expand CAE Software Capabilities," *MSN & CT*, April 1986.

[10] Rautio, J. C., and R. F. Harrington, "An Electromagnetic Time-Harmonic Analysis of Shielded Microstrip Circuits," *IEEE Trans. on Microwave Theory and Techniques*, August 1987.

[11] HFFS, Ansoft Corp, Pittsburg, PA.

[12] Penfield, P. R., R. Spence, and S. Duinker, *Tellegen's Theorem and Electrical Networks*, Cambridge, MA: MIT Press, 1970.

[13] Cuthbert, T. R., *Circuit Design Using Personal Computers*, New York: Wiley, 1983.

[14] Bandler, J. W., and C. Charalambous, "Nonlinear Programing Using Minimax Techniques," *Journal of Optimization Theory and Applications*, 1974.

[15] Brayton, R. K., and R. Spence, *Sensitivity and Optimization*, Amsterdam: Elsevier Scientific Publishing, 1980.

[16] Winter, G., J. Periaux , and M. Galan, *Genetic Algorithms in Engineering and Computer Science*, New York: Wiley, 1995.

[17] Meeham, M. D., and J. Purviance, *Yield and Reliability in Microwave Circuit and System Design*, Norwood, MA: Artech House, 1993.

[18] McFarland, A., et al., "Centering and Tolerancing the Components of Microwave Amplifiers," *IEEE MTT-S International Microwave Symposium Digest*, Las Vegas, NV, 1987.

[19] Minnis, B. J., *Designing Microwave Circuits by Exact Synthesis*, Norwood, MA: Artech House, 1996.

[20] Swanson, D. G., *Field Solver Circuit Design Cookbook*, Norwood, MA: Artech House, 2003.

[21] Sadiku, M., *Numerical Techniques in Electromagnetics*, 2nd ed., Boca Raton, FL: CRC Press, 2001.

[22] http://www.agilent.com.

[23] http://www.ansoft.com.

[24] http://www.mwoffice.com.

[25] http://www.eagleware.com.

CHAPTER 7

Passive component models

7.1 Introduction

During the development of defense electronics systems in the 1970s and 1980s, a great deal of effort was put into the modeling of active microwave devices and transmission lines. Component models were handled by teams of academic researchers and highly experienced engineers, who had extensive background of electromagnetic theory and device physics. Interestingly, commensurate attention was not given to passive RF components. Researchers evidently believed that these devices were simple enough that they would pose little challenge for designers using CAD/CAE tools. As a result, design software available at the time offered these elements in their ideal *R*, *L*, and *C* (*R*-*L*-*C*) forms. Later this was observed to be inaccurate as the need to bring commercial products to market quickly necessitated accurate models for all components, active and passive.

Ultimately all of the circuits that we design must be realized using real, physical parts. Sometimes RF circuits will employ distributed elements such as transmission lines and resonant cavities, but a large part of RF realization work (below 1 GHz) falls to lumped *R*-*L*-*C* components. In initial designs conducted using CAD models (and just as often on schematics drawn on paper napkins) we do represent these circuit elements as though they were ideal. We think first of *R*-*L*-*C* components as having frequency-independent values, and no parasitic elements, such as series inductance for a resistor or capacitor or shunt capacitance for an inductor.

This may be all right for an initial design. An ideal circuit model allows us to think about the principal functions of each circuit element. It is a simplified view of what will be built eventually. But, that is just what it is—a *simplified equivalent circuit*. Real *R*-*L*-*C* components cannot be represented by a single *R*, *L*, or *C* term. Because of losses and parasitic reactances, real resistors and inductors need at least three to four circuit elements to represent their behavior over modest frequency bandwidths. Beyond that there are the frequency-dependent characteristics of the resistive, inductive, and capacitive processes themselves, such as resistive skin effect, the change of dielectric constant with frequency, and the change of magnetic properties of inductor cores with both frequency and current magnitude. Primary and higher-order resonances further complicate the process. We eventually reach the frequencies where lumped equivalent circuits can no longer describe the behavior of our components. That is when we switch to distributed, or transmission line elements.

To gain more appreciation of the difference between performances of ideal and physical components, let us look at a very simple example. Figure 7.1 shows the transmission loss of a two-element *L-C* matching network consisting of a 4.3-pF series capacitor and an 8.2-nH parallel inductor. With ideal components the frequency response is symmetrical through the 825- to 925-MHz bandwidth, matching a 10-Ω termination to 200-Ω. When we apply two physical surface mount type components (ATC 100A4R3DW150XB and CoilCraft 0603CS-8N7XBG), the change of effective inductance and capacitance, as well as the component losses, appreciably alters the frequency response. Without considering any circuit layout related effect, the initial 0.25-dB loss at 925 MHz increased to more than 1 dB. Clearly, in a more complex circuit the changes would be proportionately more significant.

Because of these complexities, real circuit designs need to take parasitics and losses into account, as well as the frequency-dependent models of *R-L-C* variations. While the CAD programs deal with the complex mathematics of these multipart equivalent circuits, it is important for the designer to understand the nature of the equivalent circuit and what changes can be expected from the ideal estimated behavior. As is true in many pursuits, forewarned is forearmed.

An alternative approach relies on measured frequency-dependent *S*-parameter data instead of equivalent circuits. However, classical lumped circuit design techniques are based on component values rather than measured data. An *L-C* filter or matching network is specified by nominal

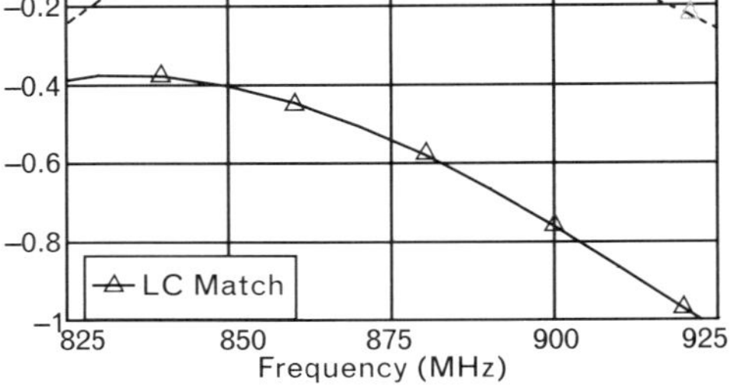

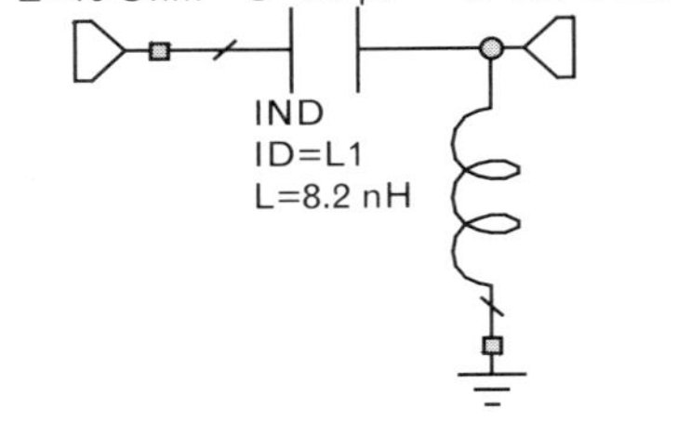

FIGURE 7.1 *Component parasitics and losses always modify the response of an RF circuit, as illustrated here for a simple two-element lumped matching network. Replacing the ideal components with surface-mount (chip) type physical elements lowers the center of the passband by about 5% and causes a 0.4 dB dissipative loss.*

components values. Substituting various measured datasets for each circuit component is a time-consuming process that generally does not lead to acceptable results. If the designer does not know how to relate the measured data to the circuit models of the components, much of the available (and expensive) software and computer power are usually wasted.

In this chapter we first explore lumped equivalent circuit models, first-order and secondary parasitic resonances, and Q-factor variations of commonly used passive components. Then we look at basic transmission line components, their discontinuities, and coupling structures. Our coverage also extends to transformers constructed with using transmission lines instead of the traditional primary and secondary windings. We also briefly touch upon measurement and test-fixture related issues.

7.2 Resistance, self-inductance, and stray capacitance of conductors

Before proceeding with the equivalent circuits for R-L-C components, let us consider three important aspects of real components: the resistive skin-effect, the self-inductance of conductors, and the stray capacitances among conductors. Although generally negligible at low frequencies, at RF we need to include these effects into our circuit simulation.

7.2.1 Resistance changes of conductors at RF

When RF signals propagate through wires and conductive traces they face frequency-dependent resistance. Let us first consider the dc resistance of a simple conductor, consisting of a cylindrical shape having cross-sectional area A, and length ℓ (Figure 7.2). The dc resistance of the conductor is given by

$$R_{dc} = \frac{\ell}{\sigma A}$$

where

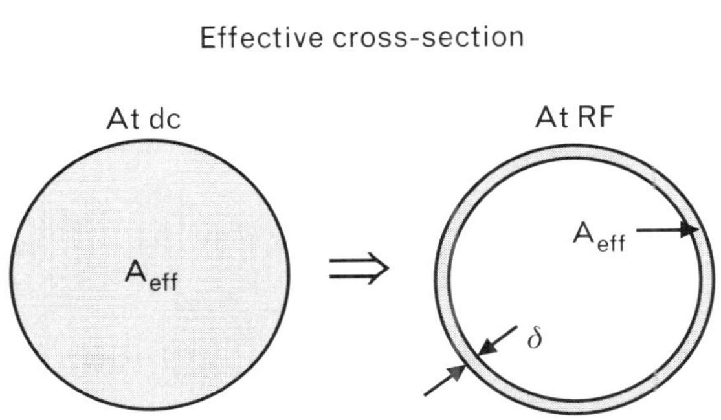

FIGURE 7.2
Cross-section of a circular conductor showing the skin depth δ. At RF the effective cross-sectional area that conducts current is a small fraction of the same at dc.

σ is the material conductivity in Siemens-centimeters.
ℓ is the conductor length in centimeters.
A is the conductor cross-sectional area in square centimeters.

The ac current in a conductor causes the magnetic field, which surrounds the current to alternate at the same rate. Magnetic fields resist being changed. They do this by inducing a voltage that opposes the change. The induced voltage is equal to the current change (di/dt) times a factor called *inductance*.

When applied to a conductor the result is that if a current within the cross-section of the conductor were to change at a high ac rate, it would generate a magnetic field whose corresponding changes would produce an opposite current change. There would also be an accompanying electrical field within the conductor. We know, however, there can be no significant electric field within a good conductor. Consequently, there can be no rapidly changing currents within a good conductor, such as copper or the other relatively high conductivity metals. Accordingly, the higher the frequency the less the current can penetrate the conductor. For good conductors, high-frequency currents exist only near the surface (within a thin "skin") of the conductor.

The resulting current density is maximum at the conductor's surface and diminishes in amplitude exponentially with distance from the surface. The depth at which the current density has decreased to $1/e$ of its surface value (where $e = 2.718$, and thus $1/e = 0.37$) is defined as the skin depth δ [1]. If one integrates the current density from the surface to the skin depth, it turns out that about 63% of the total current flow occurs in one skin depth. The skin depth in *microns* (μm) is given by

$$\delta_{\text{microns}} = \sqrt{\left(\frac{1{,}000}{\pi f_{\text{GHz}}\,\sigma\mu}\right)} \tag{7.1}$$

where μ is the magnetic permeability within the conductor in H/m and σ is the conductivity in S/m.

Many engineers approximate the effective cross-sectional area of a cylindrical conductor as being equal to the skin depth times the circumference. Clearly, this procedure is inaccurate by a factor of 100/63, since only 63% of the current is carried within the skin depth. A more accurate estimate is 1.59 times the circumference of Figure 7.2.

In any event, the skin depth is very shallow at RF frequencies and certainly at microwave frequencies (above 1 GHz). Evaluating (7.1) for copper at 1 GHz gives

$$\delta = \sqrt{\frac{1{,}000}{\pi(1)5.8\left(10^7\right)4\pi 10^{-7}}} = 2.09\ \mu\text{m}$$

Since there are 25.4 μm to the mil, *the skin depth at 1 GHz is less than 0.1 mil.* If one remembers this result at 1 GHz, the skin depth at other RF and microwave frequencies can be estimated readily by also remembering that it varies inversely as the square root of frequency. Thus, at 250 MHz it is about 4 μm, at 4 GHz it is 1 μm, and so forth.

It should also be noted that the skin depth for RF and microwave frequencies is so thin that often it is less than the typical plating thickness for most metals. Thus, if a metal is plated with gold to a depth of 0.1 mil or more (a frequently used plating to resist corrosion and provide a good contact surface), the high-frequency resistance will be determined by the conductivity of gold and not the base metal upon which it was plated. Surface smoothness is also important here because if the top of the conductor is rough, the RF surface current must travel a longer path, which translates to increased resistance.

To reduce the high-frequency resistance of metallic conductors, the formula tells us to increase the area, because high frequencies cannot penetrate very deeply at all into conductors. Thus, using larger diameter wires or wider flat strips as conductors would reduce the resistance per unit length, not because they have a greater cross-sectional area but rather because they have a *larger cross-sectional periphery*.

An interesting question that comes up frequently is: Does RF current flow on both the inner and outer sides of a hollow conductor? The answer is that as long as the conductor is much thicker than skin depth, RF current flows only on the surface, even in a hollow pipe.

We should further note that, since skin depth decreases with the square root of increasing frequency, the resistance of a conductor increases as the square root of increasing frequency, giving us a *frequency-dependent resistance*. This conductive resistance causes dissipative loss that may be significant in a long conductor. In a short conductive path, such as a plated via hole, the reactance of the element is several order of magnitudes greater than the resistance, and we can neglect the resistive portion.

7.2.2 RF considerations of resistor types

Commercial resistors may be grouped into three major categories:

1. Wire-wound;
2. Composition;
3. Thin resistive film.

Of the three forms, wire-wound types are too inductive and composition types may be too capacitive for RF applications. Frequency-independent resistors are created by using extremely thin resistive film layers. The thickness of the film must be less than the skin depth of the highest frequency of interest. Leadless, surface-mount (chip) realization minimizes self-inductance and it is the preferred form for RF applications. "Thick-film" chip resistors may be made with carbon compound, or palladium-based compounds deposited on an insulating low-dielectric base. "Thin-film" resistors generally use nichrome (NiCr) or tantalum-nitride ($TaNi_3$) layers using a sputtering process. We examine the resistor models in Section 7.3.

7.2.3 Inductance of a straight wire (far from ground and shielding)

In addition to finite resistance, the wire or ribbon leads used to connect physical *R-L-C* components within a circuit have inductances that form "parasitic reactances" in resistors and capacitors. Of course, inductors also have such parasitics, but its value usually can simply be added to the base value of inductance.

To determine the parasitic lead inductance, consider that the induced voltage in the conductor has a polarity that opposes the change of current. This effect is described as the self-inductance (L) of the wire. The inductance of any closed (two terminal) conducting path is, of course, defined by the relation

$$v(t) = L\frac{di}{dt}$$

The self-inductance of a straight wire with circular cross-section, far removed from the return path ground plane, is given approximately by

$$L(nH) = K\,\ell\left(\ln\left(\frac{4\ell}{D}\right) - 0.75\right) \tag{7.2}$$

where

ℓ is the length of the wire.

D is the diameter of the wire.

$K = 2$ for dimensions in centimeters.

$K = 5.08$ for dimensions in inches.

For a flat ribbon,

$$L(\text{nH}) = K\ell\left(\ln\left(\frac{2\ell}{W+T}\right)\right) + \frac{0.223(W+T)}{\ell} + 0.5 \tag{7.3}$$

where

W is the width of the ribbon

T is the thickness of the ribbon

ℓ is the length of the ribbon

$K = 2$ for dimensions in cm, $K = 5.08$ for dimensions in inches

Table 7.1 tabulates wire diameters of various *American Wire Gauge* (AWG) diameters [2]. The AWG standard # 50 wire is referenced to 1 mil (0.025 mm) diameter. Every drop of six in the standard numbering doubles the diameter. For example, #44 wire has diameter of 2 mil, #38 has 4 mil, and so forth.

While the inductive effects of short wires are negligible at low frequencies, they are significant at RF and very critical above 1 GHz. For example, the self-inductance of 1 cm (about 0.4 inch) of No. 26 copper wire (or any other wire material of this gauge), whose diameter $D = 0.0406$ cm (Table 7.1), is computed from (7.2) as

$$L(\text{nH}) = 2(1)\left(\ln\left(\frac{4(1)}{0.0406}\right) - 0.75\right) = 7.68 \text{ nH}$$

The inductive reactance ($\omega L = 2\pi f L$) of this 1-cm length of wire is

$$X_L = 6.28(60)(7.68)\left(10^{-9}\right) = 2.9\ \mu\Omega \text{ at } 60 \text{ Hz}$$

TABLE 7.1 DIAMETERS OF SELECTED AWG STANDARD WIRES

GAUGE AWG	DIAMETER (MILS) (1/1,000 INCH)	DIAMETER (MM)
50	1	0.025
44	2	0.051
38	4	0.102
32	8	0.203
26	16	0.406
20	32	0.813
14	64	1.626

Note: Descending gauge numbers refer to larger diameters.

$$X_L = 6.28(60)(10^6)(7.68)(10^{-9}) = 2.9\Omega \text{ at } 60 \text{ MHz}$$

$$X_L = 6.28(10^9)(7.68)(10^{-9}) = 48\Omega \text{ at } 1 \text{ GHz}$$

Note that many discrete components actually have longer combined lead length than 1 cm. But even with just 1-cm length, the inductive reactance of $j48\Omega$ is quite large at 1 GHz compared to typical desired resistive values of 50Ω or so. For this reason, components with conventional metal leads, the kind you would find in the stockroom, are not practical in circuits operating above a few hundred megahertz.

Although the inductance of a wire depends on its length and diameter, the above computed number may be used for a quick approximation as 8 nH/cm, or 20 nH/inch.

The inductance of the single conductor (far removed from ground) poses an intellectual dilemma. Inductance is only relevant when a current flows, and current flow requires a return path. Accordingly, some people may object philosophically to the concept of a straight piece of wire having a definable inductance when the return path for the current has not been specified, except to say that it is far removed from the wire. For a much deeper examination of the subject we recommend Grover's classic book on inductance [3].

7.2.4 Parallel-plate and edge-coupled capacitance

The simplest form of a capacitor is the one formed by two parallel plates separated by a dielectric material (or air or a vacuum). Additional capacitance can be formed by stacking alternate parallel conductor layers, one set connected to one terminal, the other set connected to the other terminal of the capacitor (Figure 7.3). When this is done, the total capacitance, C, of the stack is given by

$$C_{\text{pF}} = \frac{(N-1)KA\varepsilon_R}{d} FF \tag{7.4}$$

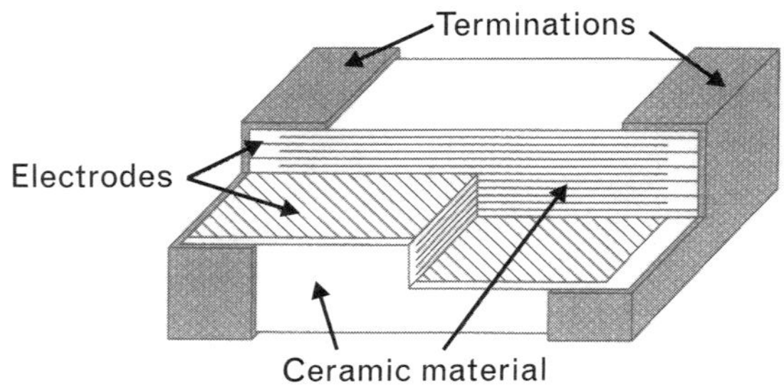

FIGURE 7.3 *Multifinger parallel plate capacitor with cutaway section to show construction. (From: [4]. ©2000 Prentice Hall, Inc. Reprinted with permission.)*

where

A is the overlapping area of one plate (in cm^2 or in^2)
N is the number of parallel plates
ε_R is the relative dielectric constant (= 1 for air or vacuum)
d is the separation of plates (in cm or inches, consistent with A)
K = 0.0885 for centimeters and 0.225 for inches
FF is the fringing factor

The relative dielectric constant, ε_R, may be frequency-dependent and may even be a complex number. Typical basic values range from 2.1 or so for Teflon, about 4.5 for FR-4 PC board, nearly 10 for alumina ceramic, 81 for water, and up to several thousand for high-value capacitors and certain ceramics used to make resonators. When one plate is much smaller than the other, as for a patch area above a large ground plane, the fringing factor exceeds 1.0, extending to perhaps 1.1 or 1.2. Fringing factor also increases as the dielectric thickness and dielectric constant are reduced.

Equation (7.4) is only valid for a lumped capacitor—that is, one whose plates and separation dimensions are less than about 5% of the operating wavelength. When this condition is not satisfied, it is necessary to treat the parallel plates as a transmission line, as will be covered later.

Let us calculate the parallel capacitance between a 50 mil by 50 mil (1.27 × 1.27 mm) component mounting pad and the ground plane for a PTFE-based multilayer printed circuit board. Use ε_R = 3.29 and a dielectric thickness of 5 mil (0.127 mm). Assume a fringing factor of 1.15.

Using (7.4), we find the capacitance as

$$C_{\text{pF}} = \frac{(N-1)KA\varepsilon_R}{d}FF = \frac{(1)0.225(0.05)(0.05)3.29}{0.005}1.15 = 0.43 \text{ pF}$$

At 1 GHz a capacitance of 0.43 pF represents $-j370\Omega$ capacitive reactance. Hence, it has noticeable effect even at a 50-Ω impedance level. If the capacitance is part of a tuned circuit, its resonant frequency can move significantly from its design value. Of course, the effect is more noticeable at higher frequencies.

EM simulation of the parallel pad capacitance gives very close correspondence to the computed value. Since it is difficult to estimate the fringing factor, use an EM simulator for the accurate solution of these types of problems, if possible.

Edge coupling becomes important in a tightly spaced layout, particularly when the impedance level between the conductors involved is relatively high. Computing the capacitance caused by the edge coupling of two conductive plates is a surprisingly difficult task, and again, EM simulation is the best tool to obtain reliable solutions.

7.3 Frequency response of physical resistors

When the physical size of a component is less than about 5% of the operating wavelength[1], it can be treated as a "lumped" element, and so it is with the resistor. However, just because we call it lumped does not mean that its equivalent circuit is simply a resistor. As we have just seen, even a short wire only 50 mil in length has 1-nH self-inductance. Regardless of its size and form, a resistor also has series inductance, L_S and has shunt capacitance, C_P. These L and C values are *parasitic*. They are not intended but are present merely because the physical resistor must have finite dimensions. The real resistor does not exist at a point of zero length, as our ideal model does.

Figure 7.4 shows two commonly used resistor types. Discrete (leaded) resistors are useful up to a few hundred megahertz only. For higher RF frequencies, surface-mount (SMD or chip) resistors are more practical due to the smaller parasitic inductance and capacitance.

Surface-mount components come in various sizes and form factors; they can handle power dissipation from a few hundred miliwatts to serveral watts. Some of the commonly used small standard sizes are tabulated in Table 7.2.

A first-level broadband model of the resistor is shown in Figure 7.5(a). This equivalent circuit with appropriate element values may be used to model discrete resistors up to a few hundred megahertz, and chip resistors to a few gigahertz. Converting the series R-L portion of the circuit to its parallel equivalent helps us to visualize the resonance between the parasitic reactances.

We see that a resistor has all three elements of a resonator in parallel R-L-C form. The parallel equivalent R and L values are frequency

FIGURE 7.4
Typical (a) discrete resistor (From: [6].© 1991 Artech House, Inc. Reprinted with permission) and (b) chip resistor (From: [4]. © 2000 Prentice Hall, Inc. Reprinted with permission.).

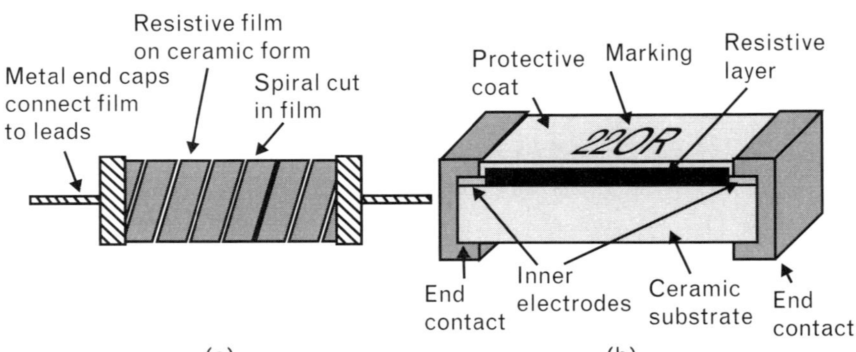

1. Digital designers generally use rise time and delay considerations and arrive to similar conclusions (see Chapter 9).

Table 7.2 Physical Dimensions of Various Surface-Mount Type Components

Chip Size Code	Width/Length (mil)	Width/Length (mm)
0402	40/20	1.02/0.51
0504	50/40	1.27/1.02
0603	60/30	1.52/0.76
0805	80/50	2.03/1.27
1206	120/60	3.04/1.52
1218	120/180	3.04/4.57

Note: The sizes are also applicable to other kinds of chip components, such as capacitors.

dependent, and there exists a resonance at which the converted parasitic inductance and the shunt parasitic capacitance resonate with each other. Converting the series R-L to parallel form yields a much higher equivalent resistance [Figure 7.5(b)].

For low-value resistors the series inductance is the prime parasitic, meaning that the total impedance of the resistor increases with frequency. At some point the parallel capacitance creates a low-Q parallel resonance with the inductance of the series branch; from that frequency the terminal impedance of the part declines. Below the resonant frequency the physical resistor behaves inductively while above resonance it looks like a lossy

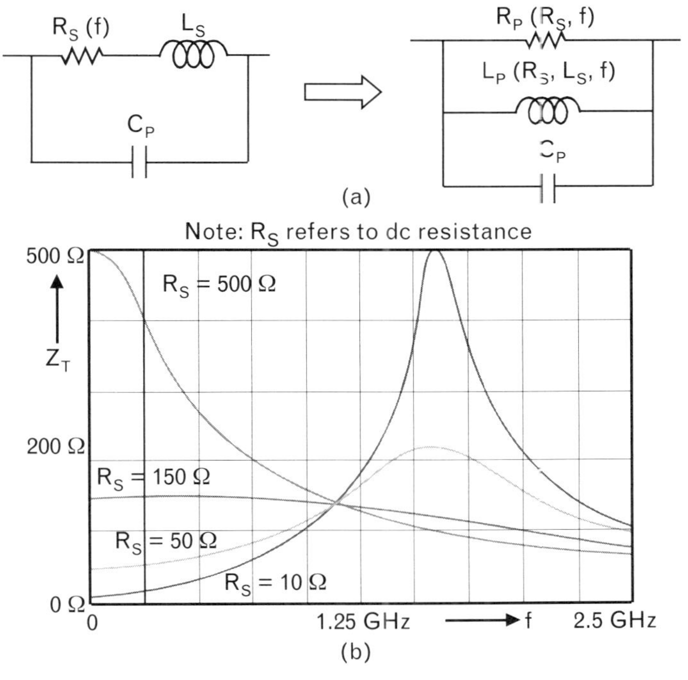

Figure 7.5 *(a) Broadband first-level model of a discrete resistor and its converted frequency-dependent parallel equivalence. (b) Terminal impedance magnitude versus frequency for discrete resistors for various nominal values between 10Ω and 500Ω.*

capacitor. For example, Figure 7.5 shows that a discrete resistor having 10Ω at dc has an impedance peak of 500Ω at its resonance.

If the nominal resistance is relatively high (several hundred ohms), the series inductance may be neglected and the parallel capacitor represents the prime parasitic. In such a case the impedance declines as the frequency increases; approximating the behavior of a parallel *R-C*. For example, Figure 7.5 shows that a discrete resistor of 500Ω at dc drops below 100Ω at 2.5 GHz. While the parallel capacitance may be relatively small (i.e., a fraction of a picofarad) its effect may need to be considered. For example, if the resistor is used as a collector-to-base dc bias resistor, even a small parallel capacitance forms significant feedback at RF.

One usually thinks of a parasitic value as something small compared to its host. However, in the case of *R-L-C* components, it is seen that the parasitic values ultimately dominate the impedance behavior of the host element. It is for this reason that is impractical to employ those lumped *R*, *L*, or *C* elements which one usually finds in an electronics stockroom in circuits operating above 500 MHz.

Surface-mount, or chip components behave similarly to their discrete equivalents, except their frequency dependencies show up at about a decade higher due to their smaller parasitics. Primary parasitics are still inductive for low-value resistors up to about 75Ω and are capacitive for resistors above 120Ω. Depending on the size, chip resistors in the range of 82–100Ω are almost purely resistive, up to several gigahertz. Hint: If you need a good 50-Ω broadband RF termination, try two parallel connected 100Ω chip resistors.

7.3.1 Fitting a model to measured resistor data

When an exact model and its element values are not available, computer optimization is very helpful in fitting a model to measured data by the following procedure:

- Select an appropriate equivalent circuit.
- Estimate the initial values for the components of the model.
- Compare the response of the model to the measured data.
- If the initial model response looks reasonable, optimize its components to duplicate the measured data.
- If the initial response is far from the target, modifying the model may help.
- Keep in mind that optimization is an iterative process and there is no guarantee of convergence. Use all available design tools to obtain ball-park initial values

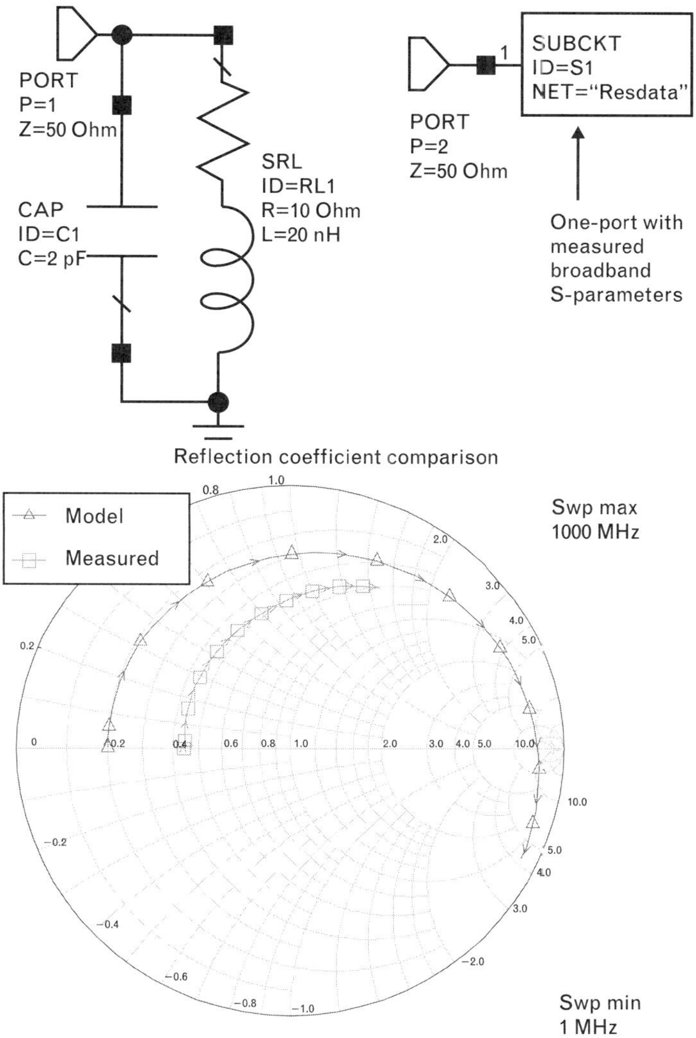

FIGURE 7.6 *Measured data (Δ-markers) remains inductive throughout the 1- to 1,000-MHz frequency range, but the model (□-markers) already shows resonance near the high end of the band, indicating too much parasitic inductance and/or capacitance. During optimization the three components of the resistor model are changed until the reflection coefficients of both one-ports are alike.*

Up to 1 GHz, relatively simple lumped equivalent circuits may be used to accurately describe the behavior of *R-L-C* components of chip and discrete types. Above that frequency, depending on the size of the components, modeling may require more detailed circuits and distributed (transmission line) elements. However, many surface-mount type components can be adequately described by the lumped models, even up to 2 GHz.

Figure 7.6 illustrates the one-port model and the measured data for a resistor between 1 MHz and 1 GHz. The model of a discrete (leaded) resistor is simulated at Port 1, and the measured one-port data is stored at Port 2. The results of the initial simulation of the one-port and that of the measured data are displayed on the Smith chart, showing significant difference between the two datasets.

The circuit optimization goal in Figure 7.6 is to minimize the error function, defined as

$$EF = \sum_{f=1\,\text{MHz}}^{f=1{,}000\,\text{MHz}} \left| 1 - \frac{s_{11\text{MODEL}}}{s_{11\text{MEASURED}}} \right|^2$$

For a simple circuit like this one, the optimization minimizes the error function in a few iterations, even without having good initial component estimates. In fact, EF was reduced from 0.51 to 10^{-5}, placing $s_{11\text{MODEL}}$ on the Smith chart right on top of $s_{11\text{MEASURED}}$. Table 7.3 shows the three component values before and after optimization.

7.4 Modeling physical inductors

Often at RF frequencies inductors are constructed as wound coils with air cores (Figure 7.7). The inductance of these *solenoidal coils* is given by

$$L(\text{nH}) = \frac{3.94 r_{\text{mm}} N^2}{0.9\dfrac{\ell}{r} + 1} = \frac{0.1 r_{\text{mil}} N^2}{0.9\dfrac{\ell}{r} + 1} \tag{7.5}$$

where r and ℓ are the coil's radius and length.

Equation (7.5) is accurate when the radius of the coil is at least 10 times the radius of the wire. For most RF coils that ratio is much smaller and (7.5) computes 10% to 20% more than the actual inductance. In small-diameter coils the precise definition of the radius r, and length ℓ, also become very important.

Equation (7.5) can be solved for the number of turns, N, needed to achieve a given value of inductance, L:

TABLE 7.3 TABULATED COMPONENT VALUES OF THE RESISTOR MODEL BEFORE AND AFTER OPTIMIZATION

	R-Ω	L nH	C pF
Initial values	10	20	2
Optimized values	21.97	10.31	0.377

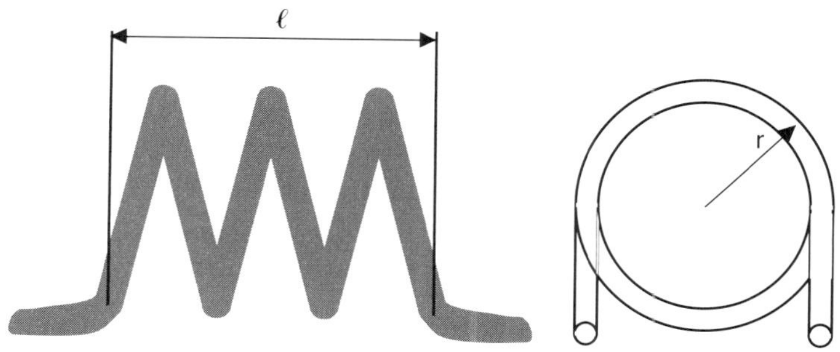

FIGURE 7.7
An RF air coil example. For optimum Q the length and diameters of the coil forms should be approximately the same.

$$N = \sqrt{\frac{L_{nH}\left(9\frac{\ell}{r}+1\right)}{r_{mil}}} = \sqrt{\frac{L_{nH}\left(0.228\frac{\ell}{r}+0.25\right)}{r_{mm}}} \tag{7.6}$$

This may be further simplified by taking into account that the highest Q is achieved when the length and diameter of the coil are *approximately* equal [3], $\ell \approx 2r$. Then,

$$N = \sqrt{\frac{28L_{nH}}{r_{mil}}} = \sqrt{\frac{0.706L_{nH}}{r_{mm}}} \tag{7.7}$$

After computing the number of necessary turns, we need to decide what wire size to select. Using the largest wire size and tight winding to fill the length, results in the lowest series resistance and best mechanical strength. Low R_S leads to higher Q. However, having the windings touch each other leads to maximum interwinding self-capacitance, which lowers

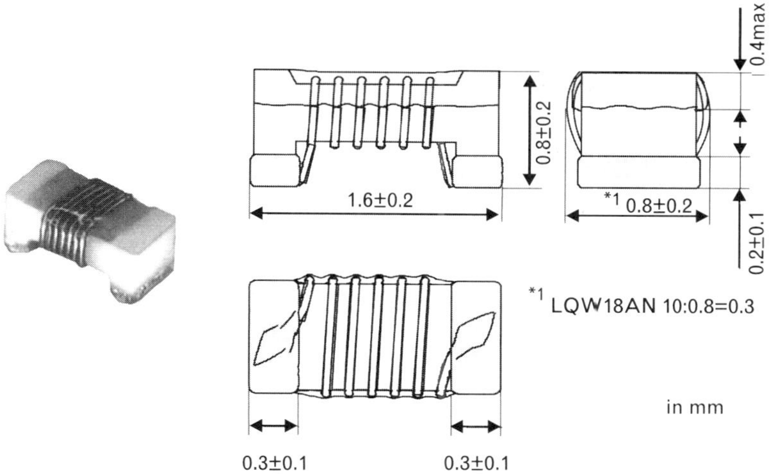

FIGURE 7.8
To improve mechanical rigidity and to offer convenient surface mounting on PC boards, RF inductors are often wound on low-loss cores. Having a core allows separation among the windings to reduce the self-capacitance of the inductor. (Courtesy of Murata Electric Co.)

f_R. Thus, there is a trade-off between optimum Q and high-resonant frequency.

Small RF air-core inductors are very sensitive to handling—a small pressure can easily change their dimensions and thereby the inductance as well. Inductors can be made more rigid by adding a low-loss ceramic core and also by placing a protective layer around the winding (Figure 7.8). Of course, now we have to include the parasitics and losses of the added core material and coating.

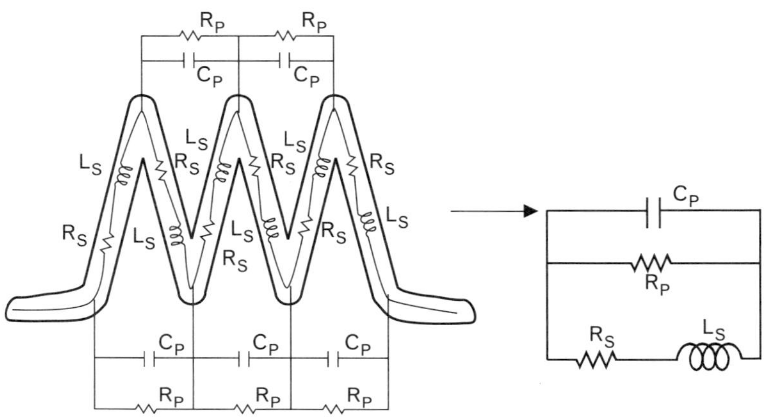

FIGURE 7.9 *The distributed parameters of a single-layer inductor modeled using a first-order lumped circuit. R_S is the conductive resistance and C_P is the self-capacitance. R_P represents all other applicable losses related to wire insulation, radiation/shielding, and coil core. Both types of losses are frequency dependent.*

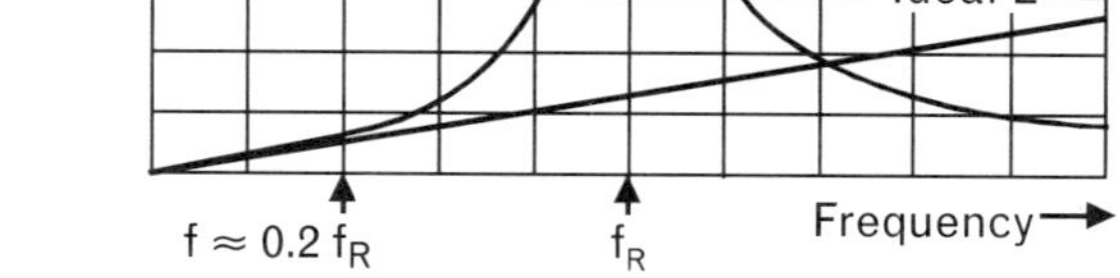

FIGURE 7.10 *Impedance behavior of ideal and physical single-layer inductor. Above its primary parallel resonance, f_R, the component is dominated by its parallel self-capacitance. Effective inductance remains relatively constant up to about 20% of f_R.*

7.4.1 Inductor self-capacitance and loss resistances

For the physical inductor shown in Figure 7.7, when the length of the wire used to wind a coil is less than 5% to 10% of the wavelength in air at the operating frequency, the distributed R, L, and C elements can be lumped into single components to model the physical inductor (see Figure 7.9).

Notice that the equivalent circuit of the inductor is similar to that for the resistor shown earlier, with the exception of the additional parallel resistance, R_p. The single-layer inductor also experiences parallel resonance (Figure 7.10), just as was incurred with the resistor. A major difference between the two resonances is that the inductor's Q is generally much higher than the resistor's Q; therefore, the inductor's resonance trace is much sharper.

A single-layer inductor does behave as a linear reactance up to about 20% to 25% of its resonance frequency. Thus, it may be modeled as a single L element, since it has nearly linear reactance behavior up to this frequency (and could even be modeled as a simple R-L without parasitic shunt C up to this frequency). At about 20% of f_R, the true inductance exceeds the nominal inductance by about 4%, but at 50% of f_R the increase in L is about 35%. When the frequency reaches 70% of f_R, the effective inductance is about twice the nominal value. At resonance the inductor looks like a high-value resistor. Above resonance the parallel self-capacitance becomes the dominant element.

7.4.2 Planar printed inductors

A square- or circular-shaped inductor printed on a dielectric board may be viewed as a transition between lumped and distributed forms, particularly if the board is thin and there is a ground-plane on its back side. A square shape is preferred because it gives more inductance for a given surface area and its artwork is also easer to generate. A lumped model [5] of a square-shape spiral printed inductor is shown in Figure 7.11.

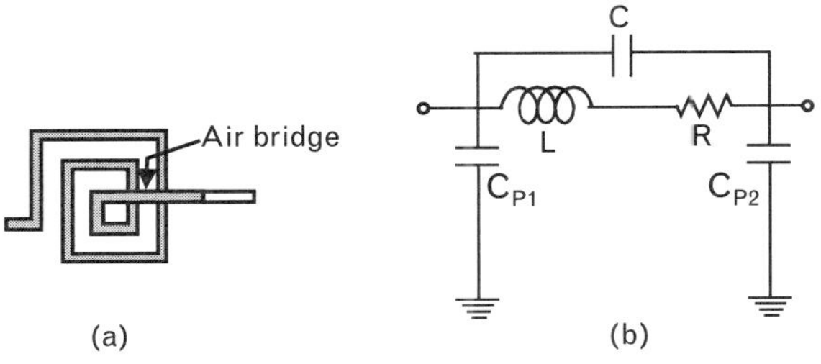

FIGURE 7.11 *(a) A spiral printed inductor and (b) its first-order lumped equivalent circuit. The parallel resistors are less significant in high-quality PC-board applications. All three resistors become important in RFIC realization, due to high conductive and dielectric losses. (From: [5]. © 1998 Artech House Inc. Reprinted with permission.)*

Exact closed form inductance computation is quite complex [6] although a simple *approximation* for square-shape inductor is given [7] as

$$L_{\text{nH}} \approx KN^{1.67}\sqrt{A}$$

where

A is the area of the square-shaped inductor in mm^2 or mil^2

N is the number of turns

K is 0.85 for dimension in mm and 0.0216 for dimension in mil

Reference [6] also has approximate formulas for other planar inductor configurations.

Commercially available RF circuit simulators provide printed-inductor models specified by physical data. Another alternative is to use an EM simulator first and pass the results to the circuit simulator.

7.4.3 Effective inductance calculations

Many designers intuitively feel that the parallel capacitance of a coil decreases the nominal inductance of the component. This belief is wrong. The parasitic capacitive reactance *reduces the total inductance susceptance* of the coil, thus making the *effective inductance larger than the nominal value*. This is obvious when we look at Figure 7.10 where the impedance (proportional to inductance) of a physical inductor is always greater than that of an ideal component.

The effective inductance of a physical inductor with parasitic capacitance increases to an infinite value at f_R, above which the parallel capacitance dominates and the actual element behaves like a capacitor. The combined effect of the losses due to R_S and R_P along with the stray (parasitic) capacitance C_P is to change the basic inductance L_0 measured at very low frequency to a frequency-dependent effective inductance, L_{EF}, computed at any given frequency f by

$$L_{EFF} = \frac{L_0}{\left(1 - \left(\dfrac{f}{f_R}\right)^2\right)} \tag{7.8}$$

where f_R is the self-resonance frequency. The frequency f_R for inductors having a Q of 10 or more can be approximated as

$$f_{R\text{GHz}} \approx 5.033\sqrt{\frac{1}{L_{0\text{nH}} C_{P\text{pF}}}} \tag{7.9}$$

For example, an inductor with a nominal value of 22 nH and 0.5 pF of stray parallel capacitance acts like a 34-nH inductor at 900 MHz—about a 55% increase above the nominal inductance. (We will also see later that the parasitic effect also increases the nominal value of a capacitor. A 10-pF chip capacitor with 1 nH total self-inductance has effective capacitance of 14.7 pF at 900 MHz, a value 47% higher than the nominal value.)

The only problem with (7.8) relating the effective inductance L_{EFF} to the zero frequency inductance L_0 is that it is not practical to measure an inductance value at zero frequency. Hence, we need a relationship between the inductance measured at some convenient frequency, f_M, and denoted by L_{EFM}, and the value that would apply at any other analysis frequency (including zero frequency), f. The relation is

$$L_{EFF} = \frac{L_{EFM}\left(f_R^2 - f_M^2\right)}{\left(f_R^2 - f^2\right)} \tag{7.10}$$

7.4.4 Q-factor calculation

An important parameter of inductors and capacitors is their quality factor, or Q. We have defined Q earlier, but let us recall this important parameter here for the purpose of describing physical components. The Q is the ratio of the stored to dissipated energy per RF cycle at a given frequency. Ideally, reactive components have no dissipation. However, their metal leads have resistance, exacerbated by skin effect, and there are additional losses associated with their operation due to radiation, insulation, shielding, magnetic core losses in inductors, and dielectric losses in capacitors. Furthermore, the effect of parasitic reactances reduces the component Q. The quality factors of the two equivalent circuits of Figure 7.12 can be expressed in series or parallel form as

$$Q_S = \frac{X_S}{R_S} = Q_P = \frac{R_P}{X_P} \tag{7.11}$$

where

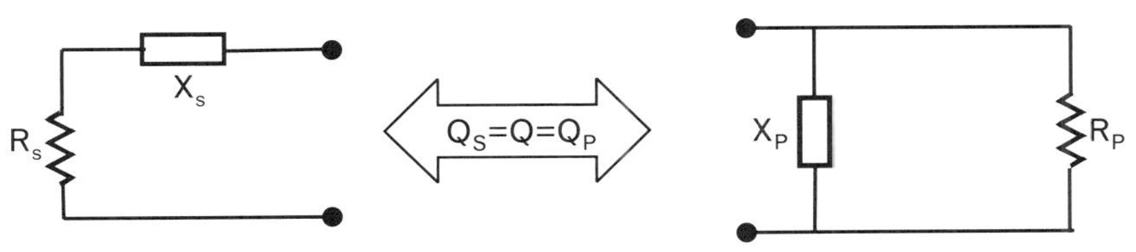

FIGURE 7.12 *Lossy reactive elements can be represented using either the two-element series or parallel form. The Q value remains the same for either representation.*

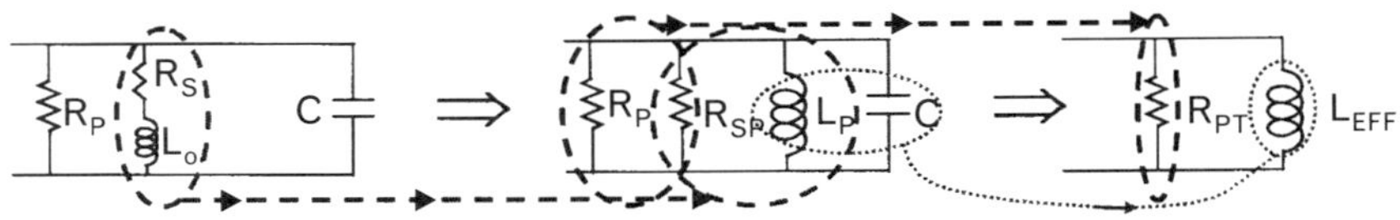

FIGURE 7.13 *Equivalent circuit of an inductor for Q calculations.*

X_S is the reactance of the series equivalent circuit.

R_S is the resistance of the series equivalent circuit.

X_P is the reactance of the parallel equivalent circuit.

R_P is the resistance of the parallel equivalent circuit.

The Q formulas in (7.11) represent a series-parallel conversion at a specified frequency, by which the Q is unchanged. Earlier we exploited the unchanging Q in the series and parallel equivalent circuits to derive the Q matching technique.

Manufacturers generally provide component Q at a relatively low frequency that does not indicate what the effective Q may be in a specific operating frequency range. We show the frequency dependency of Q for inductors next, and in Section 7.6 we show that for capacitors.

A lumped inductor model with series and parallel losses can eventually be reduced to its two-element form (Figure 7.13) for Q calculations. The effective frequency-dependent Q, Q_{EFF}, is computed by taking the ratio of the total parallel equivalent loss resistance, R_{PT}, to the effective reactance of the inductor, L_{EFF}, as shown in (7.12).

$$Q_{EFF} = \frac{R_{PT}}{X_{EFF}} = \frac{R_{PT}}{6.28 f_{\text{GHz}} L_{EFF\text{nH}}} \tag{7.12}$$

In summary, the effective inductance of a single-layer coil changes from its dc value L_0 to an effective value L_{EFF} at a given frequency f. This

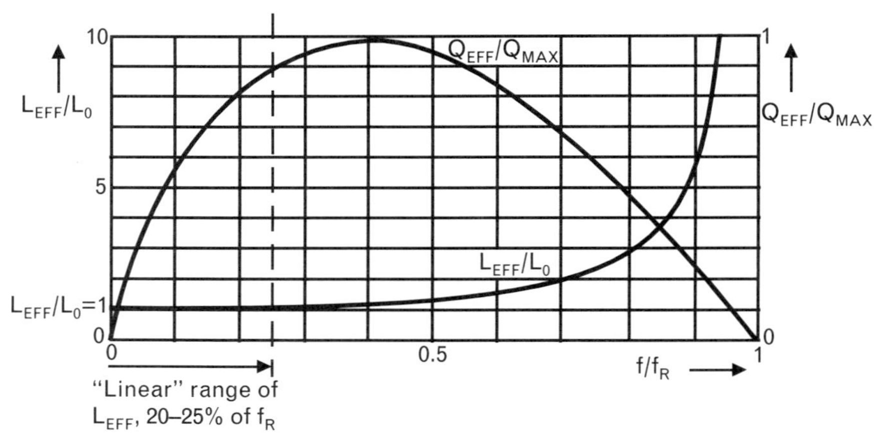

FIGURE 7.14 *Normalized effective inductance and Q variations of a typical single-layer air-core inductor. Frequency is normalized to the self-resonance frequency, f_R.*

effective inductance increases to infinity at the self-resonance frequency of the coil, f_R. Even at a frequency of only 0.1 f_R, L_{EFF} is 1% above L_0. The effective Q value of the coil is computed from the R_{PT}/X_{EFF} ratio at the frequency for which the Q is determined (Figure 7.14). More detailed modeling is available from [8, 9].

The values Q_{EFF} and L_{EFF} are both frequency dependent, with Q_{EFF} reaching its maximum value around 30% to 50% of f_R. At resonance, the effective inductance is infinite and the impedance of the component is equal to the equivalent total parallel loss resistance R_{PT}. Since parallel Q is equal to the ratio of parallel resistance to parallel reactance, the effective Q drops back to zero at f_R.

Other parameters affecting the Q are the length and diameter of the coil, wire insulation and core losses, as well as the presence of shielding created by the proximity of ground plane(s) and other conductors. Optimum Q, as we mentioned before, is reached when the length of the coil is approximately equal to its diameter.

Figure 7.14 tells us that if we want a constant frequency-independent inductance, the self-resonance of the inductor must be four to five times higher than the highest frequency of operation. Unfortunately, parts like that are not always available. Suppose we need 20-nH constant inductance at 1.9 GHz and the best available part's self-resonance is at 2.5 GHz; that places our operation at 76% of f_R. At that frequency the effective inductance is about 2.5 times of the nominal value.

One possible solution is to use a smaller nominal inductance that has an *effective* inductance of 20 nH at 1.9 GHz. This of course works only for a very narrow frequency range. However, talking with the manufacturer, we learn that if we accept five 4-nH inductors instead of one 20-nH inductor, they could make them with self-resonant frequency of 9 GHz. Then, the 1.9-GHz operating frequency is only 21% of f_R, that assures nearly constant inductance. Another alternative is to use eight 2.5-nH inductors that have even higher self-resonant frequencies, and so on. If space and cost were not important, we could consider this approach.

Clearly, using many smaller inductors in series in place of a single component *distributes* the parasitic capacitances—which takes us closer to transmission line (distributed type) component realization. When we reach the upper frequency limit where practical lumped elements can be used, we need to switch to transmission lines. We discuss distributed components in Section 7.8.

7.4.5 Multilayer inductors

For multilayer construction additional secondary resonances exist (series and parallel types), and the inductor may exhibit virtually any impedance value above the primary resonant frequency, f_R. Figure 7.15 shows the

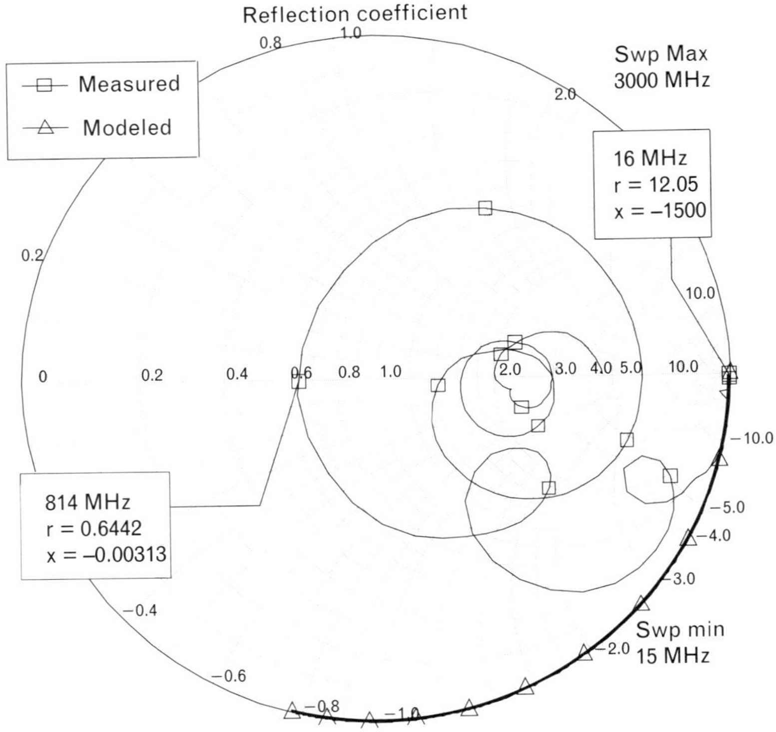

FIGURE 7.15 *What two-terminal passive component has measured impedance like the trace with square markers? Since at 15 MHz it looks nearly like an open circuit, you may think it is a low-value capacitor. Wrong—it is a multilayer inductor, shown at frequencies above its primary self-resonance of 15 MHz. Triangular markers indicate what the response would look like if the inductor had only a single 1.37-pF parallel self-capacitance.*

one-port reflection coefficient of an 82-μH discrete inductor above its primary self-resonance, that takes place at 15 MHz. The frequency response indicates six secondary resonances and a wide range of complex impedances between 15 MHz and 3,000 MHz. At 814 MHz the inductor even behaves like a 32Ω ($r = 0.64$) ideal resistor. At some other frequencies it looks like a lossy capacitor or a lossy inductor. It would be quite a challenge to model this component at RF. Later, in Section 7.6.3, we will see similar effects in multilayer capacitors.

What causes the higher-order resonances in multilayer inductors? Without going into detailed modeling that also requires EM simulation, let us place the equivalent circuit of a single-layer inductor twice in a series connection (Figure 7.16). Let us assume that the Layer 1 has lower self-

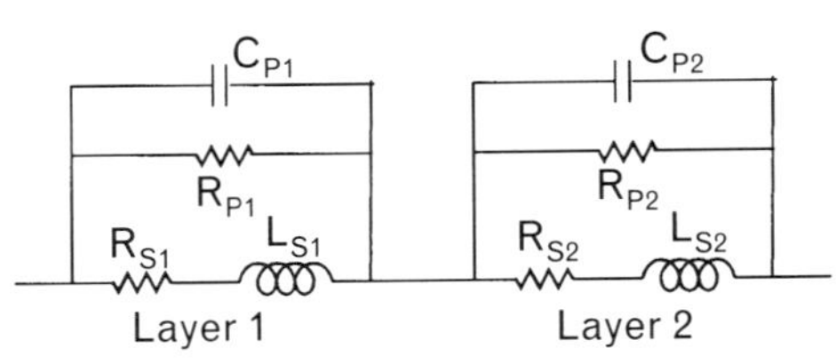

FIGURE 7.16 *Simplified equivalent circuit of a two-layer inductor, not including mutual inductance and capacitive coupling between layers.*

resonant frequency. We label the two self-resonances of the two layers as f_{R1} and f_{R2}, respectively, keeping in mind that $f_{R1} < f_{R2}$. As the frequency increases from dc, the following takes place:

- Up to f_{R1} the series combination behaves inductively.
- At f_{R1}, Layer 1 resonates and acts as high impedance. Therefore, the overall circuit impedance is also very high.
- Above f_{R1}, Layer 1 already behaves capacitively but Layer 2 is still inductive. We now have a series connection of a capacitor-like element and an inductor. The combination looks capacitive. At some frequency between f_{R1} and f_{R2} the two layers undergo series resonance. The two-layer combination has a very low impedance and gradually becomes inductive up to f_{R2}.
- At f_{R2}, the second layer goes through its parallel resonance and the overall circuit again looks like a high impedance. As we move higher in frequency, Layer 2 now acts like a capacitor.
- Since both layers behave like capacitors above f_{R2}, the two-layer inductor looks capacitive.

If both layers were lossless, the overall impedance trace rotates around the circumference of the Smith chart in a clockwise direction, passing through open and short-circuit points repeatedly. In a physical inductor, due to dissipative losses, smaller resonant loops are formed inside the chart. Additional layers, mutual inductance and capacitive coupling among layers, of course, make the impedance behavior even more complicated, as shown in Figure 7.15.

Parallel resonance may be desirable in applications for which the inductor is required only to have large terminal impedance, such as in bias chokes. However, selecting the proper inductor requires knowledge of its resonant frequency, and even then the high impedance may only be obtained over a frequency range of 5% or so bandwidth (see Figure 7.17).

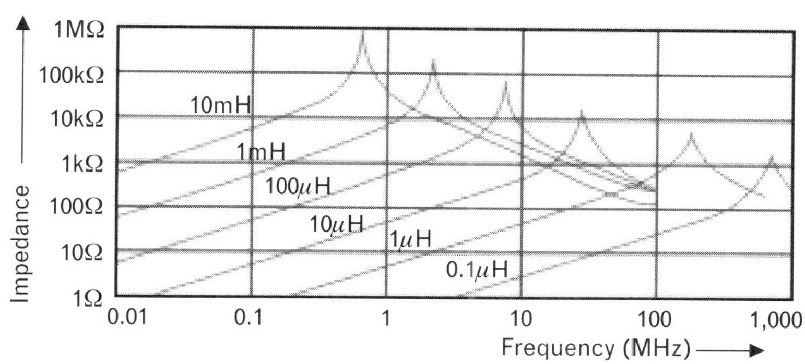

FIGURE 7.17
If you need an inductor with high impedance at 1 GHz, do not pick one in the high microhenry range because it resonates at a few megahertz and may act like a lossy capacitor at 1 GHz. (Courtesy of Murata, Inc.)

A common error is to select too large a coil inductor for a bias choke. Blindly selecting a large nominal inductance value does not assure the high impedance throughout the RF range.

7.4.6 Inductors with magnetic core

In the foregoing discussion, the coil impedance was independent of the magnitude of the voltages and currents applied to the inductor. This is usually a good approximation up to the thermal limits of the coil parts—up to the currents that would cause melting of the coil wire or the voltages, which would cause an arc over within or about the coil. Of course, the parasitic capacitances already caused changes to the effective inductance, but that is not dependent on the magnitude of the applied RF signal or dc bias current.

One way to increase inductance without increasing the size of the coil is to wind the inductor about a magnetic core material. Using a magnetic core with relative permeability μ_r $(\mu_r > 1)$ offers the following benefits:

- The effective inductance increases substantially over an air core coil of the same dimensions and turns, because the higher permeability core concentrates and increases the magnetic field intensity.
- We *may* get higher Q for a particular frequency range. The higher Q may come about as a result of the lower series (wire) resistance, since fewer turns are needed for a given L, and a larger wire diameter may be selected.
- The core can be moveable to vary the inductance.
- The core may be extended to provide shielding.
- For coupled inductors the core increases the coupling factor.

We used the word "may" when referring to an increase in Q because the presence of the core also introduces additional losses that are frequency, temperature, and current dependent. Unless the proper core material is used, the new losses can offset the improvement created by having fewer turns.

Magnetic cores also introduce other problems. Permeability varies with temperature, dc and RF signal current level applied to the inductor, and frequency. Above a specific corner frequency the permeability becomes a complex quantity[2] that requires special test equipment for characterization, such as the Agilent 4291B, 4294A, and E4991A impedance analyzers [10], that measure complex permittivity as well.

2. The complex nature of permeability is used favorably in ferrite beads (see Section 7.5).

Core losses may be combined into the frequency-dependent parallel resistor (R_p) of Figure 7.13, further complicating the effective Q calculations. Clearly, the use of a magnetic core material to increase inductance produces the onset of truly nonlinear operation and we now enter a new type of design where our modeling becomes a function of voltage and current. Since that topic is outside the scope of our textbook, we just refer the reader to specialized references on the subject [11–13]. Coils that use magnetic cores are rated for an operating temperature range and a maximum current. The current maximum is not that which necessarily would cause overheating or burnout of the inductor, but rather the maximum current for which the inductor would retain its rated inductance value.

It is also difficult to determine the effective magnetic core length of a solenoidal, rod type magnetic core inductor, since the magnetic flux lines pass partially through the air and partially through the core, as shown in Figure 7.18. The effective inductance is given by (7.13) as

$$L_{\mathrm{nH}} = \frac{4\pi N^2 \mu_i A_C}{\ell_E} \tag{7.13}$$

where

N is the number of coil turns.

A_C is the cross-section of the magnetic core in square centimeters.

ℓ_E is the effective length of the core in centimeters.

μ_i is the permeability of the core in henries per centimeters.

A more efficient coil structure is the *toroidal inductor* consisting of coil windings around a toroidal magnetic core. The toroidal geometry causes virtually all of the magnetic flux to be contained within the core. The contribution to the inductance by magnetic field lines occurring in the air

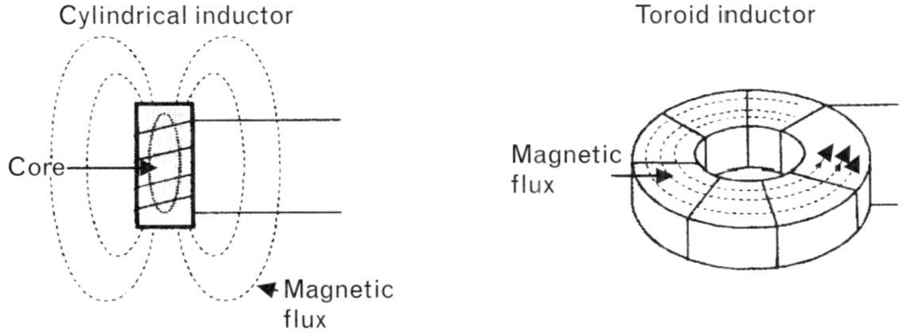

FIGURE 7.18 *A solenoidal inductor is wound around a cylindrical magnetic core of high permeability material. Since part of the magnetic flux is in the air, it is difficult to determine the exact length of the magnetic field path. A toroidal inductor is a highly efficient form since most of the magnetic flux remains inside the high-permeability core.*

surrounding the core is negligible, making calculations much simpler. The inductance is given by

$$L_{\mathrm{nH}} = N^2 A_L \tag{7.14}$$

where A_L is the inductance index of the core in (nanohenrys/turns raised to the power of 2).

Figure 7.19 shows two types of RF inductor types available to circuit designers. Tighter winding creates more inductance but also increases the interwinding capacitance, which lowers self-resonance. Careful selection of the right vendor and component is vitally important to success.

7.5 Ferrite beads

Ferrite beads provide an inexpensive and convenient way to add high-frequency isolation loss in a circuit without introducing power loss at dc and low frequencies. The beads are small and can be installed simply by slipping them over a component lead or conductor. They are most effective in providing attenuation of unwanted signals above 1 MHz. When properly used in combination with parallel capacitance, these beads can provide high-frequency decoupling and parasitic suppression.

Figure 7.20 shows a small cylindrical bead installed on a conductor and a surface-mounted bead using ribbon wire and the RF equivalent circuit. A ferrite bead is a good example of a component with an equivalent circuit model of highly frequency-dependent component values. The frequency effects, in some cases, are not easily described even with exponential terms, and may require quadratic or piecewise functions. In addition, these functions also require frequency, RF power, dc current, and temperature break points. Since modeling is quite complicated, one-port or two-port measurements are generally the most accurate component representation, at a

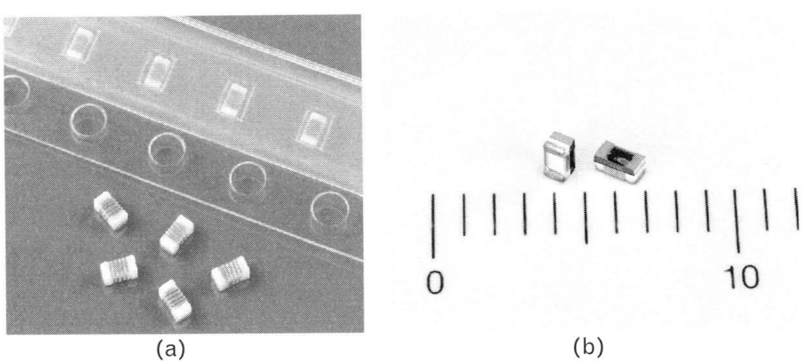

FIGURE 7.19
Small RF inductors: (a) winding with spacing between turns, and (b) tight winding for higher inductance with increased parasitic capacitance. Scale refers to millimeters. (Courtesy of Coilcraft Inc.)

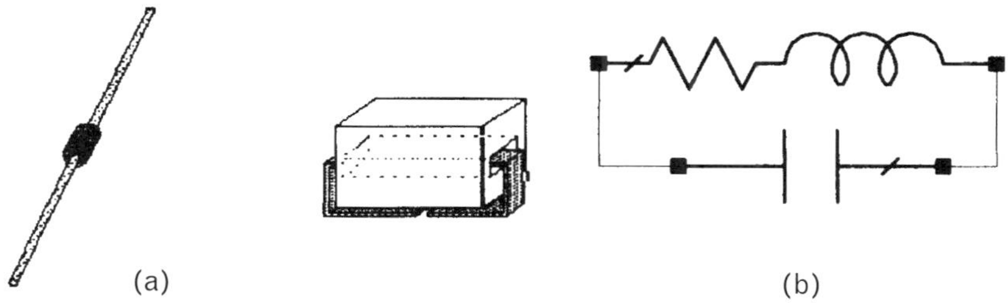

FIGURE 7.20 *(a) Ferrite sleeve on conductor wire and surface-mounted bead on flat conductive ribbon. (b) RF equivalent circuit consisting of a frequency-dependent R-L-C combination.*

specific RF power, dc current, and temperature. Figure 7.20(b) shows a lumped equivalent circuit of the component.

The value of the series resistor R_S is a function of frequency and current and the type of ferrite material used. At dc, the series resistance equals to that of the wire alone—a fraction of an ohm. The inductive portion, L_S is significant due to the high permeability of the ferrite sleeve. Due to the complex permeability nature of the magnetic sleeve, R_S increases with frequency and peaks at a specified frequency, f_P. Inductive reactance peaks

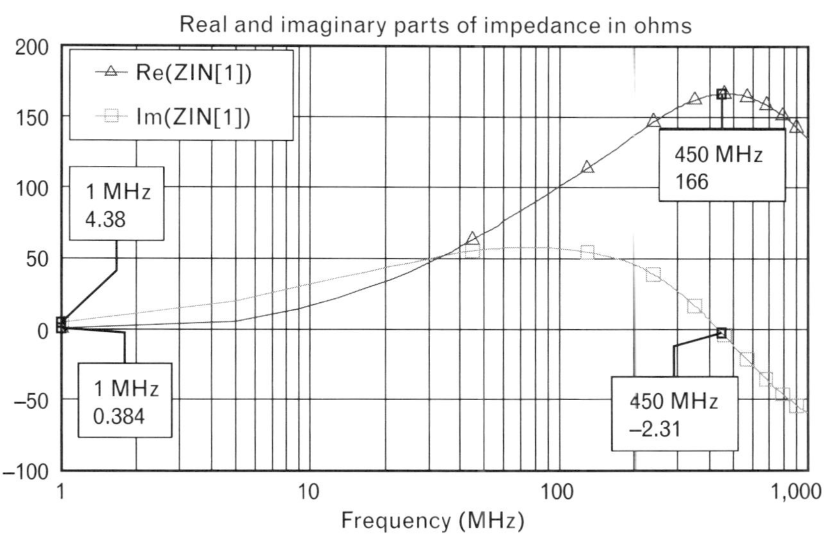

FIGURE 7.21 *Impedance behavior of a Murata BLM18AG121 ferrite bead aimed for the 450-MHz band. At 1 MHz the component has 0.38-Ω resistance and j4.38 reactance that represents L_S = 697 nH. (At dc R_S drops to an even lower value.) At 450 MHz the impedance is almost purely resistive, R_S = 166Ω.*

3. Commonly used on cables connecting printers and monitors to PCs.

much sooner than the resistance does, drops to near zero value at f_p and eventually becomes capacitive. At f_p the beads looks almost purely resistive.

Figure 7.21 displays the real and imaginary parts of a Murata ferrite bead, designed to give maximum resistance in the 450-MHz frequency range. At low frequencies the component behaves like a nearly 700-nH inductor with very low resistance. At 450 MHz it virtually looks like an ideal 166-Ω resistor. Placing this bead into the bias lead of a 450-MHz active circuit provides RF filtering without any appreciable dc voltage loss.

A benefit of the ferrite bead is very low dc resistance so it can be embedded into the dc bias circuitry without causing any significant power loss. At RF the bead acts as a resistor and it dissipates, rather than reflects, undesired signals. High-speed digital circuits and computer systems[3] also use ferrite beads to suppress unwanted signals and to meet EMI/EMC specifications [14–16].

7.6 Physical capacitor models

A lumped capacitor with series resistance R_S and dielectric loss represented by R_p can be modeled as shown in Figure 7.22. The presence of losses and the parasitic inductance change the low-frequency capacitance C_0 to a frequency-dependent effective capacitance, C_{EFF}, that may be computed at any specified frequency, using

$$C_{EFF} = \frac{C_0}{\left(1 - \left(\frac{f}{f_R}\right)^2\right)} \tag{7.15}$$

where f_R is the self-resonance frequency of the capacitor, which can also be approximated for capacitors having a Q of 10 or greater as

FIGURE 7.22 *Equivalent circuit for a lumped capacitor with conductive losses of R_S and dielectric losses of R_P. L_S is the series self-inductance and C_0 is the nominal capacitance. The reduced frequency-dependent equivalent circuit on the right consists of just the effective series resistance (ESR) and X_{EFF}, the effective series reactance. X_{EFF} is only capacitive until the self-resonant frequency f_R, and becomes inductive above that. Secondary resonances are covered later in Section 7.6.3.*

$$f_{R\text{GHz}} \approx 5.033\sqrt{\frac{1}{L_{S\text{nH}}C_{0\text{pF}}}} \tag{7.16}$$

or, if we want to compute the series self-inductance, knowing f_R and C_0,

$$L_{S\text{nH}} = \frac{25.33}{C_{0\text{pF}}f_{R\,\text{GHz}}^2} \tag{7.17}$$

For accurate modeling we need to know details of the mounting, orientation of the chip (see more on this in Section 7.6.3), and specifications of the dielectric board, since they all affect the behavior of the component.

The total effective loss of the capacitor primarily comes in two parts:

1. Dielectric loss, determined by the *dissipation factor* (DF), is caused by RF current leakage between the conductive plates. It is also called the loss tangent of the dielectric material. Dielectric loss heats up the dielectric media and in extreme cases can cause thermal breakdown that is a catastrophic failure.
2. Conductive metal losses caused by the finite resistance of the conductors and skin effect also heat the component and may contribute to failure.

Usually we do not think of a capacitor as a component dissipating RF power, since the *ESR* of a good quality capacitor is around a few tenths of an ohm in the low gigahertz frequency range. Yet, in power amplifiers, we may have tens of amperes RF current flowing through a capacitor and the I^2R losses can easily amount to several watts—enough power to damage or even destroy most small capacitors. Excessive voltage, temperature, and humidity may also affect the behavior of a capacitor.

Even the I^2R loss computation at room temperature can be misleading in high-power circuits, because the power dissipated in the capacitor heats the component, thereby increasing the *ESR*. Now the same RF current dissipates more power in the larger resistance that further increases the temperature. The cycle, called *thermal runaway*, has caused the smoky ending of many power amplifier circuits [17].

Self-inductance is unavoidable and exists even in the smallest chip capacitor. Typical inductance values range from 0.3 nH to 1.0 nH for surface-mount components and up to 10 nH to 20 nH for discrete types. The presence of series inductance *reduces* the negative reactance (not the capacitance) of the capacitor, creating resonance at frequency f_R, where the total reactance goes to zero (a short circuit). Below the resonance the reduced capacitive reactance leads to a *higher* effective capacitance value, C_{EFF}. If the effective capacitance, C_{EFM}, is measured with a capacitance meter at some frequency f_M, then the effective capacitance at any other frequency can be related to the measurement.

$$C_{EFF} = \frac{C_{EFM}\left(f_R^2 - f_M^2\right)}{\left(f_R^2 - f^2\right)} \tag{7.18}$$

Again, just as for the inductors earlier, if f_M is less than about 5% of f_R, (7.18) defaults back to (7.15).

As was noted for coils, effective capacitance increases as a result of the parasitic reactance. The effective capacitance of a capacitor increases - (Figure 7.23) to infinity at the resonance frequency, when $f = f_R$. At a frequency of $0.1f_R$, C_{EFF} is 1% above its nominal value, C_0, but at just $0.3f_R$, C_{EFF} already shows a 10% increase. Although L_S is a *parasitic* portion, it should not be considered a small perturbation to C_0. In fact, above resonance the effective reactance of the capacitor is dominated by the parasitic inductance.

Not only C_{EFF} but Q_{EFF} as well is frequency dependent. Q_{EFF} reaches its maximum value at lower frequencies, approaching infinity as frequency tends toward zero. As the frequency increases Q_{EFF} gradually drops to zero at f_R due to frequency-dependent losses and the presence of the unavoidable self-inductance.

Component Qs vary with nominal values. Generally, although not always, smaller capacitance is related to higher Q within a specific family and size of products (Figure 7.24). For example, at 2 GHz a 6.8-pF capacitor's Q is about 25 while a 2.4-pF capacitor has nearly 10 times higher Q.

Figure 7.24 tells us that paralleling two lower value capacitors of the same type gives us higher Q than using a single capacitor, which has capacitance equal to that of the parallel combination. Using two

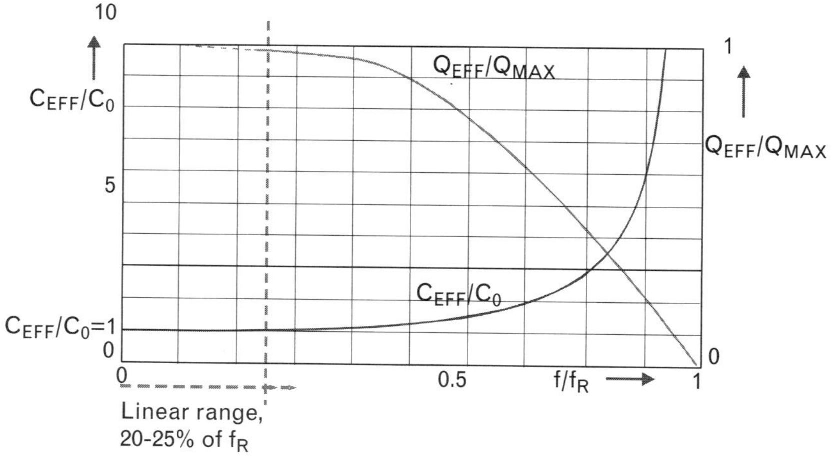

FIGURE 7.23 *Effective capacitance and Q variations with frequency. The effective capacitance is relatively constant up to about 20% to 25% of the resonant frequency. At low frequencies capacitor Q is very high but it drops quickly as we approach the series self-resonance.*

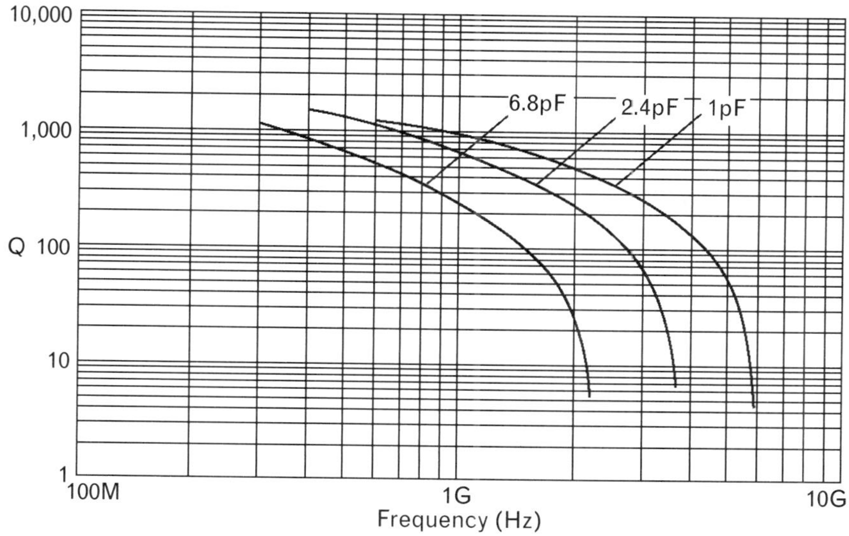

FIGURE 7.24
Q plots of the Murata GQM18 monolithic ceramic capacitors show a sharp drop of Q at higher frequencies. (Courtesy of Murata Inc.)

components also increase power-handling capability—which is an important consideration in high-power RF circuits. Of course, two capacitors cost more than one and they also take more space. Still, it is something to remember when there are no other alternatives to improve performance.

7.6.1 Interdigital capacitors

Small series capacitances on good quality dielectric boards may be created in planar interdigital form. If the board has ground plane under its surface, the fingers also create shunt capacitance and the effective model of the component is a PI-network. Computation of the capacitance is complex [18] but all the commercial RF circuit simulators offer dimension-based models for the component. EM simulation is an even better alternative to obtain accurate characterization.

7.6.2 Illustrative example of effective capacitance calculations

We need a 4.7-pF matching chip capacitor for a 1.9-GHz application. Looking at a 4.7-pF capacitor's data sheet, the specified minimum self-resonance frequency is $f_R = 3.3$ GHz. The manufacturer's test frequency is $f_M = 20$ MHz. Let us determine the effective capacitance of the 4.7-pF component. For simplicity, let us use the lossless assumption.

Since f_M is less than 1% of f_R, we can use (7.15) instead of (7.18) to compute the effective capacitance of the 4.7-pF component at 1.9 GHz. Although the resonant frequency of a typical capacitor is generally higher than the datasheet shows for minimum specification, we use the 3.3-GHz value. The effective capacitance at 1.9 GHz is

$$C_{EFF} = \frac{C_0}{\left(1 - \left(\frac{f}{f_R}\right)^2\right)} = \frac{4.7}{1 - \left(\frac{1.9}{3.3}\right)^2} = 7.03 \text{ pF}$$

which is about 50% increase of the nominal 4.7-pF value. Let us find out what kind of *nominal* value is needed to have an *effective* capacitance of 4.7 pF.

If all other nominal value capacitors would resonate at 3.3 GHz, we could simply solve (7.15) for C_{EFF} by setting C_0 to 4.7 pF. However, smaller nominal values resonate at higher frequencies, and we have to look at other capacitors' specifications to obtain that information.

Alternatively, we can make an assumption that for similar nominal values of a given capacitor family the self-inductance does not change significantly. Computing the series self-inductance[4] of the 4.7-pF component from (7.17) can give us a good estimate of the inductance of other capacitors. Since the 4.7-pF series capacitor resonates with L_S at 3.3 GHz, we find the self-inductance as

$$L_{S\text{nH}} = \frac{25.33}{C_{0pF} f_{R\,\text{GHz}}^2} = \frac{25.33}{4.7(3.3)^2} = 0.5 \text{ nH}$$

At 1.9 GHz, this inductance represents an inductive reactance of

$$X_{LS} = 6.28(f_{\text{GHz}})L_{\text{nH}} = 6.28(1.9)0.5 = 5.97\Omega$$

To have an *actual* capacitance of 4.7 pF at 1.9 GHz, we need a component with an effective capacitive reactance of

$$X_C = \frac{159}{f_{\text{GHz}} C_{\text{pF}}} = \frac{159}{1.9(4.7)} = 17.8\Omega$$

Drawing the equivalent circuit of the desired capacitor, using only the nominal capacitance and self-inductance (Figure 7.25), we show the desired effective capacitive reactance of $-j17.8$ that includes an inductive reactance of $j5.97$. The needed nominal capacitive reactance is obtained by

4. This is a good way to find self-inductance that is too low for any practical measurement.

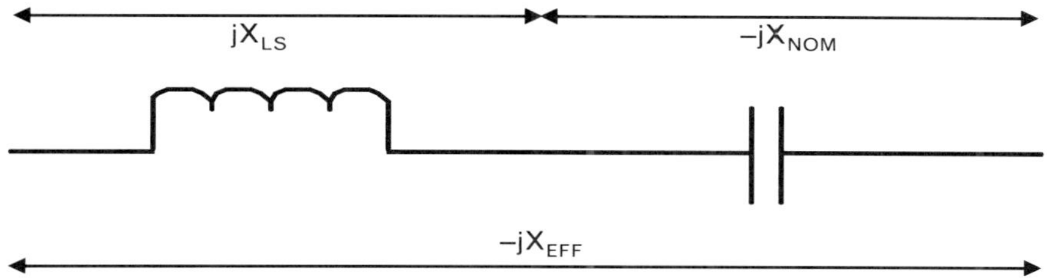

FIGURE 7.25 *Simplified schematic of a capacitor, viewing only the nominal capacitance and series self-inductance. X is the reactance of the nominal capacitor, X_{LS} is the reactance of self-inductance, and $X_C = X_{EFF}$ is the total effective reactance. At self-resonance the magnitudes of inductive and nominal capacitive reactances are equal ($X_{LS} = X$).*

subtracting the parasitic inductive reactance from the desired effective capacitive reactance.

The total effective reactance is

$$-jX_{EFF} = jX_{LS} - jX_N \tag{7.19}$$

From (7.19) we can express the nominal reactance required to obtain an effective reactance as

$$-jX_N = -jX_{EFF} - j(X_{LS}) = -j17.8 - (j5.97) = -j23.77\Omega$$

Converting the reactance back to capacitance, we get

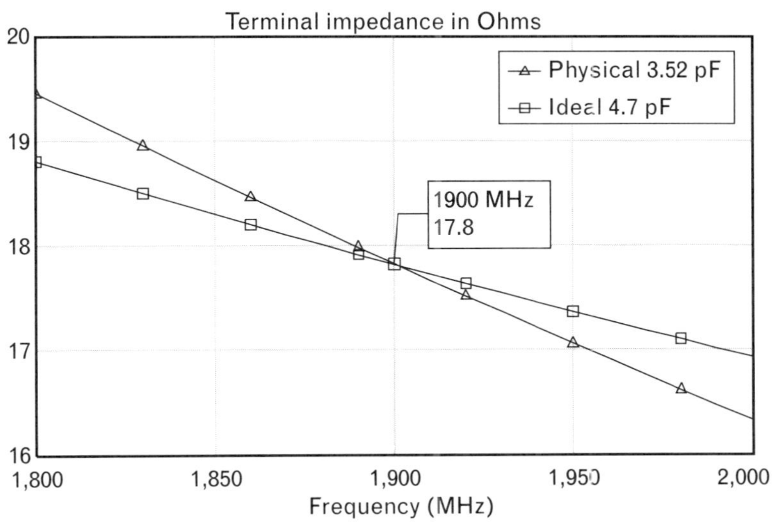

FIGURE 7.26 *An ideal 4.7-pF and a physical 3.52-pF capacitor have the same terminal impedances at 1.9 GHz, and less than 5% difference through a 200-MHz bandwidth.*

$$C_{\text{pF}} = \frac{159}{f_{\text{GHz}} X_{EFF}} = \frac{159}{1.9(23.77)} = 3.52 \text{ pF}$$

Our computation shows that in order to have an *effective* capacitance of 4.7 pF at 1.9 GHz, assuming 0.50-nH series self-inductance, we need to use a 3.52-pF *nominal* capacitance value. If the 3.52-pF valuable were available, at that frequency its effective reactance would be exactly the same as the reactance of a 4.7-pF ideal capacitor. Figure 7.26 shows that the equivalence may also be approximated through a narrow bandwidth.

In real-life applications a 3.52-pF nominal value is hard to find, and we use the nearest available smaller standard component. The difference can usually be made up by using an appropriate conductor pattern on the PC board. In the above calculations we used exact values only to illustrate the mathematical procedure.

7.6.3 Secondary resonances in multifinger capacitors

As you recall, we showed that multilayer inductors have secondary resonances that lead to eye-opening impedance behavior. Multifinger capacitors also have such problems, although generally not quite as severe as we saw for the inductors. In capacitors, the self-inductance of the contacts among the parallel fingers cause secondary parallel resonances above the primary series resonance, leading to high impedances. Depending on the way the capacitor is attached to the circuit PC board these secondary resonances can be pushed to higher frequencies by the proper orientation of the chip (see Figure 7.27). When the chip is mounted properly, secondary resonances are at least an octave above the primary resonance. Unfortunately, the form-factors of the most commonly available parts are not appropriate to take advantage of the optimum mounting orientation. Some of the capacitor makers, however, recognize this problem [19] and produce cube-like forms to help minimize the secondary resonance effects.

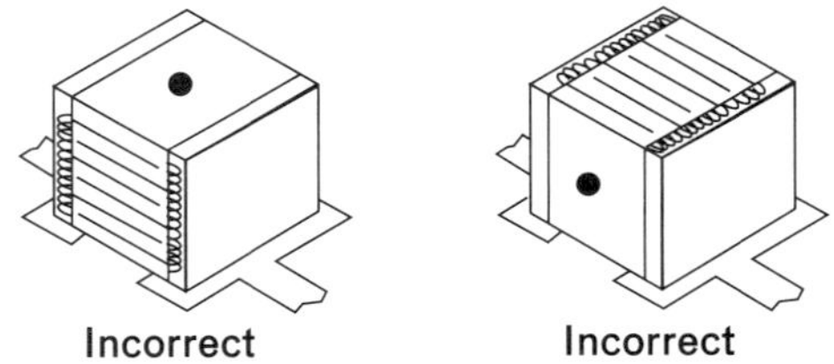

FIGURE 7.27 *PC board mounting of a typical chip capacitor, showing that current flow to the top fingers must pass through additional inductance resulting in secondary resonances. These resonances may be moved to higher frequencies by mounting the chip 90° rotated so that the shaded surfaces contact the circuit board. In the optimum position the black dot should be on the vertical side of the chip instead of the top.*

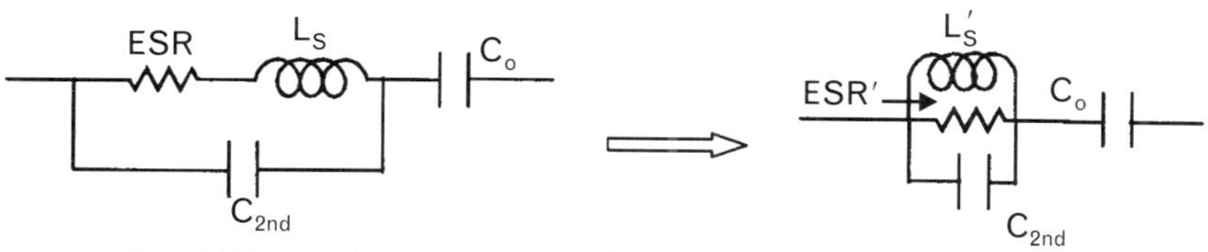

FIGURE 7.28 *Adding another capacitor C_{2nd} allow us to model the first secondary resonance that is a parallel kind, leading to high impedance. Secondary resonance frequency is affected by the positioning of the capacitor on the dielectric board.*

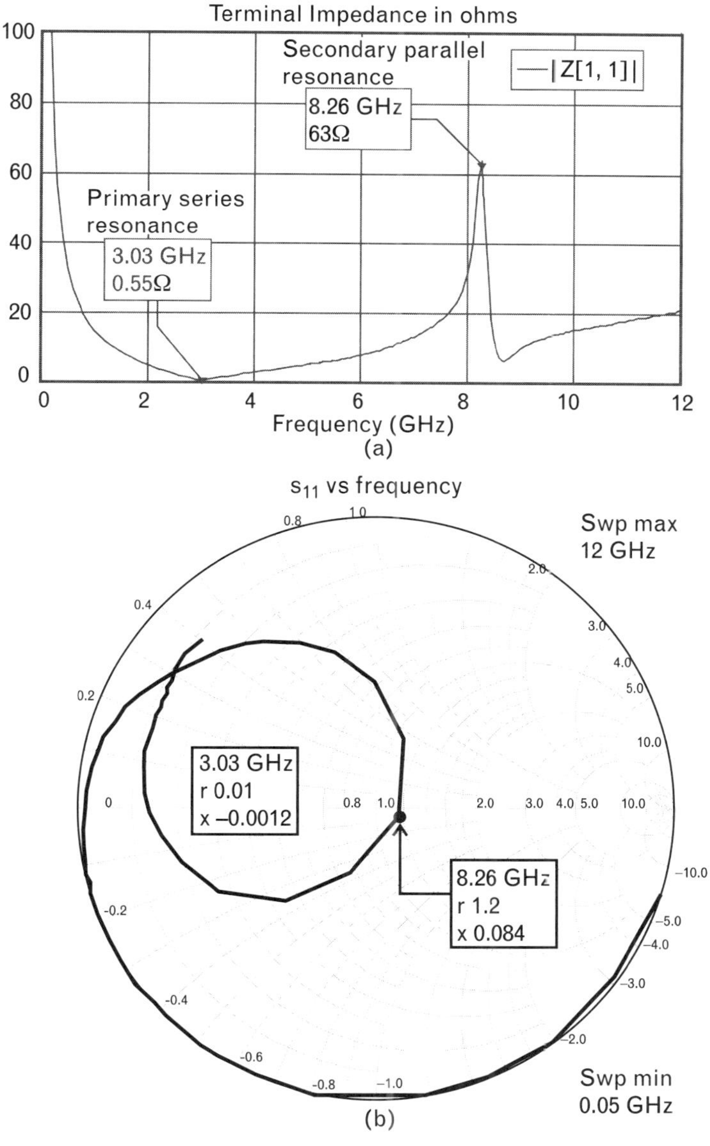

FIGURE 7.29
(a) Measured data of a 10-pF multilayer porcelain capacitor (ATC 700A100KW) shows primary series resonance at 3.03 GHz and first higher-order parallel resonance at 8.26 GHz where the capacitor behaves like a 63-Ω resistor. (b) Secondary resonances always form loops on the Smith chart. Markers on the Smith chart show resistance and reactance normalized to 50Ω.

The lumped capacitor equivalent circuit of Figure 7.22 may be modified [20] to include secondary resonance effects, as shown in Figure 7.28.

Capacitance C_{2nd}, although it is relatively small, resonates with the parallel converted inductance L_S' above the primary resonance f_R, as shown in Figure 7.29. At that secondary resonance frequency L_S' and C_{2nd} cancel each other and we get ESR' in series with the reactance of C_0. Keep in mind, however, that ESR' is quite a bit higher than ESR, due to the series-to-parallel R-L conversion.

In multifinger capacitors additional secondary resonances exist, making accurate high-frequency modeling quite difficult. Since these secondary resonances are well above the primary resonance they generally do not cause problems in narrowband communication circuits. In broadband applications, particularly if the capacitor is to provide decoupling through a wide frequency range, we need to deal with the secondary resonances.

When broadband decoupling is provided by various capacitors connected parallel, we also see parallel resonances. For example, we may use a 0.1-μF capacitor in parallel with a 30-pF capacitor and expect the larger capacitor to take care of the low frequencies and the 30-pF one to do the same in the gigahertz range. The problem is that if the 0.1μF series resonates at 20 MHz, above that frequency it behaves as a lossy inductor. (We can also expect secondary resonances above 40 MHz.) The 30-pF capacitor may have its primary resonance around 1 GHz, and up to that frequency it behaves capacitively. Now comes the bad part—between the two series resonance frequencies (20 MHz and 1 GHz), we get at least one parallel resonance also, because in that frequency range effectively there is a parallel connection of an (apparent) inductor and a capacitor. Component losses limit the impedance of the parallel circuit at resonances but can still be high enough to cause problems. Secondary resonances of the large capacitor of course can make things worse.

7.7 Via hole models

Single-ended (unbalanced) circuits require ground connections for dc current and RF signal returns. While routing of dc connections has no effect on performance, for RF considerations a precise description of the signal path is very important.

In most cases RF ground is not conveniently available PC board. RF ground may be several layers below in a multilayer board and brought to the top by a plated-through hole (Figure 7.30), called *via hole*, or sometime just via [6]. A via hole may be connected from a PC board's top to bottom, or partially through a number of layers, or it may be completely buried inside a multilayer board (called blind via). Lengths may vary from a few mils to tens of mils—a relatively short path—so why do we have to worry

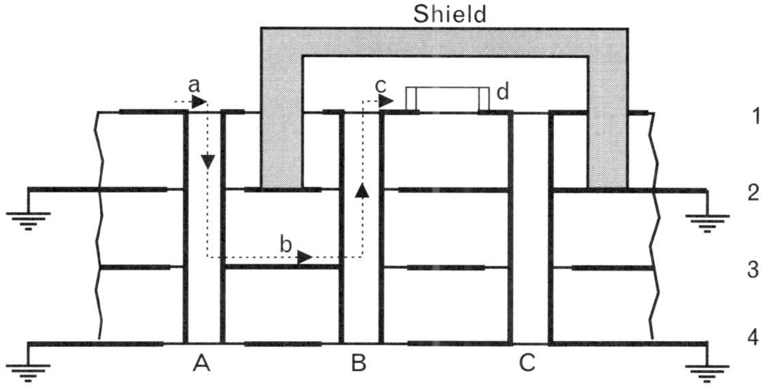

FIGURE 7.30 *Various forms of via holes in multilayer PC boards where a component placed on layer #1 requires shielding, which interrupts the conductor path of that layer. Via holes A and B connect layer #1 location a to location c, through a path of a-b-c that includes two via holes and layer #3. If cost is not important, the lower thirds of via holes A and B should not have metallization, because those sections acts as open stubs to layer #3. Via hole C brings ground to layer #1 at location d.*

about these interconnecting elements? A simple calculation reveals the problem they can cause.

Typical via hole inductances are in the 0.05- to 0.2-nH range. In the RF region the inductive reactance is significantly larger than the conductive resistance of a via hole. Therefore the resistive portion, even with skin-effect increases, can be neglected. However, the few tenths of a nanohenry inductance amounts to several ohms of inductive reactance, which is enough at RF to seriously change the behavior of our components and circuits.

In multilayer PC boards we usually want a via hole connection between the specific metal layers to be interconnected, as shown in Figure 7.30. To minimize additional parasitic capacitance, the hole could be first drilled and plated through the entire board thickness, removing the unwanted portion later. This additional process, however, adds a significant increase to manufacturing cost, and the extra section(s) of the metallized holes are generally left in the boards. For accurate modeling, particularly if the remaining sections are connected to high-impedance nodes in a circuit, we recommend that those sections are included in the circuit modeling process. In most cases they need to be modeled as open-circuited parallel stubs.

The inductance of a circular via hole, without any other added conductive path or shielding effect, is computed from [21]

$$L_{\mathrm{nH}} = K\left[H \ln\left(\frac{H + \sqrt{H^2\left(\frac{d}{2}\right)^2}}{\frac{d}{2}}\right)\right] + 1.5\left(\frac{d}{2} - \sqrt{H^2 + \left(\frac{d}{2}\right)^2}\right) \quad (7.20)$$

where

H is the thickness of the dielectric board in millimeters or mil.

D is the diameter of the hole in millimeters or mil.

K is 0.2 for dimensions in millimeters and 0.00508 for dimensions in mil.

Via hole inductance can be reduced by adding an additional closely spaced parallel hole. Intuitively we expect the total inductance of the double via to be half of that of the single hole, but in real practice the effective value is closer to 60% to 65% of the single hole, due to mutual inductive coupling. When two inductances are mutually coupled and their currents flow in the same direction, the individual inductances *increase*. Paralleling two such inductors we get half of the *increased effective* inductance [22].

The inductance is further reduced by punching and plating a rectangular hole [22] instead of using two via holes. However, the small improvement is not justified by the more difficult and less reliable process involved with the hole punching and plating of sharp edges.

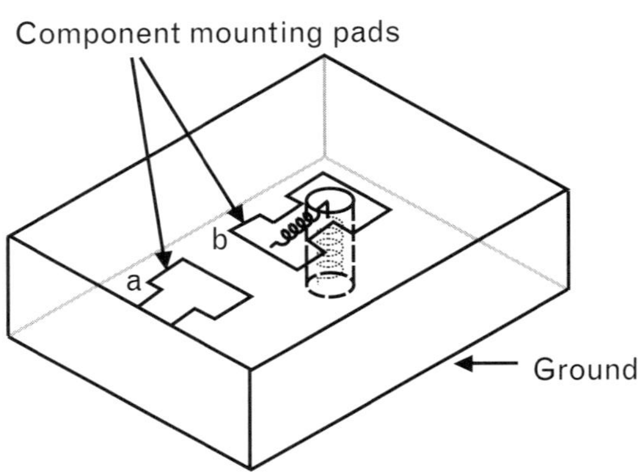

FIGURE 7.31 *When computing the effective inductance of the grounding path, we also need to include the parasitics of the surface conductive pad to which the component is attached. Above 1 GHz to 2 GHz stray coupling capacitance to ground(s) may also need to be included.*

7.7.1 Grounding-path inductance effects

A via hole providing grounding to a two-terminal passive component, such as a capacitor, just adds some series inductive reactance[5] to the component. However, even in a simple case like that, we may have to enlist the help of an EM simulator to find the proper model for the connection. Since components are placed some distance from the hole, instead of resting on the top of it (Figure 7.31), we need to consider the extra conductor path on the surface of the board for exact computation. Since RF currents do not flow uniformly through the hole, the effective inductance is always greater than the computed value derived by circuit theory [21]. Still, a small error in finding the reactance of the hole does not lead to major problems.

When the via hole is used to ground an active device or several passive components together, the finite reactance of the ground path may create feedback. Now we have a much more serious case and a relatively small error in computing the ground path can bring unpleasant results, such as unwanted oscillation and significant change of the passband frequencies and of the stopband attenuation.

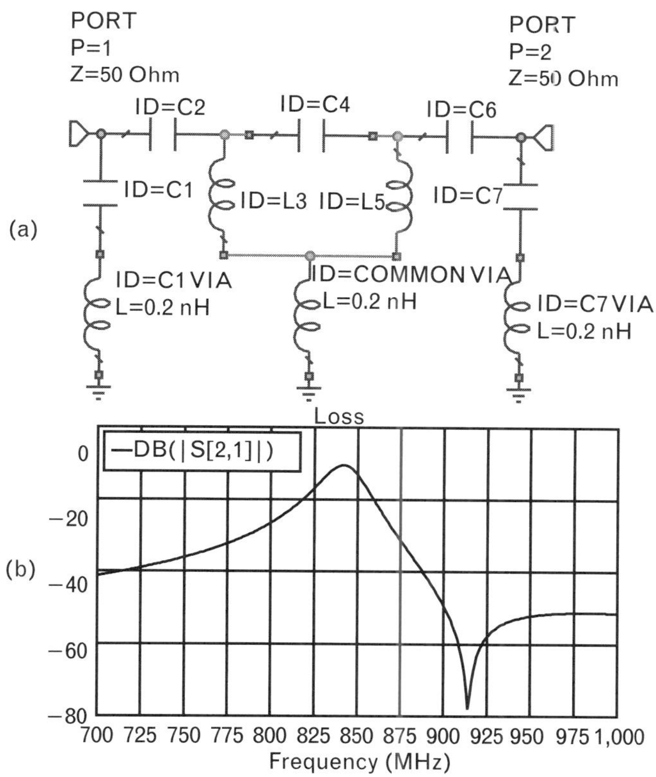

FIGURE 7.32 *Even in passive circuits, feedback caused by ground inductance can completely change the performance. (a) Instead of the ideal ground plane, inductors L_3 and L_5 are grounded through a single common via hole, labeled COMMONVIA. A 0.2-nH inductance is equivalent to only about $j1\Omega$ reactance at 850 MHz. (b) The feedback effect causes notch filtering at 915 MHz and virtually eliminated the filter's intended 830- to 860-MHz passband.*

5. Generally the reactance of the path is much greater than the resistance, so we can use a "loseless" assumption.

7.7.1.1 Illustrative example: common-mode feedback caused by via hole inductance

To emphasize our last point, let us look at an example where grounding two passive components through a single via hole made the circuit completely unusable. Figure 7.32(a) shows the initial circuit schematic of an 830- to 860-MHz bandpass filter, consisting of two capacitively coupled *L-C* resonators. The filter was to be realized with five surface-mount type capacitors and two small air-core RF inductors. Due to circuit symmetry, $C_1 = C_7$, $C_2 = C_6$, and $L_3 = L_5$. To minimize mutual inductive coupling, L_3 and L_5 were placed perpendicularly on the PC board. The lower ends of the inductors (ground sides of L_3 and L_5) were placed on a conductive pad and were grounded through a single common via hole, *COMMONVIA*. The two parallel capacitors were grounded separately. Although the circuit designer reduced nominal filter component values to compensate for

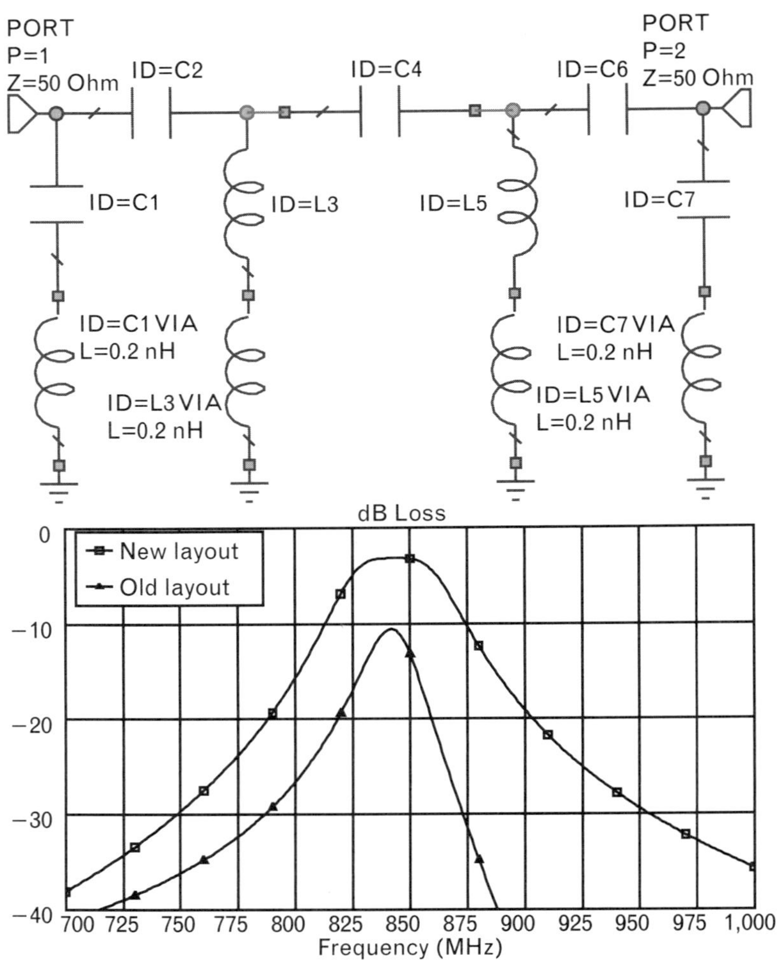

FIGURE 7.33 *After separately grounding the inductors L_3 and L_5 the intended filter response vastly improved (□-markers) compared to the common grounding case (Δ-markers). Other components still require modeling, and all layout related parasitics should also be added before going to the prototype phase. (Note: vertical scale changed from 80 dB in the previous plot to 40 dB here.)*

the presence of the ground inductance (about 0.2 nH), feedback effect was not considered.

Frequency response of the first prototype showed an unexpected dip above the passband and a very poor performance in the passband [Figure 7.32(b)]. When finally the filter was properly modeled in a circuit simulator —which should have been done before building the prototype—the effect of common-mode feedback was clearly visible. Even though we do not show here all the physical component models, the problem caused by even a minute amount of inductive feedback is obvious.

After realizing what caused the problem, the two inductors were separately grounded by the same 0.2-nH equivalent inductance and the frequency response of the following prototype was back to the expected form (Figure 7.33).

As we mentioned earlier, the circuit still needs more detailed modeling, but a simple simulation was sufficient to identify the prime reason for the poor initial performance, making the completion of the final design phase much easier. Lesson learned: Stay close to the circuit layout and add the layout related effects into the initial simulation. Do not build prototypes until the simulated performance is satisfactory. By the way, remember to include tolerance analysis also!

7.8 Planar transmission lines for RF/MW applications [23]

In the World War II era, the heavy coaxial transmission line circuitry needed for microwave circuits presented problems for engineers as well as the end users. Around 1950, the *idea* of microwave integrated circuits was considered and new types of microwave circuitry based on printed transmission lines rather than plumbing or wiring were envisioned. The microstrip, a planar[6] conductor separated from a wide metal ground-plane by a dielectric board, was first introduced in the early 1950s [24]. It was not initially accepted due to problems associated with loss, radiation, and shielding. Its application also required rigorous analytical calculations. A few years later the stripline [25] was developed, offering fewer problems and much greater predictability of performance. A stripline is a true TEM line; therefore, it has a unique characteristic impedance and constant phase velocity. Its transmission line parameters can be calculated very accurately by static techniques, such as conformal mapping. From about 1955 to 1965, stripline was the most commonly used planar line.

Microstrips came to life again in the mid-1960s, due to the newly developed higher dielectric ceramic substrates instead of plastics. Mechanically, the ceramics gave greater stability of dimensions under mechanical

6. A planar transmission line has flat conductors on a dielectric board. Other definitions may vary.

forces, temperature and chemical exposure, yielding higher reliability. Electrically, the higher dielectric constant concentrated more of the energy in the substrate, making the circuit less sensitive to nearby obstacles. It also reduced the sizes of circuits in a given frequency range with new photo-lithographic techniques. Microstrips rapidly replaced striplines as the transmission line of choice, and for the past four decades they have maintained their position as the most popular transmission medium for printed-circuit board circuits, *microwave integrated circuits* (MICs) and more recently for *RF integrated circuits* (RFICs).

There are many excellent textbooks available with detailed mathematical expressions to analyze and synthesize the various planar transmission lines and their discontinuities [26–32]; therefore, we do not repeat them in our book. We do, however, encourage you to use one of the transmission line subprograms included in all commercial RF/MW circuit simulators—some of which can even be downloaded at no charge from their Web sites [33]. These programs are constantly updated with the latest research data and offer more convenience than the alternative lengthy numerical transmission line analysis and synthesis formulas.

7.8.1 Comparison of planar transmission line forms

Let us review the most frequently used planar transmission lines shown in Figure 7.34, starting with the properties they share. All forms support a

FIGURE 7.34
Various forms of planar transmission lines.

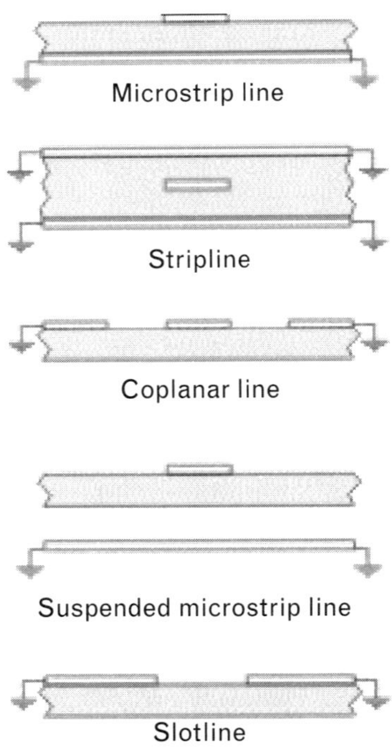

dominant mode, and that mode is characterized at a *given* frequency by a characteristic impedance, a phase constant, and an attenuation constant. With the exception of slotlines, they propagate down to zero frequency as TEM lines do and their propagation characteristics are called quasi-TEM [34]. A circuit using one of these planar lines is normally enclosed in a metal box for shielding. This box always affects the transmission line parameters to a greater or lesser degree, depending on the physical dimensions.

7.8.1.1 Microstrip

Microstrips are widely used in RF and microwave circuits. The microstrip structure has a ground plane on one side of the dielectric board with the circuit printed on the other side, and open space above. It is referred to as having an inhomogeneous dielectric because the EM waves propagate both in air and solid dielectric media. It is unbalanced with respect to ground and its physical structure is also unbalanced. Therefore, microstrip discontinuities have unequal currents above and below the strip conductor, and they have the tendency to radiate. The inhomogeneous dielectric also causes the phase-constant and characteristic impedance to have variation with frequency—a phenomena called dispersion.

Since the EM waves propagate in two different dielectric media, we need to find the effective dielectric constant, ε_{EFF}, that is between the values of the relative dielectric constants of the PC board (or substrate) and air. The effective dielectric constant varies with the conductor strip width, the thickness of the dielectric, and the spacing to the top cover.

The electromagnetic fields in microstrip are bound quite closely to the PC board, particularly for those of higher dielectric constants. As a result, hybrid elements or tuning elements can be added to the circuit without disturbing the performance of other parts.

Microstrips have been used in a wide range of applications, such as amplifiers, oscillators, mixers, hybrid couplers, and filters. Radiation and uneven phase velocities, however, pose limitations for certain types of circuits, like interdigital filters and dual-directional couplers, which are poor candidates for realization in microstrip.

7.8.1.2 Stripline

A stripline has its center conductor symmetrically located with respect to the ground planes. Although electrically it is an unbalanced line with respect to ground, it is physically balanced, and this prevents undesired radiation from discontinuities (assuming the top and bottom grounds are symmetrically spaced from the conductor). The dielectric is uniform (homogeneous) and allows true nondispersive TEM operation.

A common realization of stripline is where two dielectric boards are pressed together, both having solid conductors on the outer sides, and one of the two boards having a conductive strip on the inner side. The two outer conductors are bolted together to form the common ground. If the two boards pressed together are *soft*, as opposed to *hard* dielectrics such as ceramics, the air gap at the edges of the inner conductor becomes negligible. Striplines are difficult to connect to other planar transmission lines but they have good transitions to coaxial transmission lines.

Striplines do not work well for integrated circuitry because the solid dielectrics prevent access to the strip. We can neither add other circuit elements nor can we conveniently tune the circuit without taking its structure apart. On the positive side, it has low loss and is also suitable for broadside coupling structures while microstrip is limited to edge-coupled form only. Application of stripline has been most successful for passive circuits, such as filters, multisection couplers, and impedance transformers, laid out on soft-dielectric boards.

7.8.1.3 Coplanar waveguide

The coplanar waveguide [35] was introduced in 1969 and has found wide application, second only to microstrip. A CPW has a center conductor symmetrically placed between ground planes, all of which are on the *same side* of the substrate. Though inhomogeneous, it is less dispersive than the microstrip. Like the microstrip, it is open for addition of hybrid elements or tuning, but unlike the microstrip, the ground planes are readily accessible. Finally, the characteristic impedance of the CPW depends only on the ratio of the strip width to the ground plane spacing; therefore, a desired impedance can be achieved with a convenient ratio of those dimensions. For microstrip, once the substrate thickness is fixed, a given strip width can yield only one impedance.

The processing technology is the same as for microstrips, but the substrate requires metallization on only one side, eliminating the needs for via hole grounds. Another form, grounded coplanar waveguide also has a ground underneath to accommodate boards with bottom grounds.

Radiation at discontinuities can be a problem with CPWs, but it can be reasonably well controlled by maintaining circuit symmetry between ground planes, and by frequent use of wire bonds to keep ground planes at the same potential in the vicinity of discontinuities. The CPW offers good transition to coaxial transmission lines.

A physical and electrical dual of coplanar waveguide is the coplanar strip [36]. It consists of a pair of conductors, balanced with respect to ground, printed on one side of a substrate. Coplanar strips are more frequently used in balanced circuit applications, such as dipole antennas.

7.8.1.4 Suspended substrate

Working with high-impedance components, such as FET devices, we often need transmission lines with characteristic impedances higher than what is practically realizable in conventional microstrip form. Since the characteristic impedance is an inverse function of the spacing between the ground plane and the conductive strip, moving the ground farther away increases the characteristic line impedance [37]. The trade-off is that low-impedance lines become too wide and discontinuities are more difficult to model.

The dielectric layer may be very thin, creating very low loss transmission lines, since most of the EM wave propagation is in air. As a result, the effective dielectric constant is very low—close to unity—therefore, the physical transmission line lengths become reasonably short only at microwave frequencies. Suspended substrate transmission lines are dispersive, although not as much as microstrips.

7.8.1.5 Slotlines

EM waves can propagate along a slot cut into the metallization of a dielectric board [38], which is metallized on only one side. An inhomogeneous dielectric is necessary to bind the wave to the slot; therefore, slotlines are not TEM structures. Unlike in waveguides, slot waves have no lower cut-off frequency, but because the impedance approaches zero at low frequency, slotlines have limits to their usable frequency range.

In a practical slotline, the slot is made narrow in terms of the wavelength in order to suppress higher-order modes and radiation. The impedance of the slot is then adjusted to a convenient level by means of the slot width or the substrate thickness. The resulting phase constant describes a dispersive slow wave, comparable to that of a microstrip on the same dielectric.

Like CPW, the slotline circuit is truly coplanar, and like other planar lines it has an open surface across where other components may be connected. A slot in the ground plane couples well to the microstrip circuit on the opposite side. This allows the microstrip designer another degree of freedom. Short slotline sections cut into the ground plane of microstrip have been used to realize wideband forms of couplers and hybrids [39].

7.8.2 Coupled transmission lines

Coupled transmission lines are formed by placing two uniform transmission lines close enough to each other to have their electric and magnetic fields interact. (We already introduced this concept in Chapter 3, discussing the dual-directional coupler.) Transmission lines may be coupled in

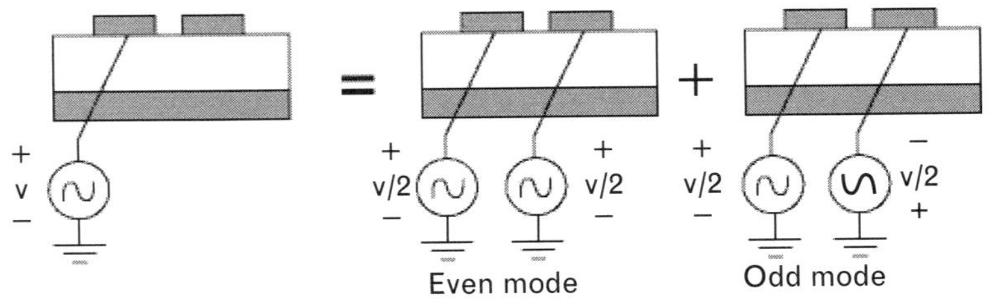

FIGURE 7.35 *An asymmetrically driven pair of coupled lines is equivalent to two separate sets of symmetrically driven coupled lines. (From: [5]. © 1998 Artech House, Inc. Reprinted with permission.)*

various ways, depending on the amount of desired coupling and bandwidth [27].

Characterization of coupled lines is somewhat more difficult than single lines because the EM fields between the two lines can have different forms. Fortunately we can apply linear superposition theory to simplify the task. When one of the two coupled lines izs driven by a signal source, the behavior of the line is equal to the sum of two separate circuit responses, as shown in Figure 7.35. When the two lines are driven by the same signal polarities, we refer to the *even mode* response, and when they see two signals out of phase, we get the *odd mode* response. Therefore, a coupled line has two sets of parameters: even and odd mode impedances, even and odd mode losses, and so forth. There is only one electrical length, however, because it does not depend on the phase of the driving signals.

The coupling between the two lines is frequency dependent, reaching its maximum when the lines are both quarter-wavelength (90°) long. Depending on the type of coupling, the practical bandwidth of a single-section coupler varies from 5% to 10% to an octave frequency range.

Unwanted coupling of parallel lines is a concern with tightly spaced circuits where conductors are closed to each other. A good rule of thumb is to consider parallel line coupling when the spacing is less than four to five times the board thickness. For example, we frequently find it necessary to "meander" long transmission lines to reduce the distance between ends. The spacing among parallel sections is often small enough that we need to include coupling effects.

As for the single lines, we again refer our readers to the various RF circuit simulators' transmission line subprograms to convert between electrical parameters and physical specifications. Discontinuities are discussed next, and some of the coupled line applications are discussed in Chapter 8.

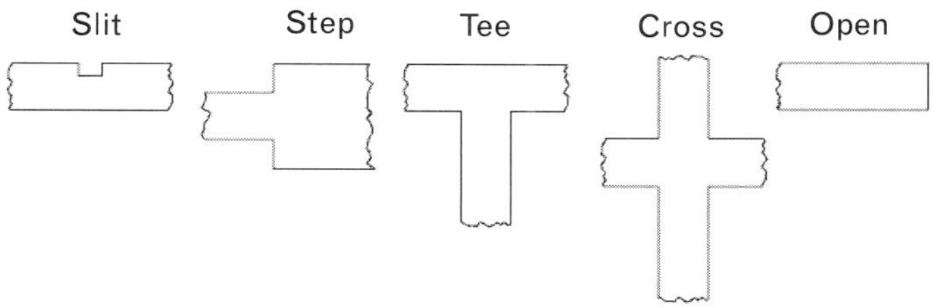

FIGURE 7.36 *Discontinuities are created when we affect the signal propagation by connecting together two or more transmission lines. Propagation changes even when we connect a 50-Ω impedance coaxial line to a 50-Ω microstrip line.*

7.8.3 Transmission line discontinuities

Every time the EM wave propagation is changed in a transmission line circuit, some disturbance is created by the RF current crowding that behaves inductively, and by the electric field fringing that has a capacitive nature. When two or more transmission lines of different types, configurations, or characteristic impedances are connected together, we experience transmission line *discontinuities*.

The most commonly occurring transmission line discontinuities are shown in Figure 7.36, with their lumped equivalent circuit models. A brief explanation follows:

- *Notch:* a cut into the transmission line. Current crowding is represented by a series inductor.
- *Step:* two transmission lines with different characteristic impedance or different modes of propagation. Current crowding takes place on both sides of the junction and the wider line's open edge creates stray capacitance.
- *Bend:* usually 90°, although variable angle and even circular forms are available. RF currents are lazy and want to flow through the shortest path, which is at the inner side of the bend. There is heavy current crowding on both sides of the corner. The outside of the corner acts as extra capacitance.
- *Tee:* two cascade lines with a parallel branch between. Current crowding occurs in the lines as they approach the junction. The presence of the junction results in an excess capacitance to ground.
- *Cross:* two cascade lines with two parallel branches between the lines. A more complex case, best described by a mixed (lumped-distributed) model.
- *Open:* unterminated transmission line, generally at the end of an open-circuited parallel stub or parallel coupled lines. A stray capacitance occurs at the open end.

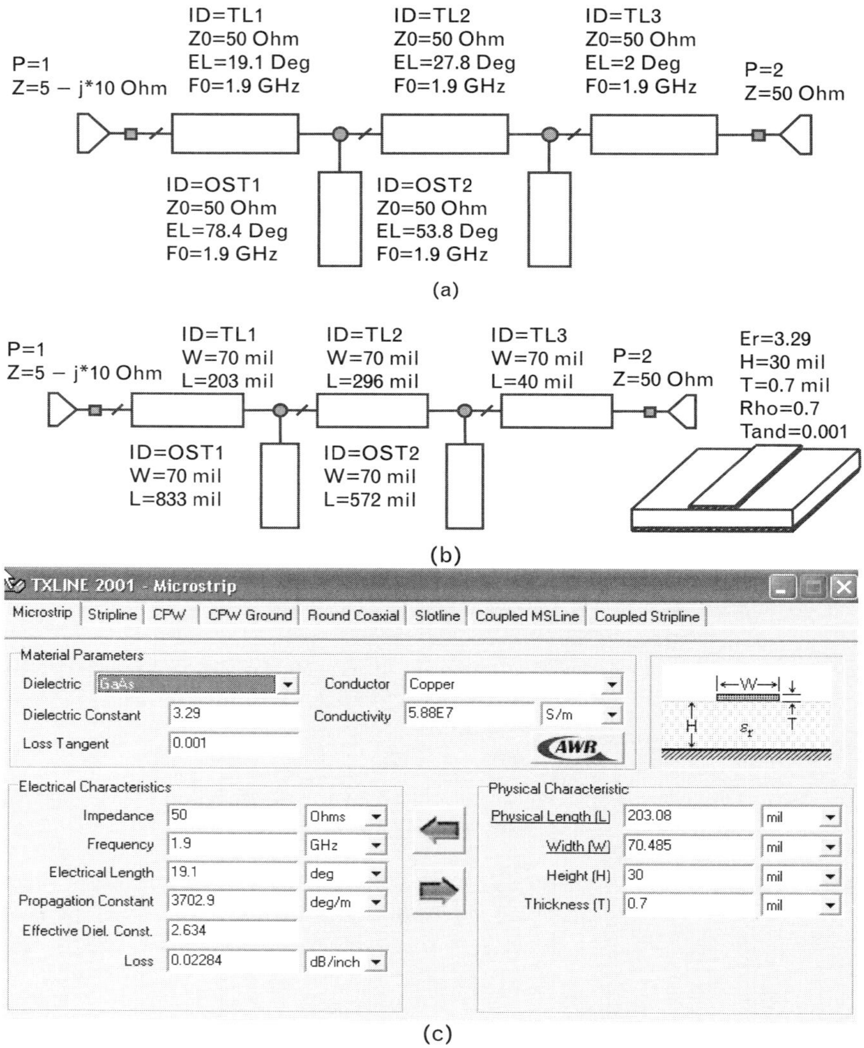

FIGURE 7.37 *Converting the four-element transmission line network to physical dimensions with TXLINE, using the PC board parameters shown above. The short 50-Ω line at the right side, TL3, is only used for physical connection. Top cover effects are not included and at this point ideal transmission line junctions are assumed. (a) Ideal electrical circuit; (b) microstrip transmission line dimensions after conversion; and (c) TXLINE window showing one of the line conversions.*

In case of discontinuities not represented in the circuit simulators, we turn to EM simulation for help. Due to the extensive research during the Cold War era, most of the standard types of discontinuities are well understood, documented [6], and properly modeled in the RF circuit simulators. For that reason we do not provide component equations for equivalent circuits.

To illustrate the importance of proper modeling, we compare the frequency response of a simple four-element transmission line circuit, with and without discontinuities. The circuit was designed using the Smith chart to match a power transistor's complex impedance of $Z_1 = (5 - j10)\Omega$ to $Z_2 = 50\Omega$ at 1.9 GHz. We used 50-Ω characteristic impedance for all lines for simplicity. We also added a short fifth transmission line (TL3) to make contact with the 50-Ω load termination. Circuit topology of the ideal transmission line circuit is shown in Figure 7.37(a).

Converting the electrical design to a physical layout, we first need to change the electrical parameters (Z_{TL} and θ) of each transmission line to a physical width and length. Transmission line calculator subprograms, such as AWR's TXLINE, Ansoft's TRL, or Agilent's LINECALC are very convenient for such task.

Dimensions of the physical lines are shown in Figure 7.37(b). Our PC board specifications for relative dielectric constant, dielectric and conductor thickness, metal resistivity, and dielectric losses are:

$\varepsilon_R = 3.29$.

$H = 30$ mil $= 0.76$ mm.

$T = 0.7$ mil $= 18\ \mu$m.

$\rho = 0.7$ (for copper, relative to gold).

$\tan\delta = 0.001$.

The next step is to combine the five transmission lines through two physical T-junctions. Adding junctions change the lengths of every line and there is no exact way to predict in advance what the effective changes are. A good estimate of compensating for the presence of a physical junction is to assume that all three lines end at the geometric center of the junction. Therefore, we initially need to

- Reduce the length of TL1 by one-half of the width of the parallel stub.
- Reduce the length of TL2 by the full width of the parallel stub since we add junctions to both ends.
- Reduce the length of the stubs by half of the width of the cascade line.

Adding three-port discontinuity junctions and optimizing the transmission line-lengths gives us a physical circuit that closely follows the performance of the ideal electrical circuit (Figure 7.38). The ideal circuit without any losses brings a perfect match $s_{21\text{dB}} = 0$ dB at band center with about 0.1 dB loss at the band edges. A physical circuit with discontinuities included has similar response at nearly 0.1 dB more loss. When the discontinuity effects are not included, we see a significant frequency shift and higher loss.

Next, we convert the simulated matching circuit to a physical layout (Figure 7.39). Most RF simulators can handle this task directly from the schematics of Figure 7.38, including the generation of the artwork needed

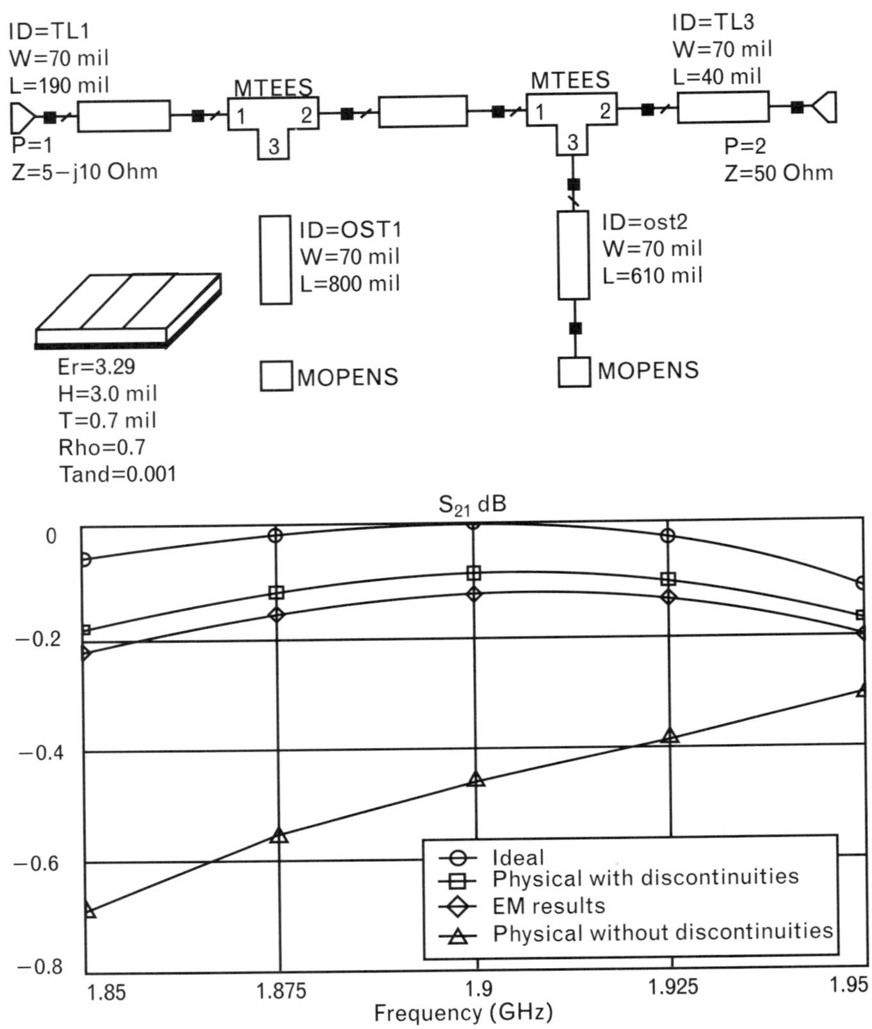

FIGURE 7.38 *Compensating discontinuities is important even in this simple matching network. The ideal lossless circuit response (o-markers) is computed using zero-size transmission line junctions without radiation at the end of the open stubs. The compensated physical circuit with schematic shown on left (□-markers) has conductive and dielectric losses and a similar response. EM simulation of the physical circuit (◇-markers) shows very good agreement with the compensated circuit simulator results. The uncompensated physical circuit (Δ-markers) shows a significant frequency shift.*

for the etching process. If the initial layout does not fit the available PC board space, the lines may be bent or meandered to reach a satisfactory size. Of course, every physical change affects the electrical performance and we may need several iterations until all requirements are fulfilled.

A final EM simulation to verify the results of the circuit simulator is again good practice, particularly when coupling among various transmission lines may be significant. In our example, the two open-circuited parallel stubs are separated from each other by 230 mil, which is nearly eight times the thickness of the dielectric board; therefore, the parallel coupling effects are negligible. When circuit and EM simulation both predict good performance (Figure 7.38), we have very high probability of getting similar results from the physical circuit also.

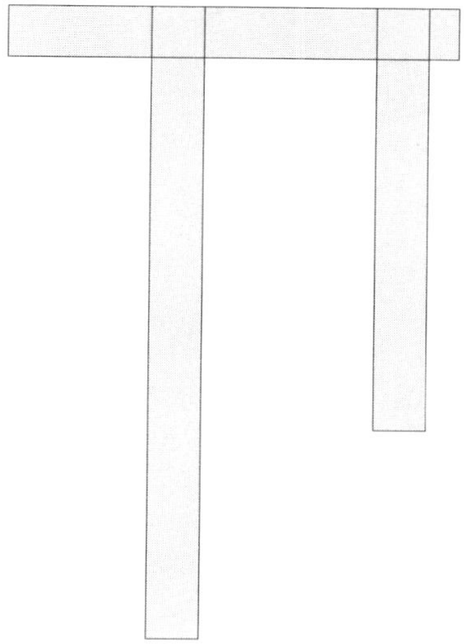

FIGURE 7.39
Physical circuit layout of the transmission line network. Modern RF circuit simulators include accurate characterizations of the various discontinuity junctions. Our circuit here uses three-port TEE-junctions to connect the three transmission lines. Each of the long open-circuited stubs can be split into two parts by using cross junctions.

The above solution is just one of the many possible options to match the two specified terminations. Another commonly used option is to use cascade transmission line(s) with parallel capacitors instead of the parallel open-circuited stub. If the necessary capacitors are realizable, the lumped-distributed combination may offer broader bandwidth and have less space requirement. Our goal here was to illustrate the conversion of electrical transmission line parameters to a physical circuit.

RF circuits and systems operating above 1 GHz to 2 GHz always include some form of transmission line elements. State-of-the-art RF circuit simulators offer models for various discontinuities that cover most of our needs. EM simulators are there to help us with the non-standard discontinuities. CAE tools readily convert transmission line designs to circuit layout and perform statistical analysis that quickly forewarn us about possible problems our products may face later in production. With the wide range of CAD tools available these days, which fit almost any budget, there is no justification for not doing a thorough circuit modeling before building the first prototype.

7.9 Dielectric board materials

At low frequencies the function of the PC board is to provide convenient printed connections and mounting to the components. At RF we have higher expectations from the board. If our circuit includes transmission line elements, the dielectric board also *becomes part of the circuit.* In that case, losses and tolerances of the board must be included in our design, including statistical considerations. Since there are several good and easily understandable publications available on this subject [40–44], we only want to raise one important issue here.

There are different opinions about the upper frequency limit to which FR-4 (epoxy-glass) material should be used, but most agree that it is somewhere between 1 GHz and 2 GHz—that happens to be where transmission lines begin to replace lumped components. FR-4 has been used widely at low frequencies and in digital circuit applications. As the frequencies of wireless communications have gradually gone higher, many designers stayed with the material they had been using. After all, FR-4 is low-cost (important consideration these days), and is available from a wide range of vendors, so why should they switch?

There is nothing wrong with FR-4 for lower frequency applications. As we get to the gigahertz range, however, the following concerns are raised:

- The relative dielectric constant varies between 3.5 and 4.5, and is frequency dependent.
- Relatively high frequency-dependent dielectric loss ($\tan\ \delta \approx$ 0.020–0.03 at 1 GHz).
- Wide dimensional tolerances, ±15%.
- Temperature and moisture absorption effects.

TABLE 7.4 TABULATED DIELECTRIC CONSTANTS AND LOSS FACTORS OF SOME HIGH-PERFORMANCE RF LAMINATES

SUPPLIER	MATERIAL	COMPOSITION	ε_R AT 10 GHz	TAN δ AT 10 GHz	WEB SITE
Arlon	25N	Thermoset/glass/ceramic	3.38±0.05	0.0024	http://www.arlonmed.com
	25FR	Thermoset/glass/ceramic	3.58±0.05	0.0035	
GIL	GML1000	Polyester/glass	3.00±0.04	0.0050	http://www.gilam.com
	GML2032	Thermoset polymer alloy	3.20±0.05	0.0029	
Rogers	RO3003	PTFE/ceramic	3.00±0.04	0.0013	http://www.rogers-corp.com
	RO4003	Thermoset/glass/ceramic	3.38±0.05	0.0027	
Taconic	TLC-30	PTFE/glass	3.50±0.05	0.0030	http://www.taconic-add.com
	RF-60	PTFE/ceramic	6.15±0.25	0.0028	

Note: Losses at 10 GHz are an order of magnitude better than FR-4 losses at 1 GHz. Dielectric constants differ, so PC board masks made for FR-4 must be changed if the circuit contains transmission lines.

PTFE-based (Teflon) laminates have been used for decades in microwave and defense applications with much better specifications—very low loss, low sensitivity to humidity and temperature changes and processing chemicals—at an order of magnitude higher price than FR-4. However, in recent years, newer dielectric materials have been introduced at somewhat higher prices than FR-4, but with considerably better performance and fewer steps during production. When these new boards become more known and accepted, their higher production volume will undoubtedly bring the cost to even more affordable levels for consumer and commercial wireless system applications.

Typical parameters of the newer dielectric boards are given in Table 7.4.

Thinner dielectric layers, multilayer boards, and smaller components enable us to reduce wireless system component size by packing more circuitry into small areas. While the closeness of cascaded components reduces the effects of interstage interaction, it increases the coupling among nearby components and conductor traces. We highly recommend EM simulation to investigate possible hidden coupling effects before building any circuitry.

7.10 Transformers

The transformer performs the electrical function of stepping up or down the voltage or current. It also provides impedance matching, isolation (no physical connection) of one portion of a circuit from another, phase inversion, and also alters ground conventions (single to differential, or balanced to unbalanced). Conventional transformers have primary and secondary

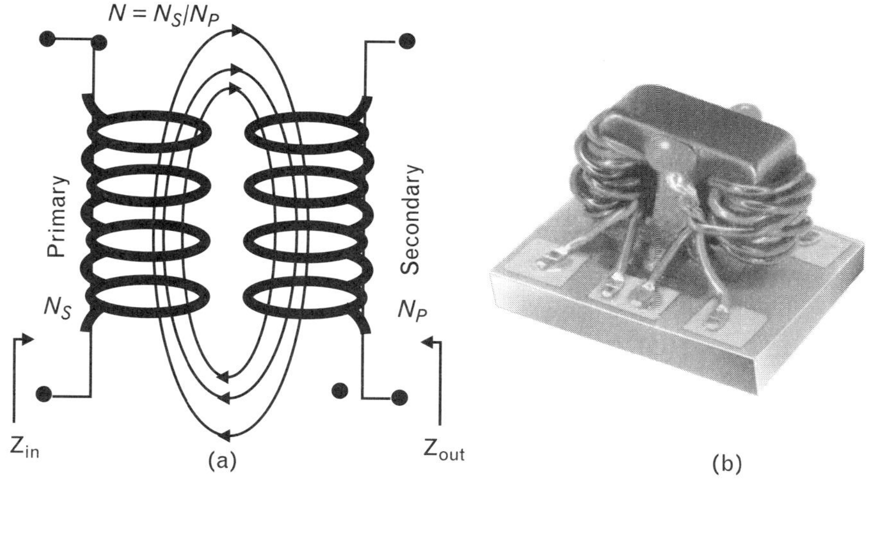

FIGURE 7.40 *(a) Ideal lumped transformer operation, and (b) toroid form realization for RF applications (Courtesy of Mini-Circuits).*

windings and establish magnetic flux linkage between the coils through a magnetic core. Ideal transformers, with the exception of dc, have no bandwidth limitations. As frequency reaches the RF region, conventional magnetic cores become extremely lossy, and the capacitive coupling between the windings makes the frequency response uneven. In the low RF range powdered iron cores are used, while at the higher frequencies we turn to ferrites (i.e., nickel-manganese and nickel-zinc compounds) to minimize core losses. Unique winding techniques can help us to maintain flat frequency response.

7.10.1 Transformer equivalent circuit with conventional windings

A four-terminal ideal transformer is shown in Figure 7.40(a). In the ideal transformer, all flux generated by the primary winding passes through the secondary. If the output is short-circuited, the input impedance is zero, and with an open-circuited load the input also looks like infinite impedance. The output voltage is a frequency-independent function of the input voltage and the turn ratio of the two windings.

Once the frequencies reach the megahertz region, the conventional primary-secondary type winding is best realized in toroid form [Figure 7.40(b)].

In a real physical transformer (Figure 7.41) a magnetization current exists, which flows in the primary winding, independent of the load. Magnetization current does not represent a power loss and can therefore be accounted for by an inductance, L_e, in parallel with the input.

Since some magnetic materials are also conductors, such as the ferromagnetic materials, the changing flux in the core sets up small current loops, called *Eddy currents*. These currents represent a power loss, which is also independent of the load. The continuous alignment and reversal of particles in the core that give the core its magnetic properties is called hysteresis loss. The total core loss, shown by a frequency-dependent resistance R_e in parallel with the primary winding, represents the sum of Eddy currents and hysteresis loss [45, 46].

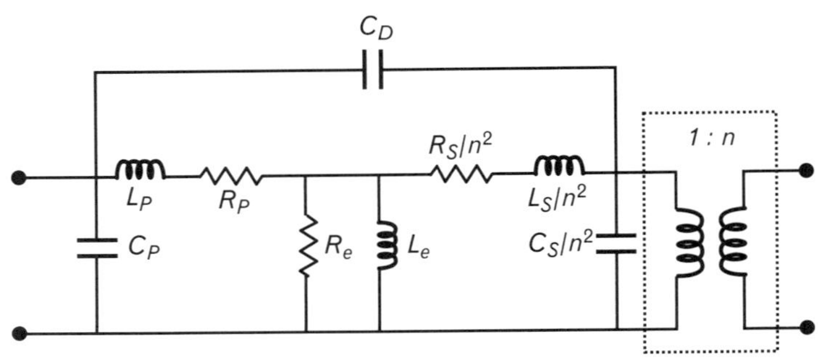

FIGURE 7.41
RF transformer equivalent circuit. The ideal transformer at the right side allows a change of output voltage's polarity.

The finite conductivity of the primary and secondary windings result in a resistance in the model associated with each. The secondary resistance may be transformed to the primary side in Figure 7.41 as R_S/n^2, where n is the turn-ratio of the transformer. The primary resistance carries not only the transformed load current but also the magnetizing and core loss currents. For this reason the primary resistance in the equivalent circuit is placed before the parallel elements as R_p.

The windings of the coils have stray capacitances similar to that of discrete inductors. Most magnetic core materials have relative dielectric constants greater than one, increasing the stray capacitance when these materials are used. Three lumped capacitors are shown in the equivalent circuit model of Figure 7.41. C_p and C_S represent lumped self-capacitances of the two windings, while C_d is the capacitive coupling between the windings.

The voltage transformation in the ideal transformer assumes that the magnetic flux links all the turns of both windings. However, even in a toroid a small amount of leakage flux exists. The presence of this leakage flux is evidenced by: (1) the failure of the inductance of the device to be exactly proportional to the square of the number of turns; (2) the failure of the voltage transformation ratio to be exactly equal to the turns ratio; and (3) the existence of a small inductance, detectable in the primary when the secondary is shorted. Two series inductors are included in the equivalent circuit model to account for self-inductance. The secondary inductance L_S is again transformed to the primary side of the ideal transformer by the $1/n^2$ factor. The primary self-inductance is L_p.

Self-inductance and stray-capacitance are truly distributed parasitic components. Lumped element approximations for these may not be representative over the frequency range desired. In addition to the operating condition limits (dc currents, ac signal level, and temperature) on the model elements, the transformer parasitics are subject to variations with frequency. Conductor loss and inductance are frequency-dependent functions, along with the magnetic material permeability and permittivity. The transformer equivalent circuit model is dependent on many factors requiring judicial user application.

If dc isolation between the source and load is not needed, a different kind of transformer may be created to transfer energy through transmission line action, instead of magnetic flux linkage. Magnetic cores may still be used to increase the low-frequency isolation between the input and output ports. RF transformers often have a usable frequency range greater than equivalent discrete (lumped) or distributed circuits. However, just like *R-L-C* elements, transformers also have parasitics that limit the performance over frequency and must be accounted for in the circuit design and analysis.

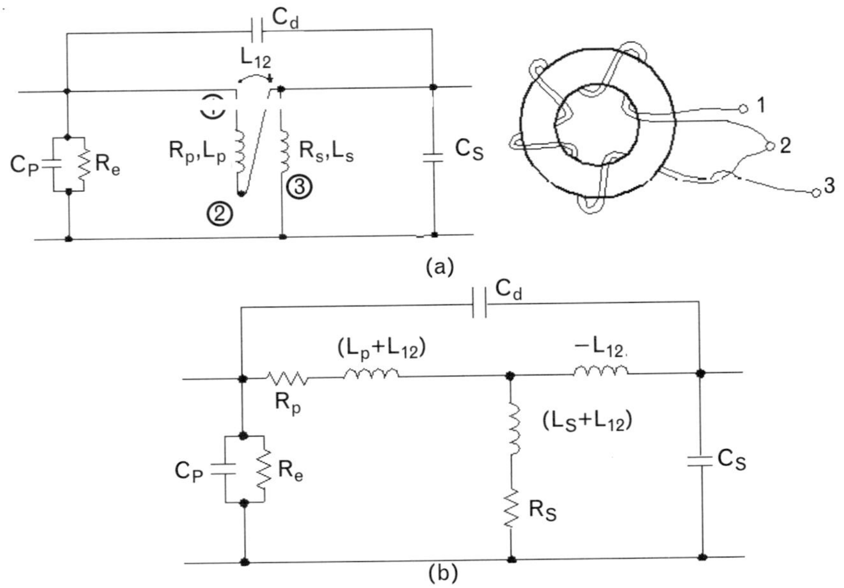

FIGURE 7.42
Electrical equivalent circuits and physical construction of a toroid RF transformer with bifilar winding. (a) Circuit with ideal transformer element; and (b) modified circuit replacing the ideal transformer element with an equivalent T-model.

Twisted-wire, or tightly held wires have been used successfully at RF to approximate twin-lead type transmission line behavior [47–49]. An equivalent transformer circuit for bifilar construction (Figure 7.42) uses mutually coupled windings with C_p, R_e, C_d, and C_S that represent inter-winding capacitances and core losses. The initial circuit may be replaced by a more detailed one shown in Figure 7.42(b), separating the inductances and the frequency-dependent conductive losses. Measured data verified the validity of this lumped model up to 600 MHz [50].

As we approach and pass 1 GHz, the twisted-wire models need to be replaced with transmission lines. Of course, transmission lines are also usable at lower frequencies, but there twisted wires offer more flexibility and easier construction. Once transmission lines are introduced, the upper limit is determined by the core material and it becomes quite difficult to extend the operation beyond 2 GHz. Coupled lines and transmission line baluns (balanced-to-unbalanced transformers) can provide some (not all) functions of transformers at those higher frequencies, but they cannot duplicate the wide bandwidth coverage of transformers.

7.10.2 Baluns

Perhaps the most frequent application of a conventional balun is to transform between balanced, or differential, and unbalanced, or single-ended operations. Two common forms of baluns that do not provide impedance transformation are shown in Figure 7.43(a).

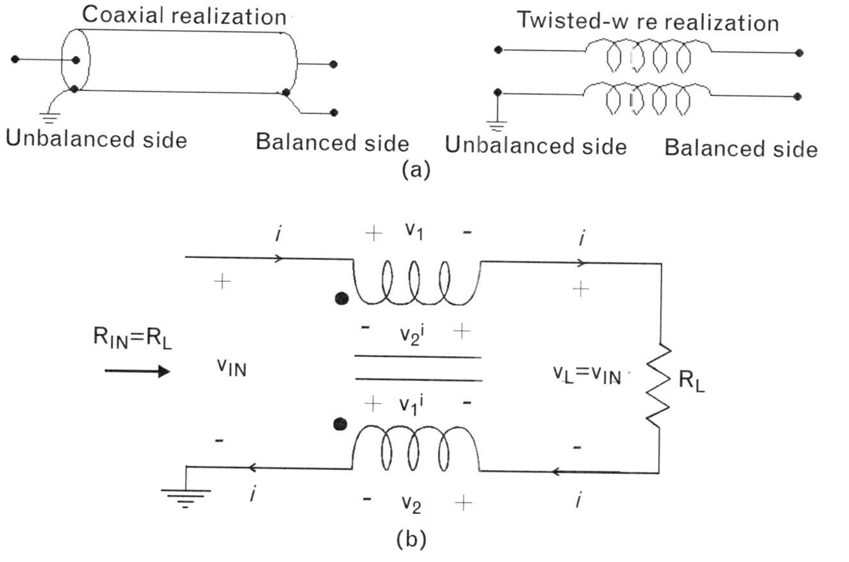

FIGURE 7.43 *(a) Two forms of baluns. Both forms are generally wrapped on a magnetic core material. (b) The current and voltage relationship shows that the unbalanced input impedance is the same as the balanced load. If we make a ground connection to the center of the load resistor, and separate the two $R_L/2$ resistors, the transformer becomes a 3-dB power divider, having 180° phase difference between the outputs.*

Since it is easier to visualize the balun's operation with the lumped model, let us use the voltages and currents of Figure 7.43(b) for our explanation.

At the unbalanced side an applied voltage v_{IN} drives a current I through both windings, creating a voltage v_1 across the top and v_2 across the lower winding, with the polarities marked. Two other voltages, v_1' and v_2', are induced by transformer action. The two windings are identical, as they are in twisted-wire form, therefore $v_1 = v_2$, and $v_1' = v_2'$. Since across each winding we have two voltages with equal magnitudes and opposite phases, they cancel each other. The effective voltage drop on the balun is zero and

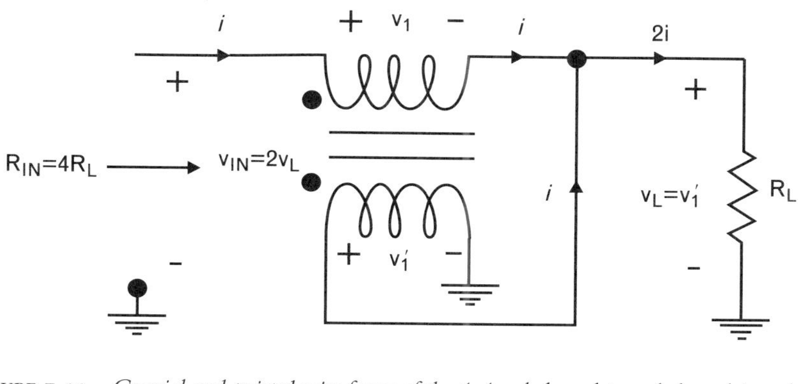

FIGURE 7.44 *Coaxial and twisted-wire forms of the 1:4 unbalanced-to-unbalanced impedance transformer. Improved accuracy may be obtained by modeling the input/output jump-wire connection as another transmission line. Core losses may be represented with a frequency-dependent lossy parallel coil at the input.*

the voltage across the floating (balanced) load is equal to v_{IN}. Input impedance at the unbalanced side is equal to the load impedance.

$$R_{IN} = \frac{v_{IN}}{I} = \frac{v_{OUT}}{I} = R_L \qquad (7.21)$$

This arrangement does not provide any impedance transformation but it changes from single-ended to balanced operation. If impedance transformation is required, additional windings are needed. For example, paralleling the inputs and series connecting the outputs of two of the 1:1 baluns of Figure 7.43 gives us a 1:4 impedance ratio between an unbalanced input and balanced output. Or, if the balun action is not required, by changing the topology of the single transformer (Figure 7.44) we get an unbalanced-to-unbalanced (*unun*) structure with a 1:4 impedance transformation.

Impedance transformation is achieved the following way. The input voltage v_{IN} drives a current i through the upper winding, creating a voltage v_1 across it. The current continues through the load, R_L. By transformer action, an equal voltage, $v_1' = v_1'$ is induced across the lower winding, which forces an equal amount of current, i, through the load. Therefore, the current through the load is twice of the input current.

The voltage drop across the load is $v_L = v_1'$, due to the direct connection between the load and the lower winding. Since $v_1' = v_1$, the voltage at the input is twice the load voltage. Comparing the input and output sides, it is seen that at the input we have twice the output voltage and half of the output current. That amounts to the input resistance being four times higher than the load resistance.

A balun has no dc isolation between the source and load. At the low end of the passband frequencies the self-inductance of the windings must be large enough to isolate the two terminations from each other. A rule of thumb is that the reactance created by the winding of the transmission line, at the lowest frequency of operation, should be three to five times greater than the larger termination.

In Chapter 4 we saw that any two resistive terminations may be matched at a given frequency with a quarter-wavelength transmission line, provided the characteristic impedance of the line is equal to the geometric mean of the two terminations. Accordingly, a 4:1 impedance transformer should be wound with a transmission line (or equivalent twisted wire) that has a characteristic impedance of

$$Z_{TL} = \sqrt{R_S R_L} = 2R_L = \frac{R_S}{2}$$

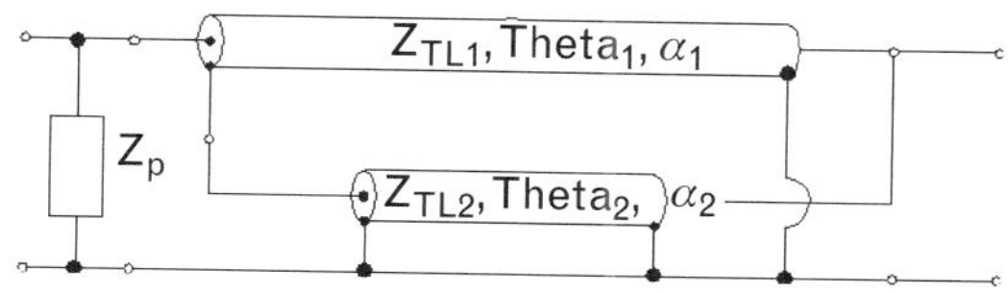

FIGURE 7.45 *Transmission line transformer model for twisted-wire realization. Losses are represented by the transmission line loss factor and the frequency-dependent parallel impedance at the input. Since the equivalent low-frequency inductance of the transmission line is directly proportional to its number of turns, the two parameters must be correlated.*

If we use such a transformer to match an unbalanced 50-Ω source to an unbalanced 12.5-Ω load, the transmission line impedance of the transformer should be 25Ω. If we want to create such a transmission line with twisted or closely held wires (or ribbons), detailed design information is available in [6, 49].

A physical model that provides an accurate broadband description of transmission line transformers [50] is shown in Figure 7.45. A four-terminal lossy coaxial line TL_1 provides input-output coupling and a high-impedance line TL_2 models the connection between the two wires. A lossy inductor, labeled Z_p, to ground is needed to limit the transformer's low-frequency response. The coaxial line losses (α_1 and α_2) represent the conductive and dielectric losses, while the magnetic core losses are incorporated into Z_p.

7.11 Crystal resonators and models

Quartz crystal resonators are indispensable to modern electronics [51]. Used for frequency generation, filtering and control, quartz crystals find applications in instrumentation, communications, computation and navigation. Navigation systems rely upon quartz crystals for precision timing. Microprocessors, computers and wrist watches use quartz crystals for synchronization and time keeping. At first glance, these useful devices seem simple, consisting of a slab of quartz with electrodes attached to each side, but their modeling touches upon almost every area of classical physics including mechanics, acoustics, wave motion, piezoelectricity, and electronic circuit theory. Crystal resonators offer precision difficult to describe in everyday language. Long-term frequency stability, referred to as *aging*, can be as small as a few parts in 10^{10} per day. A change in the earth's circumference to the same extent would amount to only 0.5 inch.

Piezoelectricity is the key property that makes quartz useful in electronics. Some crystalline materials are piezoelectric. Mechanical force generates electric charge. Electrical potential causes mechanical displacement.

A quartz plate exhibits mechanical resonances at frequencies determined by its shape, size, and mass. Alternating voltage applied to the surface of a quartz crystal slab causes vibrations. This vibration is quite small except when the electrical excitation frequency coincides with a mechanical resonance. The exchange of energy between electrical and mechanical states is extremely efficient at resonance. Very little energy dissipates in the crystal. Resonance is very sharply defined as quartz crystal resonators have quality factors (Q's) ranging from 10,000 to over 1 million. In contrast, the Q of conventional *L-C* resonant circuits is limited by the inductor to no more than 200 or 300 hundred. High quality factor is absolutely essential to precision frequency control.

7.11.1 Crystal orientation

The physical properties of crystals, unlike those of common amorphous solids, vary with the direction of measurement. Materials having direction dependant properties are anisotropic. Electrical conductivity is strikingly anisotropic in quartz, varying by a factor of 100 with direction. Because most of the physical properties of quartz affecting resonant frequency are anisotropic, the orientation of resonator blanks is critical. Although several coordinate systems are in use, the orthogonal or rectangular coordinate system is most commonly used to describe crystal piezoelectric properties.

Natural quartz crystals have the familiar shape shown in Figure 7.46. An ideal quartz crystal is a hexagonal prism with six smaller cap faces at the end. The z-axis of the orthogonal coordinate system runs along the length of a natural quartz crystal. The y-axis is perpendicular to the prism face, and the x-axis is a line through the hexagon points. A plate of quartz cut with its major surface aligned with the y-axis is called an x-cut plate. Rotating the cut 90° about the z-axis gives a y-cut plate.

Orientation of the cut, with respect to the crystal axis, determines not only the value of the physical properties of the blank, but also the temperature coefficient. An x-cut quartz plate has temperature coefficient of resonant frequency about -20×10^{-6}part/°C, which is too high for most RF

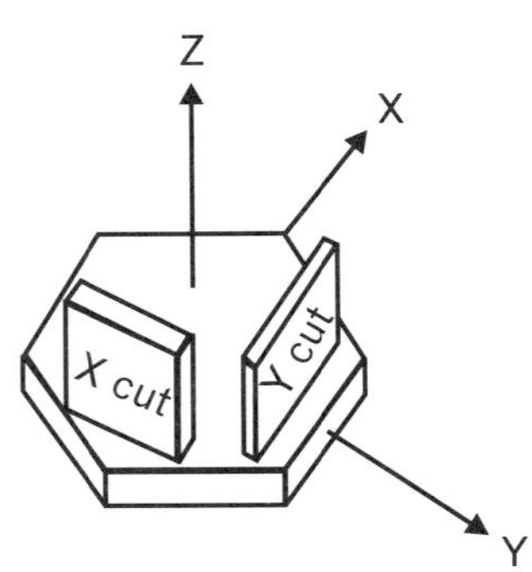

FIGURE 7.46
Quartz crystal resonator orientation. Y-cuts are perpendicular to the y-axis while x-cuts are perpendicular to the x-axis.

applications. Temperature coefficient of y-cut plates is even higher, about $+100 \times 10^6$ part/°C. Notice that a 90° rotation about the z-axis changes the frequency-temperature coefficient from negative to positive and somewhere between x-cuts and y-cuts, the plate is not temperature sensitive.

Rotating a y-cut plate 90° about the x-axis also changes the frequency-temperature coefficient sign. The resulting z-cut plate has a coefficient of –73 ppm/°C. Since a 90° rotation changes the sign again we can look for the optimum angle to minimize temperature coefficient. Actually, there are two zero coefficient angles, one clockwise and the other counterclockwise. These orientations are the AT-cuts and BT-cuts with rotation angles of approximately +35° and –49°, respectively. In practice these angles are very critical and are precisely defined with Bragg X-ray diffraction. AT-cut crystals have near-zero temperature coefficients at 25°C, and excellent frequency stability can be achieved without compensation for modest temperature excursions about room temperature.

7.11.2 Doubly rotated cuts

The majority of crystals manufactured are AT-cuts; however, doubly rotated cuts, especially IT and SC cuts, are popular in moderate and high precision applications. Singly rotated cuts are formed by aligning the saw blade with the crystal x-z plane (y-cut) then rotating the blade about the x-axis to the desired angle, θ. Preceding the x-axis rotation with a rotation about the z-axis through an angle ϕ, as illustrated in Figure 7.47, produces a doubly rotated cut. A major difficulty in the manufacture of doubly rotated resonators is the extreme accuracy required of both rotations.

7.11.3 Crystal resonator equivalent circuit

Due to the piezoelectric effect, when a voltage is applied an acoustic (shear) wave traverses the crystal between the electrodes. The amplitude of the shear wave—the crystal response—is greatest at those frequencies that excite a shear wave whose wavelength corresponds to an integral fraction

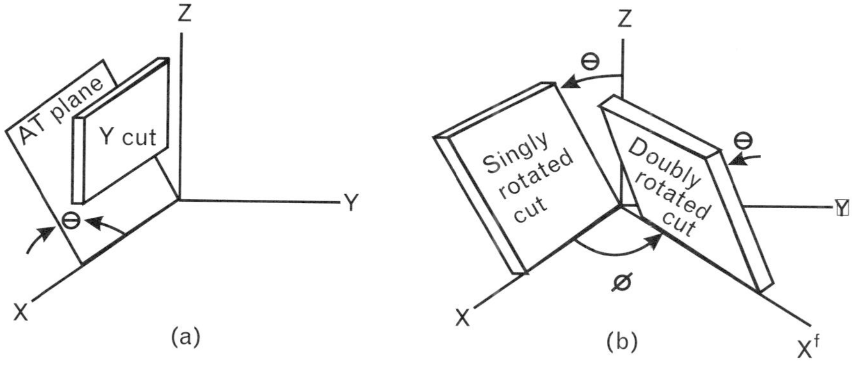

FIGURE 7.47
(a) Singly and (b) doubly rotated crystal cuts. Proper rotation leads to zero tempeture coefficient at one temperature and minimum change through a narrow temperature range.

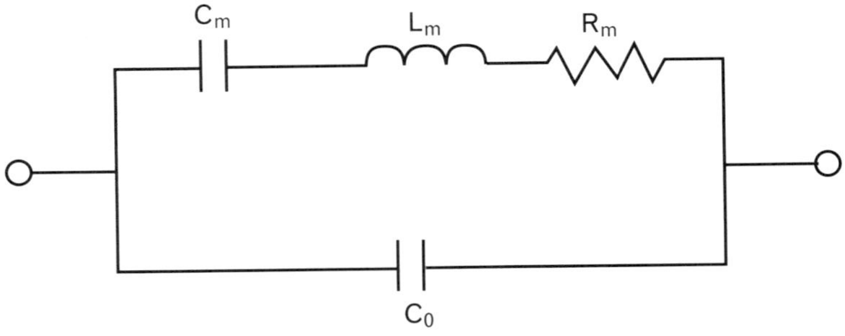

FIGURE 7.48 *Crystal resonator equivalent circuit model. Elements C_m and L_m determine the series self-resonant frequency*

of the physical distance through the quartz. The resonant frequency is inversely proportional to the crystal thickness. The first resonance occurs when the distance between the electrodes is one-half of an acoustic wavelength. Odd multiples, known as *overtones of the fundamental frequency*, also exist that satisfy the same boundary conditions as the fundamental.

The piezoelectric effect provides such a convenient and effective link to the electrical world, because a crystal resonator is actually a mechanical system. It has complementary energy storage elements, overtones, spurious responses and losses just like the pendulum it has largely replaced. In the equivalent circuit for a crystal resonator shown in Figure 7.48 the series *R-L-C* circuit represents mechanical resonance. Resistor R_m denotes the energy dissipated in the crystal and in its mounting arrangement.

Capacitance C_m comes from the elasticity of the quartz and inductance L_m is determined by the quartz mass. The last two parameters are sometimes referred to as the *motional inductance and the motional capacitance*, emphasizing their mechanical heritage. Only C_0, the electrode capacitance, is physically present as an electrical property. The electrodes, one on each side of the quartz blank, make up the plates of a capacitor with a quartz dielectric.

Resonator design sets the value of the electrical equivalent circuit parameters. Electrode capacitance, C_0, varies with electrode area. The blank thickness along with the electrode mass loading sets the resonant

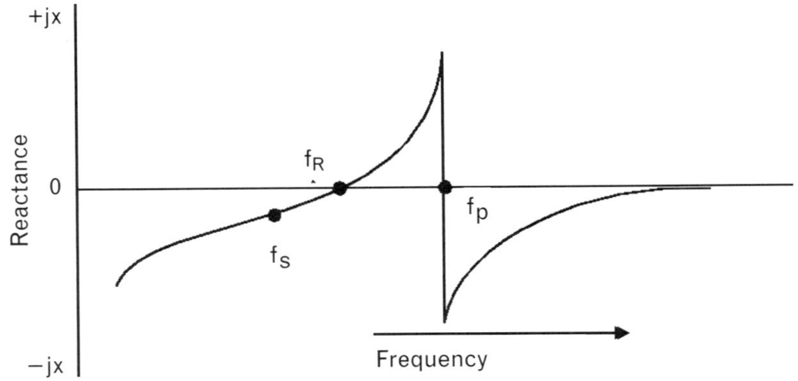

FIGURE 7.49 *Frequency response of the crystal equivalent circuit, showing both series and parallel resonances.*

frequency and determines the product of L_m and C_m. Contouring increases L_m while decreasing C_m. Blank and electrode diameter both affect resistance, with larger diameters giving lower resistance. Friction between air molecules and the vibrating blank surface increase losses significantly and lower resonant frequency. For this reason, and to achieve low aging, all but the least expensive resonators are protectively sealed.

Figure 7.49 gives a reactance plot of the crystal equivalent circuit, displaying series and parallel resonances at frequency f_s and f_p, respectively. At series resonance of the motivational arm, the crystal impedance is not purely resistive but has a capacitive component due to C_0. Above series resonance, the motional arm behaves with high-Q and it resonates with C_0 at f_P.

There are two frequencies where the crystal looks purely resistive: at f_R, that is, slightly above f_S, and at f_P. At those frequencies the phase of the impedance is zero.

The parameters of a resonator equivalent circuit are determined with a crystal impedance bridge, a vector voltmeter or a network analyzer—the latter offering many advantages in terms of accuracy, ease of use, and automation. More commonly the crystal operates somewhere between series and parallel resonance in the region where the crystal reactance is inductive. The following formulas give the resonant frequencies:

Series and parallel resonance is computed from the crystal's parameters,

$$f_S = \frac{1}{2\pi\sqrt{L_m C_m}} \tag{7.22}$$

$$f_P = f_S\left(1 + \frac{1}{2}\frac{C_m}{C_0}\right) \tag{7.23}$$

Resonator Q is determined by

$$Q = \frac{X_m}{R_m} = \frac{2\pi f_S L_m}{R_m} = \frac{1}{2\pi f_S C_m R_m} \tag{7.24}$$

Additional series R-L-C circuits in parallel to the fundamental arm model the resonance at higher overtones such as $3f_S$ or $5f_S$. Because of their high Q, they can be modeled independently and do not interact with each other.

Of the two resonances, the one due to the series arm is desired, and the second is a parallel (anti)resonance at f_P due to the interaction of the crystal inductance with its parasitic capacitance and is undesired. However, even for the desired resonant mode, the effect of the shunt capacitor will be to shift the frequency at which the net crystal reactance is zero to a frequency f_R

slightly higher than f_S. For most oscillator applications, the crystal is operated where the crystal looks inductive, at an even higher frequency, often nominally called the *load frequency*, f_L.

Due of the large stored energy in the acoustic wave at resonance, the Q of the crystal is very high, usually above 10,000. For instance, depending on the orientation of the way the crystal is cut, the Q at 100 MHz of a fundamental resonator would be between 20,000 to 50,000, or between 80,000 to 100,000 for a fifth overtone oscillator. The Q of a series *R-L-C* circuit is $\omega_S L_m / R_m = 1/\omega_S C_m R_m$, so this implies very high values of inductance (typically in millihenries) and low values of capacitance (typically in femtofarads). The small capacitance leads to a large reactance change per unit frequency near the resonant frequency

$$\left.\frac{\Delta X}{\Delta f / f}\right|_{f_S} = 2QR_m = 2\omega_S L_m = \frac{2}{\omega_S C_m} = \frac{1}{\pi f_S C_m} \tag{7.25}$$

In normal operation at the load frequency f_L, the crystal operates into a total load capacitance consisting of the shunt parasitic capacitor C_0 and an external load capacitor C_L that, for instance, a Colpitts oscillator (see Volume II, Chapter 6) imposes. The resonant frequency is essentially determined by the crystal's motional inductance in parallel with this load capacitance. From the above, we can derive the frequency at which the net reactance of the total circuit will be zero and the circuit will oscillate. This is just f_L, where

$$\frac{f_L - f_S}{f_S} = \frac{\Delta f_{LS}}{f_S} = \frac{C_m}{2(C_0 + C_L)} \tag{7.26}$$

As mentioned above, the crystal can be modeled at a higher overtone by a separate *R-L-C* motional arm set to resonate at that frequency. If N denotes the Nth overtone, then we may model that resonance with

$$C_N = \frac{C_m}{N^2} \tag{7.27}$$

for modes where the resonant frequency is determined by the thickness of crystal between the plates. It follows therefore that since the resonant frequency increases by a factor of N,

$$L_N = L_m \tag{7.28}$$

and since the motional *R-C* time constant is (approximately) independent of frequency,

TABLE 7.5 CRYSTALS TO OPERATE WITH LOAD CAPACITORS

CRYSTAL NUMBER	TYPE OF CUT	f (MHz)	C_0 (pF)	C_L (pF)	R_L Ω	R_m Ω	r	C_m (fF)	Q 1,000	M
18	AT	1	7	32	575	388	250	28	14.7	59
27	AT	5	7	32	60	40	250	28	28.4	114
29	DT	0.2	11	32	6,000	3,324	400	27.5	8.7	22
33	AT	25	12	32	17	9	2,500	4.8	147	59
36	AT	10	7	32	24	16	250	28	35.5	142
37	50X	0.25	2.5	20	5,500	4,345	120	20.8	7	59
38	NT	0.016	7.6	20	110,000	57,860	900	8.4	20.4	23
42	50X	0.25	2.5	32	5,000	4,305	120	20.8	7.1	59
57	CT	0.5	7	32	3,000	2,022	350	20	7.9	23
62	AT	20	7	32	20	13	250	28	21.9	88
68	AT	3	7	32	40	27	250	28	70	281
71	AT	5	4	32	150	119	35,000	0.11	2,340	67

Source: [52].

$$R_N = R_m N^2 \tag{7.29}$$

A variety of different crystals are summarized in Table 7.5. These crystals are specified to operate with a nominated value of load capacitor C_L, so that their frequency may be exactly specified. For frequencies above about 30 MHz, the crystal will normally be calibrated at an overtone rather than the fundamental, since above this frequency the crystal becomes too thin to manufacture to reliably support the fundamental mode. Newer crystals with an inverted mesa structure are one exception to this principle.

The equivalent series resistance, R_L, when operating with a load capacitor C_L, always exceeds the resistance of the motional arm, although not by much. Near f_L, it is close to R_m and given approximately by

$$R_L = R_m \left(1 + \frac{C_0}{C_L}\right)^2 \tag{7.30}$$

The capacitance ratio r is the ratio of the static capacitance C_0 to the motional capacitance C_m, given by C_0/C_m From (7.30), it is related to the relative frequency shift from series resonance due to the static capacitance. The value of r is usually well over one hundred because the motional capacitance will be small for a high Q crystal. As a result, the antiresonance frequency is usually very close to the series resonance frequency, since its frequency is given from (7.23) by

$$f_P \approx f_S\left(1+\frac{1}{2r}\right) \tag{7.31}$$

M is the crystal figure of merit and is defined as Q/r and is a measure of the resonant-antiresonant frequency separation. If $M = 2$, then these two frequencies coincide. As M decreases below 2, the crystal can be unusable for an oscillator since it has no inductive region above the frequency of the series resonance.

For resonators with a large figure of merit, the resonant frequency f_R is given approximately by

$$f_R \approx f_S\left(1+\frac{1}{2QM}\right) \tag{7.32}$$

Clearly, the larger the M, the more useful the crystal resonator.

Table 7.6 lists crystals designed to operate near a resonant frequency. The exact oscillation frequency might be tuned by means of a varactor

Table 7.6 Crystals to Operate Without Load Capacitors

Crystal Number	Type of Cut	f (MHz)	C_o (pF)	R_m Ω	r	C_m (fF)	Q1,000	M
19	AT	0.8	7	520	300	23.3	16.4	55
26	DT	0.3	9	4,000	400	22.5	5.9	15
45	DT	0.455	5	3,300	385	13	8.2	21
50	NT	0.1	4	60,000	900	4.4	6	7
65	AT	20	7	40	2,150	3.3	60.3	28
67	AT	30	7	40	2,200	3.2	41.4	19
74	AT	125	4.5	25	5,800	0.78	65.3	11
79	AT	10	7	30	200	35	15.2	76
80	AT	95	7	60	5,700	1.2	23.3	4
81	AT	65	7	40	2,100	3.3	18.5	9
82	AT	80	7	50	6,000	1.2	33.2	6
84	AT	60	7	40	2,300	3	22.1	10
157	AT	5	7	37	180	38.9	22.1	123

Source: [52].

capacitance. The principle difference from Table 7.5 is that for these crystals, no load capacitor is specified.

With such crystals, the frequency variation of the resonant frequency, in parts per million, is shown for the stated temperature variation.

In passing, we note that ceramic resonators can be molded to induce a piezoelectric effect, and used as low-cost, low-frequency IF filters up to several megahertz. For filters of this type, the equivalent circuit is the same as for a crystal, with the addition of a large shunt parasitic resistor to model the losses in the ceramic. Such a shunt resistor models the dc conduction path between the electrodes that can result from imperfections in the ceramic material itself. As a result, ceramic resonators have a much lower Q than quartz crystals.

7.11.4 Applications

Crystal resonators have been in use for more than 70 years. Millions are made each year for frequency generation, filtering, and timing. Frequencies span from a few kilohertz to several hundred megahertz. The majority control clock signals in digital devices ranging from quartz wristwatches to supercomputers. The least expensive resonators cost just a few cents each in large volume. Communications and navigation equipment make up the next largest use. Finally the excellent precision of the best resonators makes them valuable in electronic instrumentation and specialized sensors. Crystal characterization is difficult since the high-Q resonances are not easy to locate with low-impedance (50Ω) test equipment. Special techniques and software are available [53] for measurements.

Most crystal resonators end up in oscillators. A resonator cannot oscillate without an external active device to replace the energy dissipated by losses in the crystal. The active device provides amplification. Connecting a resonator between the amplifier output and input creates an oscillator.

7.12 Surface acoustic wave resonators

Acoustic losses that increase as the square of frequency limit the ultimate frequency of quartz crystal resonators. However it is the fragility of thin quartz plates, not excessive losses that restricts the frequency of conventional *bulk acoustic wave* (BAW) resonators. If another parameter instead of thickness determined resonant frequency, quartz could serve as a resonator material well into the lower microwave region. Surface acoustic waves enable the manufacture of high frequency quartz resonators and filters on thick substrates [54].

Anyone who has thrown a rock into a pond and watched the ripples expanding outward has witnessed surface waves. These waves carry energy outward from the initiating disturbance. Their velocity depends upon the density of the medium, and the strength of the restoring force, in this case gravity. A leaf floating on the pond surface follows an elliptical path in the vertical plane as the wave passes. Submerged objects are not much affected since surface waves are confined to a surface layer about one wavelength deep.

In the 1960s simple ways were found to generate and detect surface acoustic waves in quartz. Researchers used metallic *interdigital transducers* (IDTs) looking like interlaced combs, deposited on the surface of quartz substrates. Adjacent comb fingers, driven electrically out of phase, generate an interelectrode electric field which, because of the piezoelectric effect, deforms the surface of the resonator. Elastic restoring forces initiate two surface acoustic waves traveling perpendicular to the electrodes fingers. Electrode fingers spaced by one-half acoustic wavelength reinforce the initial disturbance. Destructive interference may occur at other excitation frequencies.

Photolithography, highly advanced by the semiconductor industry, makes possible very fine electrode spacing enabling the manufacture of ultra high frequency resonators on mechanically rugged quartz substrates. IDTs can also detect surface acoustic waves arriving from either direction. IDTs are both bidirectional and reciprocal.

Bidirectionality is a disadvantage in as much as the energy propagating in the wrong direction is wasted. Unidirectional transducers, with interlaced electrodes driven as an acoustic phased array, generate or receive surface acoustic waves in one direction. These electrodes find use in filters where the 6-dB power loss of bidirectional transducers must be avoided. Resonant gratings at the outside ends of the SAW substrate, redirect the wayward wave in another form of unidirectional transducer.

A simple SAW resonator consists of an excitation electrode, followed by a second electrode, placed along the direction of propagation. Together they form a narrow bandwidth delay line. A key advantage of SAW devices is that the propagating wave can be detected, deflected, reinforced, absorbed or otherwise manipulated anywhere along its path. Clever manipulation forms the basis for a wide array of useful RF devices including resonators, delay lines, filters, and other signal processing devices.

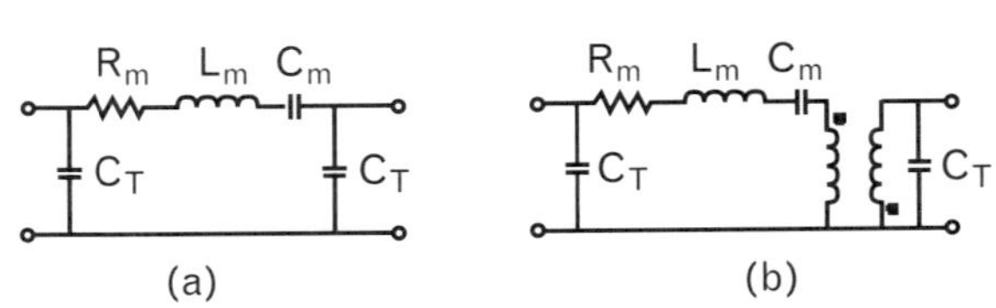

FIGURE 7.50 *Equivalent circuits for SAW resonator with (a) 0° and (b) 180° phase shift at resonance.*

Figure 7.50(a) shows a simplified equivalent circuit for a SAW two-port resonator operating near resonance. The phase shift at resonance is normally 0°; however, by reversing one pair of IDT electrode connections, the phase shift at resonance can be made 180°. Figure 7.50(b) shows the equivalent circuit for a 180° two-port resonator.

Connecting the two electrodes together gives a one-port SAW resonator. Driving the electrodes in series gives a 0°, one-port resonator, driving them in parallel gives 180° of phase shift at resonance. Two-port resonators are more common and easier to use because the transducer capacitances go to ground and can be absorbed by external input and output matching networks, instead of shunting the motional arm elements.

Performance of a differentially configured SAW filter was shown in Section 4.18.2.

7.13 Dielectric resonators

Quartz is by far the most widely used resonator material but other materials find use too. The key physical parameters that guide resonator material selection are acoustic velocity, the piezoelectric coefficient, the electromagnetic coupling coefficient, the relative permittivity, the frequency temperature coefficient and, of course, the cost. Acoustic velocity defines device geometry. Higher velocity supports higher resonant frequencies for a given thickness. Strong coupling coefficients increase the separation between series and parallel resonance, an important feature needed for moderate to wideband filters. As in quartz, these physical parameters vary with orientation and are affected by mechanical stress.

Dielectric resonators are used extensively at frequencies above 1 GHz, where their small size and tunability make them the resonator of choice [55]. With Q's of several thousand, they are able to maintain low oscillator phase noise and lock the oscillator frequency far more effectively than microstrip circuits alone, whose Q's are at most several hundred, and even less in integrated circuits.

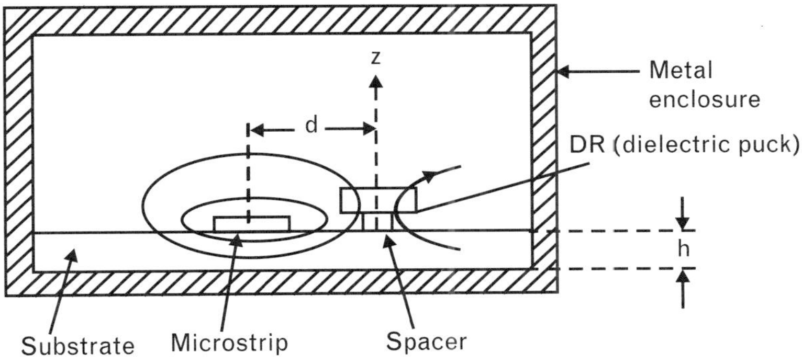

FIGURE 7.51
The coupling between a microstrip line and a dielectric resonator is controlled by their physical separation.

The resonator is a cylindrical "puck" made of high dielectric constant ceramic material. The ceramic, being low loss, ensures high Q, and its high dielectric constant ensures the size is as small as possible. Barium titanate and its variants are the most common material used, and the dielectric constant ranges from the mid-thirties to nearly 100. Dopants can be introduced to change the variation of the dielectric constant over temperature so that the resulting change in resonant frequency can compensate for opposing device effects with temperature.

The diameter of the puck determines its resonant frequency. The most common resonant mode is the $TE_{01\delta}$ mode, in which the magnetic field that encircles the microstrip line couples to the resonator along its vertical (z) axis, and excites electric field lines in the puck. There is no electric field in the direction of the z-axis (along the central axis of the cylinder). The height of the puck in the direction of the z-axis is approximately equal to its radius, and its diameter is slightly larger than one wavelength at the resonance frequency (measured in the dielectric material).

The placement of the dielectric puck relative to the microstrip, shown in Figure 7.51, determines the coupling of the resonator to the circuit. The puck is usually mounted slightly elevated from the plane of the microstrip by means of a spacer drilled through the microstrip, to prevent the resonator Q being degraded by the losses in the substrate material. In microstrip, this entails removal of a small amount of substrate material and mounting a low-loss spacer material (frequently cut from a small tube of glass) in its place. The puck is then glued on top of the spacer. The resonator is usually modeled in a metallic enclosure to establish definitive boundary conditions for the electromagnetic fields. The roof of the enclosure impacts the resonant frequency, increasingly so as it approaches the top of the puck. This is sometimes used to mechanically tune the resonator frequency using a metal plate directly above the puck. The resonant frequency increases several percent as the plate comes closer to the top of the puck.

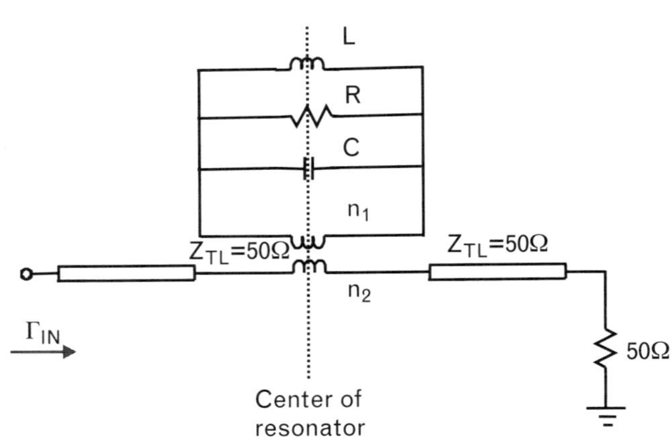

FIGURE 7.52
The equivalent circuit model of the dielectric resonator.

The equivalent circuit of the resonator is shown in Figure 7.52. It consists of a parallel *R-L-C* circuit inserted in series with the transmission line at a reference plane corresponding to the center of the puck. The *R-L-C* resonator is inserted in series with the line through a transformer whose turns ratio models the coupling. If there is no coupling, the resonator has no impact on the line so its reflection coefficient is simply that of 50Ω, that is equal to 0. As the puck is moved closer to the microstrip, its coupling to the microstrip magnetic field increases, and the resonance becomes apparent as the reflection coefficient swings around a circle of constant conductance hinged at 50Ω. At high couplings, the change in reactance around the circle is quite high for relatively small frequency changes, indicating the high Q of the circuit.

As the coupling is increased, at resonance an increasingly higher resistance (representing the loss of the puck) is added in series with the 50-Ω load. Slightly off resonance, additional shunt inductance or shunt capacitance is also added in parallel with this resistance. Well away from resonance, the parallel *R-L-C* circuit simply appears as a short circuit and adding it in series has no effect on the 50-Ω load impedance. It is apparent that used in this way the dielectric resonator has a bandstop characteristic at its resonant frequency, and the input impedance is equivalent to

$$Z = Z_{TL}\left(1 + \frac{k}{1 + 2jQ_0 \dfrac{\Delta f}{f_R}}\right)$$

where k is proportional to the degree of coupling to the microstrip, Δf is the measurement offset frequency from the resonant frequency f_R, and Q_0 is the Q of the resonator.

When used in an oscillator, the dielectric resonator can be used either as a bandstop filter to provide a resonant termination, or as a bandpass filter at the resonant frequency. As a bandpass filter, a second microstrip line on the opposite side of the puck is coupled through the magnetic field of the first microstrip, most strongly at the resonant frequency. The electrical model for the bandpass configuration adds a second transformer to the same shunt *R-L-C* circuit, and couples it to a second line microstrip line. Off resonance, the *R-L-C* circuit is virtually a short circuit and any coupling between the two lines is therefore shorted. At resonance, the high equivalent resistance of the resonator loads the two coupling transformers and power transfer can occur between the two lines. This is an effective method of providing feedback between the collector and base or drain and gate in a transistor oscillator.

Some caution is needed to model the locus of the load impedance with frequency correctly. At the reference plane corresponding to the center of the puck, the resonator appears as a parallel *R-L-C* circuit in series with the line. Approximately 90° further away along the transmission line, the circuit is rotated around the Smith chart and appears around resonance as a series *R-L-C* circuit. The resistance at resonance changes from a high value to low value, so it is very tempting to use this transformation to obtain a good value of load resistance to start the oscillator and keep it running. However, as described in Volume II, Chapter 6, this simple transformation can ruin many an oscillator. Many dielectric resonator oscillators are poorly made, with consequent mode-hopping and frequency hysteresis with temperature or tuning, because of the incorrect loading this transformation causes.

7.14 Component measurements and modeling

R-L-C components can be modeled based on one-port or two-port scattering parameter measurements. For inductors and capacitors, measuring the relatively small equivalent series loss resistances may be quite difficult in the one-port measurement format. Two-port measurements, on the other hand, can provide the necessary loss information with much higher resolution, although sometimes system impedance may need to be transformed appropriately so that the 50-Ω terminations of the network analyzer do not reduce measurement sensitivity.

In most cases the components to be measured cannot be attached directly to the calibrated network analyzer, but rather must be placed within test fixtures. As a result, the measured data includes the parameters

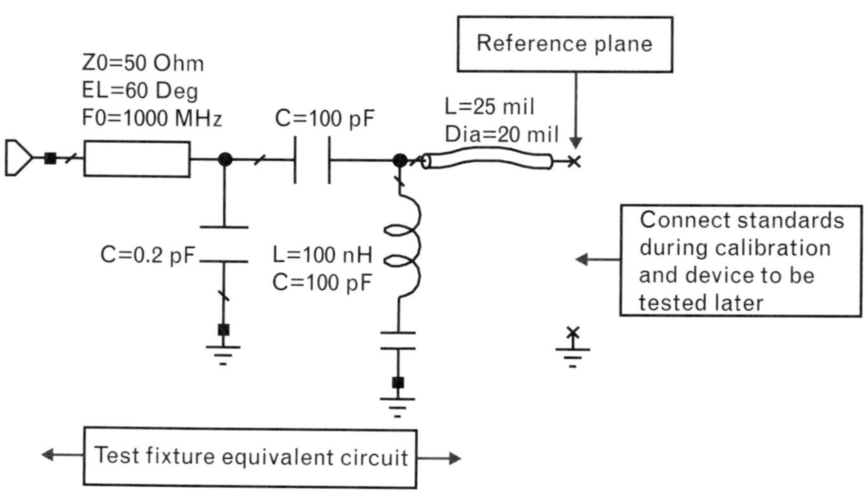

FIGURE 7.53 *Calibrating the network analyzer and a six-element test fixture together moves the reference plane to the point where the device under test is connected (marked with arrow). This procedure requires calibration standards with the same type of connectors as the components to be connected to the reference plane.*

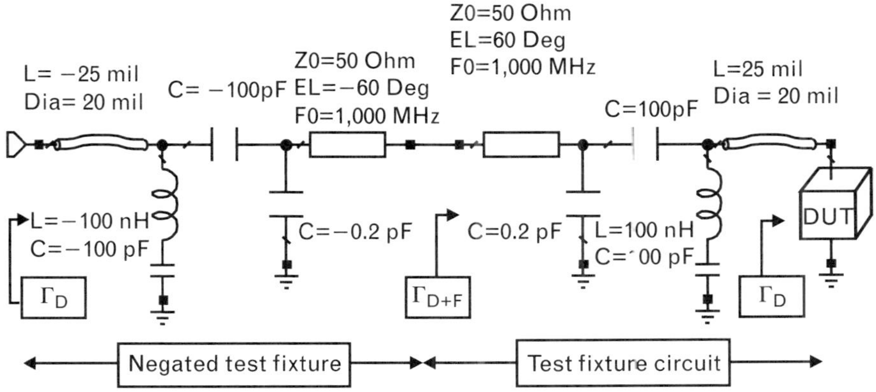

FIGURE 7.54 *The two-port test fixture is de-embedded from the measured data by cascading the fixtures' negated components to the measured data. The negated elements must be connected in a reverse order of how they appear in the test fixture. The test fixture transforms the reflection coefficient of the measured device from* Γ_D *to* Γ_{D+F}. *The two-port containing the parameters of the negated fixture cancels the effect of the fixture and enables us to display* $_D$ *at the input of the cascaded circuit.*

of both the test fixture and the device under test. By careful characterization, the test fixture parameters (or equivalent circuit) can be established and used to de-embed the device parameters from the measured data by one of the three following methods, depending on the characterization of the test fixture and the availability of calibration standards:

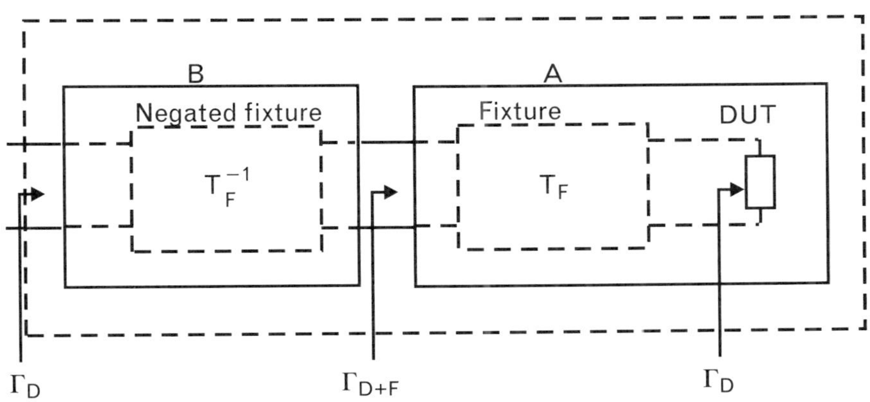

FIGURE 7.55 *One-port "A" includes the test fixture and device under test and we measure the one-port data,* Γ_{D+F}, *at its input port. When we cascade a dummy two-port "B" containing the matrix inverse of the test fixture's T-parameters to "A," we eliminate the effect of the test fixture.*

- *Standard one-port terminations (open, short, and load) are available for the side of the test fixture where the components will be connected.* Performing the regular one-port calibration with the three standards extends the measurement's reference plane to the component connection (Figure 7.53). This is the simplest of the three techniques, but high quality standards frequently do not exist, particularly for "home-made" test fixtures.

- *The circuitry of the test fixture is known.* Cascade the "negated" circuit of the fixture to the reference port of the measurement (Figure 7.54). The negated circuitry eliminates the effect of the fixture for all frequencies. For example, cascading a mathematical model of a –100-pF capacitor to a 100-pF component, results in perfect cancellation—the resultant *S*-matrix of the cascade becomes a unit matrix. Cascading the mathematical model of a transmission line of 50-Ω characteristic impedance, –60° electrical length at 1 GHz, and gain factor (remember, this is a mathematical element only) of 0.2 dB/inch to a real line of the same impedance, 60° length and 0.2 dB/inch loss, results again is a unit matrix.

- *The S-parameters of the fixture are known.* Measure the input reflection coefficient of the test fixture with the unknown device connected to the output port to obtain Γ_{D+F} (circuit "A" of Figure 7.55). Measure the two-port *S*-parameters of the test fixture and convert them to *T*-parameters by using (4.30). Compute the inverse of the *T*-matrix, T^{-1}. Create a dummy two-port "B" with T^{-1}, and cascade it to the reference port of the measurement. Cascading the two circuit blocks, matrix multiplies T^{-1} and T, canceling the effect of the test fixture. The computed input reflection coefficient of the "A-B" cascade is then equal to the unknown, Γ_D. This approach works for all frequencies where the *T*-parameters exist.

7.15 Summary

With the greater availability of computer simulation software, it is practical to predict the behavior of *R-L-C* circuits accurately, minimizing costly and time-consuming prototype cycles. However, circuit simulators can predict the outcome of the circuit with an accuracy limited by the quality of the component models provided to them. The old saying about computers, "garbage in, garbage out" applies well here. Only proper use of carefully designed test fixtures, accurate calibration, and thoughtful data extraction can provide models whose accuracy will hold up throughout the RF range. As EM simulators become more user-friendly and are integrated with circuit and system simulators, their use in practical design will increase. Of course even EM simulation can provide misleading results, particularly when the problem is not formulated properly.

Due to integration, our circuits are gradually becoming smaller. Incorporating all passive components into RFIC chips reduces cost and increases reliability. At the time of writing this book, the inherent low Qs of passive components pose severe limitations to what may be integrated. *Microelectromechanical* (MEM) processing [55–58] offers new directions to improve the quality of passive components. When this promising technology matures, we will need to add new sections to our book.

References

[1] Kurokawa, K., *An Introduction to the Theory of Microwave Circuits*, New York: Academic Press, 1969.

[2] ITT, *Reference Data for Radio Engineers*, 5th ed., Indianapolis, IN: Howard Sams & Co., 1970.

[3] Grover, F. *Inductance Calculations*, New York: Dover Publications, 1946.

[4] Ludwig, R., and P. Betchko, *RF Circuit Design*, Upper Saddle River, NJ: Prentice Hall, 2000.

[5] Maas, S. A., *The RF and Microwave Circuit Design Cookbook*, Norwood, MA: Artech House, 1998.

[6] Wadell, B. C., *Transmission Line Design Handbook*, Norwood, MA: Artech House, 1991.

[7] Liao, S. Y., *Microwave Circuit Analysis and Amplifier Design*, Englewood Cliffs, NJ: Prentice Hall, 1987.

[8] Klappenberger, A., "Designing Accurate Small Inductors for Microwave L-C Filters," *RF Design*, November 1994.

[9] Mezak, J., "Modeling Helical Air Coils for Wireless and RF Applications," *RF Design*, January 1998.

[10] Agilent Technologies, *Test and Measurement Catalog*, Palo Alto, CA, 2003.

[11] Helszajn, J., *Principles of Microwave Ferrite Engineering*, London, England: Wiley Interscience, 1969.

[12] Hayt, W. H., *Engineering Electromagnetics*, New York: McGraw-Hill, 1981.

[13] Dekker, A. J., *Electrical Engineering Materials*, Englewood Cliffs, NJ: Prentice Hall, 1959.

[14] Williams, T., *EMC for Product Designers*, Oxford: Butterworth-Heinemann, 1996.

[15] Paul, C., *Introduction to Electromagnetic Compatibility*, New York: Wiley, 1992.

[16] Ott, H. W., *Noise Reduction Techniques in Electronic Systems*, 2nd ed., New York: Wiley, 1988.

[17] Perna, V. F., *The RF Capacitor Handbook*, Huntington Station, NY: American Technical Corporation, 1994.

[18] Gupta, K. C., R. Garg, and I. J. Bahl, *Microstrip Lines and Slotlines*, Dedham, MA: Artech House, 1979.

[19] Fiore, R., "Circuit Designer's Notebook," http://www.atceramics.com., 2001.

[20] Fiore, R., "Capacitors in Broadband Applications," *Applied Microwave & Wireless*, May 2001.

[21] Goldfarb, M. E., and R. A. Pucel, "Modeling Via Hole Grounds in Microstrip," *IEEE Microwave and Guide Wave Letters*, June 1991.

[22] Swanson, D. G., "Grounding Microstrip Lines with Via Holes," *IEEE Trans. on Microwave Theory and Techniques*, August 1992.

[23] Getsinger, W. J., "Microwave Transmission Lines," *UCLA Short Course Notes*, Los Angeles, CA, 1982.

[24] Grieg, D., and H. Engelmann, "Microstrip—A New Transmission Technique for the Kilomegacycle Range," *Proc. IRE*, December 1952.

[25] Getsinger, W. J., "Coupled Rectangular Bars Between Parallel Plates," *IRE Trans. on MTT*, January 1962.

[26] Matthaei, G. L., L. Young, and E. M. T. Jones, *Microwave Filters, Impedance Matching Networks, and Coupling Structures*, New York: McGraw-Hill, 1964.

[27] Howe, H., Jr., *Stripline Circuit Design*, Dedham, MA: Artech House, 1974.

[28] Fusco, V. F., *Microwave Circuits—Analysis and Computer Aided Design*, New York: Prentice Hall, 1987.

[29] Sinnema, W., *Electronic Transmission Technology*, 2nd ed., Englewood Cliffs, NJ: Prentice Hall, 1988.

[30] Bahl, I., and P. Bhartia, *Microwave Solid State Circuit Design*, New York: Wiley, 1988.

[31] Moreno, T., *Microwave Transmission Design Data*, Norwood, MA: Artech House, 1989.

[32] Maloratsky, L. G., "Reviewing the Basics of Microstrip Lines," *Microwaves & RF*, March 2000.

[33] TXLINE software, http://www.mwoffice.com.

[34] Gupta, K. C., R. Garg, and R. Chadha, *Computer-Aided Design of Microwave Circuits*. Dedham, MA: Artech House, 1981.

[35] Wen, C. P., "Coplanar Waveguide: A Surface Strip Transmission Line," *IEEE Trans. on Microwave Theory and Techniques*, December 1969.

[36] Getsinger, W. J., "Circuit Duals on Planar Transmission Media," *IEEE MTT International Symposium Digest*, Boston, MA, May 1983.

[37] Cohn, S. B., "Slotline on a Dielectric Substrate," *IEEE Trans. on Microwave Theory and Techniques*, October 1969.

[38] Smith, J. I., "The Even- and Odd-Mode Capacitance Parameters for Coupled Lines in Suspended Substrate," *IEEE Trans. on Microwave Theory and Techniques*, May 1971.

[39] deRonde, F. C., "A New Class of Microstrip Directional Couplers," *IEEE MTT International Symposium Digest*, Newport Beach, CA, May 1970.

[40] Laverghetta, T. S., *Microwave Materials and Fabrication Techniques*, Norwood, MA: Artech House, 1991.

[41] King, J. A., *Materials Handbook for Hybrid Microelectronics*, Norwood, MA: Artech House, 1988.

[42] Kuszaj, M., and A. Aguayo, "A Substrate for All Seasons," *Printed Circuit Design*, March 2000.

[43] Guiles, C. L., "Everything You Ever Wanted to Know About Laminates," *Arlon*, 2000, http://www.arlonmed.com.

[44] Santhanam, N., "High Performance Economical Substrates: Myth or Reality?" Taconic Advanced Dielectric Division, Application Note, 2002.

[45] Losee, F., *RF Systems, Components, and Circuits Handbook*, Norwood, MA: Artech House, 1997.

[46] Abrie, P. L. D., *Design of RF and Microwave Amplifiers and Oscillators*, Norwood, MA: Artech House, 2000.

[47] Guanella, G., "Novel Matching Systems for High Frequencies," *Brown-Boveri Review*, September 1944.

[48] Ruthroff, C. L., "Some Broadband Transformers," *Proc. IRE*, December 1957.

[49] Sevick, J., *Transmission Line Transformers*, Atlanta, GA: Noble Publishing, 1996.

[50] Niclas, K. B., R. R. Pereira, and A. P. Chang, "Transmission Lines Accurately Model Autotransformers," *Microwaves & RF*, November 1992 and April 1993.

[51] Bottom, V. E., *Introduction to Quartz Crystal Unit Design*, New York: Van Nostrand, 1982.

[52] Parzen, B., and A. Ballato, *Design of Crystal and Other Harmonic Oscillators*, New York: Wiley-Interscience, 1983.

[53] Semones, T., "Characterizing Resonators with S-Parameter Network Analysis," *RF Design*, January 1989.

[54] Campbell, C. K., *Surface Accoustic Wave Devices*, New York Academic Press, 1998.

[55] Kajfez, D., and P. Guillon, *Dielectric Resonators*, Norwood, MA: Artech House, 1986.

[56] De Los Santos, H., *Introduction to Microelectromechanical (MEM) Microwave Systems*, Norwood, MA: Artech House, 1999.

[57] Richards, R. J., and H. De Los Santos, "MEMS for RF/Microwave Wireless Applications: The Next Wave," *Microwave Journal*, March and July 2001.

[58] Rebeiz, G. M., and J. B. Muldavin, "RF MEMS Switches and Switch Circuits," *IEEE Microwave Magazine*, December 2001.

CHAPTER 8

Filters and resonant circuits

8.1 Introduction

Filters play a very important role in communications systems, and their design is a highly specialized field within RF engineering. Their main function is to separate analog signals into specific, well-defined frequency bands. Circuit and system engineers learn quickly that creating a physically realizable filter, other than simple lowpass or highpass *L-C*-types, from filter tables or basic design equations, is not an easy task. Most companies have a dedicated small group, or at least a highly skilled specialist, to do challenging filter designs. Another solution is to outsource the design, or purchase the filter from a component supplier.

In my first job after college, working with active circuits, I, Les Besser, initially felt more important than my colleagues whose designs were focused on passive components. After all, my attitude was that anyone can get a set of element values from filter tables. When our project needed a 2-GHz bandpass filter, I dismissed my project leader's advice to have it done by the "passive circuit group," and volunteered to finish the design over the weekend. After a day of manual calculations, using the MYJ filter handbook [1], I felt that I was virtually through the design phase. My happiness, however, quickly evaporated after converting the electrical parameters of the transmission lines to physical dimensions, only to find out that they were totally impractical. After several unsuccessful transformations, I swallowed my pride and Monday morning quietly passed the design to the experts. Since that experience, I have a lot of respect for filter designers.

Filter design dates back nearly a century with the development of image parameter techniques. In the 1940s, exact synthesis techniques were developed that required extensive numerical computations. Since computers were not readily available, most of the designs were based on lowpass prototype tables generated by key researchers. From the 1960s the availability of computers provided a significant boost to modern filter design and led to the development of several major filter design programs [2], of which at least one is still commercially available [3].

Filters are generally designed to work with resistive terminations. Most RF system components present terminations with reactive parts, and if the filter is used in such an environment its performance will change. RF system designers need to be aware of these effects, and the solutions we described in the impedance matching section that utilized parasitic absorption may also apply here.

Since the stopband rejection is generally achieved by reflecting signals at unwanted frequencies, filters are mismatched outside of their passbands. (Exceptions exist in some cases, as we will see in the multiplexer section.) The mismatch may cause problems when components are cascaded in RF systems (see Section 4.15.1).

This chapter covers the various types of passive filters and also looks at some of the commercially available computer-aided design tools. We restrict our discussion to linear, time invariant two-port filters, containing inductors, capacitors, transmission line elements and various types of resonators.

A typical receiver system contains many filters, which serve a variety of functions. Figure 8.1 shows a generic receiver with filters for specific functions.

As outlined in Chapter 3, the RF preselector filter passes frequencies of interest and rejects others. Undesired frequencies can cause mixer spurious responses, blocking, and intermodulation. This filter also rejects the image frequency. In many cases, the bandwidth of the signal is a small fraction of the passband of the filter, so group delay is not critical. A Chebyshev or elliptic filter is typically used because it combines low passband loss with high isolation of undesired frequencies.

The image rejection filter, used after the LNA and before the mixer; rejects amplified noise at the image frequency, which is folded into the IF frequency bandwidth, increasing the system noise by 3 dB. This filter can be similar in type to the RF preselector filter, but needs to reject the image by 20 dB to 30 dB for noise figure considerations. The RF preselector and the image rejection filter together should attenuate the image by the dynamic range of the RF amplifier, typically 60 dB to 90 dB.

The IF filter passes only the frequency band of the channel selected. The signal occupies the majority of the filter's passband, and the

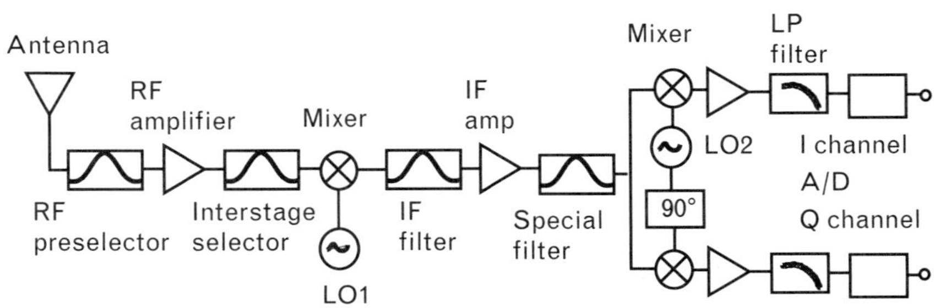

FIGURE 8.1
Part of a typical communication system receiver, showing the various types of filters used in different locations.

information must not be distorted. A filter with flat group delay and high stopband rejection is used. Accordingly, the filter elements must have very high Q to maintain the filter characteristics. Quartz or SAW filter types are generally used.

The lowpass filter at baseband provides the lowest bandwidth of the system, establishing the noise bandwidth. This filter may be of an analog or digital type.

Since the main theme of our books is RF/MW circuit design, our focus here is on filters used in the high-frequency portion of communication systems. These filters have many forms of physical realizations (Figure 8.2) and we can only cover a small part in one chapter. Our intent here is to provide a general introduction to filter design that will help you make the right decisions about *design-versus-buy* type of choices. If you choose the latter, this chapter will help you to set reasonable specifications for the purchase order.

8.2 Filter specifications

Definitions of some of the important specifications of various types of filters are as follows.

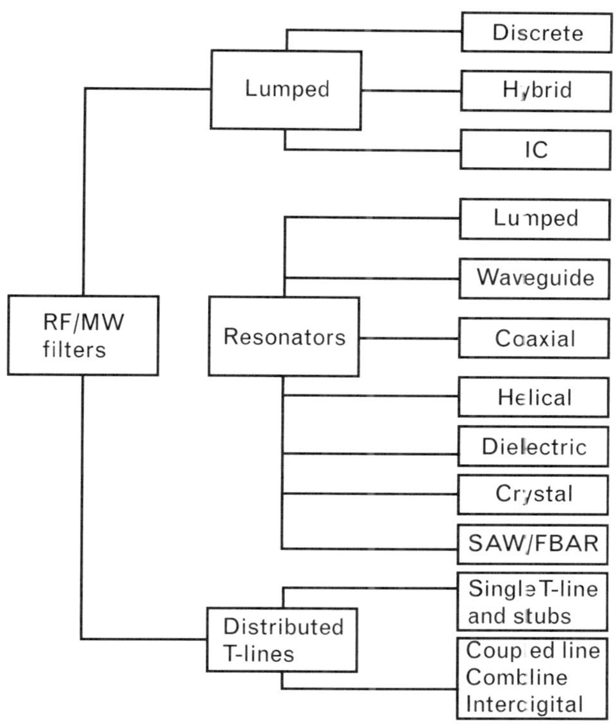

FIGURE 8.2 *Various types of physical realizations of RF/MW filters.*

Bandwidth

Many of the textbooks and filter tables specify bandwidth for lossless filters at 3-dB gain roll-off, relative to minimum attenuation. Although this is a commonly used value, it is often not very practical. We already mentioned that if the transmitted power is reduced by 3 dB, only half of the total available signal power is transmitted and the other half is reflected. When this happens the reflection coefficient magnitude is 0.707, which is equivalent to a 5.8:1 *VSWR*—not a very practical specification for an RF system block. More commonly used relative passband ripple and gain roll-off values are 0.1 dB to 0.2 dB or even less, resulting in very low input/output *VSWR* for the filter.

Transducer loss and insertion loss

Physical filters have losses that can easily amount to several decibels in the passband. For lowpass and highpass filters the element losses may change significantly throughout the passband, while the element losses of narrowband bandpass and bandstop filters are relatively uniform. There are two commonly used loss definitions, the *transducer loss* and the *insertion loss*. Although these two terms are frequently used interchangeably, the exact definition of insertion loss was given in Section 4.14. We restate both two-port expressions here for convenience. Transducer loss is the reciprocal of transducer gain,

$$L_T = \frac{1}{G_T} = \frac{\text{Power available from the source}}{\text{Power delivered to the load}}$$

$$= \frac{\left|(1 - s_{11}\Gamma_S)(1 - s_{22}\Gamma_L) - s_{12}s_{21}\Gamma_S\Gamma_L\right|^2}{\left(1 - |\Gamma_S|^2\right)|s_{21}|^2\left(1 - |\Gamma_L|^2\right)}$$

Insertion loss is given as

$$IL = \frac{\text{Power delivered to the load without the two-port}}{\text{Power delivered to the load with the two-port inserted}}$$

$$= \frac{\left|(1 - s_{11}\Gamma_S)(1 - s_{22}\Gamma_L) - s_{12}s_{21}\Gamma_S\Gamma_L\right|^2}{|s_{21}|^2\left(1 - |\Gamma_S\Gamma_L|^2\right)}$$

The total insertion loss or transducer loss in ideal filters is completely caused by reflection. In practical filters the total loss has an additional term caused by dissipative component losses. Commercial RF circuit simulators compute *transducer gain*, and the loss of a filter is treated as *negative* gain.

Passband ripple

Ripple is a measure of gain flatness, representing the difference between minimum and maximum losses in the passband. If the filter components are lossless, the passband ripple is directly related to the input and output reflection coefficients. Although component losses affect the ripple (or roll-off), particularly at the edge(s) of the passband, when the losses are moderate, they do not significantly change the reflection coefficients. In equal-ripple type filters, the order of the filter is always one greater than the total number of extremas[1] in the passband response.

Input and output match

In two-port filters the initial design value of the passband ripple, or band-edge roll-off, determines the quality of impedance match. In RF/MW filters, it is necessary to have at least 14- to 16-dB return loss at both ports. The impedance match of low-frequency filters in RF systems is less critical because it can be improved by adding matched attenuators. Table 8.1 provides a comparison of the different parameters used in specifying impedance match.

Stopband rejection

The stopband rejection in lowpass or highpass filters is the minimum acceptable attenuation at the beginning of the stopband. In bandpass and bandstop filters the rejection may be different at the low and the high end of the passband or reject band, respectively.

Shape factor

Most of the handbooks define the shape factor as the ratio between the bandwidths where the attenuation reaches certain levels, generally 3 dB and 60 dB.

Table 8.1 Conversion of Various Parameters Used for Port Impedance Specifications

Parameter Type	Parameter Values							
Ripple (dB)	0.01	0.05	0.1	0.15	0.2	0.25	0.5	1.0
Return loss	26.5	19.5	16.5	14.7	13.5	12.5	9.6	6.9
VSWR	1.1	1.24	1.35	1.45	1.54	1.62	1.98	2.65
$\|\Gamma\|$	0.05	0.11	0.15	0.18	0.22	0.24	0.33	0.45

1. Low and high values (valleys and peaks) of attenuation.

Terminations

Two-port filters are generally specified with resistive terminations at both ports, referred to as *doubly terminated* types. In special cases, one of the terminations may have an extreme value—either zero or infinity—represented by a voltage or current source, respectively. Such filters are called *singly terminated.*

Group delay

Group delay, sometimes called *envelope delay,* is related to the time a signal is delayed while passing through the filter. It is calculated by taking the derivative of the transmission phase with respect to frequency. Unequal passband group delay causes distortion in signals composed of multiple frequencies.

Phase delay

This term is usually used in lowpass delay lines and it is defined by the ratio of the phase angle in radians to the frequency in radians per second.

Group delay distortion

Filter selectivity is related to group delay distortion, as we will see in Sections 8.4 and 8.5. Compared to a maximally flat, or Butterworth filter response, an equal-ripple or Chebyshev filter response of the same order has higher selectivity, and consequently more delay distortion. Elliptical filters (equal-minima stopband) with their increased selectivity have even more delay distortion. Such comparisons however may be misleading and the group delays should be compared at equal amount of stopband rejection. For example, to achieve the specified amount of stopband attenuation provided by a fifth-order Chebyshev filter, we need an eighth-order Butterworth alternative. In the vicinity of the passband edge, the two filters' delay distortion may be very similar.

Filters can also be designed for minimum group delay distortion, but with the exception of cross-coupled filters [4] these types provide poor amplitude selectivity. Frequently the filter is first designed to meet the amplitude specifications and then one or more delay equalizer sections are added to compensate for the group delay distortion introduced by the filter. The delay equalizers are *all-pass* types of networks with resistive input and output terminations. However, when the equalizer is built with physical components, the losses may also affect the input/output impedances. When such a nonideal equalizer is cascaded with the nonideal filter, generally an overall optimization is needed to meet the specifications.

Intersymbol interference

When rectangular pulses pass through a band-limited channel, the pulses spread into time intervals of succeeding symbols causing *intersymbol interference* (ISI). This interference increases the probability of error during symbol detection. It is easier to shape the signal spectrum at baseband frequencies than at RF, and a *raised cosine* type filter is ideal for that task.

8.3 Various filter types

Functionally filters can be classified into one of four groups, with an addition fifth group of special filter types listed:

1. *Lowpass filters.* As its name implies, a lowpass filter passes low frequencies and gradually rejects signals above a specified corner frequency. Transmission zeros of lowpass filters are located at infinite frequency, and sometimes also at finite frequencies, but not at dc. Physical lowpass filters always have transmission zeros at finite frequencies, due to resonances caused by component parasitics, as we will see later. Our ears have "built-in" lowpass filters that cut off around 16 kHz (much sooner for senior citizens).

   In its simplest form a passive lowpass filter is realized as a ladder network with series inductors and shunt capacitors. At low frequencies the series reactance and the shunt susceptance of the components are insignificant, and all signal power available from the source is delivered into the load. As the frequencies increase the reactances and susceptances become more significant and the frequency response gradually rolls off. If the source and load resistances are equal, a filter built with ideal (lossless) elements has zero transmission loss at dc.

   When the terminations are not equal, their impedance difference create mismatch losses that apply may throughout the entire frequency range. Such applications are mainly used in the low-frequency portion of a system. It is possible however to match unequal terminations through a specified passband with a quasi-lowpass design that uses lowpass topology yet performs as a bandpass filter.

   Ideally, we want a frequency response that has zero loss until a specified corner frequency and infinite loss from there on. In filter design we approximate this response with various types of rational functions to provide a *gradual* attenuation above our corner frequency. The selectivity of the frequency response just above the corner frequency depends on the order of the filter, and the type of filter response we choose.
2. *Highpass filters.* If we replace the series inductors with series capacitors and the shunt capacitors with shunt inductors, the response of the filter changes from a lowpass to a highpass type. Using lossless components, the attenuation is zero at infinite frequency and

gradually increases below a specified corner frequency. Otherwise, everything we stated for the response of the lowpass filter type is reversed with mirror image symmetry about the corner frequency. Highpass filters do not have transmission zeros at infinite frequency. A waveguide is a good example of a highpass filter, since it does not pass low frequencies.

3. *Bandpass filters.* When our filter structure includes components with transmission zeros both at dc and infinity, our filter will pass a specified frequency range and reject both low and high frequencies below and above the bandpass range. The simplest form of a bandpass filter consists of cascaded series and parallel resonators, while more complex filters can be designed with inductively or capacitively coupled resonator sections. Bandpass filters are very important in communication systems and they are realized in a wide variety of forms.

4. *Bandstop filters.* If our goal is to block transmission through a specified group of frequencies we use a bandstop filter, which has a response that is just the opposite of the bandpass filter discussed above.

5. *Special filter types.* Two additional groups are worth mentioning here. One is used for equalization; the other is a multiport circuit that passes different frequency bands to different load terminations.

   - *All-pass filters.* Circuits that pass all frequencies with equal amplitudes but lead to a predictable phase shift are called *all-pass networks.* We mentioned earlier that the group delay of a component indicates how long it takes the signal to pass through the two-port. For accurate reproduction of the signals passing through the system, all components should have a flat group delay, meaning that through the passband the signals should take equal time to pass through the component. Filters are designed to introduce sudden changes in frequency response, which causes nonlinear transmission phase. Since the group delay is a derivative of the phase, a nonlinear phase response leads to different group delays at different frequencies. Components that are specially designed to compensate the group delay distortion of filters are called delay equalizers. These circuits do not affect the magnitude of signals of various frequencies, but they compensate the group delay distortion introduced by a filter network. Some of the filter synthesis programs also allow us to design group delay equalizers with the appropriate number of sections, to offset the delay distortion of a specified filter.

- *Multiplexers.* The simplest form of a multiplexer is the diplexer (sometimes called duplexer) filter that is made up by connecting together one end of two filters, such as a lowpass and a highpass, or perhaps two bandpass filters. These filters can be designed to present constant, or near-constant impedance at the port where they are connected, through the passbands of both filters. Connections between the filters are generally in parallel for *L-C* elements and in series for waveguides.

  If the 3-dB corner frequencies of the individual filters are the same, the filters are called *contiguous*, and if the frequencies are not the same they are referred to as *noncontiguous*. Special design techniques allow us to design these filters individually and then connect them together to achieve their final forms.

8.4 Low-frequency versus RF/MW filters

Based on their applications in RF and microwave systems, there are two major categories: baseband [5] and RF filters. Although the focus of this book is on RF-related techniques, we also include a brief review of filtering that takes place in the low-frequency section of a wireless communication system.

8.4.1 Baseband filters

Ideally, we would like filters to pass all frequencies within a given band and to reject all others. The ideal lowpass filter of Figure 8.3 strictly band-limits the signal to all frequencies below some cutoff frequency, $f_C/2$. The bandpass equivalent of this filter, when shifted to some center frequency, has the same bandwidth of $f_C/2$. The transfer characteristic for signals within the passband has a magnitude 1.0 (i.e., the output amplitude is equal to the input amplitude). The output impulse response consists of the classic "sinc" or $\sin(x)/x$ shape of Figure 8.3, with precursors before and trailers after the main impulse has passed through the filter. If we define a bit period as $T_b = 1/f_C$, then the output response to an input impulse will pass through zero at all unit intervals of the bit period except the instant that the impulse reaches the output (defined as $t = 0$). This implies that a succession of impulses can be detected by sampling the output at successive bit periods, since the response from all previous and future impulses will be zero, and only the impulse at the corresponding time instant at the input will contribute. There is no ISI since each pulse in the output stream can be sampled independently of the value of pulses around it.

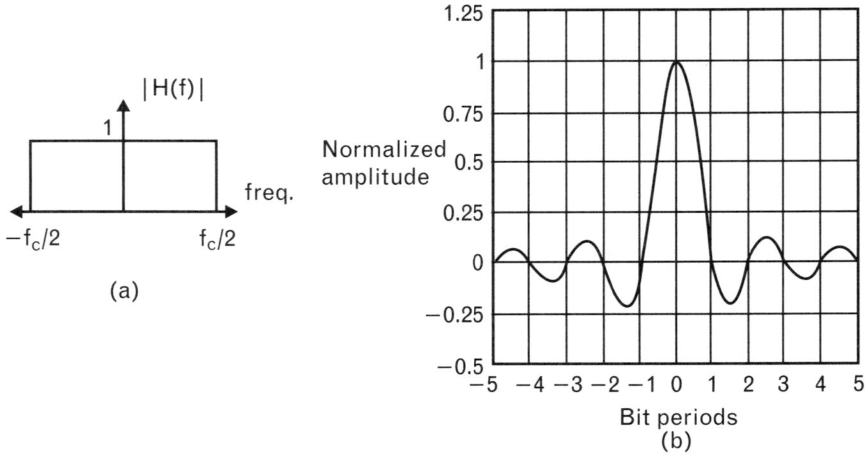

FIGURE 8.3
The (a) frequency and (b) impulse responses of the ideal brick-wall filter. (From: [5]. © 1996 Artech House, Inc. Reprinted with permission.)

8.4.1.1 Raised-cosine filters

Ideal responses such as this are noncausal, as they extend infinitely in time. They cannot be realized, only approximated. A class of filters known as *Nyquist filters* preserves the ISI-free property of the ideal filter. A realizable Nyquist response is shown in Figure 8.4. In such a filter, the frequency response tapers off slowly to zero outside the passband and follows a cosine roll-off in frequency given by

$$\begin{aligned} H(f) &= 1 \quad \text{for} \quad |f| \le \frac{1}{2T_b}(1-r) \\ H(f) &= \cos^2 \frac{\pi T_b}{2-r}\left(|f| - \frac{1-r}{2T_b}\right) \text{for } \frac{1}{2T_b}(1-r) < |f| \le \frac{1}{2T_b}(1+r) \end{aligned} \tag{8.1}$$

This response is sometimes referred to as a raised-cosine spectral shape. $T_b = 1/f_C$ is the bit period, and the roll-off factor r, where $0 \le r < 1$, defines the frequency during which the frequency response is tapered. As r increases from 0 to 1, the frequency response changes from a brick-wall filter with cutoff frequency $f_C/2$ to a squared-cosine roll-off, beginning at

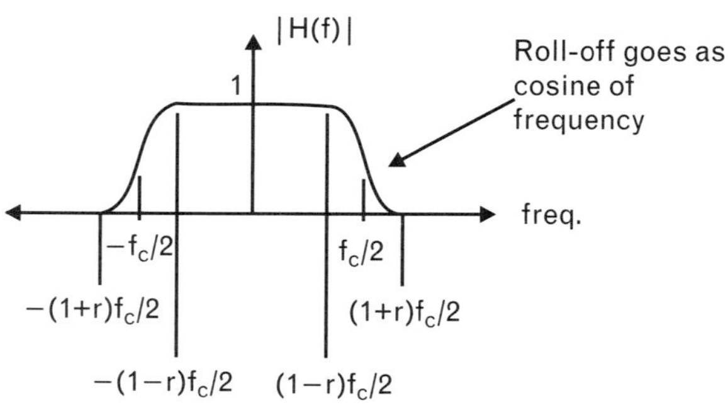

FIGURE 8.4
A realizable Nyquist filter response. (From: [5]. © 1996 Artech House, Inc. Reprinted with permission.)

dc and tapering off to zero at f_C. The total bandwidth occupied thus increases from $1/(2T_b)$ Hz to $1/T_b$ Hz (i.e., the spectral occupancy reduces from 2 symbols/sec/Hz to 1 symbol/sec/Hz as r is increased).

Figure 8.5 shows that the ISI-free property of the ideal filter is preserved in the Nyquist filter, as the impulse response passes through zero at all integral values of the sampling period T_b except the origin. As the bandwidth of the filter is increased by increasing r, the impulse response tapers off more rapidly with time. The classic sinc pulse shape is modified by a cosine term,

$$h(t) = \frac{\cos \pi rt}{T_b\left[1 - \left(\frac{2rt}{T_b}\right)^2\right]} \frac{\sin(\pi t/T_b)}{\pi t/T_b} \tag{8.2}$$

Although Nyquist pulses are perfectly band-limited, we now lose the constant envelope properties of the original modulation. We will see that this turns out to cause major problems in keeping the receiver and transmitter operating in their linear regions and will require careful attention to ensure that any nonlinear distortion is minimized.

The impulse response is one way of analyzing the effect a filter has on a single pulse. It shows the filter's response time and roll-off time, and it is useful to predict if ISI will result. However, an easier way to visualize the impact of filtering on a baseband waveform is through an eye diagram, reviewed in Chapter 2. For instance, in the case of a radio transmitter, the eye diagram would repetitively plot the baseband signal that enters the

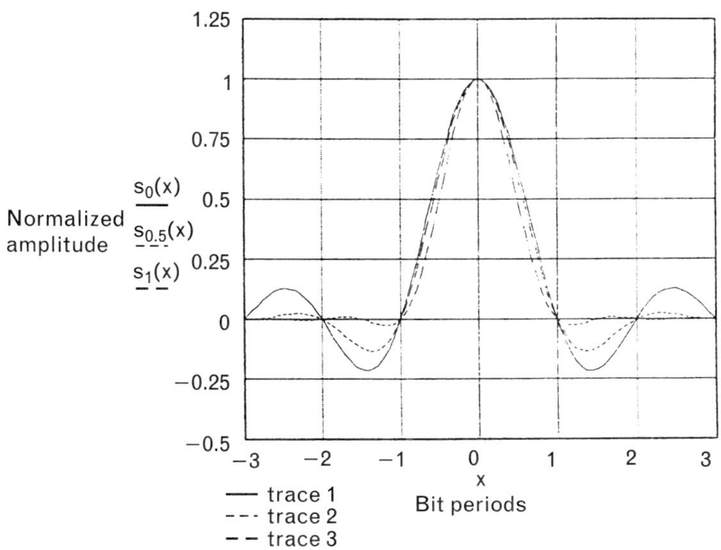

FIGURE 8.5 *The output impulse response from a Nyquist raised-cosine filter for r = 0, 0.5, and 1.0. (From: [5]. © 1996 Artech House, Inc. Reprinted with permission.)*

modulator to produce the IF signal. In an amplitude-modulated system, it will plot the multilevel sequence of symbols; with phase-modulation, it will plot the in-phase (I) and out-of-phase (Q) symbol streams separately.

Figure 8.6 shows an example of a raised-cosine Nyquist filter's eye diagram. There is no observable intersymbol interference, even though the bandwidth of the waveform has been restricted by passing the waveform, with its instantaneous transition between symbols, through a filter of limited bandwidth.

The reason for filtering the baseband signal becomes obvious when we look at the system of Figure 8.7, and the output IF spectrum resulting after the first upconversion. The two spectra shown correspond to a filtered and an unfiltered baseband input. The modulator is a biphase modulator that changes the phase of the 80-MHz LO signal by 0° if a "1" symbol is applied and 180° if a "–1" symbol is applied. In both cases, the IF spectrum is centered at 80 MHz and the first spectral null occurs at 10 MHz on either side of the center frequency. However, the baseband signal that has been filtered prior to the modulator is much more spectrally efficient. Its spectral sidebands are all 50 dB below the main lobe, instead of 20 dB as for the unfiltered signal. The channel spacing in the former case can be 20 MHz without appreciable power from one channel appearing in the adjacent channel. We need to maintain spectral efficiency and minimize adjacent channel interference, without introducing intersymbol interference and compromising correct detection of the signal, which is a great challenge of filter design. In modern radios, filters are optimized to achieve this for the system's digital modulation format.

Strictly speaking, raised-cosine filters are not realizable because of their preringing and postringing extending out infinitely in time. They can be

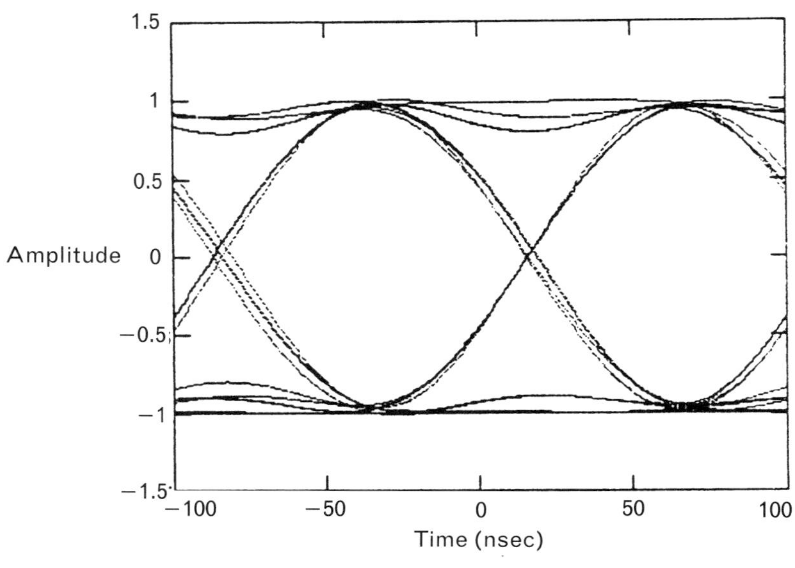

FIGURE 8.6
The eye diagram of random nonreturn to zero (NRZ) data after raised-cosine filtering. (From: [5]. © 1996 Artech House, Inc. Reprinted with permission.)

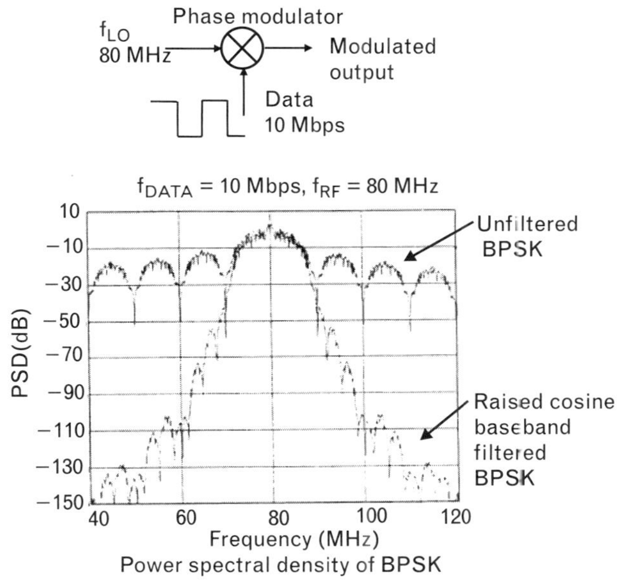

FIGURE 8.7 *A biphase modulator system, with f_{LO} = 80 MHz and data rate of 10 Mbps. (From: [5]. © 1996 Artech House, Inc. Reprinted with permission.)*

approximated, however, for instance with phase-equalized elliptic filters of high order, or digitally at baseband, as in Figure 8.8. Digital filters [6] implement the filtering function by combining time samples of the input waveform with appropriate weight and delay. Such filters are known as *FIR filters*—finite impulse response, or transversal filters [7]—their output is computed from the weighted sum of the present input and a finite number of past inputs. They can be designed with linear phase. *Infinite*

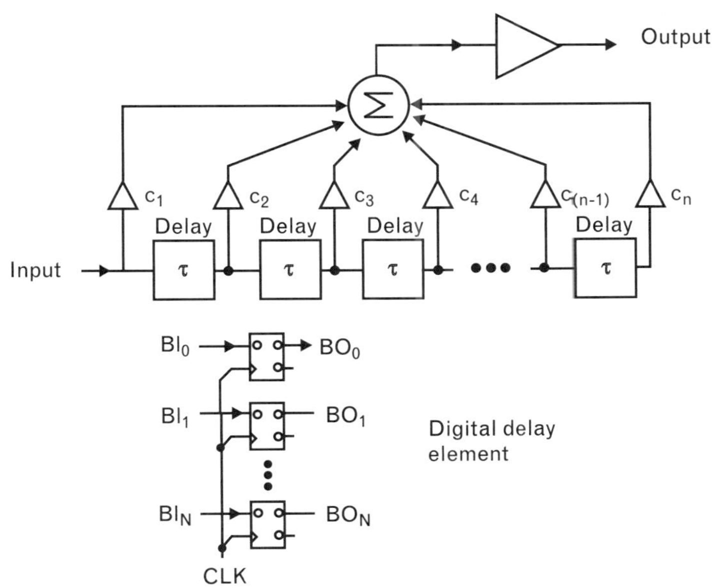

FIGURE 8.8 *A raised-cosine FIR filter. (From: [5]. © 1996 Artech House, Inc. Reprinted with permission.)*

impulse response (IIR) filters are similar, but their output terms can include unlimited number of past outputs and can thereby generate much sharper cutoff characteristics [7]. However, unlike FIR filters, an IIR filter cannot be made with linear phase response, so it is less suited to communications applications. Although FIR filters and their delays and multipliers are implemented digitally at baseband frequencies, RF filters based on surface acoustic wave technology, so-called SAW filters, are an analog equivalent. With SAW filters, an acoustic wave is launched on a quartz substrate, and the wave is detected by a series of interdigital fingers printed on the quartz. The delay is proportional to the physical distance between adjacent fingers, and the sampling coefficients are proportional to the length of each finger, since a longer finger will contribute a stronger component to the total.

As discussed above, the characteristic of a Nyquist filter is that it gives steep frequency roll-off with no ISI. The eye diagram of a 22-tap raised-cosine lowpass filter is shown in Figure 8.9. The input baseband waveform is square, since the transition between symbols is instantaneous; filtering preserves the clean nature of the transition, but by rounding the square edges and restricting the transition rise time, the resulting reduced bandwidth after filtering is evident. The eye diagram again has a "wide-open" eye, with previous or future transitions having no impact on the state of the pulse at the detection instant. With a cutoff frequency of $f_C/2 = 5$ MHz, the zero crossings of the impulse response occur every $T_b = 1/f_C$ or 100 ns. This corresponds exactly to the bit period of the data rate, which is 10 Mbps.

8.4.1.2 Gaussian filters

A Gaussian filter is another example of a Nyquist filter. It is a simple *L-C* filter with very poor frequency roll-off but a good phase response [8]. A Gaussian filter has a frequency response of the form

$$H(f) = \tau\sqrt{2\pi}\exp\left(-0.5(\tau f)^2\right) \tag{8.3}$$

Its step and impulse responses have no overshoot, and it has low rise time and group delay. However, its selectivity is worse than filters without these characteristics.

An example of the baseband response of the Gaussian filter, excited under similar conditions to the raised-cosine filter above, is shown in Figure 8.10. The Nyquist nature of the response is again evident in that the eye diagram is very clean, with almost no closure of the eye. This allows the detector to make a straightforward decision as to which symbol has been transmitted, and reduces the criticality of sampling at the precise sampling point. The *BT* product of bandwidth and bit period in this example is 0.5, corresponding to a 5-MHz filter and a 100-ns bit period. The Gaussian filter used in the common GSM cellular system has a *BT* product of 0.3.

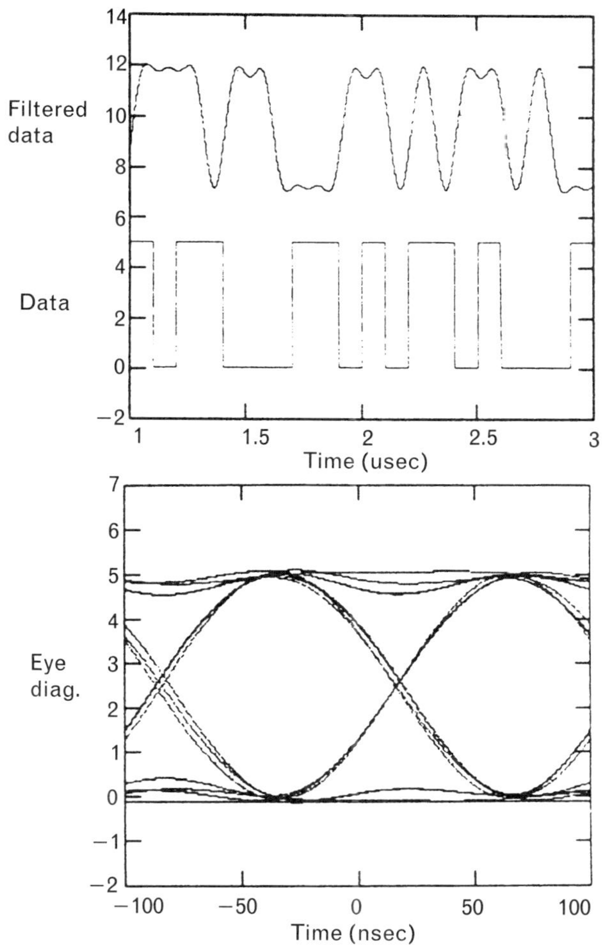

FIGURE 8.9 *Time-domain response and eye diagram of a 5-MHz raised-cosine lowpass filter. The input nonreturn to zero data rate is 10 Mbps. (From: [5]. © 1996 Artech House, Inc. Reprinted with permission.)*

The advantage of using a raised-cosine filter at baseband should be apparent, because not only does such a filter limit the spectral occupancy of the IF and RF channel, it also minimizes the impact of adjacent symbols on each other in the time domain (ISI).

8.4.2 RF filters

At RF and microwave frequencies, analog filters must pass the signal with minimum phase and amplitude distortion. Synthesis techniques for the design of filters are discussed in Chapter 6, but to understand the rejection of spurious responses and the selection of the desired IF channel in a receiver, we will briefly cover some of their characteristics in this section.

The baseband filter is the narrowest filter in the receiver chain and it sets the overall noise bandwidth. Because its bandwidth is as narrow as possible, its delay characteristics will be comparable to those of the modulation it is

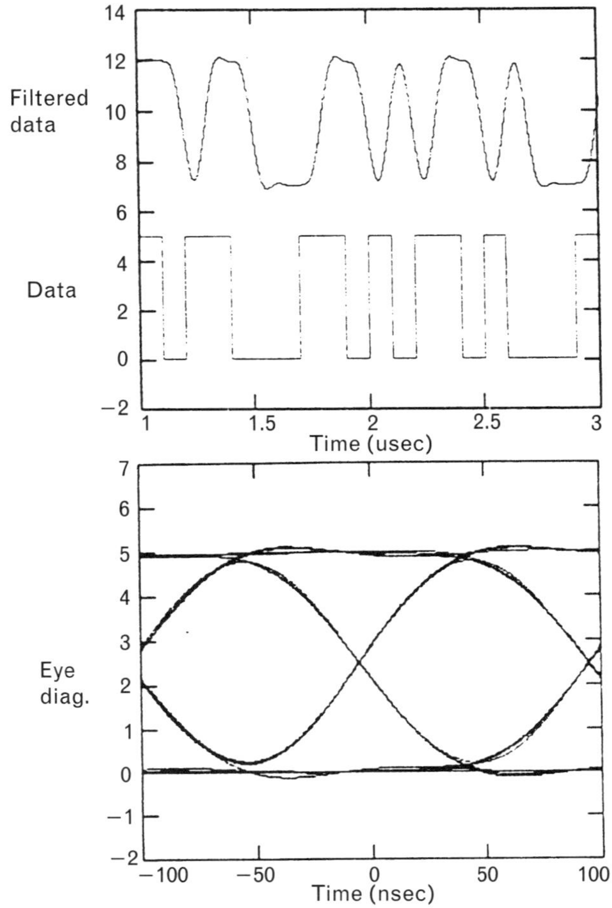

FIGURE 8.10
The time-domain response and eye diagram of a 5-MHz Gaussian filter. The input nonreturn to zero data rate is 10 Mbps. (From: [5]. © 1996 Artech House, Inc. Reprinted with permission.)

passing. For this reason, eye diagrams that convey time and amplitude information are important analytical tools to show the impact of the filter on the signal and its modulation envelope. However, at RF the bandwidth corresponds to the entire passband, which is likely to be much wider than the modulation content of a single signal. Therefore, the time delay of an RF filter is likely to be insignificant compared to the modulation time scale, and we work more with the overall amplitude and phase response in the frequen cydomain, or transfer characteristic, of the filter.

The transfer characteristic of a filter, the ratio of its output voltage to its input voltage, is usually expressed as a Laplace transform. This is essentially the same as its frequency response but with $s = j\omega$. The roots of the denominator of the transfer characteristic define the poles of the filter, and by adjusting the topology and the value of the components that make up the filter, the poles can be positioned in such a way as to produce a desired response. CAD tools are available to automate this process for a desired transfer response. We list here the most commonly used filter response types that are described by various analytical functions.

Before looking at the various types of filter responses, we want to point out that parasitic capacitances and inductances of lumped *L-C* components lead to strange frequency responses at RF. Since parasitics always increase the effective component values, a shift of the passband always takes place. Component self-resonances create finite transmission zeros and make a Butterworth or Chebyshev filter behave somewhat similarly to an elliptic filter (covered in Section 8.4.2.3). Above the self-resonant frequency, a lowpass filter begins to look like a highpass filter, because inductors act capacitively and capacitors are dominated by their self-inductances. Accurate component modeling therefore is vitally important in RF filter design.

8.4.2.1 Butterworth (maximally flat) filter

In a Butterworth filter the poles are positioned so that the maximum number of derivatives of the amplitude versus frequency response are zero at the center frequency of the filter. A typical attenuation response is shown in Figure 8.11, where the flatness of the response in the passband of the filter is evident. The order of the filter, n, is determined by the number of poles of the transfer characteristic (number of nonredundant transmission zeros); as it increases, the attenuation increases proportionally. The attenuation of a Butterworth lowpass filter at a frequency f is given by

$$A_{\mathrm{XdB}} = 10\log\left[1+\left(\frac{f}{f_C}\right)^{2n}\right] \tag{8.4}$$

where f_C is the 3-dB passband corner frequency.

From (8.4) we can find the necessary 3-dB corner frequency if a lowpass filter is to be designed for a specified passband edge roll-off, r, in dBs, at a frequency f_r. That is,

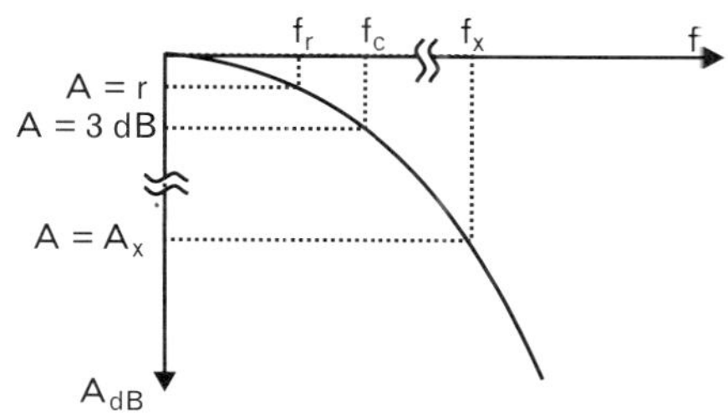

FIGURE 8.11 *Attenuation response of a lowpass Butterworth filter.*

$$f_C = f_r\left(10^{\frac{r}{10}} - 1\right)^{-\frac{1}{2n}} \tag{8.5}$$

The Butterworth filter has substantial ringing in its response to both an impulse and to a step function, as shown in Figure 8.12. The group delay will be compared with other filters later in Figure 8.20. Ideally, the group delay should be constant with frequency so that all modulation components go through the filter with the same delay. If it is variable, as it is in the Butterworth near the band edge, linear distortion will occur since modulation components of the signal extending to the band edge suffer increased delay relative to others.

Figure 8.13 shows the ISI of a baseband signal passing through a Butterworth lowpass filter. It is obvious that each symbol is affected by both prior and following symbols, and that the eye is much more closed than with the Gaussian filter of Figure 8.10. The detection decision becomes much harder even in the absence of noise, as the step response of the filter has substantial ringing that results in varying amplitude and timing levels for the same symbol. However, in spite of its poor delay and phase linearity, the Butterworth filter has the smoothest passband of the most common RF filters, low VSWR in the passband, and least sensitivity to component Q and changes of its element values.

8.4.2.2 The Chebyshev (equal-ripple) filter

Various English spellings are used for the name of the man who formulated the approximation (Tschebyscheff, Tsebysheff, and so forth), but the most interesting one I, Les Besser, have encountered is my wife's. After listening to my dictation she transcribed his name as "Chubby Chef." I hope the

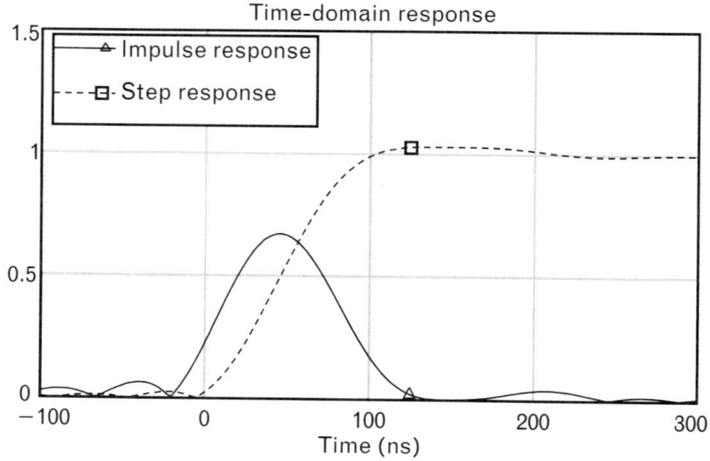

FIGURE 8.12
Impulse response and step response of a Butterworth filter.

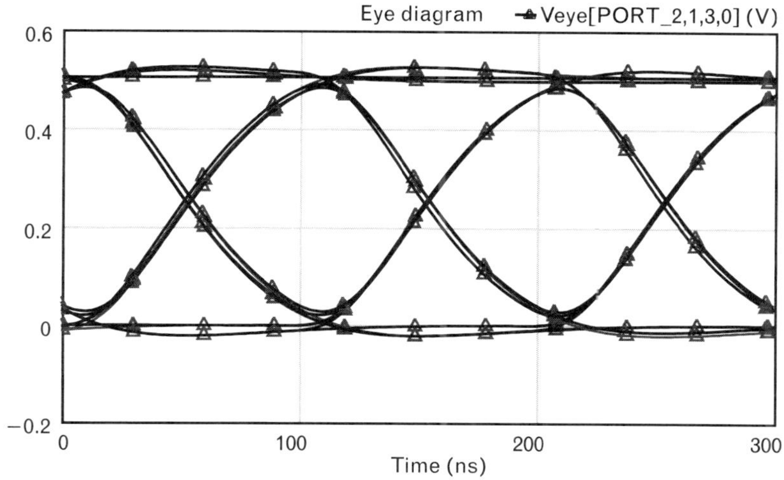

FIGURE 8.13 *The eye diagram of a 5-MHz Butterworth lowpass filter. The data rate is 10 Mbps. (From: [5]. © 1996 Artech House, Inc. Reprinted with permission.)*

spirit of the great Russian mathematician was as amused as I was when I read it.

Poles of a Chebyshev filter are placed to provide rapid roll-off outside the passband of the filter at the expense of some ripple within the passband. The magnitude of the allowable ripple is specified as some value, *r*, typically a fraction of a decibel. Increased ripple brings increased selectivity. Among the monotonic stopband type of filters, Chebyshev provides the greatest attenuation outside the passband. A lowpass filter's attenuation is given by an expression of the form

$$A_{\text{dB}} = 10\log\left[1 + \varepsilon C_n^2(\omega)\right] \tag{8.6}$$

where $\varepsilon = \sqrt{10^{\frac{r}{10}} - 1}$

C_n is the *n*th-order frequency-dependent Chebyshev polynomial.

The relationship between the 3-dB corner frequency, f_c, and the equal-ripple lowpass frequency, f_r, shown in Figure 8.14 is defined as

$$f_C = 0.5 f_r \left[x^{\frac{1}{n}} + x^{\frac{-1}{n}}\right] \tag{8.7}$$

where

$$x = \frac{1}{\varepsilon} + \sqrt{\frac{1}{\varepsilon^2} - 1}$$

Tabulating (8.7) with passband ripple values of 0.01, 0.1, and 0.2 for filter orders from one to five, allows us to have a quick comparison

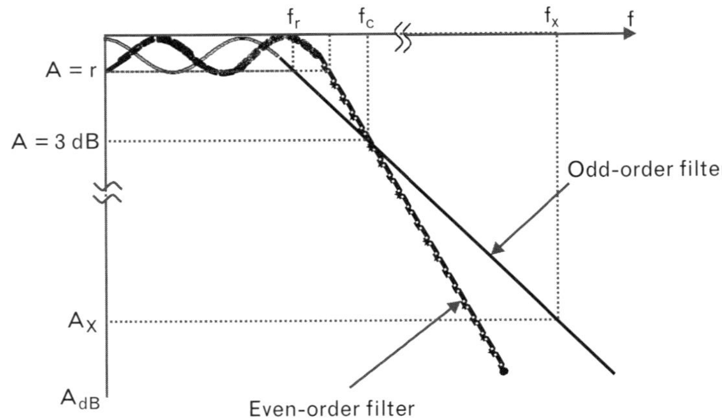

FIGURE 8.14 *The attenuation characteristics of a lowpass Chebyshev filter with passband ripple of r. Even-order filters, with maximum number of passband extremas, cannot be fitted to the Chebyshev polynomials with equal source and load resistances and therefore require unequal terminations. Frequencies f_r and f_x refer to the odd-order response.*

between f_c and f_r (see Table 8.2). We mentioned earlier that the 3-dB corner frequency is not very practical for RF applications, due to the large mismatch at the passband edge. To avoid poor impedance match, we need to design filters that maintain a small equal-ripple performance throughout the entire passband. If we select lowpass element values manually from filter tables *that are normalized to the 3-dB corner frequency*, then f_c must be specified properly above the desired f_r. For example, if we want a third-order lowpass filter with $r = 0.1$ dB to $f_r = 1$ GHz, then we need to specify $f_c = 1.389 f_r = 1.389$ GHz. Some of the filter tables normalize components *directly to the equal-ripple corner frequency*, so this extra step is not required. There are also differences in reference termination: some table values are based on the source normalized to 1Ω and others reference everything to a 1-Ω load, so you need to be very careful to keep everything in proper

TABLE 8.2 RATIO OF THE 3-DB CORNER FREQUENCY, f_c, TO THE EQUAL-RIPPLE CUTOFF FREQUENCY, f_r, FOR VARIOUS CHEBYSHEV FILTER SPECIFICATIONS

	f_c/f_r Ratio		
Order, n	For $r = 0.01$ dB	For $r = 0.1$ dB	For $r = 0.2$ dB
1	20.852	6.628	4.714
2	3.305	1.953	1.690
3	1.878	1.389	1.290
4	1.467	1.215	1.160
5	1.292	1.136	1.101

order. Computerized filter synthesis programs generally ask for r, f_r, *and the actual terminations*, so the 3-dB corner frequency may not even show up in the specification window.

In baseband filters, steeper roll-off outside the passband and higher adjacent channel rejection are achieved at the expense of group delay and intersymbol interference. This is evident in Figure 8.15, which shows substantial overshoot and undershoot during the transition between symbols, and a marked deterioration in the eye diagram. Due to its better selectivity, the Chebyshev filter is a more economic solution than a Butterworth type. Group delay response will be shown later in Figure 8.20.

FIGURE 8.15
(a) The time-domain response and (b) eye diagram of a 5-MHz Chebyshev lowpass filter. The data rate is 10 Mbps. (From: [5]. © 1996 Artech House, Inc. Reprinted with permission.)

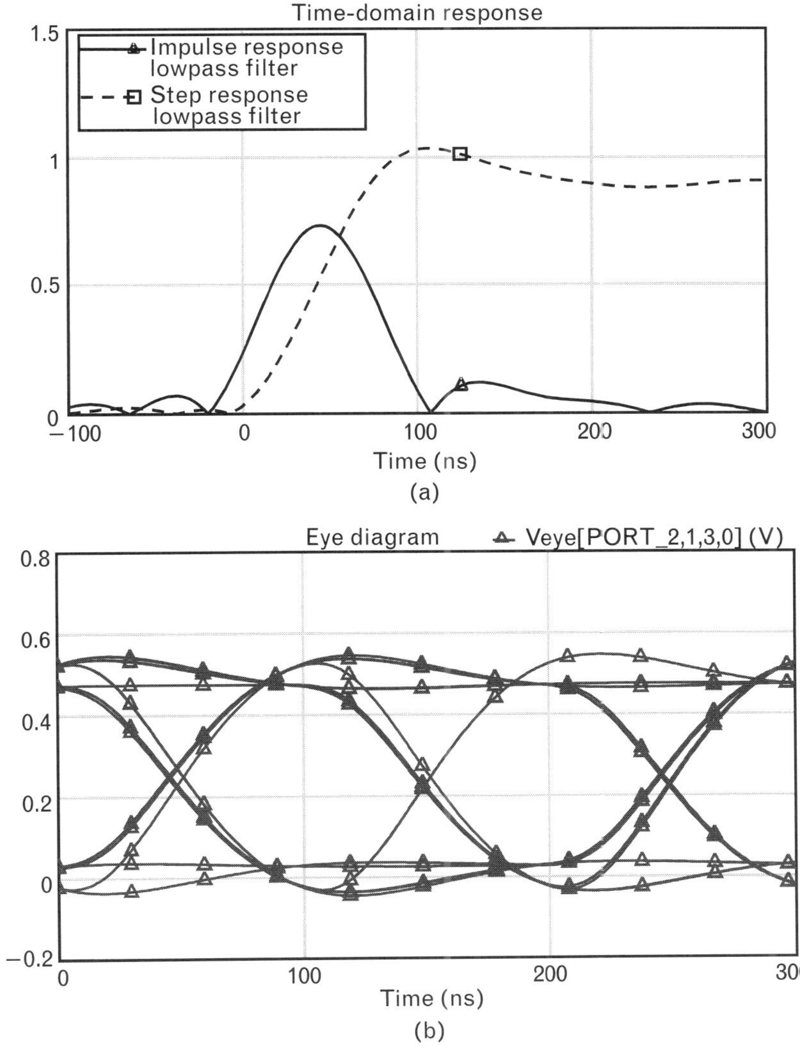

8.4.2.3 The elliptic (Cauer) filter

Filters with equal-ripple passband and equal-minima stopband attenuation provide improved selectivity over Chebyshev types. Elliptic filters include resonant circuits which create transmission zeros at finite frequencies. For a given filter order, the designer has the choice of placing the finite transmission zeros close to the passband edge and settle for lower minimum stopband attenuation, or move the zeros farther to obtain greater stopband attenuation. Most of the microwave bandpass filters with highly demanding specifications use cross-coupled elliptic filters where finite resonances are achieved by coupling among nonadjacent resonators [9–11].

Figure 8.16 shows how the finite transmission zeros are selected for two different minimum stopband attenuation requirements. Specifying 20-dB minimum attenuation places the first pair of finite transmission zeros to 560 MHz, providing 20-dB loss at 550 MHz. The second filter with 40-dB stopband attenuation specification has its lower transmission zeros at 730 MHz, creating 40-dB loss at 710 MHz. The plots also show that sharper selectivity leads to increased group delay distortion, but we need to

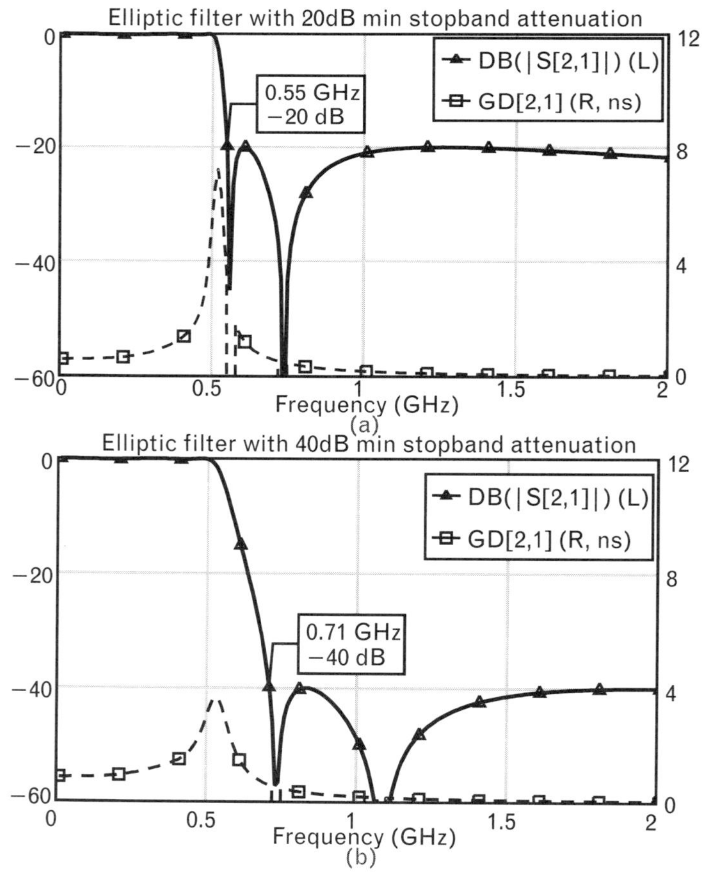

FIGURE 8.16 *Comparisons of two fifth-order elliptical lowpass filters, both having 0.1-dB equal-ripple passband. (a) With 20-dB specified minimum attenuation we can have the stopband starting at 550 MHz, which is only 10% above the passband. (b) Increasing the minimum attenuation to 40 dB moves the stopband edge up to 708 MHz. Group delay scale on the right side is in nanoseconds.*

interpret them carefully. After all, we would expect the filter with 40-dB minimum stopband loss to be more selective above the passband, but it is just the opposite.

If we only need to maintain 20-dB minimum attenuation, it can be reached with a fifth-order filter at only 10% above the 0.1-dB equal-ripple passband cutoff frequency. To meet the 40-dB minimum stopband attenuation with the same filter order, the roll-off above the passband is *less selective* because it must stay down below the 40-dB level at all stopband frequencies.

The attenuation of an elliptic filter is given by

$$A_{\text{dB}} = 10\log\left[1+\varepsilon^2 Z_n^2(\omega)\right]$$

where $Z_n(\omega)$ is an nth-order elliptic function.

Since resonant circuits require additional components, the elliptic filter requires more elements than an equal order Chebyshev filter. For example, a fifth-order Chebyshev lowpass circuit has five components, while a fifth-order elliptic filter has seven (Figure 8.17). Designing with 0.1-dB equal-ripple passband to 500 MHz, the fifth-order elliptic filter with 40-dB minimum stopband loss outperforms the fifth-order Chebyshev for selectivity (708 MHz versus 1,100 MHz), but it also has higher group delay distortion. Comparing the selectivity of the same elliptic filter with a seventh-order Chebyshev still favors the elliptic (708 MHz versus 788 MHz for 40-dB attenuation), which also has lower group delay distortion. Of course, we need to keep in mind that the elliptic filter gives only a minimum of 40-dB loss above 708 MHz, while the attenuation of the Chebyshev continuously increases above 788 MHz.

If the fifth-order elliptic is designed for 20-dB minimum attenuation, its skirt selectivity is much better than the Chebyshev's (550 MHz versus 625 MHz), but the elliptic has considerably more delay distortion. Of course, when the ideal elements are replaced with physical models, the performances can be quite different. The important factor is to carefully compare the physical circuit responses side-by-side before deciding the best option for a specific application.

It is also fair to mention that the superior selectivity of elliptic filters is not obvious in low-order filters. As the orders increase, the differences are very significant. For example, a specified stopband rejection at only 5% above the passband may be fulfilled with a tenth-order elliptic filter, but with a Chebyshev type, it requires an order greater than 30.

We mentioned earlier that due to physical component resonances, even Butterworth and Chebyshev filters frequency responses show finite transmission zeros when built with lumped *L-C* components. Although those circuits do not have equal-minima stopbands, they resemble the

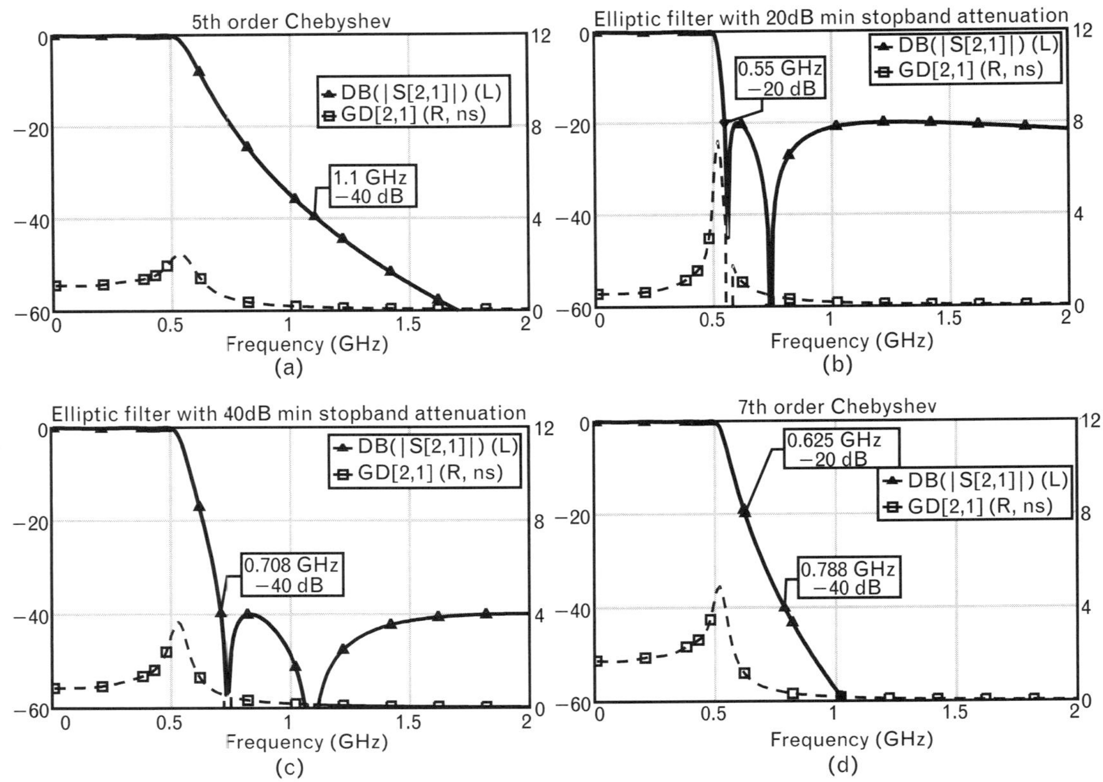

FIGURE 8.17 *Frequency response comparison of (a) fifth-order Chebyshev, (b) fifth-order elliptic with 20-dB minimum stopband loss, (c) fifth-order elliptic with 40-dB minimum stopband loss, and (d) seventh-order Chebyshev lowpass filters. All four filters have 0.1-dB equal-ripple passbands of 500 MHz. Dashed lines show group delay.*

behavior of elliptic filters at RF. We will look at an example of the resonance effects in Section 8.7.2.

8.4.2.4 The Bessel filter

The Bessel filter is similar to the Butterworth filter in that it seeks to maximize the flatness of the group delay by setting to zero as many derivatives of the phase of s_{21} (i.e., of $d\theta/df$) as possible. The transfer function of the filter is then $\exp(-s\tau)$ where τ is the delay. The resulting impulse response is shown in Figure 8.18. The attenuation outside the passband is not nearly as steep as the Butterworth or Chebyshev filters, which are optimized for attenuation rather than phase response. To obtain any reasonable selectivity, the passband corner-frequency attenuation needs to be specified for a relatively large ripple, generally 3 dB. Due to the large mismatch loss and unique rounded frequency response, the passband VSWR is much worse than those of the above-mentioned filters. As a result, Bessel filters are generally not used at RF where impedance matching is an important factor.

An interesting observation for Bessel filters is that after reaching an order of 5 to 6, the *selectivity changes very little* for larger orders. Increasing the order, however, increases the absolute value of the group delay and *widens the constant-delay bandwidth.*

The flat delay and good impulse response with no ringing result in low intersymbol interference. The fact that the attenuation response is not as good as with other filters will result in poorer adjacent channel interference performance. The Gaussian filter discussed above is related to the Bessel filter, since its transfer characteristic has the Gaussian form of $1/\exp-((\omega - \omega_0)/\omega_{pass})^2$. As with the Bessel filter, the poles are located by expanding the denominator and designing the component values appropriately.

8.4.2.5 The equiripple linear phase filter

The equiripple linear phase filter is to the Bessel filter what the Chebyshev filter is to the Butterworth. By accepting a small amount of equal-ripple delay error, selectivity of the equiripple filter is better than that of Bessel filter for a specific equal-ripple delay bandwidth. A maximum phase ripple ε may be specified to determine the poles of the filter. Figure 8.19 shows the attenuation and equiripple group delay. The equal-ripple group-delay bandwidth increases with the specified delay ripple. Changing from the maximally flat delay to a 0.05% equal-ripple delay increases the delay bandwidth nearly 40%. Moving up to 0.5% delay ripple extends the equal-ripple delay bandwidth about 60% over the maximally flat delay response.

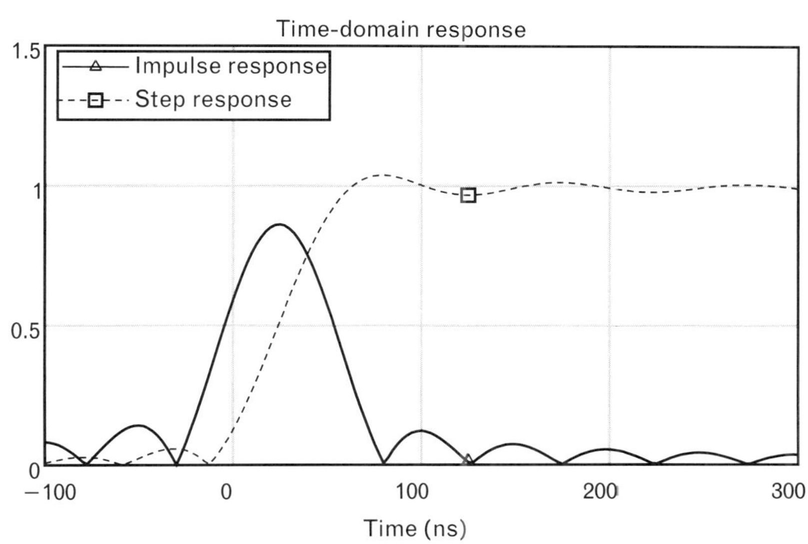

FIGURE 8.18
Impulse response of a Bessel filter.

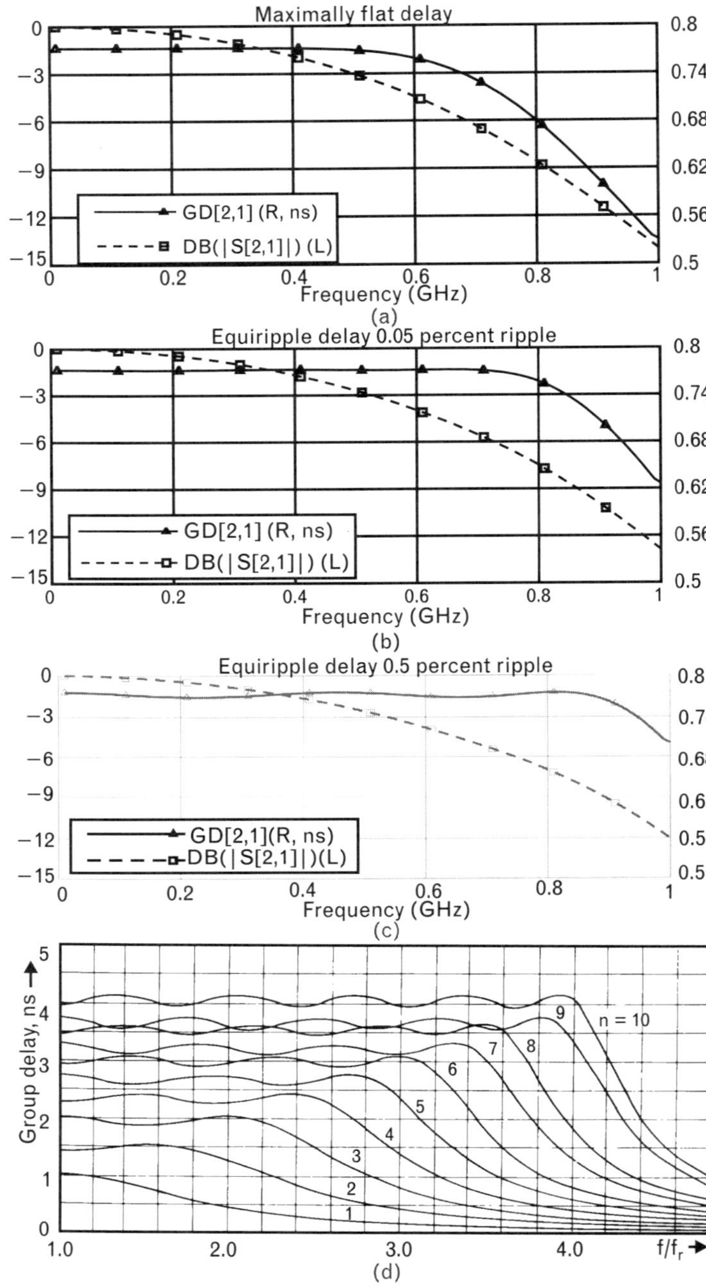

FIGURE 8.19
(a) Attenuation and group delay of the maximally flat and equiripple filters with (b) 0.05% and (c) 0.5% delay ripple, showing the increased delay bandwidth. (d) Equal-ripple delay bandwidth variation versus filter order for 0.5° phase error. Numbers on the horizontal axis refer to the quantity of f/f_r, so the left side of graph indicates the passband corner frequency, f_r. [Figure 8.19(d) is after: [8].]

Figure 8.19(b) shows how the equal-ripple delay bandwidth increases with the order of the filter for a specified delay ripple.

Just as we mentioned for the Bessel filter, the passband VSWR of the equiripple delay filter is also poor. Neither of these filters is very practical for RF and MW applications.

8.5 Comparison of filter responses

A side-by-side comparison of four different filter types—Butterworth, Chebyshev, elliptic, and Bessel—is shown in Figure 8.20. Comparing filters of equal (fifth-) orders, we can rank the elliptic as best for selectivity but poorest for group-delay distortion. The maximally flat delay Bessel filter is at the other extreme—best delay distortion and worst selectivity.

Comparing the four filter types with compatible 40-dB stopband edge attenuation at $f_x/f_c \approx 1.8$ (with respect to the 3-dB corner frequency) shows that three of them have a similar amount of delay distortion. We used an eighth-order Butterworth, fifth-order Chebyshev, fourth-order elliptic, and eighth-order Bessel type. We should point out that with the Bessel we could not reach the specified selectivity and stopped at an eighth-order filter. Above that order the selectivity did not improve significantly.

Out of the four filters, the elliptic filter has the widest 0.1-dB passband, steepest selectivity and surprisingly good group-delay distortion, while the Bessel filter shows poorest selectivity and best delay performance. Note: An even-order elliptic filter is not always easy to realize because it may require coupled inductors.

We need to understand that phase response and group-delay characteristics are related to the selectivity of the filter. With the exception of the previously mentioned cross-coupled filters, to achieve a specified amount of selectivity, we have to accept the corresponding delay distortion.

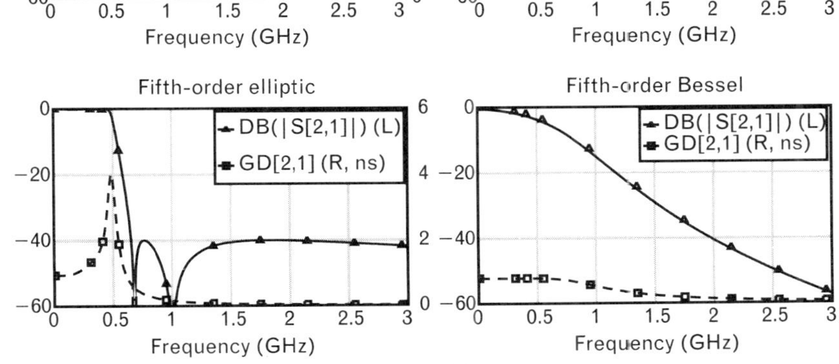

(a)

FIGURE 8.20 *Attenuation and group delay of four lowpass filter types: Butterworth, Chebyshev, elliptic, and Bessel. (a) All filters are designed to have 3-dB ripple at 500 MHz and the elliptic type shows much higher group-delay variation. (b) When the filters of different orders are compared for nearly equal selectivities, with the exception of the Bessel, their delay distortions are similar.*

FIGURE 8.20
Continued

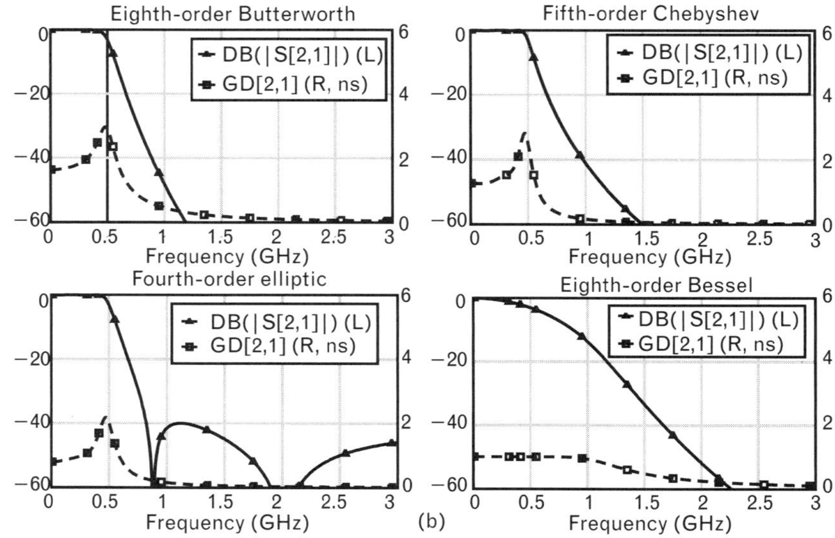

8.6 Multiplexer filters

In a contiguous diplexer filter, two individual filters, such as a lowpass and a highpass, may be designed for the same 3-dB corner frequency. Combining the circuits provides constant-resistance port impedance at all frequencies, at the side where the two-port filters are connected together. A typical application is shown in Figure 8.21, illustrating a cable television line amplifier. Another common use for a diplexer is at the antenna of a transceiver, to split the transmitted and received signals.

In a cable television system the cable carries *downstream* UHF and VHF channels to the subscribers. Low-frequency *upstream* signals, such as polling, voting, and buyer subscription, are sent in the reverse direction back to the cable provider. Due to cable attenuation, the signals need to be amplified repeatedly in their respective frequency range and direction.

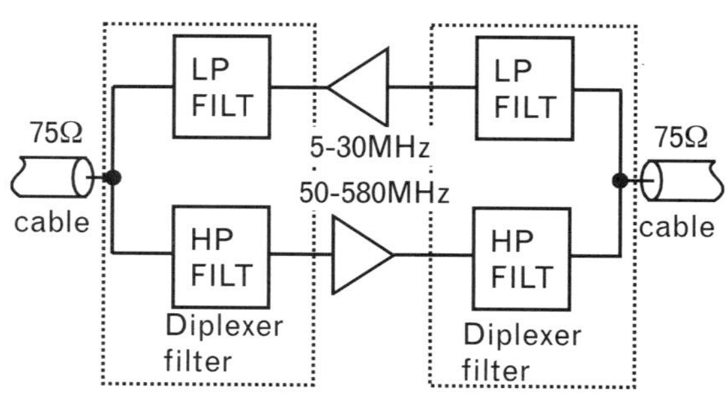

FIGURE 8.21 *Dual-directional cable-television line amplifiers boost signals into two directions at different frequencies. Diplexer filters separate the frequency bands and also present proper terminations to the transmission lines.*

The cable must be properly terminated in both frequency bands to avoid reflections, which can create undesirable ghosting for television viewers. By placing two sets of complementary filters around the amplifiers, the signals can be routed into the proper direction and the combination of the filters provides the proper resistive terminations to the cable at both sides.

To maintain near constant-resistance at the common port at *all frequencies*, the individual filters should:

- Have the same basic topology, and 3-dB corner frequency;
- Have complementary response specifications, such as 0.5 dB Chebyshev for both filters;
- Be designed using a *singly terminated* approach, using an extreme termination at the common side.

In real-life applications the constant-resistance input may not be required at all frequencies and a pseudo-complementary performance is acceptable [12]. Even if the above rules are followed, parasitics and losses apply to the two filters differently. The important point is to control the interaction between the filters to achieve their individual passband specifications.

Another commonly used diplexer application in RF transceivers is where the transmitter and receiver share the same antenna (Figure 8.22). For example, in the CDMA IS-95 cellular telephone system described in Chapter 3, the signals transmitted are at 824 MHz to 849 MHz, while the receiver works in the 869- to 894-MHz band. Two highly selective bandpass filters connect the transmitter and receiver sections to the antenna in their respective passbands, and also provide isolation from each other. In such application, since size and cost are also very important considerations, generally, coupled resonators (Section 8.10) are used.

The combined input impedance of narrowband filters with widely separated (i.e., an octave separation) passbands looks purely reactive outside of the passbands. Therefore, with a segment of cascade transmission line one input may be transformed to act as an open circuit in the second filter's passband. This way, when the two bandpass filters are connected in parallel at their input ports, their interaction is reduced. As the passbands get closer to each other, this task becomes increasingly more difficult.

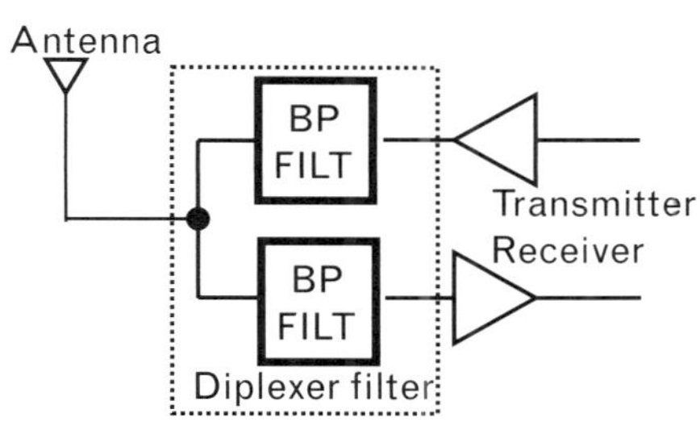

FIGURE 8.22
Partial block diagram of an RF transceiver, showing the transmitter, receiver, and the antenna they share through a diplexer filter.

8.7 Filter design outline

Several good textbooks cover manual filter design based on tabulated prototype values. We do not wish to duplicate their work because now every major RF circuit simulator has at least a basic filter synthesis routine to provide the element values, already frequency and impedance scaled. For transmission line filters, the simulators also perform the conversion from electrical parameters to physical dimensions for any specified dielectric media. We advocate computer-aided design and our recommendation is to use an automated filter synthesis program rather than filter tables to obtain the component values. Since in RF circuit design the initial ideal components need to be replaced with physical models, computer-aided simulation is a must. Still, it may still be useful to some readers to review the manual approach—if for nothing else, to appreciate the computerized alternative. Our review is limited to lowpass filter design and some of the circuit transformations.

8.7.1 Lowpass filter design using filter tables

The required manual steps to design a lowpass filter are as follows.

8.7.1.1 Determine filter order

Finding the order of the filter by solving closed-form expressions, like (8.4) and (8.5) for a Butterworth response, may be a tedious procedure, and particularly for even more complex filter types. The alternative is to use charts or nomographs provided by filter handbooks, such as the ones we show later for Butterworth, Chebyshev, and elliptic filters. Use of these nomographs is simple, as shown in Figure 8.23. We need to specify the ratio of the specified stopband attenuation frequency and passband corner frequency, f_X/f_r, the passband decibel ripple (or roll-off at band edge), r, and the minimum required decibel stopband attenuation, A_{MIN}. If the exact answer is a fractional number, since the filter order must be an integer, the common practice is to round it up to the next higher integer. Actually, adding parasitics to lumped components improves the selectivity of the filter in general, and almost always the rounding may be performed to the lower rather than to the higher order, as we will see in Section 8.7.2. Filters with distributed components behave just the *opposite way*, and the initially computed order should be increased.

Use of the illustrative nomograph is summarized below. Notice that the true passband corner frequency, f_r, is used in the selection, rather than the 3-dB corner frequency, f_c. The required steps are as follows:

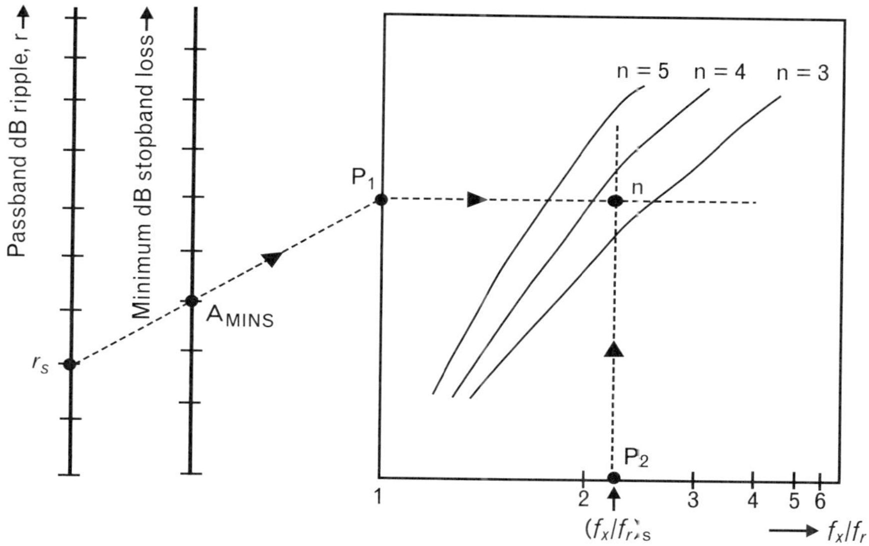

FIGURE 8.23 *Summary chart illustrating the use of lowpass filter order nomographs, showing curves for third-, fourth-, and fifth-order filters. User inputs are r_S, A_{MINS}, and $(f_x/f_r)_s$. (After: [8].) (Specific scale values are shown in Figures 8.24 through 8.26.)*

1. On the far left-side vertical scale, mark the specified passband decibel ripple or roll-off, r_s, allowed to the upper passband edge, f_r.
2. Mark the minimum specified stopband decibel attenuation, A_{MINS}, required at the frequency f_x on the next vertical scale.
3. Draw a straight line from r_s thru A_{MINS} until the left side of the rectangular chart is intersected. Label the point as P_1.
4. Draw a horizontal line from P_1 through the rectangular chart.
5. Mark the value of ratio of specified $(f_x/f_r)_s$ on the scale provided on the lower side of the rectangular chart and label it as P_2.
6. Draw a vertical line from P_2 upward in the rectangular chart. The intersection of the horizontal and vertical lines gives the exact value of the filter order, which in most cases is somewhere between two integer numbers. Round the answer to the next higher or lower integer, based on your past experience.

To illustrate this procedure for Butterworth, Chebyshev, and elliptic filters, we marked on Figure 8.23 the three required lines for 0.1-dB ripple (roll-off factor), 40-dB minimum acceptable stopband attenuation and f_x/f_r = 2.25. The intersection of lines drawn from P_1 and P_2 gives us the order between third and fourth. As a general practice, most RF filter designers choose a fourth order in such a case, although a third-order lumped-element filter may also meet the specifications once the component parasitic effects are included. Even with lumped components, however, selecting the larger order can provide safety margin, considering the fact that the exact synthesized element values are often not available. Component tolerances and temperature effects also degrade performance, so it is good to be conservative at this step.

Three useful nomographs are shown next for the Butterworth, Chebyshev, and elliptic filters. They are used to find the necessary filter order for a specified stopband attenuation. The use of these charts is similar to that in Figure 8.23.

To compare the selectivity of the filters, we applied identical specifications to the Butterworth, Chebyshev, and elliptic filter nomographs (Figures 8.24 through 8.26), as follows:

- Passband corner frequency, f_r, for $r = 0.1$-dB ripple: 400 MHz;
- Minimum stopband attenuation above $f_x = 900$ MHz: 40 dB.

Normalizing f_x to f_r gives us $f_x/f_r = 2.25$. Looking at the specific charts we see that it would take an eighth-order Butterworth, a fifth-order

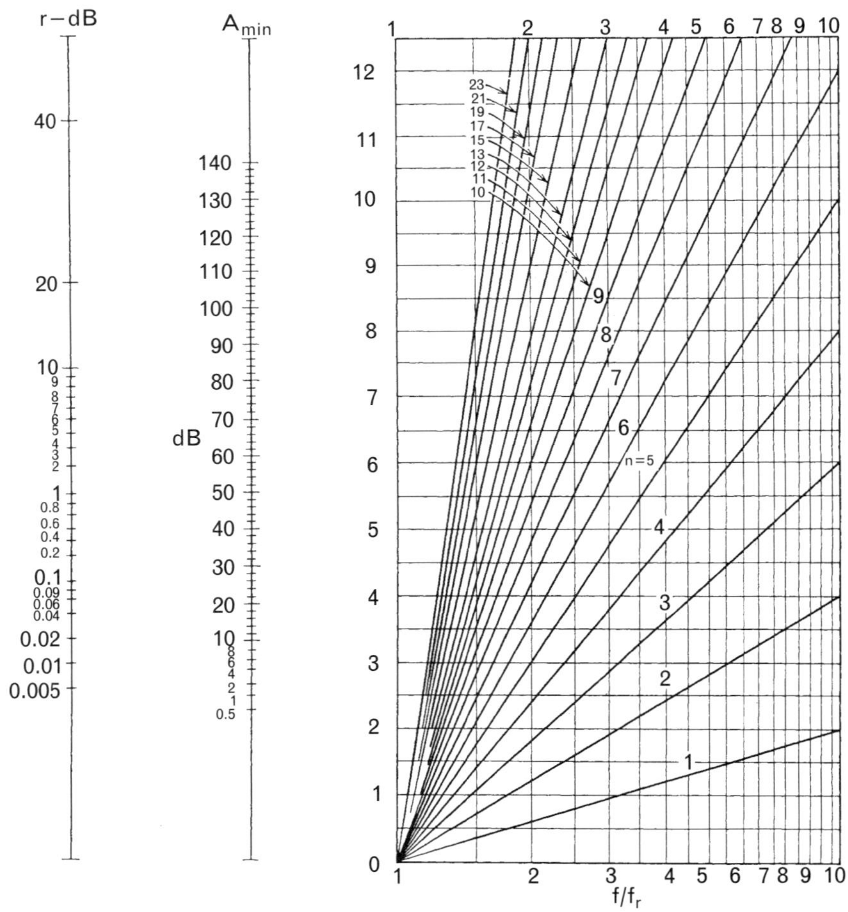

FIGURE 8.24 *Nomograph to determine the order of a Butterworth filter as a function of the passband roll-off at f_r, the minimum stopband attenuation starting at f_x, and the f_x/f_r ratio.* *(After: [8].)*

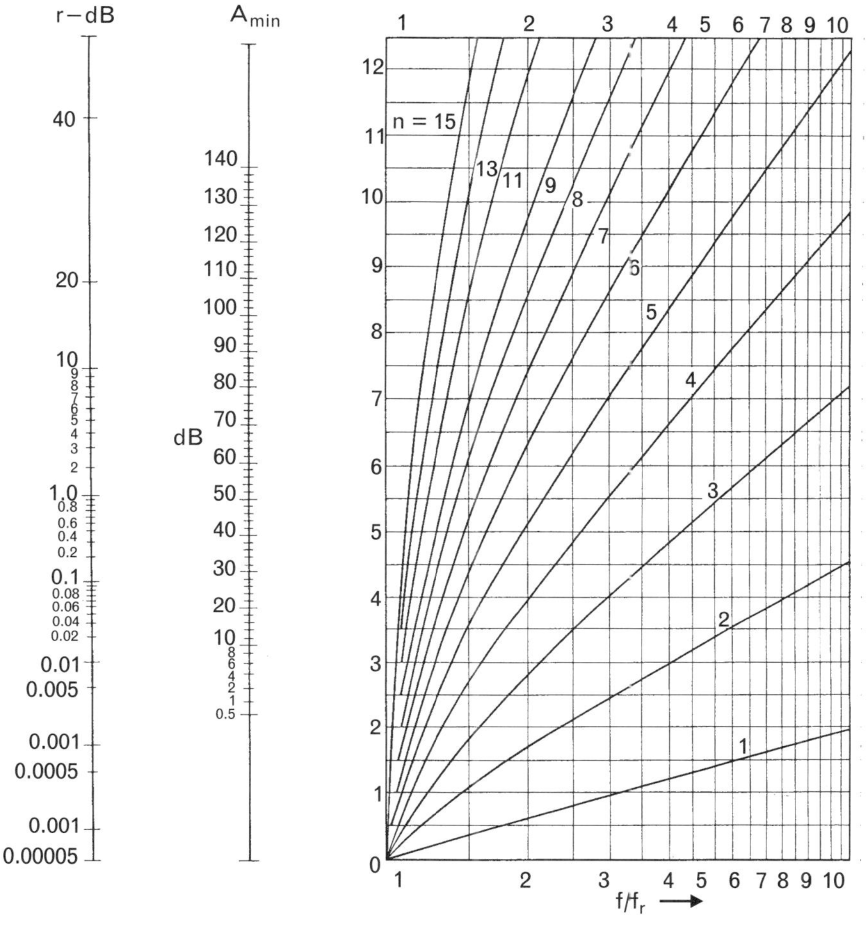

FIGURE 8.25
Nomograph to determine the order of a Chebyshev filter as a function of the passband ripple at f_r, the minimum stopband attenuation starting at f_x, and the f_x/f_r ratio. (After: [8].)

Chebyshev, and a fourth-order elliptic filter to meet a minimum of 40-dB stopband attenuation at $f_x/f_r = 2.25$. We already saw in Figure 8.20 that the delay characteristics of these filters are similar. Generally, for low-order filters the Chebyshev requires the fewest number of components, and we will choose that filter later for our illustrative example. Keep in mind, however, that in high-selectivity applications where the filter order may exceed 8 to 10, the elliptic filter becomes the superior performer.

8.7.1.2 Find the prototype component values

Once the order is determined, we can read the normalized prototype element values from the appropriate filter table. There are, however, different forms for the tables published by various sources. Some of the filter tables are normalized to 1-Ω source while others to 1-Ω load termination. Frequency normalization in some tables is referenced to the 3-dB corner

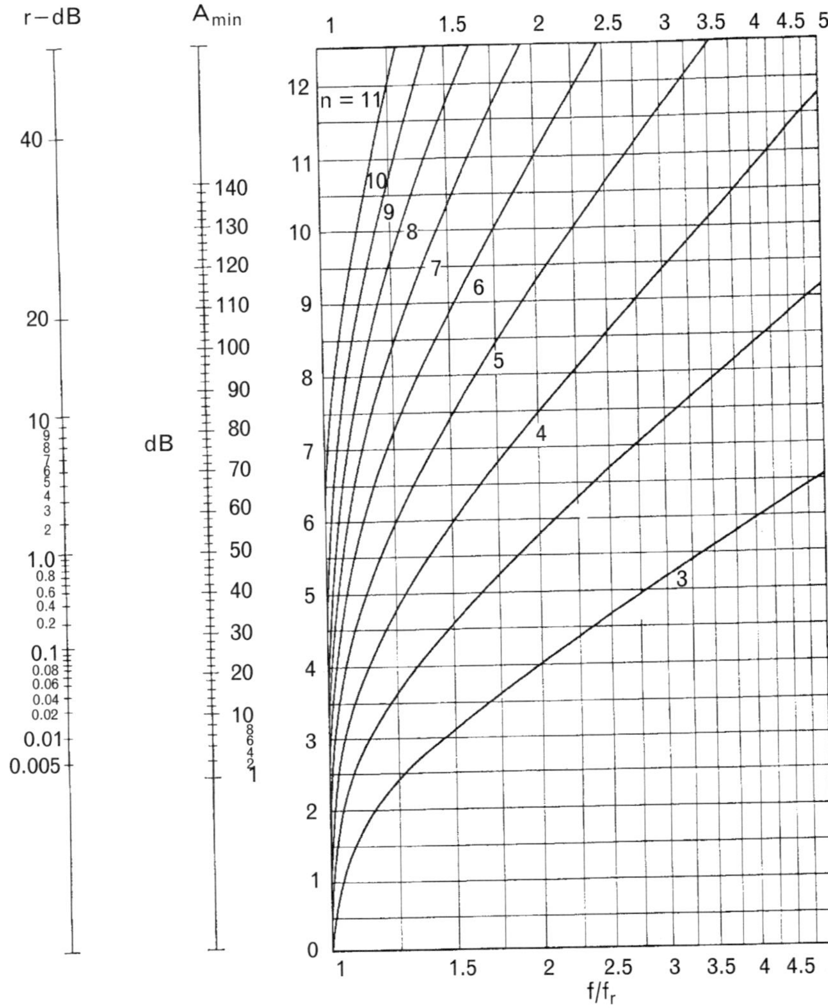

FIGURE 8.26 *Nomograph to determine the order of an elliptic (Cauer) filter as a function of the passband ripple at f_r, the minimum stopband attenuation starting at f_x, and the f_x/f_r ratio. (After: [8].)*

frequency, f_c, and in some cases to the passband ripple frequency, f_r, and we need to be careful to recognize these differences.

Filters are generally designed with equal source and load resistance. For example, if our filter faces equal terminations, such as $R_S = R_L = 75\Omega$, then we use filter tables referring to unity termination ratio, $R_S/R_L = 1.0$. In case of uneven terminations, we need to find a table appropriate for that *termination ratio* In our example we only show element values for two R_S/R_L ratios, one for unity and another for infinity. As we stated, most doubly terminated filters fall into the $R_S/R_L = 1.0$ category, but we also want to mention the singly terminated case where $R_S/R_L = \infty$, or $R_S/R_L = 0$.

Figure 8.27 shows two sets of normalized lowpass prototype element values for 0.1-dB equal-ripple fifth-order Chebyshev lowpass filters, with two possible topologies. Component labels in the top row refer to element

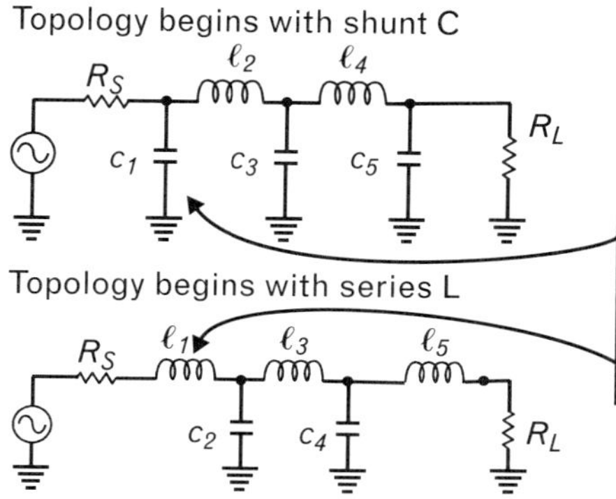

R_S/R_L	c_1	ℓ_2	c_3	ℓ_4	c_5
1.000	1.147 *(1.301)*	1.371 *(1.556)*	1.975 *(2.241)*	1.371 *(1.556)*	1.147 *(1.301)*
∞	1.376 *(1.561)*	1.592 *(1.807)*	1.556 *(1.766)*	1.249 *(1.417)*	0.573 *(0.651)*
R_L/R_S	ℓ_1	c_2	ℓ_3	c_4	ℓ_5

FIGURE 8.27 *Tabulated normalized L-C element values for fifth-order Chebyshev lowpass filters with 0.1-dB passband ripple, normalized to the equal-ripple passband frequency, f_r. For doubly terminated cases: $R_S/R_L = 1$; and for singly terminated design: $R_S/R_L = \infty$ or $R_L/R_S = \infty$, (Italicized values in parentheses are normalized to 3-dB corner frequencies, f_c instead of f_r.)*

values for the R_S/R_L termination ratio, where the circuit topology begins with a parallel capacitor. In equivalent dual circuits, the termination ratio refers to R_L/R_S. Then, a series inductor is used for the first element, and the bottom row indicates the element values for the alternative topology. The first set of components is applicable to the doubly terminated case with $R_S = R_L = 1$. Singly terminated filters are designed either with $R_S/R_L = \infty$ (current source) or $R_L/R_S = \infty$, which is equal to $R_S/R_L = 0$ (voltage source), using the second set of element values.

With the exception of even-order filters with equal terminations (not shown here), the tables always provide two possible sets of component values: one for a circuit starting with a parallel reactive element and another one with a series component, as shown at the top and bottom rows of Figure 8.27. If the desired topology is the one shown at the top, the numerical values in the second column refer to R_S/R_L. On the other hand, if we use the alternative topology at the bottom, the ratios refer to R_L/R_S.

For singly terminated filters, the ∞ termination ratio may be realized in two ways:

1. Upper topology, $R_S/R_L = \infty$ => Current-generator source (open-circuit) and a finite load resistance;
2. Lower topology, $R_L/R_S = \infty$ => Voltage-generator source (short-circuit) and a finite load resistance.

Circuit diagrams of singly terminated filters are shown in Figure 8.28. Extreme terminations are used in special cases, such as diplexer filters, and we refer readers to [13, 14] for their designs.

The inductors and capacitors in Figure 8.28 have large component values, but keep in mind that we are dealing with prototype filters, normalized to $R_L = 1\Omega$ and 3-dB corner frequency, $f_r = 1$ radian $= 0.159$ Hz. The actual element values are determined next, using frequency and impedance scaling.

8.7.1.3 Frequency and impedance scaling

Each of the normalized component values need to be scaled to the proper corner frequency and load termination by using the following expressions. Each capacitor, c_n, and each inductor, ℓ_n, will be scaled as

$$C_n = \frac{c_n}{(2\pi f_r)R_L} \tag{8.8}$$

and

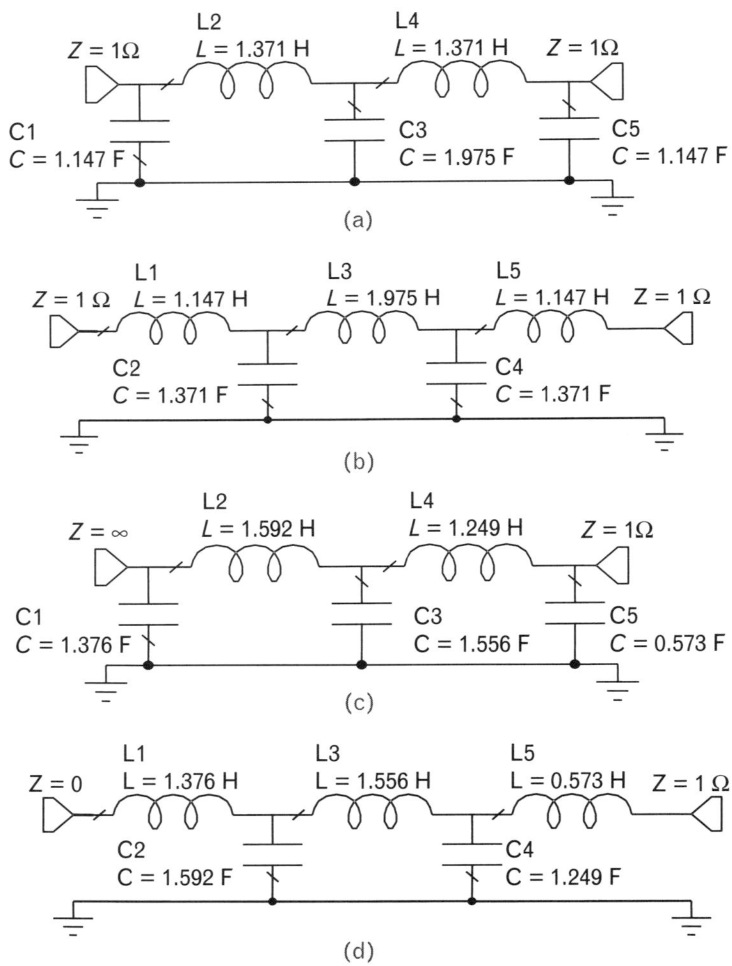

FIGURE 8.28 *Four possible realizations of the normalized Chebyshev lowpass prototype filter. (a,b) Filters to work with $R_S = R_L = 1\Omega$. The 0.1-dB passband corner frequency of the normalized filter, f_r = 1 radian (Note $f_r \neq f_c$). Singly terminated filters to work with (c) current source ($R_S = \infty$) and (d) voltage source ($R_S = 0$) into a 1-Ω normalized load.*

$$L_n = \frac{\ell_n \; R_L}{(2\pi f_r)} \tag{8.9}$$

where

f_r is the *desired* equal-ripple passband cutoff frequency in hertz (or f_c in some tables).

R_L is the *actual* load resistance of the filter in ohms. Note: Some filter tables are normalized to the *source termination*.

After the impedance and frequency scaling the circuit will perform exactly as specified as long as all the components are ideal and the terminations are resistive. Real-life components and real-life terminations unfortunately do not meet these criteria. We will show the effects of component losses later in Section 8.7.2 during the illustrative exercises.

8.7.2 Illustrative exercise: 400-MHz Chebyshev filter design with lumped components

Design a Chebyshev filter with 0.1-dB equal ripple passband of 400 MHz, 50-Ω source, and load terminations with a minimum number of inductors in the filter. Minimum required stopband attenuation is 40 dB at 900 MHz. Simulate the circuit response with ideal components and find the nearest standard components with ±5% tolerances. Check the response of the circuit by using discrete components where each capacitor has 10-nH self-inductance and the inductors have 1.0-pF capacitance. Then, replace the discrete components with surface mount types, reducing the series self-inductance of the capacitors to 0.5 nH and the self-capacitance of the inductors to 0.3 pF.

Solution

Our first step is to determine the order of the filter. Normalizing the low edge of the stopband frequency, f_x = 900 MHz to the equal ripple passband corner frequency, f_r = 400 MHz, we get

$$\frac{f_x}{f_r} = \frac{900\ \text{MHz}}{400\ \text{MHz}} = 2.25$$

Apply the 2.25 normalized frequency ratio, 0.1-dB passband ripple, and 40-dB minimum attenuation to the nomograph Figure 8.25, we find that the required filter order is just below five. We select the next higher integer order of five.

Our next task is to obtain the normalized lowpass prototype values for the components.

From Figure 8.27, we will read off the five component values for a fifth-order filter. The component values we get from the tables are in farads for capacitors and henries for inductors, but keep in mind that this is a very low-frequency prototype circuit that will be frequency scaled later. We have two alternative topologies: one is to use three parallel capacitors and two series inductors and the dual solution uses three series inductors and two parallel capacitors. Since inductors are more troublesome to handle, and our specifications stated a minimum number of inductors, we choose the first option that starts with a parallel capacitor. The normalized prototype element values are

$$c_1 = c_5 = 1.147F$$
$$\ell_2 = \ell_4 = 1.371H$$
$$c_3 = 1.975F$$

The lowpass prototype filter with 1-Ω normalized source and load terminations is shown in Figure 8.29.

The next step is to apply frequency and impedance scaling with the desired 3-dB corner frequency and load termination of 50Ω. The circuit topology is symmetrical, and we only need to calculate the first three elements, C_1, L_2, and C_3, since $L_4 = L_2$, and $C_5 = C_1$. The element computations, using $f_c = 400$ MHz are as follows:

$$C_1 = C_5 = \frac{c_1}{2\pi f_r R_L} = \frac{1.147}{2\pi(0.4)10^9(50)} = 9.11 \text{ pF}$$

$$L_2 = L_4 = \frac{\ell_2 R_L}{2\pi f_r} = \frac{1.371}{2\pi(0.4)10^9} = 27.3 \text{ nH}$$

$$C_3 = \frac{c_3}{2\pi f_r R_L} = \frac{1.975}{2\pi(0.4)10^9(50)} = 15.7 \text{ pF}$$

The impedance and frequency scaled circuit is shown in Figure 8.30(a). Without component parasitics or losses, the circuit has exactly 0.1-dB equal ripple response to 400 MHz, dropping to 3 dB at 454.4 MHz, and give us slightly over 40-dB attenuation at 900 MHz. (Note that we are just barely able to meet the minimum required attenuation at 900 MHz, but let us not change the filter until we will see the effects of the component parasitics.) Return loss is greater than 16.5 dB in the passband. As the frequency response in Figure 8.30(c) shows, the filter also meets our passband specifications since we used ideal components.

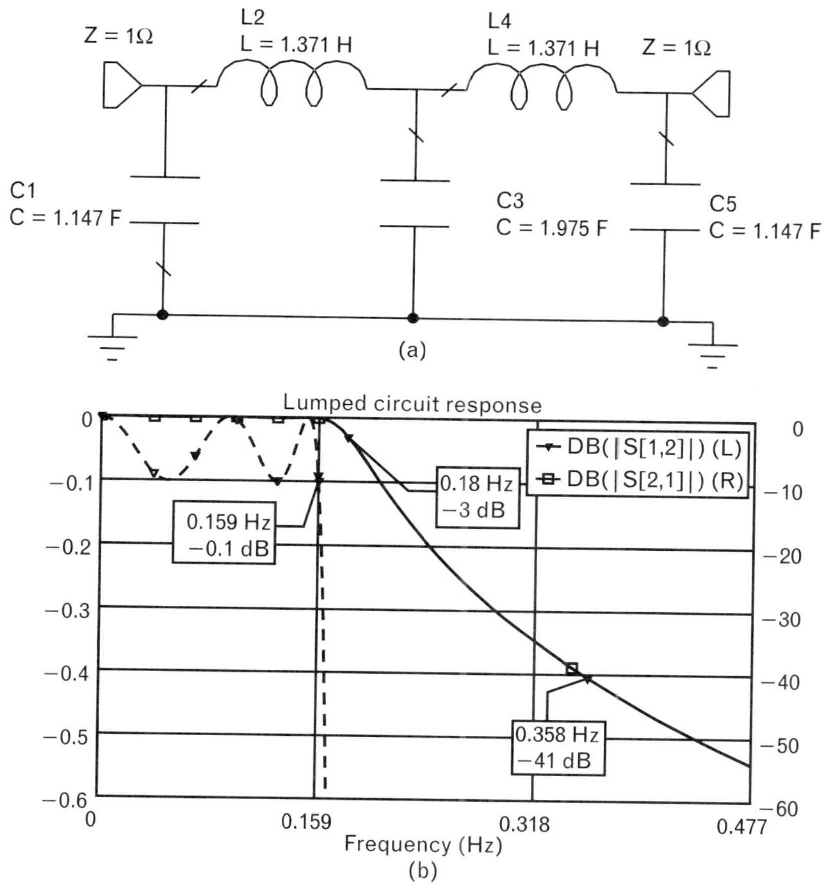

FIGURE 8.29 *(a) Circuit configuration with a minimum number of inductors for a fifth-order lowpass prototype Chebyshev filter. (b) Frequency response is equal-ripple 0.1 dB to f_r = 1 radian = 0.159 Hz. At 2.25 times the passband corner frequency, 0.358 Hz, the minimum attenuation of 40 dB is (barely) satisfied. The solid line refers to 10-dB steps on the right-side scale, and the dashed line refers to the 0.1-dB steps on the left-side scale.*

Our next step is to investigate the effect of finite component Qs and component parasitics. First, we see how the circuit performs with discrete components where capacitors have a 10-nH self-inductance and the inductors have a 1-pF self-capacitance. Figure 8.31(a) shows the response of the physical circuit with parasitics, and Figure 8.31(b) with component losses added, using component Qs of 100 for all five components. Note the three resonances occurring at finite frequencies, caused by the parasitic inductances and capacitances. Another interesting observation is that the passband of the filter shifted to a lower frequency since the parasitics increase the effective values of the components, thereby pushing the passband to a lower frequency. Although by using smaller component values we could improve the filters performance, with these large parasitics the response is not easy to compensate. Instead, we look at surface-mount components, which have significantly lower parasitic inductance and capacitance.

The five discrete components self-resonate with their parasitics, giving us three sets of finite transmission zeros. Since the circuit has symmetry,

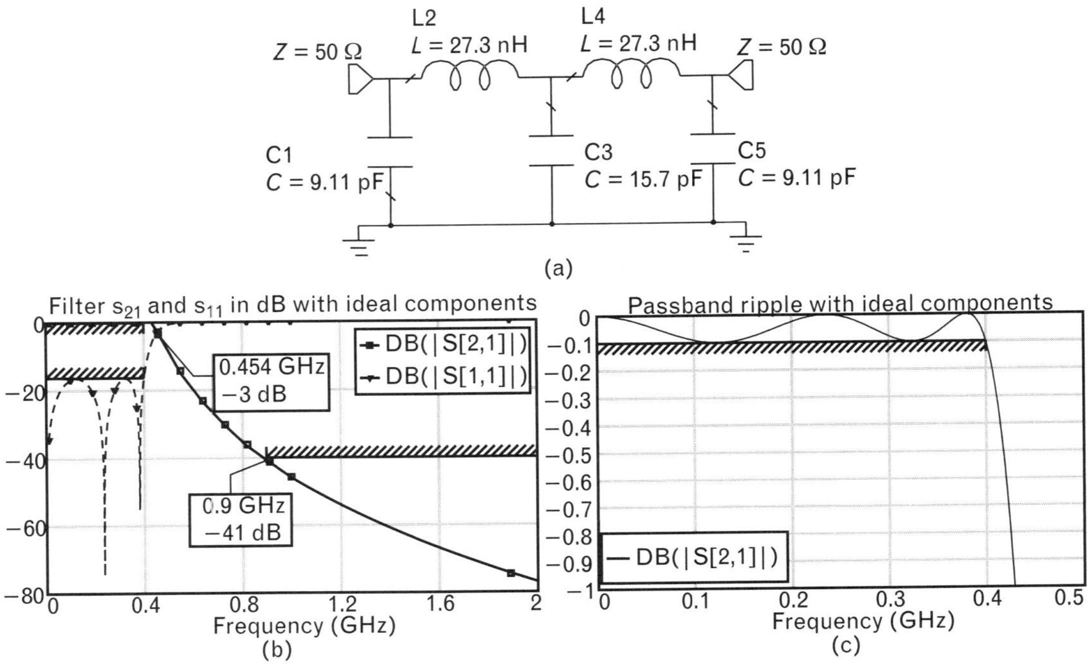

FIGURE 8.30 *(a) Schematics of the fifth-order filter after frequency and impedance scaling. (b) Response of the scaled filter, using ideal components. At f_c = 454 MHz, where $|s_{21}| = |s_{11}|$, the transmitted signal portion becomes equal to the reflected part. Forward-slanted attenuation limits show design specifications. (c) The passband frequency response shows that the synthesized filter meets the 0.1-dB equal-ripple specification.*

both C_1 and C_5 self-resonate at 527 MHz. L_2 and L_4 also have identical self-resonances at 963 MHz. Capacitor C_3 is unique and its resonance is at the lowest of the three frequencies, at 402 MHz. The effect of the component losses is a gradual roll-off of the frequency response.

Above the highest self-resonance frequency of 963 MHz, the discrete filter's stopband performance deteriorates and a second passband opens up. Once the components go through their primary self-resonances, their parasitics take over—capacitors behave like inductors and inductors are controlled by their self-capacitances. At that point the lowpass filter becomes a highpass circuit, as we mentioned earlier in Section 8.4.2.

Figure 8.32 shows the performance of the circuit with surface-mount components where the self-inductance of the capacitors is 0.5 nH and the self-capacitance of the inductors is 0.3 pF. Once again we are showing the circuit with and without component losses. The frequency shift of the response is less with these components compared to their discrete equivalents. (Recall that component parasitics increase the effective values of the components and lower the frequency response of the filter.)

Optimizing the exact synthesized element values to the nearest available lower standard components with ±5% tolerances shifted the frequency

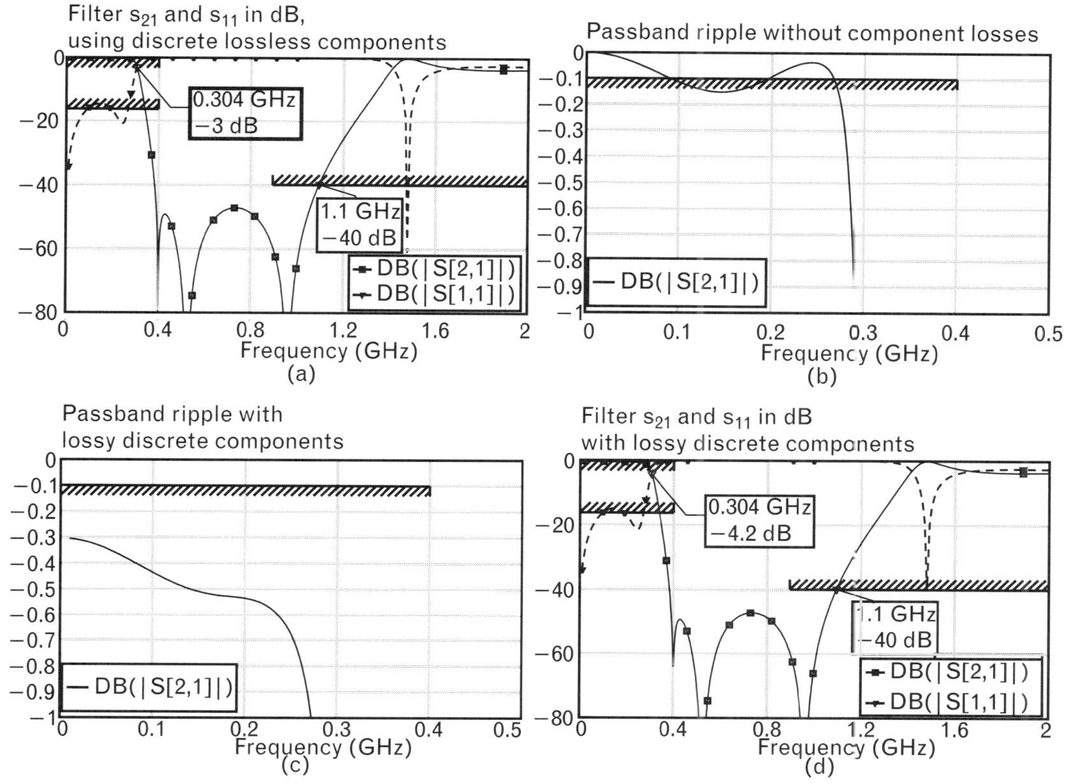

FIGURE 8.31 *(a, b) Frequency response of the filter with discrete component parasitics, without losses. The response shows the component self-resonances and frequency shift. A second passband also exists above 1.1 GHz. Adding the component losses primarily affects (c) the passband, without any significant change in (d) the stopband.*

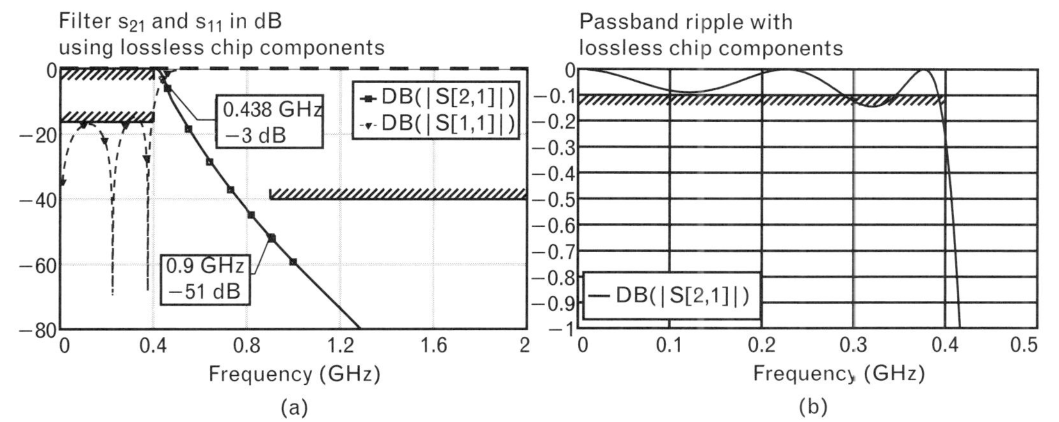

FIGURE 8.32 *(a) Since surface-mount components have lower parasitic inductance and capacitance the frequency-shift of the passband is less than the response shown in Figure 8.31. Self-resonances moved to higher frequencies, and the filter exceeds the stopband requirements. (b) The passband ripple is slightly greater than 0.1 dB.*

response upward to meet the original specifications, as shown in Figure 8.33(a, b). Note that we no longer have an exact equal-ripple passband since the element values no longer follow the requirement of Chebyshev polynomials. Finally we added component Qs to model the loss of the filter using Q values of 100 at 400 MHz. The result is a gradual roll-off toward the high end of the passband, giving 0.5-dB loss at the band-edge. The stopband response is not affected significantly, and we easily meet the 40-dB minimum attenuation at 900 MHz. The final response with standard component values is shown in Figure 8.33(c, d).

The statistical analysis and the yield optimization of a similar filter, with higher passband ripple and corner frequency, was shown in Chapter 7. The same process could be applied to this filter to check the initial yield and optimize it if necessary.

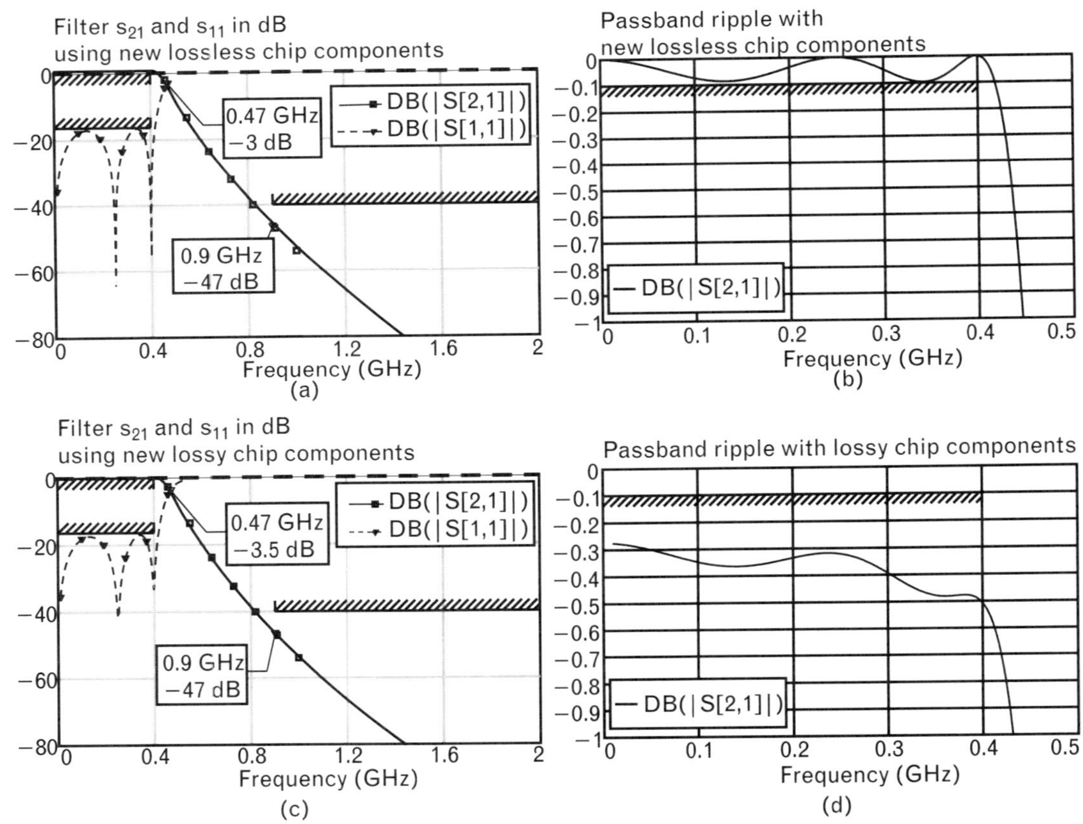

FIGURE 8.33 *Final analysis of the 400-MHz Chebyshev filter with reduced values of physical surface-mount components, using $C_1 = C_5 = 8.2$ pF, $L_2 = L_4 = 25$ nH, $C_3 = 14$ pF. Frequency response of (a, b) lossless circuit and (c, d) lossy circuit with component Qs equal to 100 at 400 MHz.*

8.8 Transmission line (distributed-element) filters

As the frequencies reach the region where lumped elements cannot be practically realized, we need to build filters with transmission line components. Much of the theory used for low-frequency filter design is also applicable to microwave filters except different elements are used to realize the filters. Richards' transformation [15] allows us to convert the lumped elements to transmission line forms, changing the linear frequency dependency of lumped L-C components to tangent functions of open and short-circuited series or parallel transmission line stubs.

Richards' transform applies to commensurate[2] networks by mapping the complex s-plane into a new complex Ω-plane (no relation to ohms). The transformation moves infinite frequency of the s-plane to the quarter-wave length frequency of the transmission lines used in the filter. The forms of the two complex variables are

$$s = j\Omega$$
$$\Omega = \tan\theta$$

where θ is the commensurate length of all transmission lines in the network at the passband cutoff frequency of the filter.

The transformation allows us to convert the lumped prototype inductor and capacitor values, ℓ_n and c_n, with transmission line stubs. Inductors are replaced with short-circuited transmission line stubs, where the impedance-scaled characteristic impedances of the shorted stubs are [16]

$$Z_{SS} = \frac{R_L \ell_n}{\tan\theta_{SSc}} \tag{8.10}$$

Capacitors are replaced with open-circuited transmission line stubs, where the impedance-scaled characteristic impedances of the open stubs are

$$Z_{OS} = \frac{R_L \tan\theta_{OSc}}{n} \tag{8.11}$$

where

R_L is the actual load resistance of the filter.

2. Equal electrical length.

$\theta_{SSc} = \theta_{OSc}$ is the commensurate electrical length of the short-circuited and open-circuited transmission line stubs.

8.8.1 Illustrative example: converting a lumped filter to distributed type using Richard's transformation

Until now, we only used transmission line stubs in parallel configuration because series stubs are difficult to realize in planar structures. They are frequently used, however, in other physical forms, such as coaxial structures. In the design phase of distributed filters, series stubs can play important roles, *even when the final circuit is created without them.* As we will see in the next section, during network transformations filters may go through *intermediate topologies* with series stubs, before the final realizable circuit is obtained.

Using Richards' transform, the five-element lumped L-C lowpass filters of Figure 8.29 can easily be transformed to five-element transmission line circuits. Parallel capacitors are converted to parallel open stubs and series inductors to short-circuited series stubs. Our goal is to maintain a 0.1-dB equal-ripple response to 2 GHz.

The characteristic impedances transmission line stubs in Figure 8.34 are computed with (8.10) and (8.11), by setting the commensurate electrical length arbitrarily to 45° at 1 GHz in both filters. We will see later that moving the commensurate length closer to 90° at the corner frequency increases the selectivity, but also narrows the stopband.

We are now ready to compute the characteristic impedances in Figure 8.34. The distributed element circuit has three parallel open-circuited stubs and two series short-circuited stubs with 50-Ω source and load. Since the topology is symmetrical, we only need to compute the characteristic impedances of the first three elements. That is,

$$Z_{OS1} = Z_{OS5} = \frac{R_L \tan\theta_{OSc}}{c_{CP1}} = \frac{(50)\tan 45^\circ}{1.147} = 43.6\Omega$$

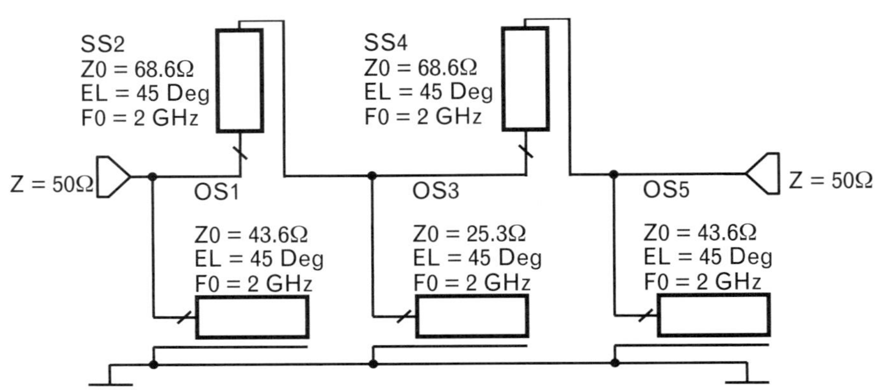

FIGURE 8.34 *A fifth-order Chebyshev lowpass filter converted from a lumped L-C prototype circuit to its transmission line equivalent using Richards' transformation. All transmission lines have the same electrical lengths of 45° at 2 GHz.*

$$Z_{SS2} = Z_{SS4} = \frac{R_L \ \ell_{LS2}}{\tan\theta_{SSc}} = \frac{(50)1.371}{\tan 45^\circ} = 68.6\Omega$$

$$Z_{OS3} = \frac{R_L \tan\theta_{OSc}}{c_{CP3}} = \frac{(50)\tan 45^\circ}{1.975} = 25.3\Omega$$

The characteristic impedance of the third element, Z_{OS3} may be too low for physical realization. We could increase the commensurate line length, which increases the impedances of Z_{OS1} and Z_{OS3}, while decreasing the impedance of Z_{SS2}. For example, increasing the commensurate line length to 60° at 2 GHz gives us

$$Z_{OS1} = Z_{OS5} = 75.5\Omega$$
$$Z_{SS2} = Z_{SS4} = 39.6\Omega$$
$$Z_{OS3} = 44.3\Omega$$

which are in a better range for microstrip realizability. However, this is not the final form of the filter, and we will modify the circuit in Section 8.9.3.1.

The frequency response of the distributed filter (Figure 8.35) meets the 0.1-dB equal-ripple specification from dc to 2 GHz, just as the lumped L-C filter did up to 400 MHz. Outside of the passband, however, the similarity ends because the distributed circuit does not really behave as a true lowpass filter. Ideal inductors and capacitors change their reactances linearly with increase of frequency while transmission line stubs follow a tangent function, which has a cyclical behavior. For example, an open-circuited stub acts capacitively as long as its electrical length is less than 90°. Above the

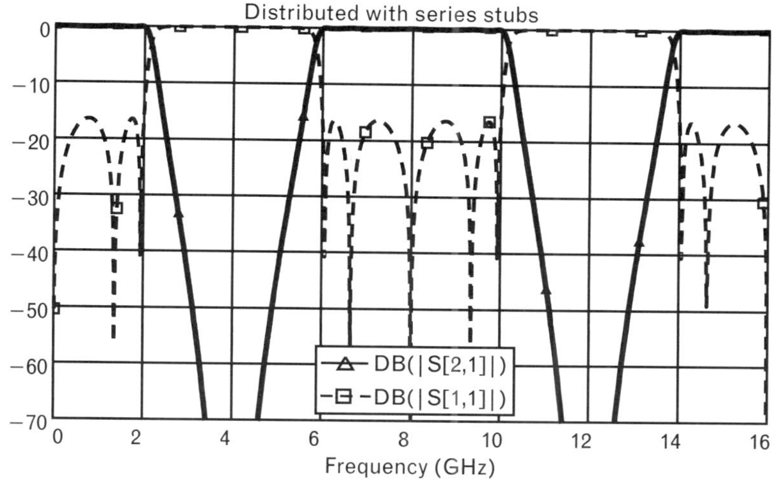

FIGURE 8.35 *Frequency response of the lowpass filter with distributed components. Additional passbands are centered at multiples of 8 GHz. These passbands can be shifted to higher frequencies by reducing the commensurate transmission line lengths. Traces with dashed lines show return loss.*

quarter-wavelength frequency, the stub behaves inductively, until it reaches a frequency where the electrical length is 180°. Then the stub becomes capacitive again, and the cycle continues forever—assuming ideal components.

Whereas in the stopband the frequency response of the lumped lowpass Chebyshev prototype filter rolls off monotonically, the response of the distributed filter rolls off rapidly to reach infinite attenuation at the frequency where the transmission lines are quarter wavelength long. Since our commensurate length is 45° at 2 GHz the quarter-wavelength frequency is twice that or 4 GHz. If the attenuation would remain high, that would be great for lowpass filter designers. Unfortunately, above 4 GHz the attenuation begins to diminish and there is a second equal-ripple passband, beginning at a frequency of $3f_r$. This symmetrical response second passband has a center frequency of 8 GHz and upper corner frequency at 10 GHz, which is equal to $5f_r$. Then the cycle repeats itself giving us a new passband between 14 and 18 GHz, infinite attenuation at 20 GHz, and so on. If the transmission lines were ideal this cycle would repeat to infinity frequency.

An interesting observation is that it is not possible to make a perfect lowpass filter with just transmission line stub components. The unavoidable passbands can be moved higher by reducing the commensurate length of the transmission lines, which sets the quarter-wavelength frequency higher. For example, if instead of 45° we set our commensurate length to 30° at 2 GHz, then the quarter-wavelength frequency is three times higher, or 6 GHz. Now the lowpass filter would roll off and reach infinite attenuation at 6 GHz only, forming the second bandpass between 10 GHz and 14 GHz. However, the price of moving the quarter-wavelength frequency higher is decreased selectivity and more extreme element values.

A second concern with our distributed circuit is that series stubs are not realizable conveniently in planar circuits. In addition, in physical form, transmission line interconnections require elements with finite dimensions. The interconnecting links also need to be modeled in distributed form.

A possible solution to both problems is to add cascade transmission lines[3] to the filter, and with the help of Kuroda transformations eliminate the series transmission line stubs, as shown in Section 8.9.3.1.

8.9 Network transformations

Various types of network transformations exist. Some of them change the response type of a filter, others change circuit topologies, and/or

3. Filter literature refers to cascade transmission lines as *unit elements*. Sometime they are also called *series lines*.

component values to obtain physically realizable filters. Let us look at the most commonly used forms in three major groups.

8.9.1 Transformations to change the filter's response

Lowpass prototype filters can be transformed to highpass, bandpass, and bandstop types with simple mathematical operations and component modifications. Since even the basic synthesis programs provide direct solution to bypass the transformations, we briefly review them, and refer to Zverev's book [8] for details.

8.9.1.1 Lowpass-to-highpass transformations

A highpass filter can be transformed from the lowpass prototype before impedance and frequency scaling. The lowpass filter must be first designed for mirror image symmetry around the passband corner frequency, f_r, of the highpass filter (Figure 8.36), with reciprocal stopband attenuation requirements, where

$$\frac{f_{XL}}{f_{rL}} = \frac{f_{rH}}{f_{XH}} \quad \text{and} \quad f_{rL} = f_{rH}$$

The basic circuit topology remains but each parallel capacitor is replaced by a parallel inductor and each series inductor by a series capacitor. The new element values are the reciprocal of the lowpass prototype values, as shown in Figure 8.37.

8.9.1.2 Lowpass-to-bandpass transformation

Once again we start with a lowpass filter designed for the same bandwidth as the targeted bandwidth of the bandpass filter (Figure 8.38). After the lowpass filter's transformations are completed, we add into each of the branches a resonating element (i.e., to a parallel capacitor we add a parallel inductor, and to a series inductor we add a series capacitor) and set the resonant frequency at the center of the desired passband, f_0. Adding these new elements will transform the lowpass filter to a bandpass response. For narrowband

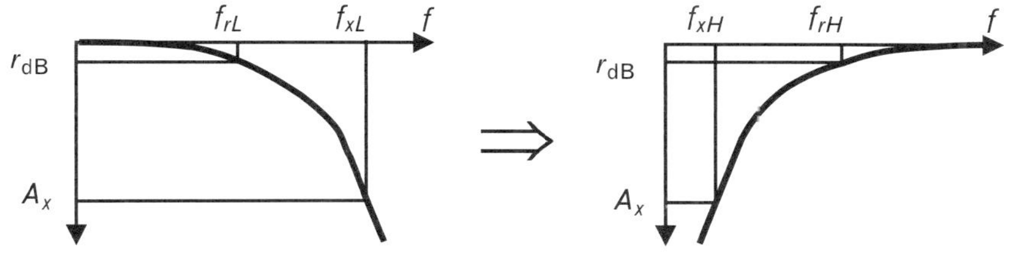

FIGURE 8.36 *A lowpass prototype filter can be transformed into a highpass circuit about its f_r corner frequency.*

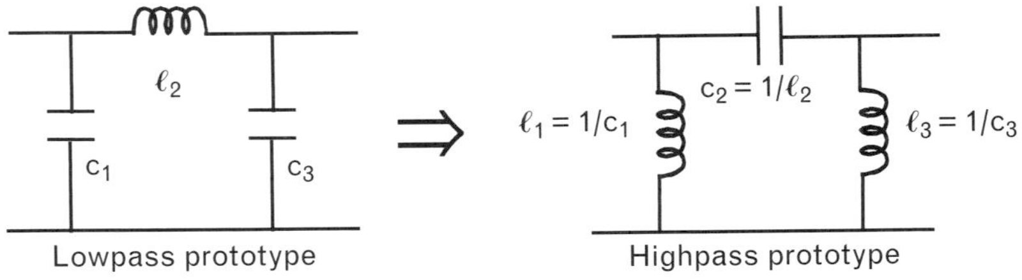

FIGURE 8.37 *Transformed elements of the highpass prototype filter are computed from the lowpass prototype circuit.*

filters the transformation is nearly identical, but as the bandwidth increases, the width of the stopband, BW_2, changes during the transformation.

8.9.1.3 Lowpass-to-bandstop transformation

The technique is similar to what is used in the lowpass-to-bandpass transformation, but here the resonating elements are to block transmission at resonance. To accomplish this we add to each parallel capacitor a series inductor and to each series inductor a parallel capacitor (Figure 8.39). Once again the resonant frequency is selected to be at the center of the desired stopband. The original lowpass filter is to be designed for the same bandwidth as the bandwidth of the bandstop filter.

Although the lowpass-to-highpass transformation always brings physically realizable topologies and element values, the bandpass and bandstop transformations work only for large fractional bandwidths. For narrowband designs, we find that the element values are not realistic and a different approach is used [8] for physical realizable circuits.

8.9.2 Transformations to change termination ratio or element type

Filter synthesis frequently provides unrealizable element values, circuit topologies, or improper source-to-load termination ratios. Norton and Kuroda transformations can be very helpful in such cases, and we provide a brief review of their various forms here. Norton's transforms are shown in Section 8.9.2.1 for lumped *L-C* networks while commensurate-length transmission line transformations are illustrated with Kuroda's techniques in Section 8.9.3. We will also review delta-to-star, also called PI-to-TEE transformations, for lumped inductors and capacitors.

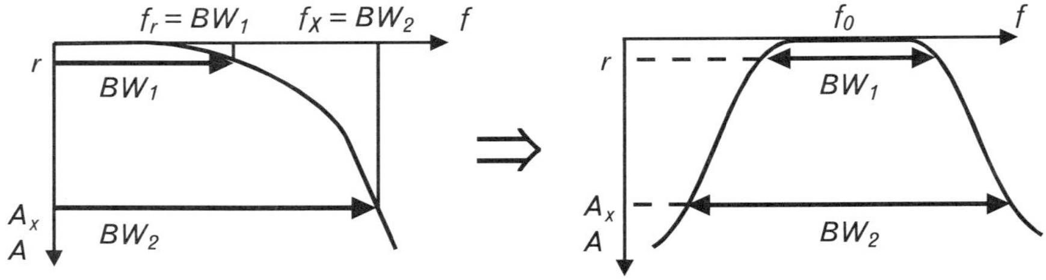

FIGURE 8.38 *A filter with lowpass bandwidths of BW_1 and BW_2 can be transformed into a bandpass filter with nearly identical bandwidths.*

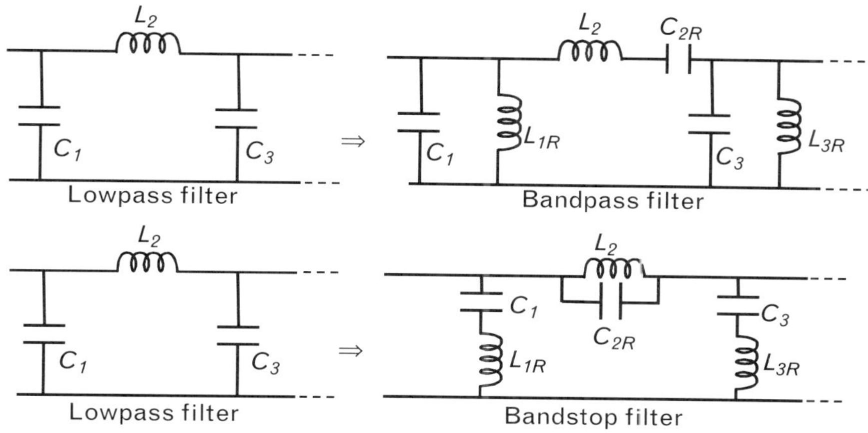

$$C_{iR} = \frac{25.33}{L_i(f_{0GHz}^2)}, \qquad L_{iR} = \frac{25.33}{C_i(f_{0GHz}^2)} \qquad \text{where } i = 1, 2, 3, \ldots$$

FIGURE 8.39 *For bandpass and bandstop transformation, we add an element, C_{iR} or L_{iR}, to resonate each lowpass filter branch. Resonance is set at the center frequency of passband or stopband.*

8.9.2.1 Norton transformations

Through impedance scaling (Section 8.7.1.3), we can always change the impedance level of any passive two-port *L-C* ladder network simultaneously at both sides. Norton impedance transformations may be used to change the termination ratio of a filter or (matching network), and also to *rearrange circuit topology, without changing the order and the response* of the circuit. The transformation is performed by an appropriate pair of adjacent branches of a ladder network. In the simplest forms, the branches may contain only a pair of single capacitors or inductors. In more complex forms, the branches may be made up by several components Depending on the configuration of the component pair, the transformation may be used to increase or decrease the impedance level at one side of the circuit, without affecting the impedance at the second side—a very useful technique in filter and matching network design.

The Norton transformation to increase the load termination is illustrated by a simple two-element circuit in Figure 8.40. Let us assume that the synthesized "left-handed" second-order *L*-network meets certain specifications when operating between a 50-Ω source and a 12.5-Ω load. However, our actual load is four times higher, 50Ω. If we impedance scale the circuit by a factor of 4.0, we arrive to the right load of 50Ω but the source becomes 200Ω. Using the Norton impedance transformation [16], the circuit topology and the element values can be changed to work into an $n = 9$ times larger load of 112.5Ω [since 9(12.5) = 112.5Ω], *without changing the order of the network or the frequency response*. The load impedance transformation factor, n, comes from a relationship that is a function of the two-element values of the circuit

$$n = \left(\frac{L_S + L_P}{L_P}\right)^2 = \left(\frac{10+5}{5}\right)^2 = 9 \tag{8.12}$$

where L_S and L_P are the inductances of the two circuit elements.

While the above-shown transformation is useful if we need *exactly* a 9:1 increase of the load impedance, it does not help us since our desired load is greater than 12.5Ω but less than 112.5Ω. Our actual load is 50Ω, and we want only a 4:1 impedance transformation, not 9:1. The good news is that by adding an extra element, the impedance transformation may be performed to increase the load by any factor k, between 1.0 and n (in this case: $1.0 < k < 9$).

By splitting one of the circuit elements into two parts before the transformation, we can arrive at any transformation ratio between the value of 1.0 and n. For example, let us split the 10-nH series inductor of Figure 8.40 into two equal parts, as shown in Figure 8.41(a). Then, the 5-nH parallel inductor, L_p, and the adjacent series inductor L_{sb} can be used to give us a *partial impedance transformation* ratio of k = 4:1. Inductor L_{sa} remains unchanged in the transformed circuit.

$$k = \left(\frac{L_{sb} + L_P}{L_P}\right)^2 = \left(\frac{5+5}{5}\right)^2 = 4$$

By having three components in the transformed circuit [Figure 8.41(b)], we reached the desired 50-Ω load impedance. *The order of the network has not changed*, even though we have an extra element—the order is still 2. The frequency response of the circuits before and after transformation is exactly the same from dc to infinity.

A second option to obtain a 50-Ω load is to split the 5-nH parallel inductor, L_p, into two equal parts of $L_{pa} = L_{pb}$ = 10 nH each. Now we perform the impedance transformation with the original series inductor, L_S, and L_{pa} to reach a 4:1 transformation ratio (Figure 8.42). Our new circuit again has three components, but it is still a second-order network. Even

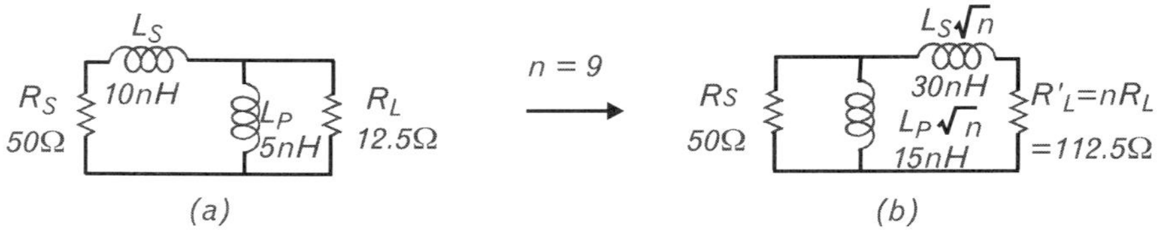

FIGURE 8.40 *Two second-order networks with identical frequency responses from dc to infinity. (a) Original network and (b) results of the Norton transformation to nine times higher load impedance.*

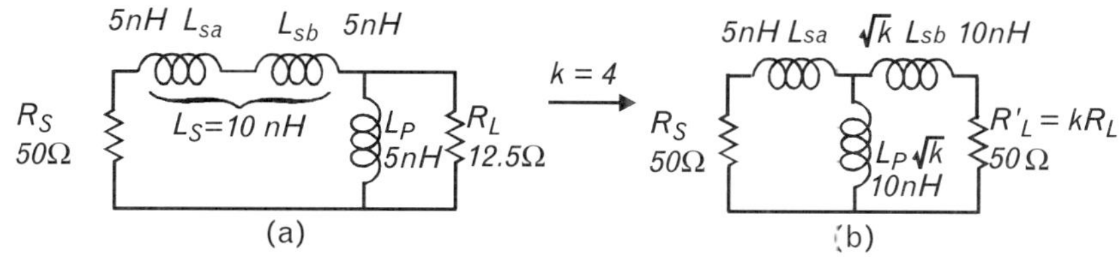

FIGURE 8.41 *(a) Using only part of the series inductor for the transformation allows us to increase the load by a factor of $k = 4.0$ (instead of the original transformation with $n = 9.0$). (b) The final circuit is a TEE-section with three components.*

with the new element values and topology, the frequency response will be exactly the same as it was before.

Two inductors or two capacitors in a *left-handed* L-network configuration may always be used to provide a full or partial transformation to a *higher* load impedance. If we want to *reduce* the load termination, we need two elements that form a *right-handed* L-network, as shown in Figure 8.43. For this circuit the maximum amount of down-transformation, $1/n$, is also computed from the two element values using (8.12).

$$n = \left(\frac{L_S + L_P}{L_P}\right)^2 = \left(\frac{200 + 50}{50}\right)^2 = 25$$

A full transformation reduces the load from 200Ω to 200/25 = 8Ω.

If instead of the full transformation we only need a partial transformation, we can again split the series or the parallel component into two parts, as shown in Figure 8.44, to give us partial lower transformation with the ratio $k = 4$ for a 50-Ω load. Again we end up with three-element networks but neither the order nor the frequency response changes from the initial design.

Figure 8.45 summarizes the two-element Norton transformations to increase load termination with a left-handed *L*-network, and Figure 8.46

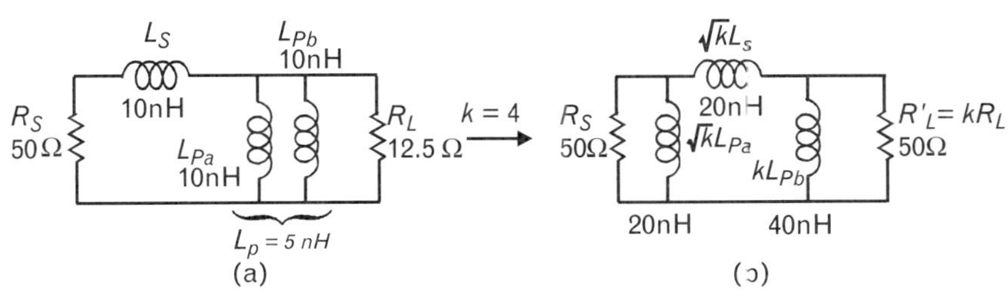

FIGURE 8.42 *(a) Transforming through the series inductor, L_S, with L_{pa} also leads to a $k = 4$ times increase of the original 12.5-Ω load. (b) The result is now a Pi-section. Notice that inductor L_{pb} also increases by a factor of $k = 4.0$.*

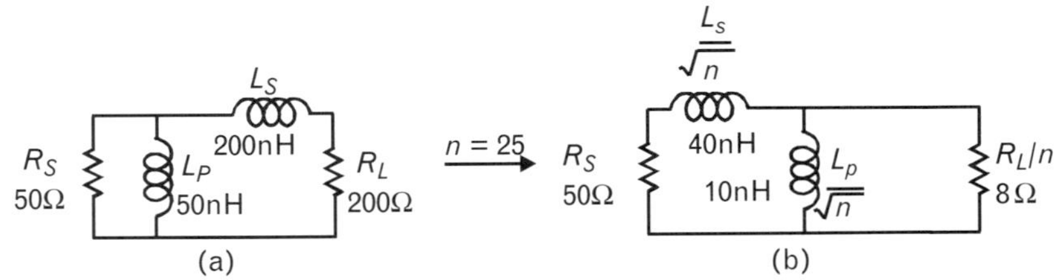

FIGURE 8.43 *(a) A right-handed L-network with two inductors or two capacitors can be used to lower the load impedance. (b) The transformed circuit has new element values.*

provides the formulas for lowering loads with right-handed *L*-networks. In the case of the full transformation the ratio *n* is used. Full transformation does not increase the number elements but interchanges the positions of the transformer section's components. For partial transformation of ratio *k* the intermediate parameters *m* and *p* are introduced, and the initial two-element transforming section changes to a three-element PI- or TEE-section.

Notice that in all cases, the reactances of any additional component placed between the transforming circuit section and the load is also transformed by the same factor as the load. Also note that since the transformation requires a bandpass section of two identical types of components (two capacitors or two inductors), it is *not applicable to lowpass or highpass circuits.*

An example of the Norton transformation is shown in Figure 8.47, applying it to the result of a synthesized fifth-order bandpass circuit. A given set of specifications are fulfilled with the synthesized circuit operating between $R_S = 50\Omega$ and $R_L = 12.5\Omega$, as shown in the schematics. The actual load ($R_L = 50\Omega$) is four times greater than the synthesized result.

From Figure 8.45 we compute the transformation ratio available with the two 10-pF capacitors, to increase the load by a factor of *n*.

FIGURE 8.44 *For load reduction by $k < n$, we also have two options, leading to (a) a TEE-network and (b) a PI-network.*

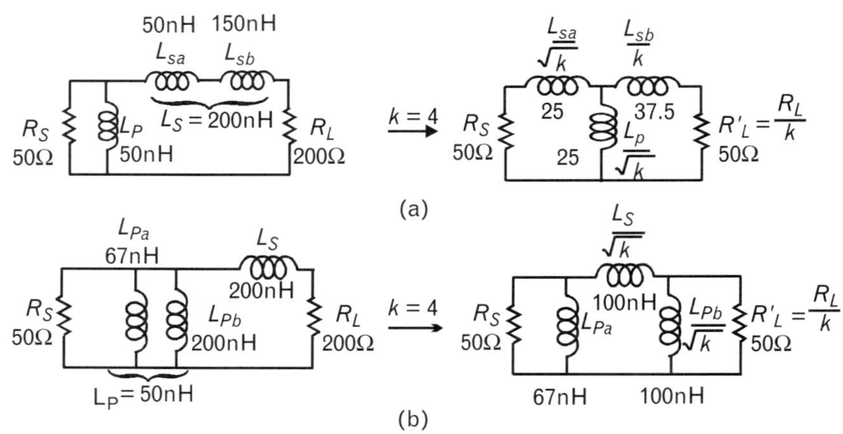

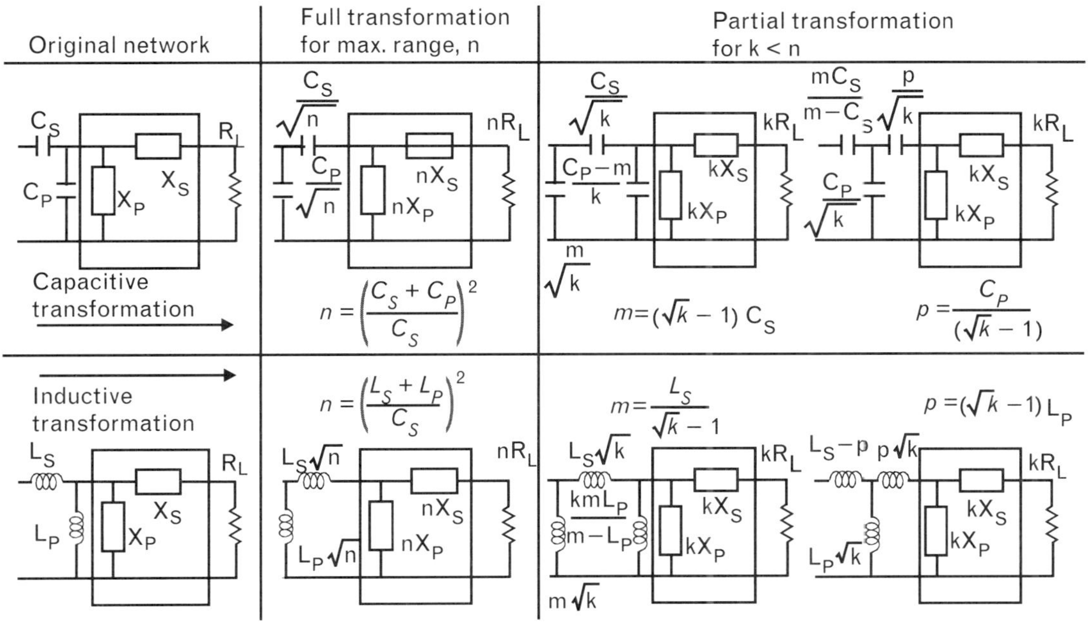

FIGURE 8.45 *Summary of Norton impedance transformations to increase the load (right-side) termination by a factor of n for full and by a factor of k for partial transformation.*

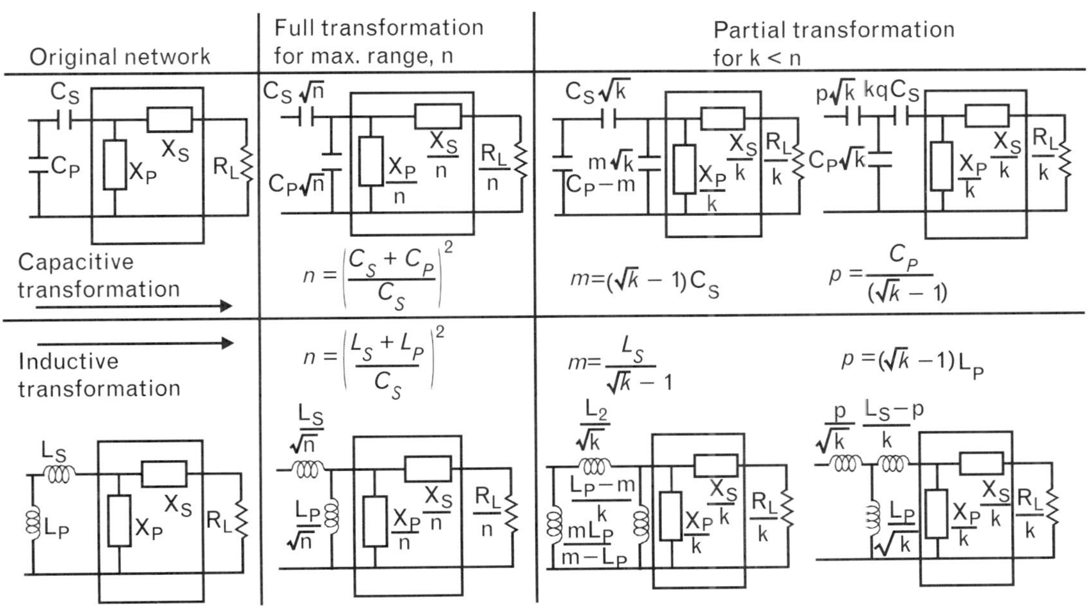

FIGURE 8.46 *Norton impedance transformations to lower the load (right-side) termination by a factor of n for full and by a factor of k for partial transformation.*

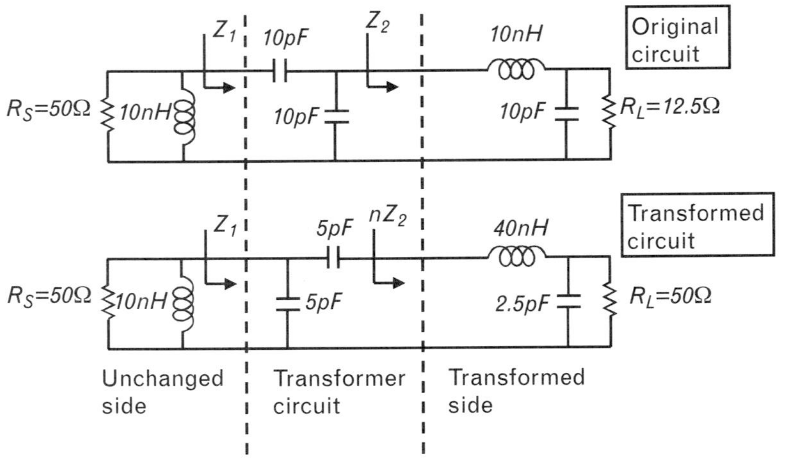

FIGURE 8.47 *Norton transformation enables us to increase the load impedance of the initial circuit from 12.5Ω to 50Ω without changing the frequency response of the circuit. All components in the transformed side of the circuit increase their impedances by a factor of four. The left side of the circuit remains unchanged.*

$$n = \left(\frac{C_S + C_P}{C_P}\right)^2 = \left(\frac{10+10}{10}\right)^2 = 4$$

Applying a 4:1 up-transformation with the two 10-pF capacitors to the circuit of Figure 8.47, changes the topology and some of the element values. Note that at the left side of the transforming section, elements R_S and the 10-nH parallel inductor are *unchanged*. All elements on the right side of the transforming section increased their impedances by a factor of n. We should point out that increased reactance leads to a *larger* inductor and *smaller* capacitor.

Several additional forms of Norton transformations exist with more complex branches as well as circuit topologies, and we refer interested readers to [16]. These transformations are commonly used to obtain more physically more realizable topologies and/or component values.

8.9.3 Transformations to change termination ratios and/or circuit topologies of transmission line networks

8.9.3.1 Kuroda identities (transformations)

For commensurate-length transmission line circuits the various types of Kuroda identities [17] are very useful. We show 10 of them in Figure 8.48. In the first two [Figure 8.48(a,b)] the transformation changes the topology,

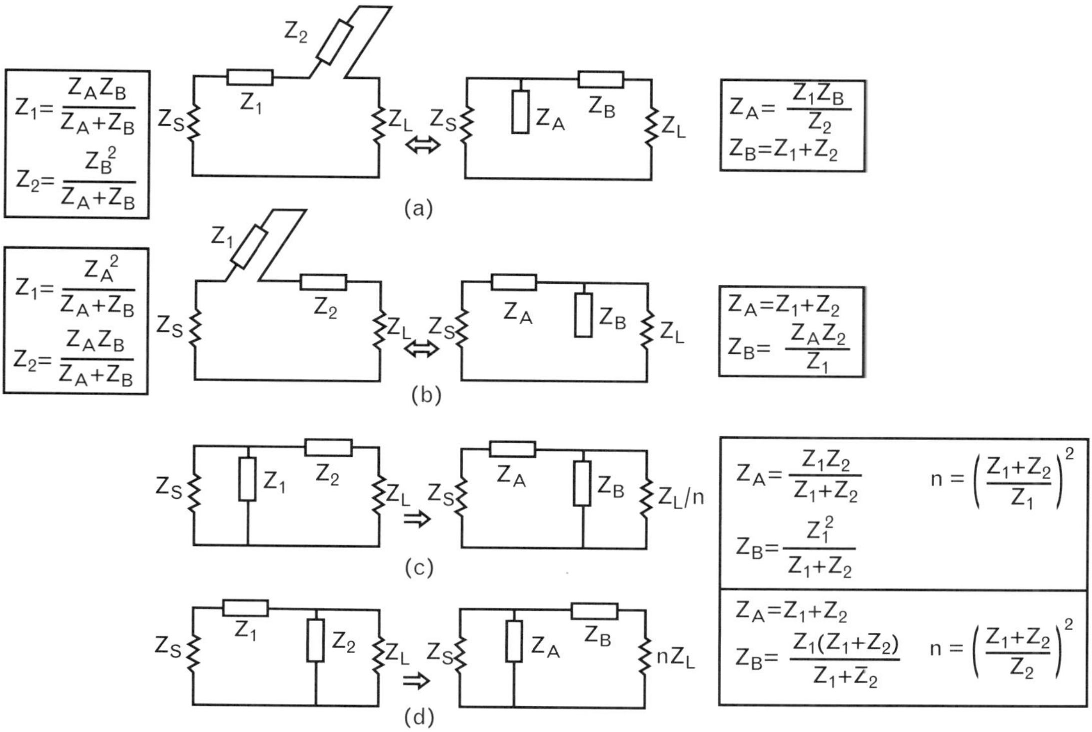

FIGURE 8.48 *Various Kuroda transformations and related impedance values for commensurate-length line circuits. (a) and (b) do not affect the termination ratios, while (c) through (j) are similar to the Norton transformations since they change the termination ratios by a factor of n.*

the type of the transmission line elements, as well as their impedances, *without affecting the termination ratio*. The electrical lengths of both transmission lines remain the same after the transformation. These transformations are used when the element type attained by the filter synthesis is a series short-circuited stub, that is not physically realizable in planar circuits. It may also be used as an intermediate step in a multistep transformation.

The remaining eight of Kuroda identities change the configuration of the transmission lines, the impedance values, and the termination ratios. These transformations are used when the initial source-to-load ratio is improper. Just as in the case of the Norton transformations, the Kuroda transformations do not affect the order of the network or the frequency response.

8.9.3.2 Illustrative example: eliminating series stubs with Kuroda transformations

If we add four commensurate length 50-Ω lines to the filter of Figure 8.34(a), with four consecutive Kuroda transformations we can change the

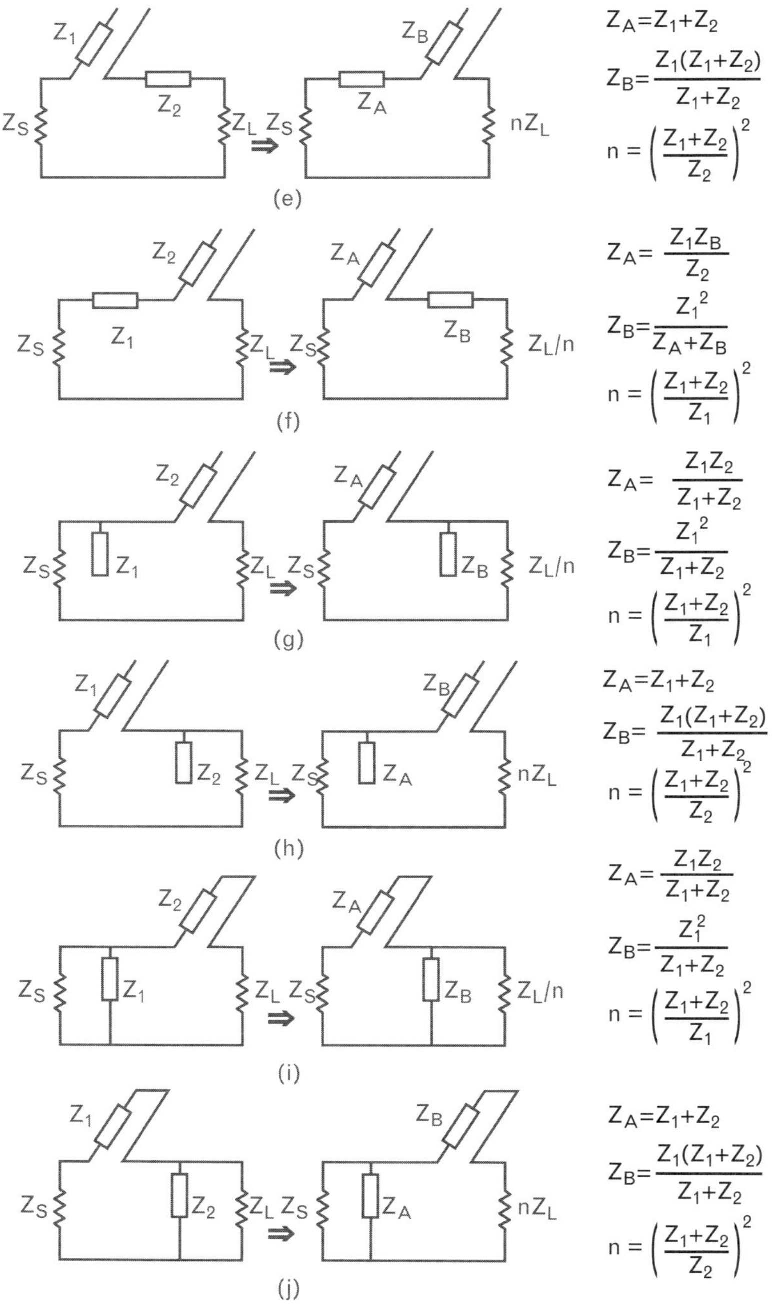

FIGURE 8.48
Continued.

circuit topology to an *easily realizable physical* form. The final transformed circuit has five parallel open-circuited stubs, separated by four cascade transmission lines. In addition to eliminating the series stubs, the cascade lines are also useful to establish well-modeled physical discontinuities with the parallel open-circuited stubs. The results of the Kuroda transformations are shown in Figure 8.49. Since the 50-Ω unit elements added to the 50-Ω terminations at both sides are redundant types, they do not change the order or the frequency response of the circuit.

The transformations of the circuit shown in Figure 8.49 are done in the following sequence:

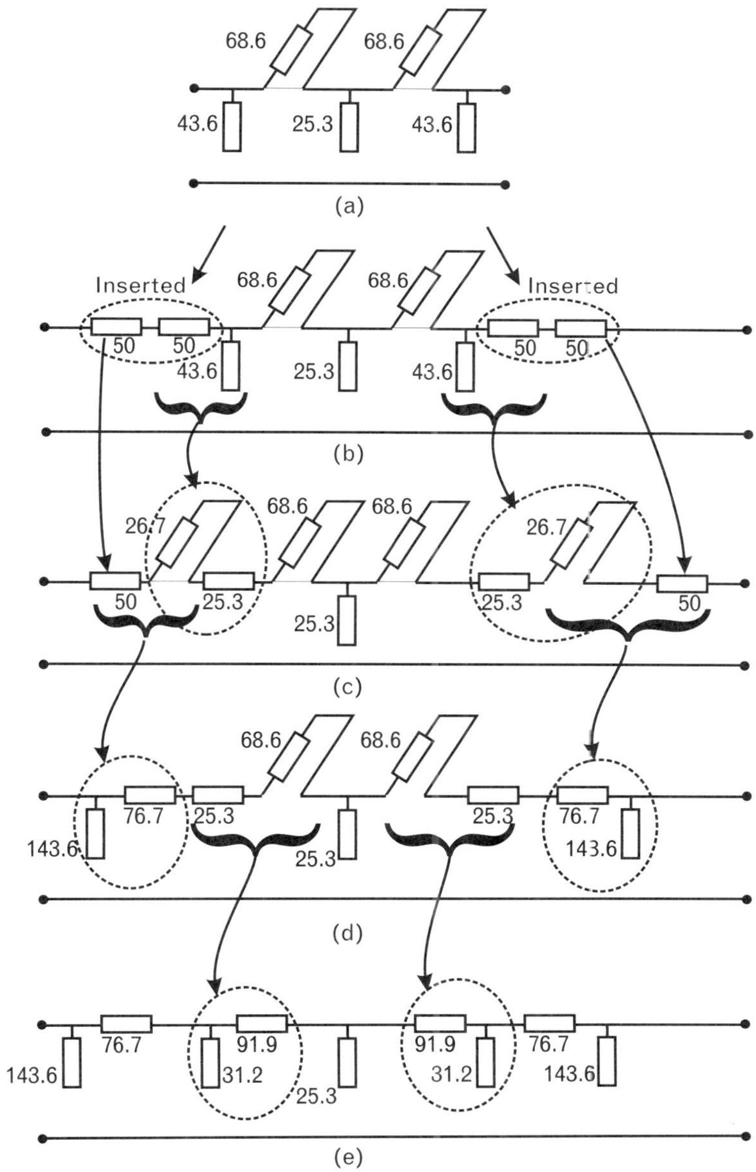

FIGURE 8.49 *Kuroda transformations convert (a) the five-element transmission line stub filter of Figure 8.34 to (e) a new nine-element circuit, which can be realized in planar form. Although we added four more elements to the initial circuit in steps (b) through (d), neither the order of the circuit nor the frequency response is changed. Transmission line lengths are 45°at 2 GHz. Numbers on the schematics refer to characteristic line impedances.*

1. The original fifth-order distributed filter network [Figure 8.49(a)] has three parallel open-circuited stubs and two series short-circuited stubs. The series elements are not realizable in planar form and the circuit topology assumes zero spacing among the elements. All five transmission lines have *equal electrical lengths*. Keep in mind that if the circuit is realized in a nonhomogeneous dielectric media where $\varepsilon_{EFF} \neq \varepsilon_R$, the physical line lengths are not equal.

2. Adding two commensurate 50-Ω cascade lines to each sides of the circuit does not change the order or the frequency response [Figure 8.49(b)].

3. The Kuroda transformation is first applied to a 50-Ω cascade line – 43.6-Ω parallel open stub combination at each side, as indicated in Figure 8.49(b). The circuit in Figure 8.49(c) results. Now we have four series short-circuited stubs, so we seem to be making things worse.

4. Transforming the first two elements 50-Ω cascade line – 26.7-Ω series shorted stub, at both sides [as indicated in Figure 8.49(c)] changes the two series shorted stubs to parallel open stubs, as shown in Figure 8.49(d). Now we only have two series stubs left.

5. Finally, transforming the third and fourth elements, 25.3-Ω cascade line – 68.6-Ω series shorted stub at both sides as indicated in Figure 8.49(d) eliminate the series stubs and now we have a planar circuit [Figure 8.49(e)]. The 25.3-Ω parallel stub's characteristic impedance at the center may be too low for microstrip realization, but we can replace it with two equal-length open stubs of 50.6-Ω impedance[4]. If needed, the same can be applied to the 31.2-Ω impedance stubs. The only remaining problem is presented by the 143.6-Ω impedances at the ends that lead to narrow physical dimensions. However, we still need to add the discontinuity junctions to the circuit (see discussion on discontinuities in Chapter 7). After that step, a slight reoptimization can help to increase the width of the high-impedance lines.

In Section 5.1 we introduced the concept of transmission zeros that helped us to establish the order of ladder networks. With lumped elements, we can always find the order by simple physical inspection of the circuit topology. We simply added the number of nonredundant transmission zeros at dc, infinity, and finite frequencies. In transmission line circuits, as long as we have only open- and short-circuited stubs in series or parallel

4. Although a mathematically correct procedure, it can lead to frequency-response problems in bandstop filter [1].

forms, counting the transmission zeros again is relatively easy. A nonredundant series short-circuited or parallel open-circuited stub blocks transmission at 90°, 270°, and every additional 180° electrical length. Even though each stub creates a multiplicity of transmission zeros at various finite frequencies, it only adds one to the order of the network. Nonredundant stubs also affect selectivity.

We emphasize the term *nonredundant*, because we mentioned earlier that a parallel open stub may be replaced by two equivalent stubs. Obviously the order of the network does not increase after such a procedure, and the newly created stub is treated as a *redundant element* for order determination.

When the network also includes cascade transmission lines, finding the order takes more work. Each *nonredundant* cascade transmission line adds one to the order of the network and increases the selectivity of the filter. Redundant cascade lines only affect the transmission phase, not the order or the selectivity. However, in most cases we cannot tell by observation only if a cascade line is redundant or nonredundant. Obviously, if we have a 50-Ω source and there is a 50-Ω transmission line at the input side of the filter, the line is a redundant element; we can remove that line from the circuit without affecting the order and magnitude of the frequency response. On the other hand, when a cascade line is placed between two stubs, forming a PI-network, we cannot tell just by observation whether it is redundant or nonredundant—even when its characteristic impedance happens to be equal to the source or load termination. To determine whether the line is redundant or not, we need to perform a Kuroda transformation to move the transmission line in both directions, toward the load and toward the source, to see if the transformed line impedance is equal to one of the terminations. Only after these tests are made can we be certain about the nature of the unit element. Therefore, we should not attempt to establish the order of a distributed network filter merely by observation of the circuit topology. To prove our point, let us look again at the example shown in Figure 8.49.

At the beginning of our Kuroda transformations, we had a fifth-order network [Figure 8.49(a)] to which we added four 50-Ω redundant cascade transmission lines [Figure 8.49(b)]. The final circuit, shown in Figure 8.49(e), has nine elements, yet the frequency response remained unchanged by the transformations. The four cascade lines are *redundant elements*. Therefore, even though we increased the number of filter components from five to nine, the final circuit is still a fifth-order circuit. If you only had access to the final circuit, it requires a fair amount of work to find the network order. Assuming that it has an order of nine is obviously a wrong conclusion.

8.9.4 Transformations to change element values

These network transformations allow us to increase and decrease element values through topology changes.

8.9.4.1 Delta-*Y* (Delta-Star, PI-TEE) transforms

Three-element inductive (or capacitive) PI and TEE-networks can always be interchanged with each other, often bringing an order of magnitude change to element values. They can be very helpful when component values are approaching the high or low end of the physically realizable range. The relationships between the forms are given in Figure 8.50.

Viewing the forms of the transform equations in Figure 8.50 tells us that if the three capacitors have *similar values*, changing from PI to TEE configuration the capacitor values generally increase, and decrease when going from TEE to PI. To show a simple example, let us use a PI-circuit with three equal 10-pF capacitors. The equivalent TEE-network has three 30-pF components. If our PI-network is realized with three 10-nH inductors, the effect of the transformation is just the opposite—we get a TEE-circuit with 3.33-nH inductors.

If our initial components have a wide range of values, the above-mentioned rules may not hold. For example, starting with a capacitive PI-network where $C_1 = C_2 = 1$ pF and $C_3 = 10$ pF, the transformed *TEE*-network has $C_A = C_B = 21$ pF, but $C_C = 2.1$ pF. In this case only, the first and third elements (C_A and C_B) increase and the middle one (C_C) actually decreases. We will apply one of these transformations later in Section 8.10.1.2 to an actual circuit to illustrate their use.

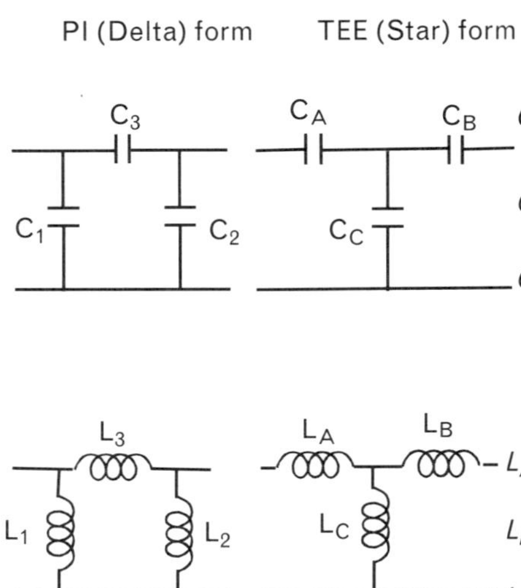

$$C_1 = \frac{C_A C_C}{C_A + C_B + C_C} \qquad C_2 = \frac{C_B C_C}{C_A + C_B + C_C} \qquad C_3 = \frac{C_A C_B}{C_A + C_B + C_C}$$

$$C_A = \frac{D}{C_1} \qquad C_B = \frac{D}{C_2} \qquad C_C = \frac{D}{C_3} \qquad D = C_1C_2 + C_1C_3 + C_2C_3$$

$$L_1 = \frac{T}{L_A} \qquad L_2 = \frac{T}{L_B} \qquad L_3 = \frac{T}{L_C} \qquad T = L_AL_B + L_AL_C + L_BL_C$$

$$L_A = \frac{L_1L_3}{L_1 + L_2 + L_3} \qquad L_B = \frac{L_2L_3}{L_1 + L_2 + L_3} \qquad L_C = \frac{L_1L_2}{L_1 + L_2 + L_3}$$

FIGURE 8.50 *PI-TEE network transformation forms and related equations. Transforming from PI to TEE generally lowers the element reactances, leading to larger capacitor and smaller inductor values. The opposite is true for a TEE-to-PI transform.*

8.10 L-C resonant circuits in filter design

We already saw in Section 5.2.1 that a simple series L-C network placed between two resistive terminations acts as a bandpass filter. The 3-dB bandwidth of this bandpass filter is computed from the loaded Q of the resonant circuit. The same applies for parallel resonant circuits as shown in Figure 8.51 where two resistive terminations are separated by a parallel L-C network. At the resonant frequency, the susceptances of the inductor and capacitors cancel each other and the source works directly into the load. If the terminations are not equal, there is a mismatch loss that is directly computed from the resistances of the two terminations.

The total parallel loading at resonance, R_{PT}, is presented by the parallel connection of the two terminations,

$$R_{PT} = \frac{R_S R_L}{R_S + R_L}$$

At resonance, the magnitudes of the two reactances are equal, $X_L = X_C$. The loaded Q of the resonant circuit, denoted by Q_{RES}, is computed by taking the ratio of the total loading resistance and either one of the two reactances.

$$Q_{RES} = \frac{R_{PT}}{X_L} = \frac{R_{PT}}{X_C}$$

Knowing the loaded Q, we can compute the 3-dB bandwidth at a known resonant frequency, f_0,

$$BW_{3\,\mathrm{dB}} = \frac{f_0}{Q_{RES}} \tag{8.13}$$

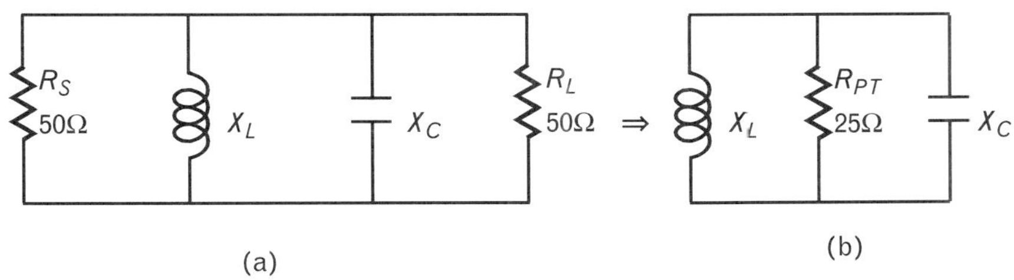

FIGURE 8.51 *(a) Adding a resonant L-C section between two resistive terminations creates a bandpass filter. (b) The loaded Q of the resonator determines the bandwidth.*

8.10.1 Illustrative example: bandpass resonant circuit design

Design a bandpass filter with a single resonant circuit for a center frequency of 500 MHz and desired 3-dB bandwidth of 50 MHz. Use a 50-Ω source and 50–Ω load terminations and initially ideal circuit elements. Next, show the changes caused by finite component Qs of 80 at 500 MHz, and plot the frequency response through the 450- to 550-MHz range.

Solution

To design a bandpass filter with a resonant circuit, we first need to find Q_{RES}. From (8.13) we compute

$$Q_{RES} = \frac{f_0}{BW_{3\,\text{dB}}} = \frac{500}{50} = 10$$

The effective total loading with ideal resonator components is found from the two terminations.

$$R_{PT} = \frac{R_S R_L}{R_S + R_L} = \frac{50(50)}{50 + 50} = 25\Omega$$

From the loaded Q and the total effective loading resistance we can compute the individual reactances and component values at the center frequency.

$$X_L = X_C = \frac{R_{PT}}{Q_{RES}} = \frac{25}{10} = 2.5\Omega$$

$$L = \frac{0.159 X_L}{f_{0\text{GHz}}} = \frac{0.159(2.5)}{(0.5)} = 0.8 \text{ nH}$$

$$C = \frac{159}{f_{0\text{GHz}} X_C} = \frac{159}{(0.5)2.5} = 127 \text{ pF}$$

The circuit schematics and frequency response are shown in Figure 8.52. The element values are, however, not practical since the inductance is too low, and the capacitance is too large at 500 MHz, but we will deal with that problem later. Simulating the circuit response from 450 MHz to 550 MHz gives us the expected bandpass shape, and the 3-dB bandwidth is exactly 50 MHz. Although for narrow bandwidth the frequency response looks nearly symmetrical, the frequency symmetry is geometric rather than arithmetic.

The circuit shown in Figure 8.52 has two problems. First, the element values are not practical for that frequency range, and second we use ideal

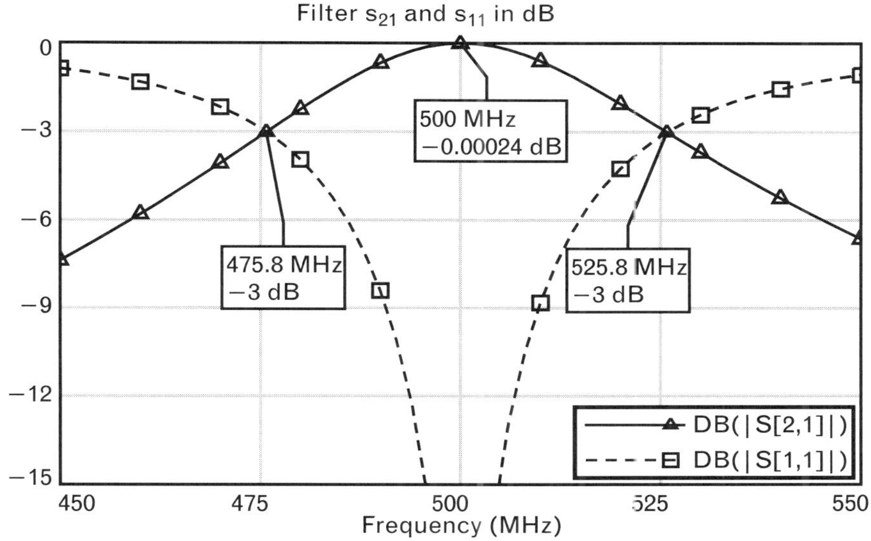

FIGURE 8.52 *Bandpass resonant circuit centered at 500 MHz with 3-dB bandwidth of 50 MHz. Note that the 3-dB frequencies of transmission and reflection coefficients are equal in a passive lossless two-port that operates between equal terminations.*

components without losses. Resimulating the circuit using $Q_L = Q_C = 80$ at 500 MHz for both components shows the effect of the dissipative circuit losses. To compute the effect of inductor and the capacitor losses on the bandwidth, we added two parallel resistors, R_{PL} and R_{PC}.

$$R_{PL} = Q_L X_L = 80(2.5) = 200\Omega$$

$$R_{PC} = Q_C X_C = 80(2.5) = 200\Omega$$

Looking at the revised schematics (Figure 8.53) the total loading on the resonant circuit is now the parallel equivalent of four resistors,

$$R_{PT} = R_S \,//\, R_{PL} \,//\, R_{PC} \,//\, R_L = 20\Omega$$

The new loaded Q of the resonator is now

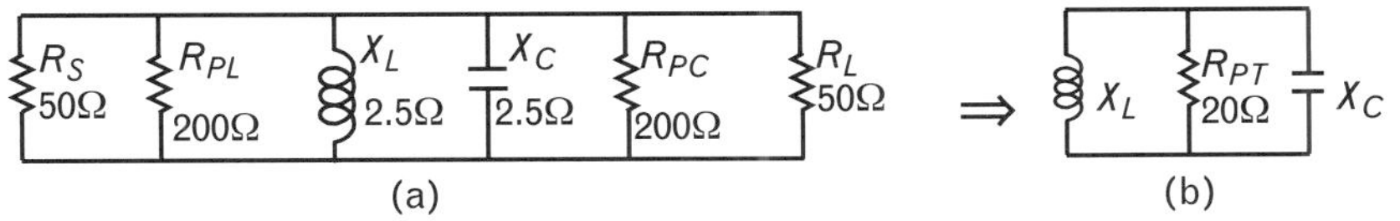

FIGURE 8.53 *Equivalent circuit of the bandpass filter with lossy components. (a) The two additional 200-Ω parallel resistors, R_{PL} and R_{PC}, represent the losses of the inductor and capacitor. Both component have unloaded Qs of 80 at 500 MHz. (b) Equivalent circuit used for bandwidth computation.*

$$Q_{RES} = \frac{R_{PT}}{X_{PL}} = \frac{R_{PT}}{X_{PC}} = \frac{20}{2.5} = 8.0$$

The new 3-dB bandwidth has increased to

$$BW = \frac{f_0}{Q_{RES}} = \frac{500\text{ MHz}}{8.0} = 62.5\text{ MHz}$$

The frequency response of the resonant circuit with and without the losses is shown on Figure 8.54. Finite component Qs cause 2-dB loss and poorer selectivity.

The original 50-MHz bandwidth could still be achieved by using lower reactances to increase the loaded Q back to 10. The new reactances,

$$X_L = X_C = \frac{R_{PT}}{Q_{RES}} = \frac{20\Omega}{10} = 2.0\Omega$$

The corresponding new inductance and capacitance values would be even *less practical* to realize at 500 MHz. We would also have to change the two resistor values, which represent the component losses, to maintain the Q values of 80. The resistor change then affects the total loading on the circuit, so we have an iterative problem to work out. We leave that as an exercise to interested readers.

The concern regarding impractical element values could be solved by impedance scaling if we could change the source and load terminations. For example, using the initial component values, impedance scaling the

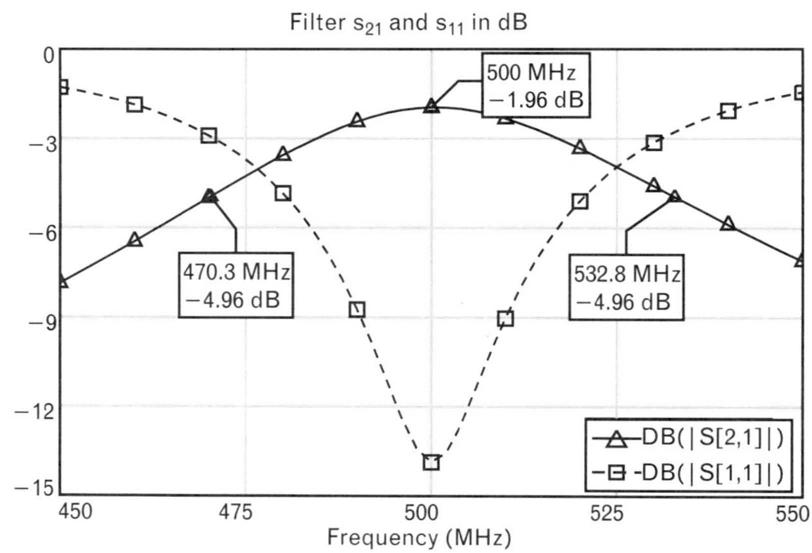

FIGURE 8.54 *The lossy resonant circuit has close to 2-dB total loss at the center frequency and the 3-dB bandwidth, relative to the attenuation at 500 MHz, increases to 62.5 MHz. The 3-dB frequencies of s_{11} and s_{21} are no longer equal to each other.*

circuit upward by a factor of 10 increases the inductance to 8 nH and lowers the capacitance to 12.7 pF, making them reasonable values for the 500-MHz frequency range. However, impedance scaling produces 500Ω source and load, which are 10 times larger than what we actually have in our system. Since we only have one inductor and one capacitor, none of the impedance transformations are available to reduce the terminations back to the actual values. The main problem is that the existing 50Ω terminations are too low, presenting too much loading for the resonant circuit. However, there is a way to *increase the effective resistance* seen by the resonator by *tapping* the inductor or the capacitor. Tapping the inductor is the more complicated of the two options since it also involves mutual coupling, so we will use the second alternative for illustration[5].

To increase the effective resistance of both terminations, we first split the parallel resonator's capacitor into two parts, C_1 and C_2. Then, we replace the individual parallel capacitors with parallel-series capacitor combinations to effectively uptransform the source and load to higher impedances, as shown for the source side in generalized form in Figure 8.55. Even though the external terminations are still at the original 50-Ω values, as far as the loading of the resonant circuit is concerned, the resistances are uptransformed by a factor of 10 from both sides. The values of the series and parallel capacitors are computed with (8.14) through (8.18). Details of the circuit example are shown in Figure 8.56.

$$C_{P(\text{pF})} = \frac{159}{f_{\text{GHz}} X_P} \tag{8.14}$$

$$C_{S(\text{pF})} = \frac{159}{f_{\text{GHz}} X_S} \tag{8.15}$$

where

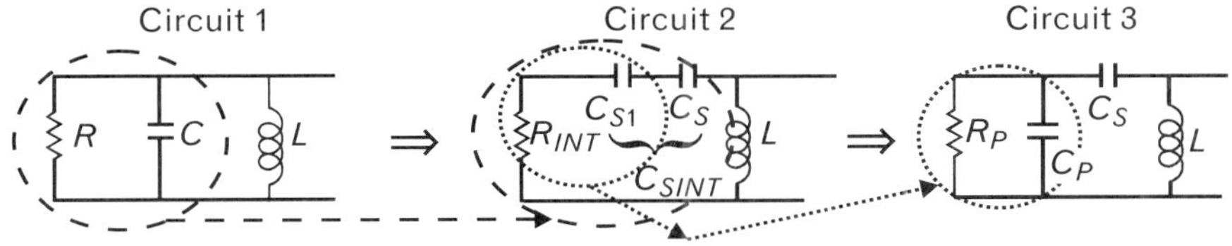

FIGURE 8.55 *Impedance transformation through C_P and C_S to lessen the loading effect of the terminations to the resonant circuit. Through a specified narrow frequency range, the loading effect of circuit section R_P-C_P-C_S is the same as that of R-C. The desired termination value, R, is chosen above the actual termination, R_P. Components R_{INT} and $C_{S,NT}$ are only intermediate values, not used in the final circuit.*

5. Interestingly, in distributed networks, just the opposite applies and we tap the parallel short-circuited stubs.

$$X = \frac{159}{f_{\text{GHz}} C_{(\text{pF})}} \tag{8.16}$$

$$X_P = \frac{1}{\sqrt{\frac{R}{R_P}\left(\frac{1}{R^2} + \frac{1}{X^2}\right) - \frac{1}{R_P^2}}} \tag{8.17}$$

$$X_S = \frac{R^2 X}{R^2 X^2} - \frac{R_P^2 X_P}{R_P^2 + X_P^2} \tag{8.18}$$

R_{SYS} is the *actual* system termination that is transformed to a desired higher value, R, through C_P and C_S.

R is the desired (transformed) *effective resistance* seen by the resonator.

C is the capacitor of the *initial* resonator.

C_P and C_S *replace* C and also *transform* R_{SYS} to R through a narrow passband.

Applying these formulas to our initial component values gives us the circuit shown in Figure 8.56. These transformations are frequency dependant but they hold up quite well for a 5–10% frequency range in a bandpass circuit. Comparison of the initial and the final circuit is shown in Figure 8.56(a–d) in four steps, where

(A) The initially computed circuit with unrealizable resonator elements (L = 0.8 nH, C = 127 pF);

(B) Initial circuit after impedance scaling all components by a factor of 10, decreasing C to 12.7 pF;

(C) Splitting the new resonator capacitance, C, into two parts: $C_1 = C_2$ = 6.35 pF;

(D) Transforming both terminations back to 50Ω through the parallel-series capacitor combinations. Element values are computed from (8.14) and (8.15).

Computations of the two capacitors, C_P and C_S, shown in Figure 8.56(d), are shown next. First we compute the reactance of the 6.35-pF parallel capacitor, C, from (8.16)

$$X = \frac{159}{f_{\text{GHz}} C_{(\text{pF})}} = \frac{159}{0.5(6.35)} = 50.1\Omega$$

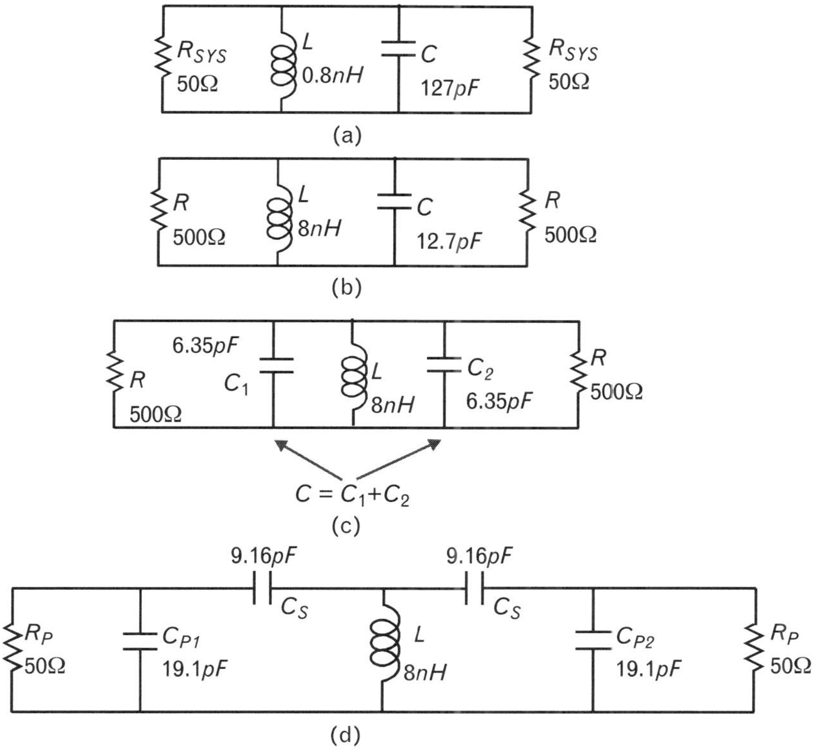

FIGURE 8.56 *Initial to final circuit transformations leading to realizable resonator elements. (a) Initial filter with unrealizable element values at RF. (b) Results of impedance scaling. (c) Splitting the capacitor of the resonant circuit. (d) Transforming back to the original 50-Ω source and load.*

Next, the reactances of the two new capacitors, C_P and C_S, are found from (8.17) and (8.18),

$$X_P = \frac{1}{\sqrt{\frac{R}{R_P}\left(\frac{1}{R^2}+\frac{1}{X^2}\right)-\frac{1}{R_P^2}}} = \frac{1}{\sqrt{\frac{500}{50}\left(\frac{1}{500^2}+\frac{1}{50.1^2}\right)-\frac{1}{50^2}}} = 16.6\Omega$$

$$X_S = \frac{R^2X}{R^2+X^2} - \frac{R_P^2X_P}{R_P^2+X_P^2} = \frac{500^2(50.1)}{500^2+50.1^2} - \frac{50^2 16.6}{50^2+16.6^2} = 34.7\Omega$$

Finally, the capacitor values of C_P and C_S from (8.14) and (8.15),

$$C_{P(\text{pF})} = \frac{159}{f_{\text{GHz}}X_P} = \frac{159}{0.5(16.6)} = 19.1\,\text{pF}$$

$$C_{S(\text{pF})} = \frac{159}{f_{\text{GHz}}X_S} = \frac{159}{0.5(34.7)} = 9.16\,\text{pF}$$

Frequency responses of the two circuits shown in Figure 8.56(a, d) track very closely through a 100-MHz bandwidth centered at 500 MHz, that is, a 20% fractional bandwidth.

Most of the challenging narrowband filters using communication systems are designed with resonant circuits. For highly selective filters two or more resonators may be coupled inductively or capacitively to give us an equal ripple passband and increased selectivity. Inductively or capacitively coupling two or more resonant circuits improves selectivity. The amount of coupling is very important—tight coupling detunes the resonators and insufficient coupling leads to high loss. Between the overcoupled and undercoupled case there is a critical coupling for the lowest equal-ripple passband and maximum selectivity [18]. To minimize the loading of the terminations on the resonant circuits, we can again use the inductive or capacitive step-up approach shown in Figure 8.55.

8.10.2 Illustrative exercise: synthesis and transformations of a capacitively coupled resonator filter

To demonstrate the use of various network transformations we synthesized a capacitively coupled *L-C* resonator filter with 0.1-dB equal-ripple passband between 820 and 860 MHz. The schematic of the original filter is shown in Figure 8.57(a). Synthesis and transformations were performed by Filsyn, a general-purpose filter program [3]. The various steps are as follows:

1. The resonators of the synthesized circuit have unreasonable component values [Figure 8.57(a)]. A 0.987-nH inductor is not practical at the 850-MHz frequency range, and a 32.2-pF capacitor is greatly affected by its self-capacitance, as we discussed in Chapter 7. The ratio of largest and smallest capacitors is nearly 14:1, which is also too high. We need to increase the inductor to a more reasonable value, to at least 5 to 10 nH and also to reduce the extreme value range of the capacitors. We will aim at 8.2-nH inductance, which is a reasonable value for realization at 850 MHz.
2. Impedance scaling the circuit by a factor of n increases the reactances of all components, including the terminations by the same ratio. Choosing $n = 8.2/0.987$ *increases* the original synthesized inductors to 8.2 nH, and *reduces* the resonator capacitors to 3.88 pF, making both elements more realizable [Figure 8.57(b)]. However, now the series capacitors are unreasonably small, and in addition the source and load terminations are increased to 415.3Ω.
3. Next we perform Norton impedance transformations at both sides of the filter to lower the terminations back to 50Ω. The transformations are performed by the series parallel combinations (0.48 pF

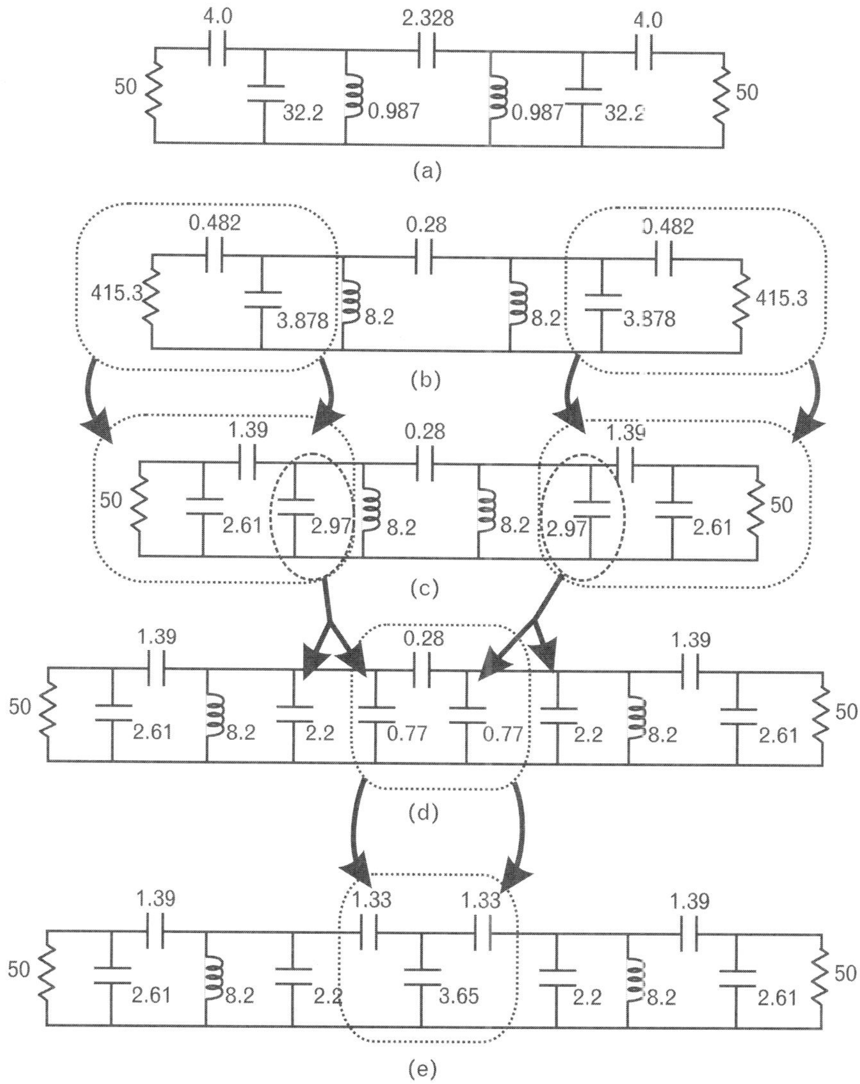

FIGURE 8.57 *Transformations of the synthesized bandpass filter helped to obtain realizable component values. (a) Initial circuit is transformed in four steps, (b) through (e), to reach practical element values. All circuits have the same order and frequency response. Element values are given in ohms, nanohenries, and picofarads.*

and 3.88 pF) of the two capacitors next to terminations. After the Norton transformations we have all components in reasonable ranges—with the exception of the 0.28-pF coupling capacitor between the two resonators—and we restored our original 50-Ω terminations [Figure 8.57(c)].

4. Now we need to do something about the 0.28-pF series capacitor. We can interchange the 2.97-pF capacitor and the 8.2-nH inductor on both sides to form a capacitive PI-network at the center of the filter, as shown in Figure 8.57(d). This move prepares for a PI-to-TEE transformation, which can help to increase the value of the series component.

The series and parallel capacitors of the PI-network have nearly a 10:1 ratio and unless we reduce that ratio, after the transformation we maintain comparable ratios in the new TEE-network. By splitting the 2.97-pF parallel capacitors into two parts, one of 2.2 pF and the other one of 0.77 pF, gives us PI-networks with two parallel capacitors of 0.77 pF and series capacitor of 0.28 pF.

5. Converting the newly formed PI-network to its TEE equivalent gives us three new capacitors with realizable values. The final circuit is shown on Figure 8.57(e). In our original circuit the ratio of minimum and maximum capacitors were about 14:1. In the final circuit we reduced this ratio to less than 3:1. The 8.2-nH inductor is a reasonable element for this frequency range. However, we are paying the price by the increased number of components.

None of the final components have standard values, and at this point we also need to examine the effects of physical models. In finding the optimum element values, we want to use as many standard component values as possible. As a general practice, a few tenths of a nanohenry or a few tenths of a picofarad can always be fabricated on the PC board during the circuit layout. For example, a short conductor trace may be a reasonable substitute for a series inductance, and small addition to the component-attach pads for parallel capacitance. Buying nonstandard elements costs extra money, and is usually provided by one vendor—neither of which are desirable.

The purpose of this exercise was not necessarily to create a state-of-the-art filter, since we used discrete elements and only two resonators. Rather, we wanted to show the power and the flexibility of the various network transformations that are extremely useful in creating realizable filter structures.

8.11 Other forms of resonators

Inductors and capacitors have limitations in resonant circuit design due to their relatively low Qs. To obtain high selectivity we need resonators with loaded Qs of thousands, which is totally impractical to realize with L-C components. Instead, at low frequencies crystal SAW types of filters are used while at microwave frequencies dielectric and waveguide resonators are favored. These types of resonant circuits may be realized with extremely high Qs. Microstrip and stripline realizations also have Q limitations, although the latter can have *Qs* of a few hundred in the RF-MW region. High-quality coaxial resonators have Q values around 1,000, while the Qs of waveguide resonators with air dielectrics may be in the 10,000 to

20,000 range. Table 8.3 provides a description of the various resonators used in RF/MW design.

8.11.1 Illustrative example: coupled transmission line filter synthesis using CAD

A 2.2- to 2.3-GHz coupled transmission line filter was designed with four different commercially available CAD programs: Agilent's ADS, AWR's Microwave Office, Eagleware's MFilter, Forem's ECM, and Optotek's MMICAD programs. Specifications of the filter are:

- Passband: 2,180 MHz to 2,320 MHz (2,200 MHz to 2,300 MHz with 20-MHz "guardbands" at the edges);
- Passband ripple: 0.05 dB;
- Stopband rejection: 35-dB minimum at 2.0 GHz and 2.5 GHz.

The filter is realized in microstrip form on a ceramic substrate with the following specifications:

Table 8.3 Various Resonator Elements and Their Related Characteristics

Type	Frequency Range (GHz)	Size	Cost	Q	Bandwidth	Long-Term Stability	Initial Frequency Set Tolerance
L-*C* lumped	<20	Medium	Low to medium	<50	Wide	Fair	±2%
Helical	<20	Large	Medium	<1,000	Moderate	Good	±1%
Waveguide TE10	<200	Very large	High	$\frac{18,500}{\sqrt{f(\text{GHz})}}$	Moderate	Excellent	±0.01%
Coax	<100	Medium	Low	<1,000	Wide	Excellent	±0.1%
Stripline	<100	Medium	Low	<500	Wide	Excellent	±0.1%
Microstrip	<100	Medium	Low	<200	Wide	Excellent	±0.1%
Dielectric puck (TE01δ)	<50	Medium	Low	$\frac{42,000}{\sqrt{f(\text{GHz})}}$	Wide	Excellent	±0.1%
High K, $\lambda/4$ ceramic	<20	Small	Medium	<1,000	Very low	Excellent	±0.5%
Quartz	<0.10	Very small	Medium	$\frac{1,000}{\sqrt{f(\text{GHz})}}$	Low	Excellent	±1%
Ceramic	<0.02	Very small	Low	$10/f$ (GHz)	Very low	Excellent	±0.001%
SAW	0.1 to 4	Miniature	High	<14,000	Low	Excellent	±0.5%

$\varepsilon_R = 10.0$

$H = 25$ mil (0.635 mm)

$\tan\delta = 0.0001$

$t = 0.05$ mil (1.27m)

$\sigma = 4.7 \times 10^{-7}$

The above specifications require a fourth-order edge coupled filter with five coupled line sections. Since the circuit has symmetry, the dimensions are tabulated for the three unique elements in Table 8.4.

All five filters met the minimum required stopband specifications, but their passbands showed differences when simulated with another CAD program. An explanation may be that while linear simulation of passive *L-C* circuits lead to the same results by all simulators, microstrip coupled lines are modeled in various ways and therefore their results are often different. Therefore, our comparison is not quite fair because we did not apply the discontinuities uniformly to duplicate the treatment of each individual circuit simulator. We did, however, submit all five designs to the Sonnet EM simulator and extracted two-port *S*-parameters from that program. Results of the individual filter's performances are shown in Figure 8.58.

The conclusion of this exercise shows that while only one of the programs met both the passband and the stopband specifications, all of them provided initial results for possible optimization. Depending on the difference between initial simulation and desired specification, the optimization may be straightforward or difficult, keeping in mind that a careful selection of weighting factors of the error function (see Chapter 6) is very important during the optimization.

TABLE 8.4 PHYSICAL DIMENSIONS OF THE COUPLED FILTER BY VARIOUS COMMERCIAL FILTER DESIGN PROGRAMS

PROGRAM	SECTION 1, $i = 1$ SECTION 5, $i = 5$			SECTION 2, $i = 2$ SECTION 4, $i = 4$			SECTION 3, $i = 3$		
	W_i	S_i	ℓ_i	W_i	S_i	ℓ_i	W_i	S_i	ℓ_i
ADS	17.28	8.00	515.3	23.28	40.37	501.3	23.46	55.75	501.3
ECM	18.12	9.41	503.9	22.60	37.89	499.6	22.74	45.71	499.8
MFILTER	22.01	7.02	508.6	23.12	38.34	500.1	23.20	46.76	499.6
MMICAD	22.20	7.70	501.0	28.50	37.10	495.6	28.50	45.70	495.4
MW Office	18.87	9.54	508.5	23.27	38.37	498.2	23.41	46.72	497.5

W_i = line width, S_i = spacing between coupled line segments, and l_i = coupled line length, where i = 1, 2, and 3.

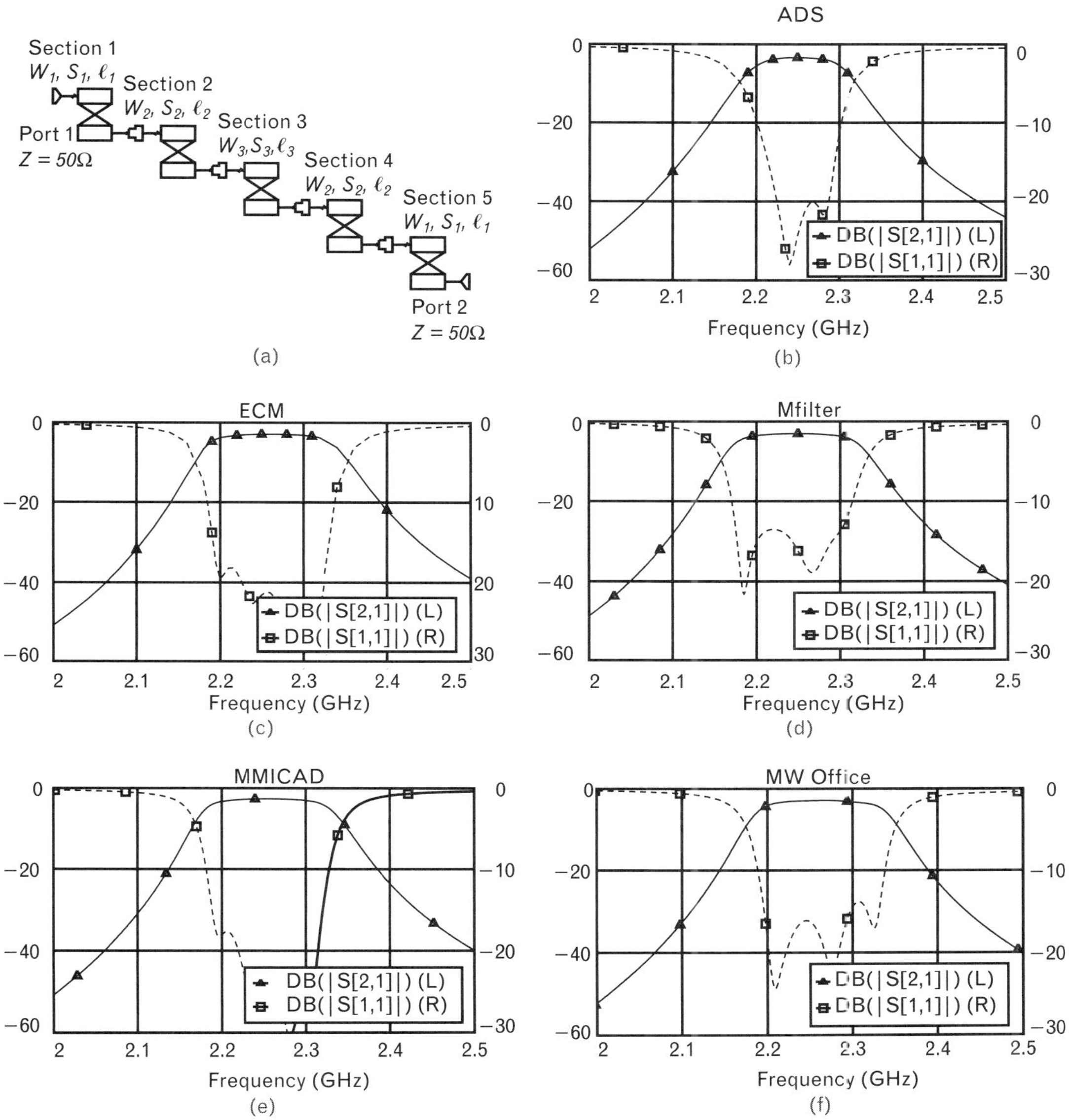

FIGURE 8.58 *(a) Circuit schematics of the fourth-order coupled-line filter showing symmetrical topology. (b–f) The comparison of transmission loss and return loss of the various synthesis programs [(b) ADS, (c) ECM, (d) Mfilter, (e) MMICAD, and (f) MW Office] through their two-port S-parameters obtained from the Sonnet EM simulator.*

8.11.2 Dielectric resonators

Dielectric resonators are formed using high-Q, high-permitivity, and highly temperature stable dielectric materials in various forms. Their physical shapes can be rectangular, circular, ring-type, or a disc (pill) type. The advantage of the high dielectric constant is reduced size because the effective wavelengths in the dielectric material are reduced by the square root of the effective dielectric constant. Therefore, if we can produce a dielectric resonator with effective dielectric constant of 80, the wavelength is reduced by approximately a factor of nine,[6] which is a very significant improvement for small portable applications. Some of the manufacturers provide the loss tangent (tan δ) for the material, which is the reciprocal of the Q-factor. Table 8.5 provides a list of manufacturers producing high Q dielectric materials for resonator applications.

8.11.3 Crystal resonators

Because the piezoelectric effect provides such a convenient and effective link to the electrical world, it is easy to overlook the fact that a crystal resonator is actually a mechanical system. It has complementary energy storage elements, overtones, spurious responses and losses. The equivalent circuit for a crystal resonator is shown in Figure 8.59 where the series *R-L-C* circuit represents mechanical resonance. Resistor R_m denotes the energy dissipated in the crystal and in its mounting arrangement. The capacitance C_m comes from the elasticity of the quartz, and the inductance L_m is determined by the quartz mass. The last two parameters are sometimes referred to as the *motional inductance* and the *motional capacitance*, emphasizing their mechanical heritage where springs and storage tanks are used to bounce energy between them. Only C_0, the electrode capacitance, is physically present as an electrical property. The electrodes, one on each side of the quartz blank, make up the plates of a capacitor with a quartz dielectric.

The resonator design sets the value of the electrical equivalent circuit parameters. The electrode capacitance C_0 varies with electrode area. The blank thickness along with the electrode mass loading sets the resonant

TABLE 8.5 SAMPLE OF DIELECTRIC MATERIALS AVAILABLE COMMERCIALLY

MANUFACTURER	MODEL NO.	ε_R	Q_U	@	f_{GHz}
Tekelec Components	E200 series	37	950		5.0
Tekelec Components	E9000	37	40,000		1.0
Dielectric Materials	Sapphire	9.394	450,000		13.2
Dielectric Materials	Sapphire	9.394	650,000		9.0
Alpha/Trans-Tech	83000 series	35–36.5	28,000		0.85
Alpha/Trans-Tech	86000 series	80	3,000		3.0
Murata Manufacturing	Resomics-U	36–40	6,000		7.0
Murata Manufacturing	Resomics-Tem	37–39	300		2.5

6. Although the puck size is reduced, the overall size of the resonator may be controlled by the enclosure.

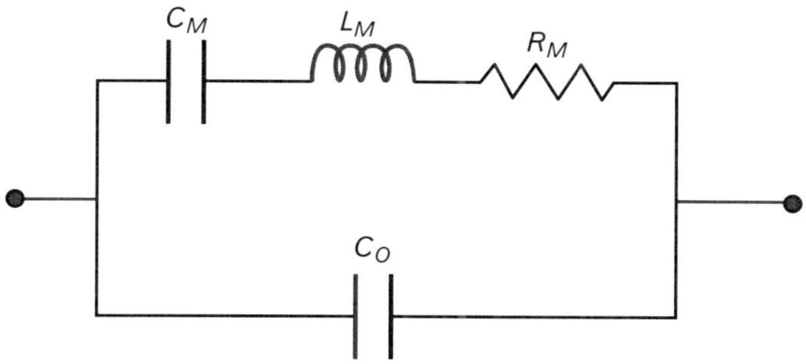

FIGURE 8.59
Crystal resonator equivalent circuit model.

frequency and determines the product of L_m and C_m. Blank and electrode diameter both affect resistance, with larger diameters giving lower resistance. Friction between air molecules and the vibrating blank surface increases losses significantly and lowers the resonant frequency. For this reason, and to achieve low aging, all but the least expensive resonators are hermetically sealed.

The frequency response shown in Figure 7.49 gives a reactance plot of the crystal equivalent circuit, displaying series and parallel resonances at frequency f_S and f_p, respectively. There are two frequencies where the crystal looks purely resistive: at f_R (where f_R is slightly above f_S) and at f_P. At those frequencies the phase of the impedance is zero.

The formulas shown in Chapter 7 give the resonant frequencies based on the equivalent circuits. In filter design the crystal replaces one or both elements of the synthesized resonator components.

Application of a crystal resonator is shown in Volume II, Chapter 6.

8.12 Summary

As we mentioned in the beginning, filter design requires a fair amount of experience, imagination, and assistance from the appropriate CAD tools. While the latter is absolutely necessary to simulate the performance with actual physical models, it may not always provide the appropriate network topology and component values. This is where experience and creativity become extremely helpful.

A successful filter designers' CAD library must include an electromagnetic simulator, regardless of the form in which the circuit is realized. Most likely there is sufficient interaction and coupling among the components that the true performance cannot be achieved by circuit simulation alone. The latest generation of circuit and system simulators includes at least a 2.5D EM simulation routine for convenient cosimulation. Filter synthesis programs are also becoming standard attachments in the simulators, and we encourage you to become familiar with them.

If your filters are realized in distributed form, we recommend that you verify the performance of a few existing circuits already built in the same form as the new filter to be designed. Do not start any new project until you find the right combination of CAD tools that provides accurate and reliable simulation.

Most filter synthesis books are highly mathematical, written for people who already know the material. Two of the few exceptions are recommended to those who are relatively new to the subject [19, 20]. An additional recommendation is a new book on EM simulation [21].

References

[1] Matthaei, G., L. Young, and E. M. T. Jones, *Microwave Filters, Impedance Matching Networks, and Coupling Structures*, New York: McGraw-Hill, 1964.

[2] Szentirmai, G., (Ed.), *Computer-Aided Filter Design*, New York: IEEE Press, 1973.

[3] Filsyn Program Manual, ALK Engineering, http://www.alkeng.com.

[4] Levy, R., and R. V. Snyder, "Design of Microwave Filters," *IEEE Trans. on Microwave Theory and Techniques*, March 2002.

[5] Larson, L. E., (Ed.), *RF and Microwave Circuit Design for Wireless Communications*, Norwood, MA: Artech House, 1996.

[6] Chen, W. K., (Ed.), *The Circuits and Filters Handbook*, Boca Raton, FL: CRC Press & IEEE Press, 1995.

[7] Korn, I., *Digital Communications*, New York: Van Nostrand Reinhold, 1985.

[8] Zverev, A. I., *Handbook of Filter Synthesis*, New York: Wiley, 1967.

[9] Levy, R., "Direct Synthesis of Cascaded-Quadruplet (CQ) Filters," *IEEE Trans. on Microwave Theory and Techniques*, December 1995.

[10] Levy, R. "Corrections to 'Direct Synthesis of Cascaded-Quadruplet (CQ) Filters'," *IEEE Trans. on Microwave Theory and Techniques*, December 1996.

[11] Hershtig, R., R. Levy, and K. A. Zaki, "Synthesis and Design of a Cascaded Trisection (CT) Dielectric Resonator Filters," *European Microwave Conference Digest*, Jerusalem, September 1997.

[12] Wenzel, R. J., "Some General Properties of Commensurate Line-Length Complementary and Pseudo-Complementary Microwave Filters," *IEEE Trans. on Microwave Theory and Techniques*, March 1968.

[13] Wenzel, R. J., "Applications of Exact Synthesis Methods to Multi-Channel Filter Design," *IEEE Trans. on Microwave Theory and Techniques*, January 1965.

[14] Wenzel, R. J., "Wideband High-Selectivity Diplexers Utilizing Digital Elliptic Filters," *IEEE Trans. on Microwave Theory and Techniques*, December 1967.

[15] Richards, P. I., "Resistor-Transmission Line Circuits," *Proc. of IRE*, February 1948.

[16] Medley, M. W., Jr., *Microwave and RF Circuits: Analysis, Synthesis, and Design*, Norwood, MA: Artech House, 1993.

[17] Schiffman B. M., and L. Young, "Design Tables for an Elliptic Function Bandstop Filter," *IEEE Trans. on Microwave Theory and Techniques*, October 1966.

[18] Weinberg, L., *Network Analysis and Synthesis*, New York: McGraw-Hill, 1962.

[19] Minnis, B. J., *Designing Microwave Circuits by Exact Synthesis*, Norwood, MA: Artech House, 1996.

[20] Rhea, R. W., *HF Filter Design and Computer Simulation*, Atlanta, GA: Noble Publishing, 1994.

[21] Swanson, D. G., *Field Solver Circuit Design Cookbook*, Norwood, MA: Artech House, 2003.

Selected Bibilography

Budimir, D., *Generalized Filter Design by Computer Optimization*, Norwood, MA: Artech House, 1998.

Giacoletto, L. J., (ed.), *Electronics Designers' Handbook*, 2nd ed., New York: McGraw-Hill, 1977.

Rhodes, J. D., *Theory of Electrical Filters*, New York: Wiley, 1976.

Temes, G. C., and J. W. LaPatra, *Circuit Synthesis and Design*, New York: McGraw-Hill, 1977.

Temes, G. C., and S. K. Mitra, *Modern Filter Theory and Design*, New York: Wiley, 1960.

CHAPTER 9

Similarities and differences of RF and high-speed digital designs[1]

In this chapter, we give a brief overview of the major differences and similarities between RF and high-speed digital designs. This chapter targets RF design engineers who may have no knowledge of digital systems. Space does not permit detailed explanations or derivations of the several aspects of digital designs. Rather, this chapter is a collection of opinions and statements for RF designers who are interested in what is common and what is different between today's analog and digital worlds. Interested readers needing more details of high-speed digital designs are referred to the suggested list of readings at the end of the chapter. Section 9.1 gives a brief historical overview of analog and digital designs, illustrating why and how the two, initially very different, design concepts are now merging. Section 9.2 describes why time-domain parameters are the primary factors in digital signaling. Section 9.3 shows why the electromagnetic coupling among interconnects of digital systems is usually an unwanted and dreaded side effect. Section 9.4 highlights the modeling differences of discrete *R-L-C* elements in RF and digital circuits. Section 9.5 illustrates why and how electrical loading, vias, and discontinuities have a different impact on RF and digital designs. Section 9.6 shows when time-domain and frequency-domain methods are preferred in digital designs. Finally, Section 9.7 compares simulation and measurement approaches used in RF and digital designs.

9.1 Historical perspective of analog RF and digital designs

Soon after the discovery of radio wave fundamentals, experiments began that harnessed the new medium to transmit information. There were two major application areas under consideration for the practical use of RF signals: wireless transmission of information, pioneered by Marconi and others; and wireless transmission of power, pioneered by Tesla [1]. Although

1. This chapter was contributed by Istvan Novak.

the wireless transmission of power had been investigated in different forms, so far no commercial application had been found. Transmitting information without wires, however, soon became a hot new development area, which dominated a large part of the twentieth century.

Digital communications started out wideband, with spectrum starting at zero. Analog communications started out narrowband, where the envelope was slowly changing with respect to the carrier. Absolute bandwidths for both types of communications have gradually been increasing.

Figure 9.1 illustrates the spectral contents of digital and RF signals with typical numbers from about 20 years ago. On the left, the spectrum envelope of a bus signal from an early personal computer is shown. Assume 5 MHz for the highest fundamental frequency of data. A 5-ns transition time of the signal, as was shown in Chapter 2, produces 65-MHz upper bandwidth for the signal. The lower end of the spectrum extends to dc, because the digital bus is assumed to be able to switch with any arbitrary bit pattern, including continuous steady low or high. The absolute bandwidth is $BW =$ 65 MHz – 0 = 65 MHz; the relative bandwidth is high, because the signal spectrum extends down to 0 Hz.

On the right, the spectrum of an FM broadcasting signal is shown around a 108-MHz carrier. The typical bandwidth is 0.2 MHz; in this case the relative bandwidth is a fraction of a percent.

Figure 9.2 shows the spectra of digital and RF signals in use today. On the left, the spectrum envelope of the signaling of the emerging InfiniBand architecture is shown. The InfiniBand architecture uses a 2.5-Gbps signaling rate, which means that the signal can change at every 400 ps, creating a 1.25-GHz bit-rate frequency. The 8B/10B coding limits the spectrum at the low end to about one-fifth of the bit-rate frequency. A typical 100-ps transition time of the signals creates a 3.1-GHz upper edge of the bandwidth. The absolute bandwidth is $BW =$ 3.1 GHz – 0.25 GHz =2.85 GHz.

The relative bandwidth is still high, but the spectrum does not extend down to dc. On the right, the spectrum envelope of a wideband RF transmission is shown. The absolute bandwidth is $BW =$ 2.6 GHz – 2.1 GHz = 0.5 GHz. The relative bandwidth becomes 0.5 GHz/2.35 GHz = 21.2%.

There are major differences between digital and RF analog designs:

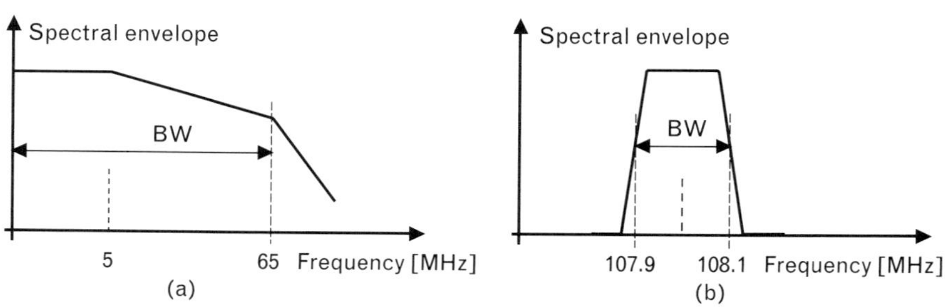

FIGURE 9.1 *Spectral contents of (a) digital and (b) RF signals from the 1980s. Bandwidths indicate 3-dB roll-off frequencies.*

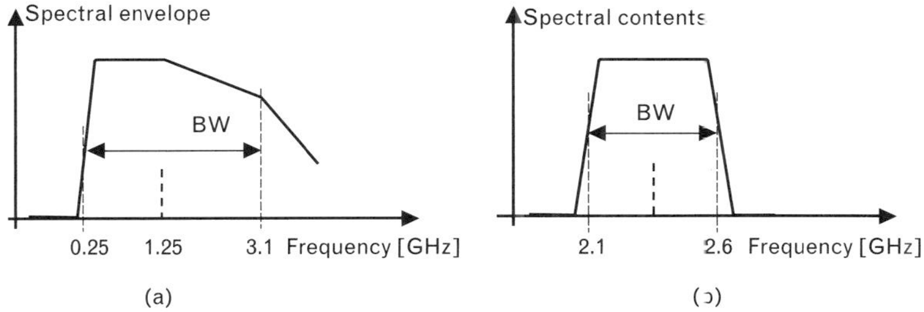

FIGURE 9.2 *Spectral contents of (a) digital and (b) RF signals from 2000.*

- *Bandwidth:* Traditional RF applications tend to be high frequency but narrowband. Digital applications traditionally are lower frequency, but in the majority of applications, where dc also needs to be transmitted, the relative bandwidth is high.

- *Input power:* RFICs tend to have matched inputs and outputs. RF input power is needed for proper operation, and there is a large variation in input output voltage/current/power levels.

- *Interconnect density:* In contrast to high-density digital circuits, RF and analog circuits have lower circuit and trace density, lower number of layers on printed-circuit boards, and the majority of printed circuit board interconnects are microstrips or coplanar; there is little need for signal vias. In lower-density RF interconnects, traces tend to be wider than traces on digital boards, and therefore, dielectric loss matters more.

- *Discontinuities:* In RF circuits, reactive loading and vias are eliminated or minimized. The remaining imperfections are bends, junctions, and surface roughness of conductors.

- *Circuit schematic:* In digital circuits, schematic does not show interconnect as separate circuit elements; they may need to be added and their delay and dispersion considered separately in prelayout and postlayout analysis. In RF, layout becomes an integral part of circuit schematic because many circuit elements are formed of printed circuit board patterns.

- *Emc:* in RF systems, there are large differences in input/output power levels of RF cells. Therefore, RF circuit subsystems require good shielding to avoid interference within systems. This protective shielding reduces the susceptibility and radiation. In contrast, digital levels are standardized within a narrow range, so interference protection is rarely needed within a system. This comes with a higher chance of radiation, which is harder to stop on the box level, as every part of the box may become radiator.

The above major differences are not without exceptions, and as time goes by, some of these differences are becoming smaller and in some cases become blurred. These days we witness the blending and merging of analog RF and digital systems: analog transmissions go very wideband with significant portion of digital baseband and digital signal processing of intermediate-frequency signals.

Digital signaling, on the other hand, began many years ago to use coding techniques to limit the relative bandwidth of the signals by eliminating the lower-frequency portion of signal spectrum. The most popular scheme is the 8B/10B coding, which transforms each 8-bit digital word into a stream of coded 10-bit series [2], and by doing so the maximum number of zeros or ones in one block is limited to no more than five. Digital high-speed systems recently started to implement multilevel signaling. While pushing the signaling speeds to the gigabit per second range, the interconnect path is allowed to create higher losses and distortions on the signal, because the receiver inputs are becoming more sensitive.

9.2 Time-domain and voltage-current parameters (transition times, delays, skew, and signal levels)

RF and microwave circuits traditionally require a certain amount of signal input power to work properly. It is also the goal that the circuit's input impedance be real and should match the transmission medium (matching for maximum available power). This is because in wireless applications the large and distance-dependant open-space attenuation creates a huge range of possible input powers that the circuits have to process properly. The input power has to overcome the atmospheric, man-made, and thermal noise, but depending on the bandwidth and frequency range, very low input powers can be used for RF inputs. For example, in the MW broadcasting band, the atmospheric and man-made noise dominates. At the 1-MHz amplitude-modulated broadcasting frequency, assuming 10-kHz bandwidth (in ITU Region 1, the MW broadcasting channel spacing is 10 kHz and in Region 2 to 3 it is 9 kHz), the average received signal may need to be 1 mV or higher for an acceptable (>26 dB) signal-to-noise ratio. In the VHF and UHF bands, the limiting factor is the thermal noise of the receivers. With a few decibels of noise figure and a 0.1-MHz bandwidth, FM broadcasting receivers may be able to work with 1 μV or less input voltage to produce an acceptable signal-to-noise ratio. The limited gain of a single amplifier stage necessarily results in a series of cascaded amplifiers in such applications. And as a result, each amplifier stage has to be able to process a different level of input voltage and power.

In contrast to RF and wireless, digital circuits were first defined and used in situations where almost all of the interconnect paths were possible to be considered lossless. At the beginning, most of the digital circuits were physically much smaller than the wavelength of the highest frequency of interest. Therefore, the interconnects were rightfully treated as lumped, ideal, lossless, and delayless elements. Voltage swing and voltage levels at logic outputs and logic inputs were similar. With the widespread use of binary signaling, it was easy to standardize on the input and output voltages. In the early years of *transistor-transistor-logic* (TTL), the margin created by the difference of minimum output voltage and maximum required input voltage was set to overcome static fan out losses[2] and power supply noise. With the introduction of *complementary metal-oxide silicon* (CMOS) circuits and at higher speeds, the same ratio now serves as the noise safety margin against crosstalk, simultaneous-switching noise, power noise, and EMI-induced noise. Standard logic inputs today have very high resistance in parallel to a small (few picofarad typical) capacitance. This high impedance creates full reflection when connected to transmission lines, such as printed-circuit-board traces and/or cables; therefore, it is increasingly common to create a matched termination resistor on the digital integrated circuit itself.

However, inputs of early logic families do not have on-chip termination; therefore, the real input power is virtually zero, because the input impedance is mostly reactive. Therefore, instead of the input power, the voltage is specified and standardized at digital inputs and outputs. To save current and power, and to reduce simultaneous-switching noise problems, the voltage swings are constantly dropping as new signaling and logic families arise. At the same time, the input sensitivity of digital receivers is improving, allowing designers to allocate a bigger part of output voltage swing to conductive and dielectric loss of traces, reflection, crosstalk, and other noise contributors. This trend is captured in Table 9.1, summarizing a few standard input-output voltages.

With binary signaling, there are two steady states of signals: low and high. The highest possible operation frequency is achieved when the transition between the two states is made short, thus maximizing the steady low and steady high portions of the waveform. The shape of the transition is often approximated with a piece-wise-linear straight edge, mainly in simple simulations. The transition time is separately defined for the two transitions: it is called rise time for the signal going from steady low to steady high, and it is called fall time for the signal going from steady high to steady low. Figure 9.3 defines the rise and fall times on an ideal piece-wise-linear waveform.

2. The TTL was first defined and built with bipolar transistors, where the generic input was an emitter of an NPN transistor. When the input was pulled to high, the base-emitter junction became reverse biased, and required literally no current to hold it at that level. When the input was pulled to low, the input had to supply the sum of base and collector currents of the first stage, approximately 1.4 mA in the standard TTL cells. When a TTL output had to drive several parallel-connected TTL inputs, the static output current capability of the driver had to match or exceed the sum of all input currents.

TABLE 9.1 TYPICAL INPUT-OUTPUT VOLTAGE RANGES FOR SOME OF THE WIDELY USED DIGITAL SIGNALING CONVENTIONS

	TTL	5-V CMOS	3.3-V PECL	BTL	LVDS
Supply voltage V_{dd}	5V	5V	3.3V	3.3V	2.5V
Minimum output voltage range (V)	0.4–2.5	0.1–(V_{dd}-0.1)	2.0–3.2	1.1–1.9	0.4 V_{pp}
Maximum input voltage range (V)	0.8–2.0	0.3–0.7 V_{dd}	2.3–2.9	1.47–1.62	0.2 V_{pp}

CMOS: complementary metal-oxide silicon; PECL: pseudo (or positive) emitter-coupled logic; BTL: backplane transceiver logic; LVDS: low-voltage differential signaling

Note: Values shown are approximate, as these signaling standards have several flavors. All of the above signaling schemes can be used single ended or differential. Values shown are for single-ended implementations or for each leg in a differential implementation.

The nonlinear drivers and inherent band limitations of circuits and interconnects, however, necessarily create rounded corners and nonstraight edges, when the 0 to 100% points cannot be defined. For these situations, it is better to specify the rise and fall times as the transition times between predefined percentage values of the steady-state differences, called voltage swing. For wave shapes, which can be well approximated with one-pole exponential functions, the 10% to 90% transition time definition is common. For band-limited signals with more rounding near the corners, the 20% to 80% transition time definitions became popular. These options are illustrated in Figure 9.4. As shown in Figure 9.5, real waveforms may be even more complex, containing nonmonotonic sections and/or distinctly different slew rate portions.

Synchronous digital systems work with timed edges. Unlike in most RF systems, the voltage uncertainty of a digital receiver input is a noticeable portion of the voltage swing. As illustrated in Figure 9.6, faster edges help to reduce the timing uncertainty (also called *skew*) due to the input receiver threshold uncertainty.

In RF applications, the receiver sensitivity may be limited by phase noise of local oscillator signals, and it is described in the frequency domain.

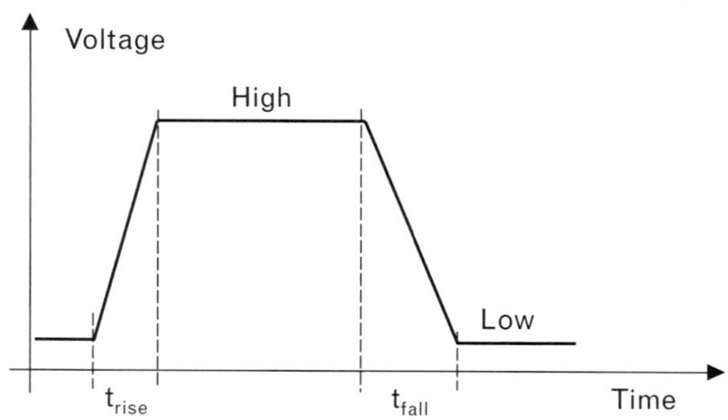

FIGURE 9.3 *Definitions of rise and fall times on a piece-wise-linear signal shape.*

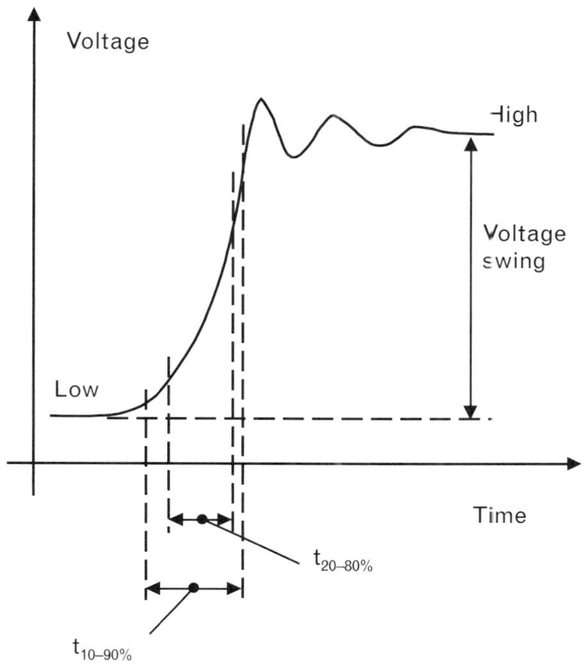

FIGURE 9.4
Definition of 10% to 90% and 20% to 80% transition times on band-limited edges. The low and high values are measured in steady state, after the settling of transients.

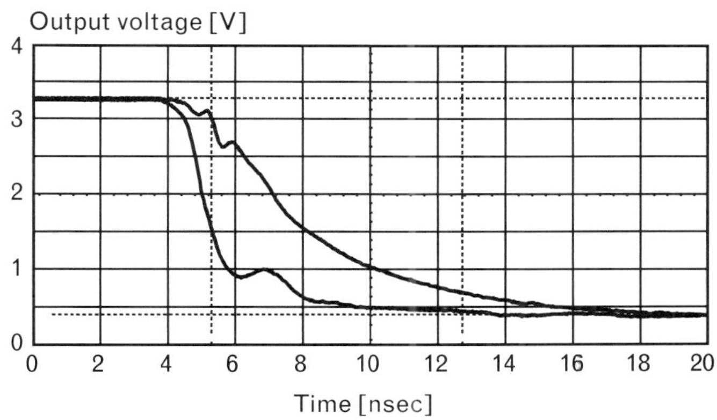

FIGURE 9.5
Illustration of real-life waveforms with non-monotonic edges and different transition times. The two measured traces show two separate clock output waveforms in a two-bank clock-driver chip with 20 individual outputs.

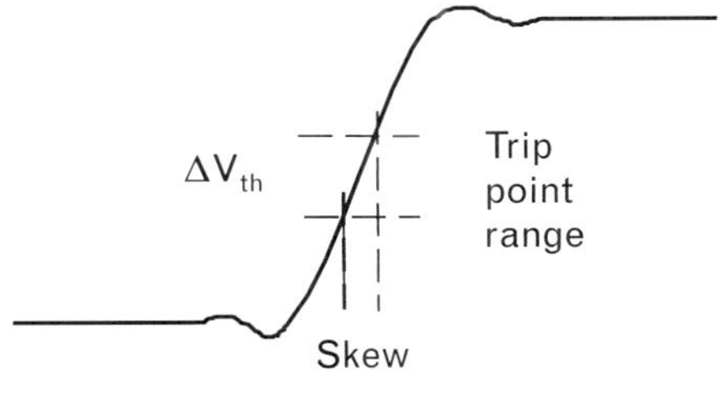

FIGURE 9.6 *Definition of skew due to the finite input trip-point range. Note that to provide noise margin, the input trip-point range is usually smaller than the 10% to 90% or 20% to 80% transition time of the received signal.*

The corresponding counterpart in digital applications is jitter. The spectrum of a digital clock source with significant amount of jitter is shown in Figure 9.7. Digital clocks with harmonically related output frequencies often exhibit a form of deterministic jitter, called modal jitter. It is illustrated in the time-domain picture of Figure 9.8. Random jitter is the result of stochastical variations of the edge, as shown in Figure 9.9.

9.3 Crosstalk versus coupling

Coupling among various pieces of interconnects in digital circuits is called crosstalk. Crosstalk in digital circuits is an undesirable side effect, to be kept under a few percent of the active signal. In contrast, in RF circuits,

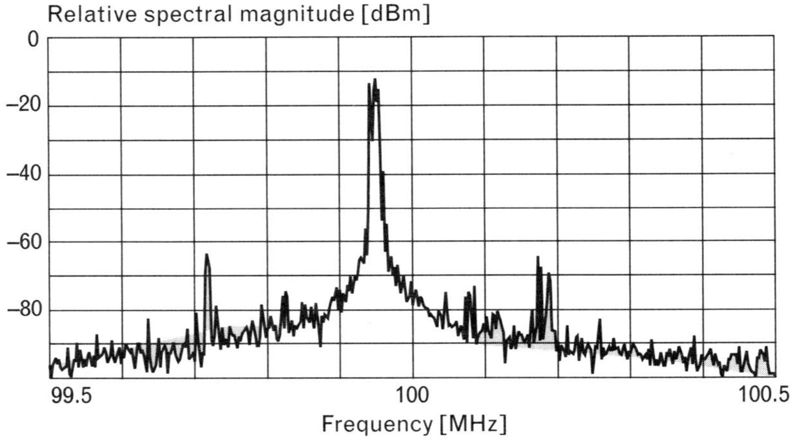

FIGURE 9.7
Illustration of phase noise of a digital clock source.

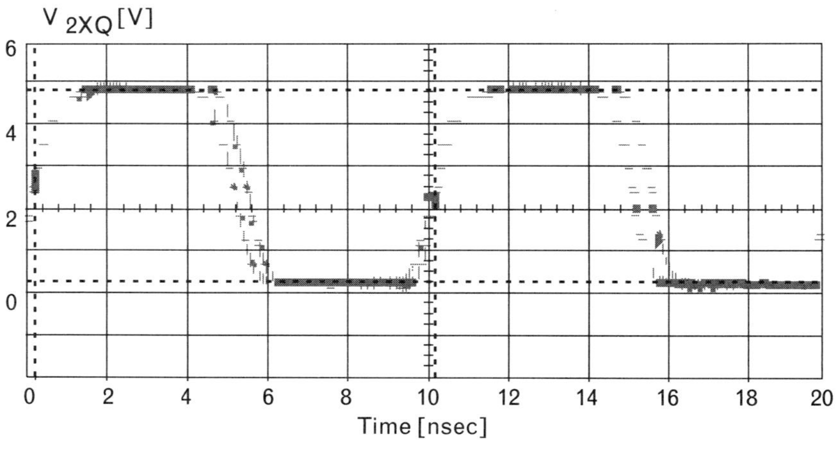

FIGURE 9.8
Time-domain deterministic jitter of a clock source.

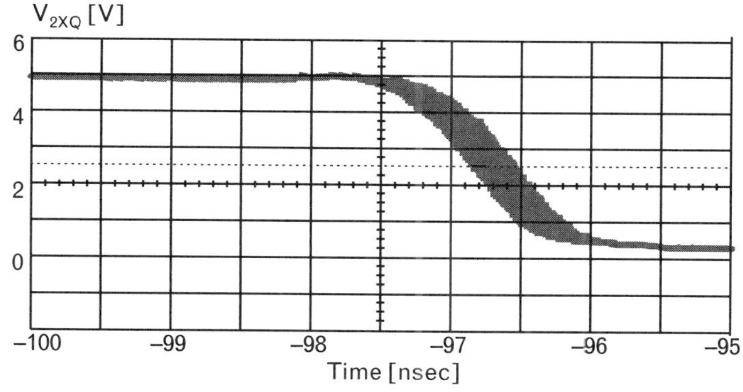

FIGURE 9.9
Random jitter on a digital clock edge.

coupling provides a useful and desired function in directional couplers, which are frequently used to separate incident and reflected waves. Directional couplers are inherently band-limited, and therefore cannot be used in wideband digital systems.

The definitions of crosstalk terms used by digital designers are shown in Figure 9.10.

The trace carrying the active signal is also called aggressor; the victim line is also called quiet line. A signal appearing on the victim line at the node close to the active source is called *near-end crosstalk* (NEXT); at the node opposite to the source it is called *far-end crosstalk* (FEXT). Note that in digital circuits, strongly nonlinear and nonreciprocal materials, like ferrites, are not typical in the signal paths.

One exception, when crosstalk (or coupling) is used on purpose between traces, is the coupled implementation of differential pairs. Edge coupled and broadside-coupled differential pairs are possible. Edge-coupled pairs can be implemented as single or asymmetrical dual stripline layers or microstrips. Broadside-coupled traces may be implemented as

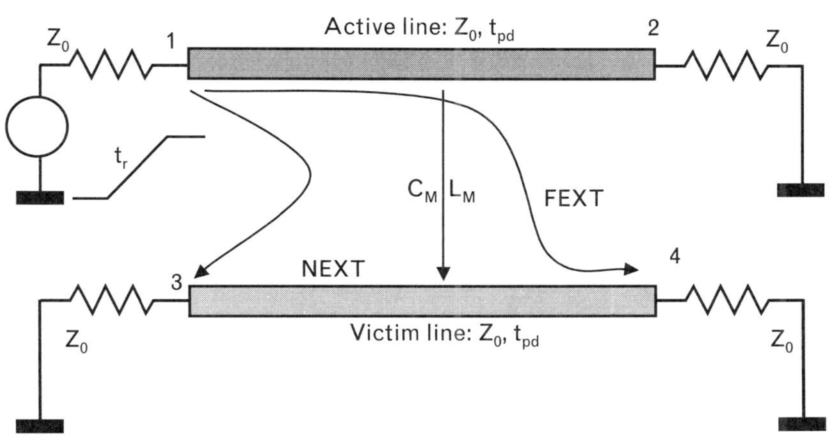

FIGURE 9.10
Definition of terms related to coupling in digital circuits.

dual stripline layers. The various coupled-line realizations are illustrated in Figure 9.11.

Broadside-coupled traces are used when the stronger coupling is required to increase the ratio of common-mode and differential-mode impedances. However, in multilayer high-density boards, strong coupling is not always practical, as the trace width would need to be reduced below the manufacturability limit. Low-cost manufacturing of large rigid boards today requires at least 0.1-mm (4-mil) trace width, which requires approximately 1.75 mm (70 mil) or more plane-to-plane separation of a broadside-coupled pair in FR4 dielectric material.

In digital applications, sources are time-domain pulses. The crosstalk waveforms in the time domain [3] are illustrated in Figures 9.12 (for microstrip construction) and 9.13 (for stripline construction). The equivalent-circuit parameters of the coupled trace pair, and the time-domain waveforms were generated with Applied Simulations Technology's ApSimRLGC and ApsimSPICE software, respectively.

In digital circuits, multiline crosstalk in busses with many parallel lines is of special interest, as it determines the worst-case noise. In RF applications, the lower trace density usually allows wider spacing and the neighbor traces have negligible couplings.

Analog circuits rarely have standardized levels, so the required isolation always depends on the specifics of the circuit and system. Also, because of the narrower-band nature of analog circuits, subsystems working in different frequency bands and ranges may tolerate much more interference and therefore may require much less isolation. High-power transmitters and sensitive receivers working in the same frequency band require the highest isolation and are usually isolated by separate metal enclosures.

Digital circuits operate with standardized levels, and the system noise margin determines the permissible level of crosstalk. The bothersome crosstalk coefficient has a narrow range: strong crosstalk may prevents the digital circuit from operating, on the other hand, levels below 1% usually do not matter.

Use of microstrip and stripline constructions is also different in analog RF and digital circuits: in multilayer digital circuits most of the layers are necessarily inner layers and therefore become stripline layers. Interconnect

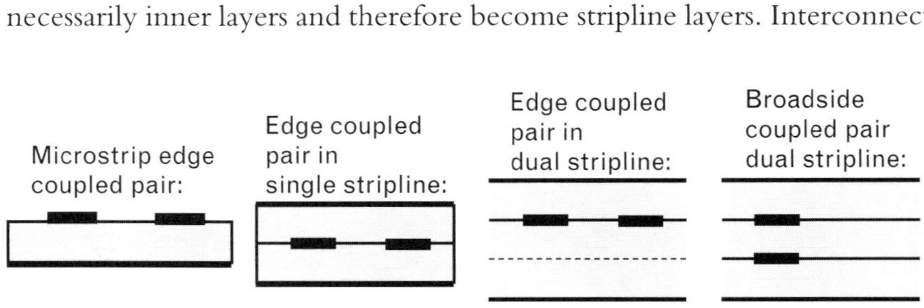

FIGURE 9.11 *Stack up implementation illustrations of differential pairs with coupled traces of single and multilayer printed-circuit boards.*

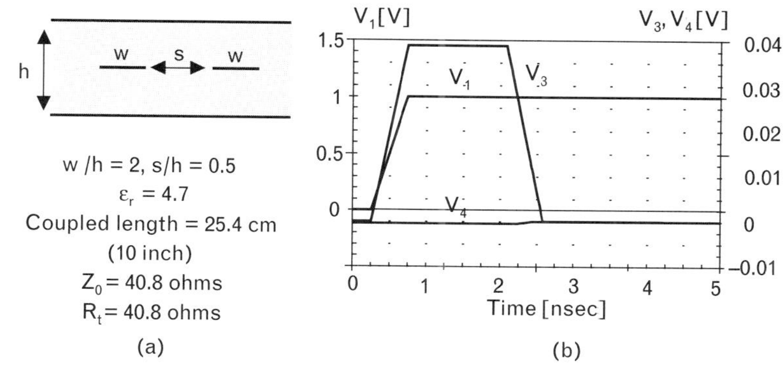

FIGURE 9.12 *Crosstalk example in a microstrip-coupled trace pair. (a) cross-section geometry and simulation parameters and (b) simulated waveforms. Node labels are according to Figure 9.10 .*

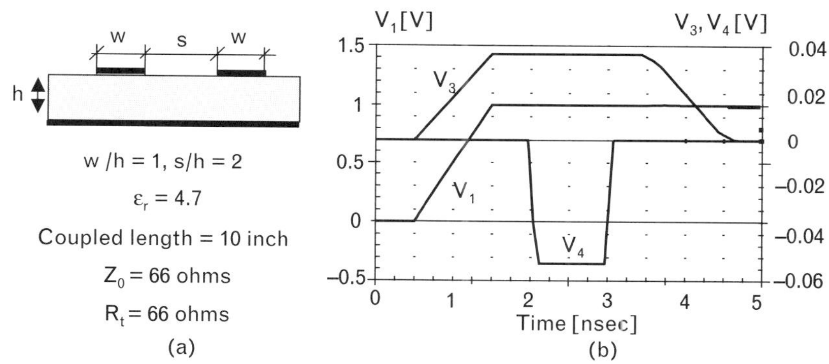

FIGURE 9.13 *Crosstalk example in a stripline-coupled trace pair: (a) cross-section geometry and simulation parameters and (b) simulated waveforms. Node labels are according to Figure 9.10.*

density and routing as well shows characteristic differences: digital buses may be several hundred bits wide, which is not used in analog circuits. The crosstalk and signal interactions in wide digital busses often create a *simultaneous switching noise* (SSN) scenario, a typical degrading factor in high-speed digital circuits.

9.4 R-L-C models for digital applications

9.4.1 Resistors

Resistors in high-speed digital applications fall into two categories. In the first category, which is the most typical usage, resistors are used to terminate signal traces or occasionally power planes. Resistors may also be used to set voltage references for inputs (*V*ref), to set impedance reference in *high-speed transceiver logic* (HSTL), and/or to set frequency of clock oscillations. Applications in this second category, though, are usually accomplished with only dc or low-frequency signals applied to the resistors, and therefore, we do not consider those further.

For signal termination applications, the main difference between RF and digital applications is that traces carrying digital signals come with a somewhat wider range of impedances. For instance, trace impedances with the reactive loading of daisy-chained inputs taken into account may be as low as 20Ω. Some BTL applications assume 27-Ω loaded trace impedance. Rambus, for instance, requires 28-Ω termination. Average lightly loaded point-to-point traces may have 50-Ω or 60-Ω impedances. Fibre Channel and Gigabit Ethernet traces need 150-Ω differential impedance, which can be implemented with uncoupled or loosely coupled 75-Ω single-ended traces. Single-ended SCSI traces require 92-Ω nominal impedance. Legacy partial-terminated buses may specify termination resistors in the hundreds of ohms range, such as the VME buses 330-Ω pull-up and 470-Ω pull-down resistors.

Power planes should have low impedance; therefore, terminating power planes require resistance values in the low ohm or subohm range.

As shown in Section 7.3, the typical inductance and capacitance associated with resistor bodies will result in a peaky impedance profile for resistance values below about 100Ω, because the inductive reactance will show up first as frequency increases. For a given body and connection inductance, the corner frequency is lower as the resistance value goes down. The capacitance of the equivalent circuit in Figure 7.4 refers to the static capacitance of the terminals of the resistor body. In RF applications, chances are good that the resistor connects to surface traces, where the trace width may be comparable to the width of the resistor, introducing little extra discontinuity. In dense digital applications, however, narrow stripline traces may need to be connected to the resistor on the surface through vias and pads. Figure 9.13 shows a typical geometry for resistor connection in digital applications. The resistor size is determined by the dissipation requirement, available space, cost, and accuracy. Handheld digital circuits frequently use the smallest-size components available, which is a 20 × 10-mil footprint (0201 case size) at the time of writing. Large-size digital boards, on the other hand, typically use bigger components, because larger resistors are cheaper and easier to assemble. Figure 9.14 shows a 60 × 30-mil (603 case size) surface-mount component with its recommended pad geometry. Compared to the connecting 4-mil-wide traces, the 50-mil-wide and 30-

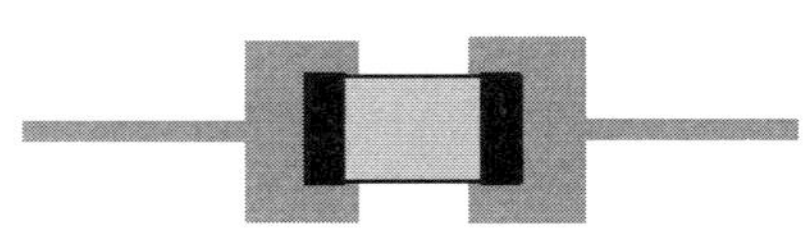

FIGURE 9.14 *Surface-mount 0603-size component on PCB pads, connected to 4-mil wide traces (approximately to scale).*

mil-long pads are big, and they create an approximately 0.45-pF/pad extra capacitance.

If detailed modeling becomes necessary, the large difference between the trace width and pad requires the pad to be considered in the equivalent circuit as separate capacitor connected from the resistor terminal to ground, as shown Figure 9.15. As the pad capacitance is usually in the order of a picofarad or so, it has to be taken into account only at gigabit per second signaling speeds. The via connection model is considered separately in Sections 7.7 and 9.5.

In applications where one side of the termination resistor is connected to a supply voltage, V_{dd} or V_{tt}, the low impedance of the supply rail shunts out the pad capacitance, and only the pad capacitance on the signal side needs to be considered. In simulations, this pad capacitance on the signal side may be combined with the via capacitance, or the package-pin capacitance connected to the same location. Typical supply-rail impedance values depend on the voltage and current rating, and on the maximum allowed noise on the rail: it is in the milliohm range for low voltages and high currents, and it could be several ohms for higher-voltage low-current applications. For instance, a 2.5-V rail supplying 50-A maximum current may require a few milliohms impedance; a low-power 5-V rail supplying a maximum of 10-mA current may require a rail impedance in the order of tens of ohms.

9.4.2 Inductors

While resistor applications are somewhat similar in RF and digital applications, inductors and capacitors are typically used in more differing ways. In digital circuits inductors are very seldom used connected directly to high-speed signals. One of the few exceptions is the compensation of capacitive loading of inputs, by adding properly sized series inductors in front of the load capacitors. The typical usage in digital circuits is to provide decoupling of supply rails between circuits and subsystems. On the other hand, supply-rail decoupling in RF circuits is the secondary application and is less demanding, because the supply noise current may be narrower band. The primary application of inductors in RF circuits is to create filters and other

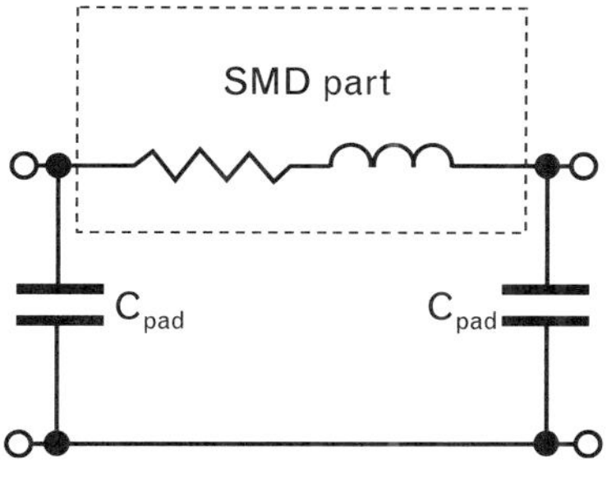

FIGURE 9.15
Equivalent circuit of resistor in digital applications.

tuned circuits in the signal path. In filters and tuned circuits the quality factor (Q) of the inductor becomes of primary concern: even if the loaded Q of the circuit is low, providing a wideband network, the unloaded Q of the components should be as high as possible. This requires the lowest possible losses in the inductors; moreover, the inductance value also should be stable with respect to frequency, temperature, and aging.

In contrast to RF applications, in digital circuits the supply-rail decoupling is practically the only use for inductors. As it was shown in Sections 7.4 and 7.5, in digital applications accuracy and stability of the inductance matter less, there is no need for high Q, and in fact usually the lowest possible Q is preferable. To maintain the required series isolation of different supply-rail points, a minimum specified series impedance value should be maintained over the specified frequency range and dc bias conditions.

The inductance of a coil with magnetic core is linearly dependent on the core's permeability. Permeability does depend on the dc bias conditions: as the dc current through the coil increases, the B-H curve gradually goes into saturation, reducing the permeability and hence reducing the inductance and inductive reactance. The impedance curves with three different dc bias current measured on a multihole ferrite bead are shown on Figure 9.16. This particular ferrite bead has been widely used in digital applications. In supply-rail decoupling applications we can assume that the ac ripple current is much less than the dc bias current, and therefore, saturation due to the ac contents is usually not a concern.

To complete the supply-rail decoupling, bypass capacitors are needed on both sides of the inductor. In low-power applications, composite capacitor-inductor arrays can also be considered. The outline and the equivalent circuit of a three-terminal filter are shown in Figure 9.17.

It should be kept in mind that these T-filter components are meant for impedance-matched applications, typically working between 50-Ω terminations. The supply-rail impedance is often much lower, and therefore, the transfer function of the T-filter may become peaky, boosting rather than attenuating certain noise components.

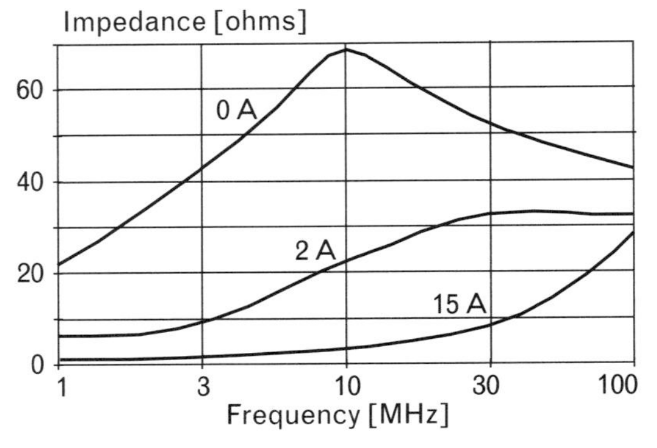

FIGURE 9.16
Impedance versus frequency of a ferrite bead, with dc bias current as a parameter.

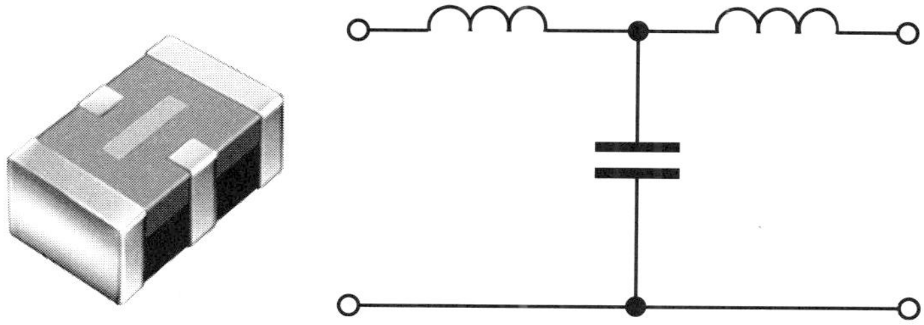

FIGURE 9.17
Three-terminal inductor-capacitor array and its equivalent circuit. (Courtesy of TDK Corporation.)

9.4.3 Capacitors

In RF circuits, the most typical application of capacitors is in filters and tuned elements; a secondary application is bypassing the supply rail against the narrower bandwidth noise. In digital applications, these two roles reverse: the primary application is providing clean supply power for the active devices by bypassing the wideband noise, and a more rare, though not negligible application, is using capacitors directly connected to high-speed signals in form of RC terminations or dc blocking capacitors. In RF applications, the quality factor in filter and tuned-circuit applications matters a lot, and therefore, the goal is to use low-loss capacitors. For the same reason, the stability and tolerance of capacitance itself is also of primary interest.

In digital applications, even when connected directly to signals, neither the capacitance value nor the loss of the capacitor is critical. In ac termination applications, the capacitor is in series to a resistor of 20-Ω to 200-Ω value, so the subohm *effective series resistance* (ESR) of the capacitor is hardly a problem. The capacitance value sets the time constant of the ac termination, but it may easily vary 30% to 50% without impacting the circuit operation. Similarly, in dc blocking applications, the ESR value does not matter as long as it is much lower than the characteristic impedance of the trace. There is a minimum requirement for the capacitance, and the maximum capacitance usually does not matter, leaving a large range for possible capacitance tolerances.

In ac termination and dc blocking applications, high Q is not required, but if the capacitor has high Q, it is not a problem either. In the much more typical bypass applications, however, high Q of the capacitor usually becomes a problem. For power-distribution applications, the goal is to provide a predetermined impedance profile. Depending on the applications, the required impedance value may vary with frequency. The typical requirement is flat, resistive impedance over a wide frequency range.

To create the flat impedance response over a wide frequency range, bypass capacitors of different values may be connected in parallel. Larger capacitance values will provide the required impedance at lower frequencies, but they tend to be physically bigger, so their inductance tends to be bigger, limiting their effectiveness at high frequencies. To provide the required impedance at higher frequencies, capacitors have to have sufficiently low inductance. This in turn requires a physically small body; hence, their capacitance may be lower. If these paralleled capacitors have high Q and widely separated series resonance frequencies, their inductances and capacitances may create significant antiresonant peaks in the impedance profile. Figure 9.18 shows such a situation, where C1 = 100 μF, ESR1 = 0.1Ω, ESL1 = 50 nH, C2 = 1 μF, ESR2 = 0.1Ω, ESL = 10 nH, C3 = 0.01 μF, ESR3 = 0.05Ω, ESL3 = 5 nH. Csum is the overall impedance of the three capacitors in parallel, neglecting interconnects. C1 and C2 have resonance frequencies of 71.2 kHz and 1.59 MHz, and a Q of 0.224 and 1.0, respectively. These moderate Q values assure that between their series resonance frequencies the resulting parallel impedance curve does not significantly exceed the lower boundary of the two capacitors' impedance profiles. C3 has a resonance frequency of 22.4 MHz and a Q of 14.1, and therefore, between the second and third series resonance frequencies, there is an antiresonance with a 3-Ω peak, whereas the lower boundary of the C2 and C3 impedance curves would be less than 0.7Ω.

Figure 9.19 is a measured illustration of this antiresonance on two parallel-connected capacitors: a 220-μF aluminum bulk capacitor with ESR = 0.23Ω and ESL = 4.3 nH, and a 4.7-nF ceramic disc capacitor with ESR = 25 mΩ and ESL = 2 nH. The leads of capacitors were soldered directly together, with no printed circuit board etch or plane. Note the 18-MHz antiresonance with a magnitude of about 2Ω.

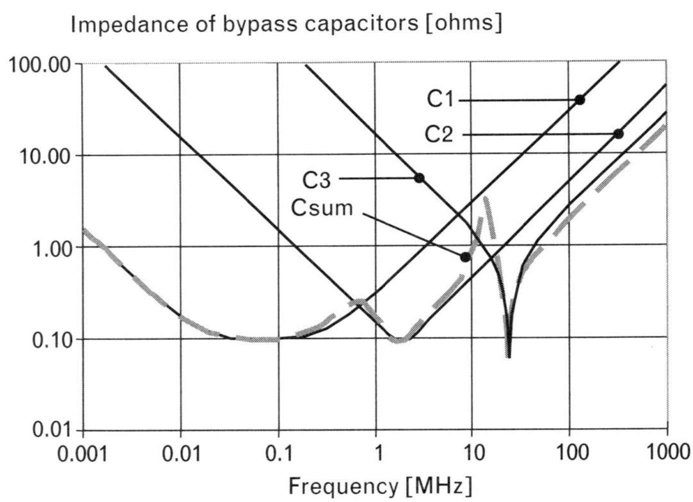

FIGURE 9.18
Impedance of three capacitors in parallel.

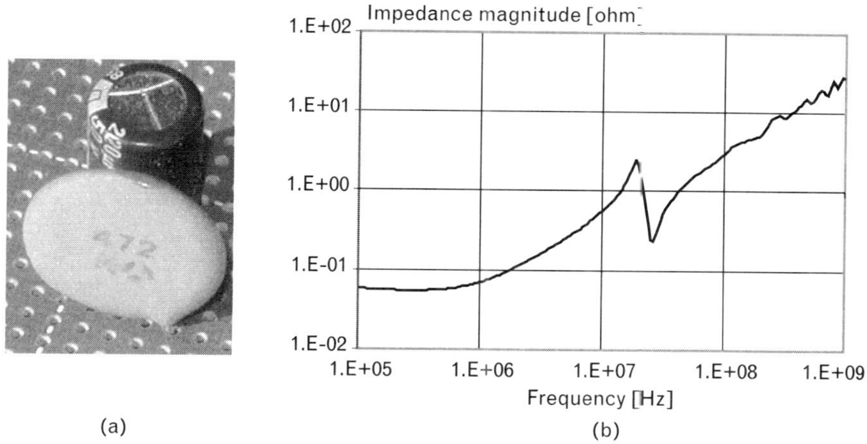

FIGURE 9.19 *Parallel antiresonance of a 220-µF electrolytic bulk aluminum and a 4.7-nF ceramic bypass capacitor: (a) capacitors and (b) their measured impedance magnitude.*

9.5 Parasitics of passive interconnects, loading, vias, and losses

The all-pass nature of ideal interconnects are modified by several factors:

- Electrical loading of daisy-chain connected inputs and outputs;
- Via- and through-hole discontinuities;
- Bends and junctions of interconnects;
- Losses and dispersion of interconnects.

In RF circuits, daisy-chain connection and buses[3] are rare. Via and through-hole connections are traditionally much fewer than in dense digital circuits. Therefore, in RF circuits, the major limiting factors are usually the interconnect parasitics, dispersion, as well as discontinuities and junctions.

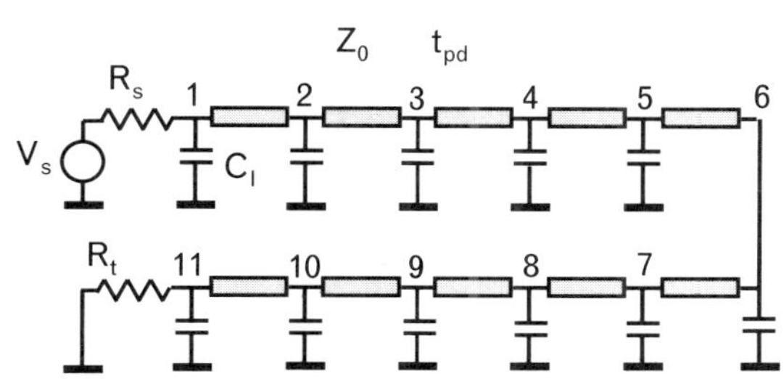

FIGURE 9.20 *Topology of example schematic with daisy-chain capacitive loads.*

3. Daisy-chain or bus connection is when a signal is routed sequentially to more than one input. In daisy chain, the spacing and delay between successive inputs are not necessarily the same. Buses, often used for plug-in card connections, tend to be more regular, with a fixed spacing and delay between connection slots.

In a typical digital system all four of the above limiting factors exist. At low signaling speeds, the biggest disturbance is caused by the input capacitance in daisy-chain I/O connections [4], which may be up to 10 pF/input. The second biggest discontinuity is the pad and via capacitance with up to about 1 pF/via. The smallest disturbance comes from bends and junctions, with an order of magnitude lower equivalent capacitance.

Figure 9.20 shows a 10-inch-long lossless trace, loaded with daisy-chain connected 10-pF capacitors at 1-inch intervals. Capacitors represent the inputs of I/O circuits at each node.

Voltage transfer function magnitudes at various nodes as a function of frequency are plotted in Figure 9.21. The uniform capacitive loading along the trace creates a band-limited transfer function. For light loads, the –6-dB frequency point can be approximated with a simple formula:

$$f_{-6\,\mathrm{dB}} = \frac{1}{2\pi\sqrt{L_{trace}C_{load}}} \tag{9.1}$$

The band-limited transfer function creates ringing on the steady-state digital signals, introduces rounded corners, and most importantly, decreases the equivalent characteristic impedance and increases the propagation delay. The equivalent characteristic impedance and loaded propagation delay of the trace with uniform daisy-chain loading can be approximated as

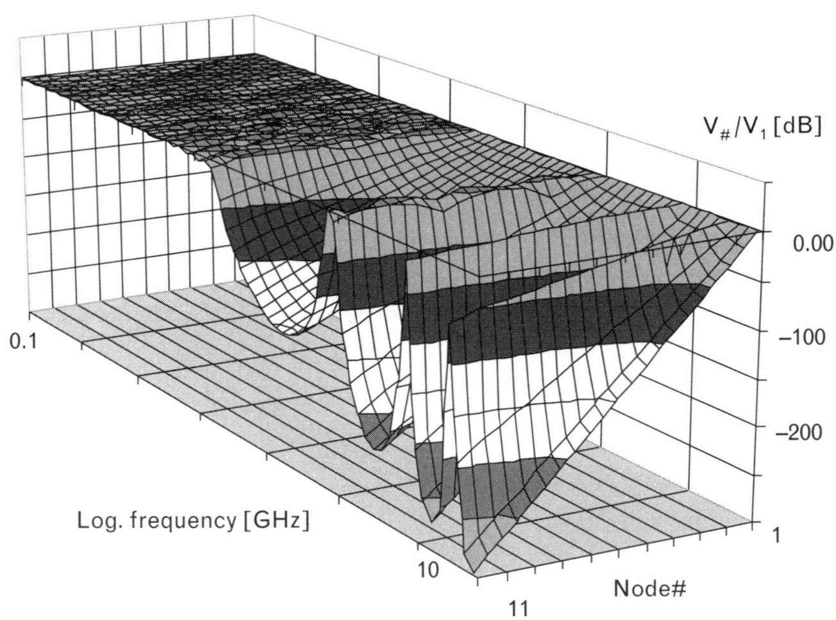

FIGURE 9.21 *Voltage transfer function versus frequency of a trace with 10-pF/inch daisy-chain load.*

$$Z_{o\_loaded} = Z_o \frac{1}{\sqrt{1 + \frac{C_{load}}{C_{trace}}}}; \quad t_{pd\_loaded} = t_{pd}\sqrt{1 + \frac{C_{load}}{C_{trace}}} \tag{9.2}$$

Figure 9.22 shows the waveform at the far end of a loaded trace with different load capacitance values.

The large I/O capacitance in daisy-chain connections not only reduces the characteristic impedance and increases the propagation delay, but also creates a large uncertainty of timing because the input capacitance of digital devices are not tightly specified. This potentially large skew is one strong driving force in high-speed digital systems to gradually move away from daisy-chain connections and towards using point-to-point interconnects.

After the active circuit's input capacitance, plated-through holes and vias are the next largest discontinuities in digital circuits. Typical equivalent circuits are shown in Figure 9.23. Below 1-Gbps signaling speeds, it is sufficient to approximate the via impedance with a single capacitance (primarily representing the pads). If the time of flight through the vertical metal cylinder of via has to be also captured, a *C-L-C* PI-network can be used. In dense digital circuits, and with narrow traces, impedance-matched vias are not common.

Finally, as long as the load capacitance of active circuits is in the several picofard range, bends and junctions in digital circuits do not represent the primary limitations.

Interconnect losses are due to resistive losses in the conductors and dielectric losses in the supporting insulation material. The series resistance

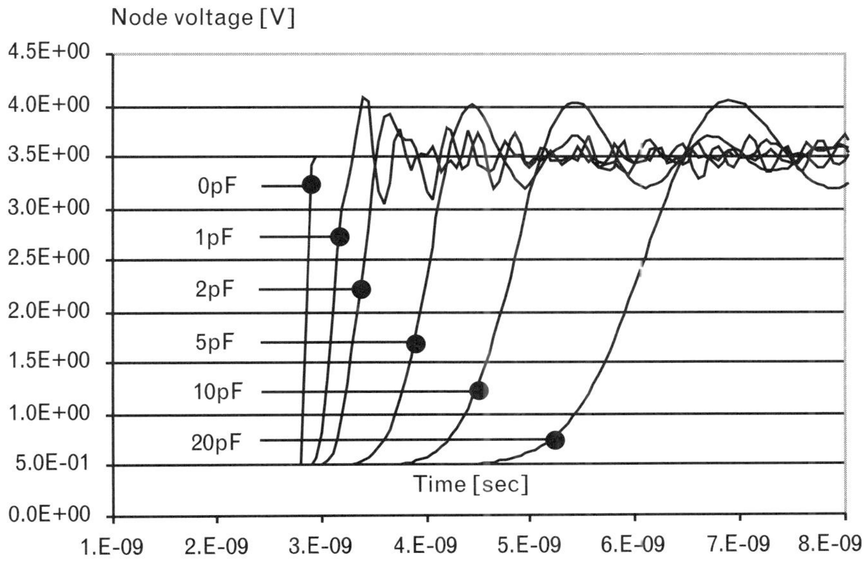

FIGURE 9.22 *Far-end time-domain waveforms of a trace with daisy-chain load capacitance of various uniform values.*

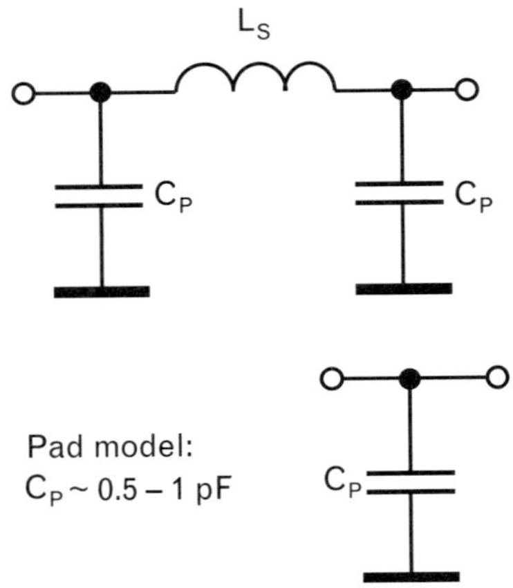

FIGURE 9.23
Equivalent circuits for vias in digital circuits.

of conductors and parallel conductance of dielectric materials can be approximated with (9.3):

$$R(f) = R_{DC} + R_s\sqrt{f};$$
$$G(f) = G_o + G_{df} \tag{9.3}$$

With matched termination, the series and parallel losses create attenuation as a function of frequency,

$$A(f)^{\text{dB}} = 4.35\left[G(f)Z_0 + \frac{R(f)}{Z_0}\right] \tag{9.4}$$

The conductive losses vary with approximately the square root of frequency, whereas the dielectric losses increase linearly with frequency. This means that beyond a crossover frequency the dielectric losses dominate. Below the crossover frequency, skin loss is dominant. The crossover frequency is lower for wider traces, and higher for narrow traces. For this reason, RF circuits with wide traces are mostly limited by dielectric losses, and digital circuits with narrow traces are mostly limited by skin loss. Figure 9.24 shows the approximate loss contributions in a 4-mil-wide microstrip trace with FR4 dielectric, assuming a loss tangent of 0.02. Below 5 GHz, the skin loss dominates the attenuation. This explains why high-speed digital circuits are less sensitive to dielectric losses, and conventional dielectric materials may work well up to a couple of gigabit/second speeds.

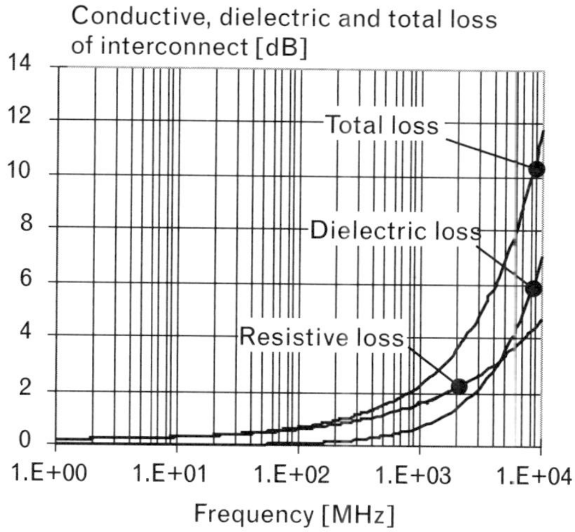

FIGURE 9.24
Resistive, dielectric and total loss of a 4-mil-wide microstrip trace on FR4 dielectric material.

9.6 Frequency-domain versus time-domain considerations

Due to the fact that digital signals are specified by their voltage and current waveforms, time-domain simulations and measurements of digital systems are the primary tools of design and verification. There are parts of digital systems, however, where the frequency-domain approach is preferred. The fixed spectra of clock signals, for instance, make them better suited for steady-state, frequency-domain analysis. Legal regulatory requirements limit the emissions from electronic apparatus, and the allowed maximum radiation values are also given in the frequency domain. Another example is the design of *power-distribution networks* (PDNs), where usually the time-domain transient noise excitation is not known; therefore, a convenient approach is to make the design and verification in the frequency domain.

9.6.1 When frequency domain is essential: clock networks

Clock signals of synchronous digital circuits are similar to the local oscillator signals in heterodyne receivers: they determine the timing. Digital clocks are square waves, though theoretically the zero crossings of a sine wave could also be used. In digital applications, either the rising edge or the falling edge, or both, are used to trigger the electronics. As only the edges are used, the monotonicity of edges is critical: any glitch within the inputs' threshold range may create double or false triggering. Spectra of clock signals have steady spectral lines at integer multiples of the fundamental frequency. A constant clock frequency also means that any intersymbol interference may synchronously build up and may create severe ringing or distortion on the edges.

Because in clock networks the steady-state matters, it is important to run simulations in the time domain until the steady state is reached. Figure 9.25 shows a daisy-chained clock network simulated for the first period versus the fifth period. Note that the ring backs during the first period leave a 1V opening between the minimum value of high signal and maximum value of low signal, whereas in steady-state the opening is only 0.4V. The waveform data was generated with Berkeley SPICE.

9.6.2 When frequency domain is useful: power distribution

Both analog and digital systems used to have just one or a few supply voltages. With the continuously shrinking chip sizes, digital systems today may have several different supply rails, each consuming sometimes hundreds of watts by high-speed digital devices. As digital buses may draw current at any arbitrary frequency below the clock, the typical requirement for the PDN is to provide a flat impedance response within the full operating bandwidth of the system. The actual limit of PDN noise is specified by the maximum tolerable noise voltage on the supply rails. To simulate or

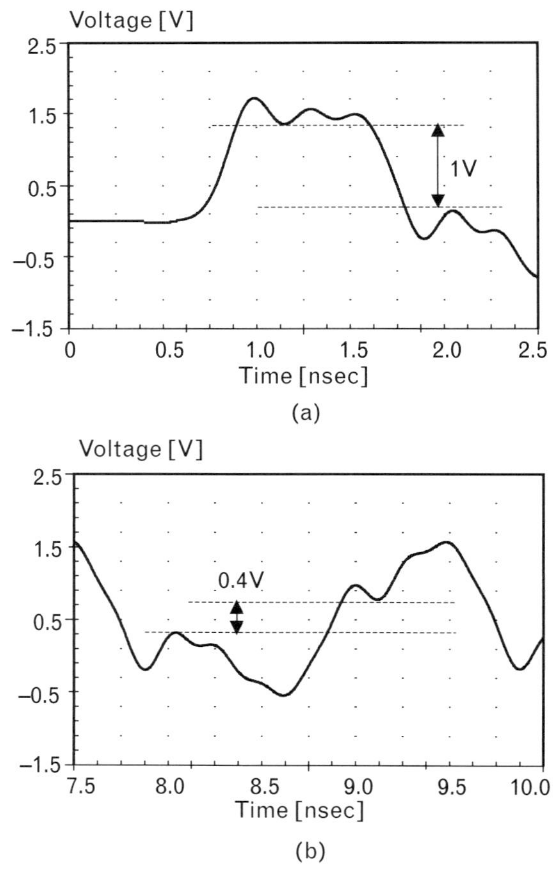

FIGURE 9.25 *Simulated response of a clock network during its (a) first and (b) fifth periods.*

measure the time-domain noise, we have to know both the noise current and the impedance profile of the PDN. Since the actual noise current may be hard to estimate accurately, a convenient middle ground is to specify the impedance provided by the supply rail.

To cover a wider frequency range, it is customary to connect several, sometimes many, bypass capacitors in parallel, arranged into some number of capacitor banks. A capacitor bank is a given number of capacitors with the same nominal value. Figure 9.26 shows the impedance profile created by three capacitors banks, all with low-Q capacitors. The first capacitor bank has five pieces of 1,500-μF, 0.02-Ω, 5-nH bulk capacitors. The second bank has 10 pieces of 10-μF, 0.035-Ω, 1.2-nH ceramic capacitors, and the third bank has 25 pieces of 0.22-μF, 0.1-Ω, 0.5-nH ceramic capacitors. The impedance profile was simulated assuming negligible interconnection delay among the capacitors—an assumption that is valid as long as the capacitors are much closer than the shortest wavelength of interest. Note that in this case, with the proper selection of capacitance, ESR and ESL values, the impedance profile is smooth and relatively flat up to at least 50 MHz. The transient noise simulated with a 10 App noise current is shown in Figure 9.27. It has a peak-to-peak value corresponding to the peak impedance times the peak current.

If, for any reason, the impedance profile is not smooth, the peak magnitude may create higher transient noise voltages [5]. For instance, the impedance profile in Figure 9.28 was simulated assuming lower ESR (higher Q) for the ceramic capacitors. The bulk capacitors stayed the same, in the second and third capacitor banks ESR was reduced from 0.035Ω to 0.01Ω and from 0.1Ω to 0.01Ω, respectively. All other parameters and conditions remained the same. Note that the impedance profile now has two pronounced minima at the series resonance frequencies of the ceramic capacitors, and a 12-mΩ (versus 5-mΩ) peak impedance magnitude

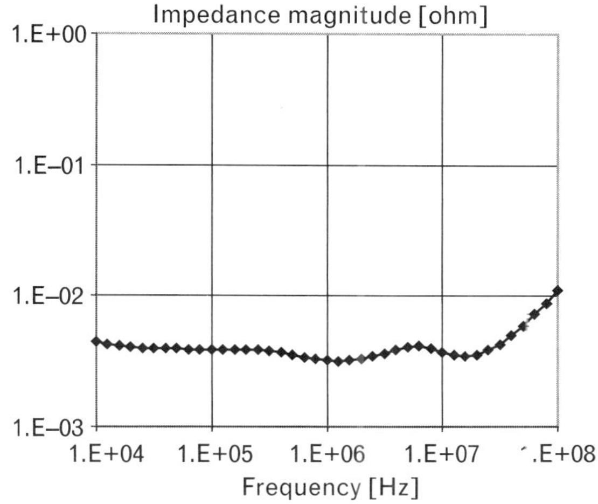

FIGURE 9.26 *Simulated impedance profile created by three low-Q capacitor banks.*

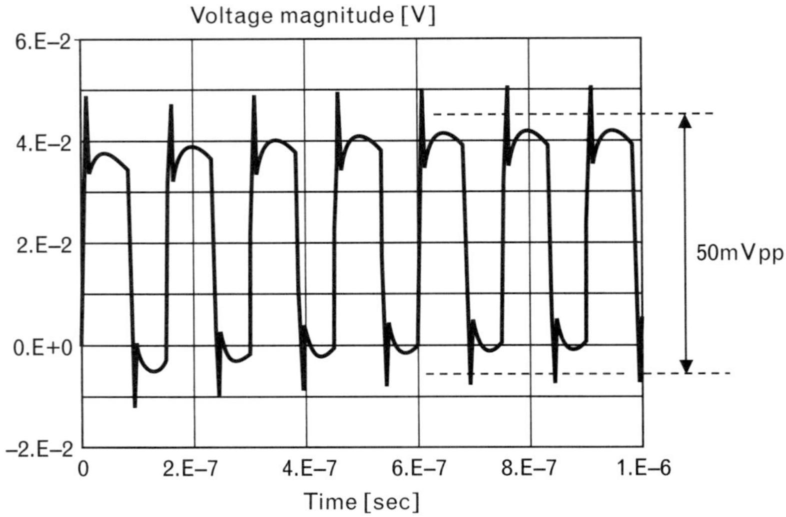

FIGURE 9.27 *Simulated transient noise on the PDN of Figure 9.26 with 10-App repetitive noise current.*

between the two minima. The simulated transient noise in Figure 9.29 shows a correspondingly higher 120-mVpp noise on the supply rail.

9.7 Measurement and simulation considerations

For traditional narrowband RF designs, the simulation and measurement requirements are mostly in the frequency domain. Total signal power and spectral purity are the two most important measures. The typical instrumentation is vector network analyzers to measure voltage reflection coefficients and transfer functions, and spectrum analyzers to measure power, spectral contents, and spectral purity. Similarly, RF CAD packages

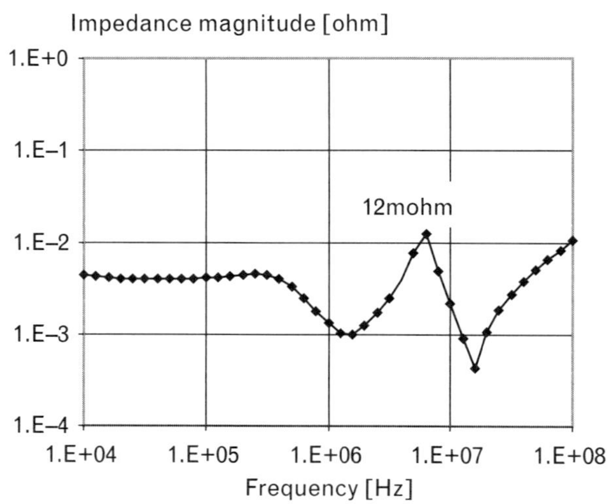

FIGURE 9.28 *Simulated impedance profile with three medium-Q bypass capacitor banks.*

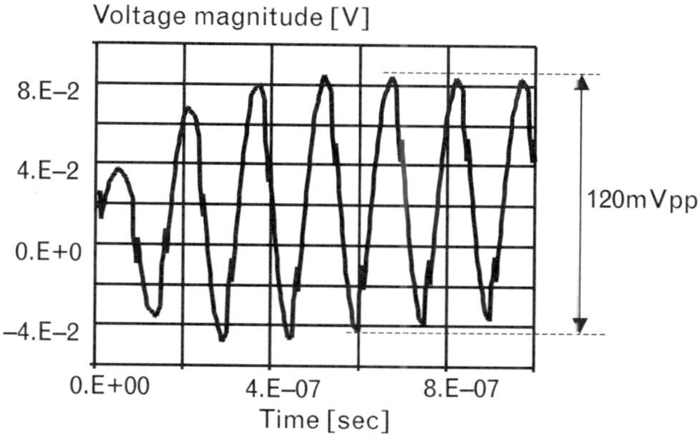

FIGURE 9.29
Simulated transient noise voltage on the PDN of Figure 9.29 with 10-App repetitive transient noise current.

primarily target the frequency domain, and nonlinearity is taken into account as a perturbation.

In digital systems, the time-domain waveform details are specified, so those must be checked by simulations and verified by measurements. To visualize and measure the signals versus time, the primary measuring instruments are oscilloscopes and time-domain reflectometers. Special-purpose oscilloscope-like instruments can measure time intervals and jitter on clock and data signals. CAD packages for digital circuits usually target the time domain, and frequency-dependent behavior (dispersion) is taken into account as perturbation.

The potentially fast edges of digital signals create a huge demand toward the bandwidth of measuring instruments. As illustrated in Figure 9.30, to capture the waveform details with sufficient detail and accuracy, we usually require the bandwidth of measuring chain (probes, cables, instruments) to be three to five times wider than the bandwidth of digital signals. With 100-ps transition times becoming common, the signal bandwidth is in the order of 3 GHz, requiring a measuring bandwidth of 10 GHz or more. While the analog front ends and the sampling aperture of digital oscilloscopes could provide this bandwidth, the bottleneck is in the probe connections and in the processing speed of captured data. Today, commercially available digital sampling oscilloscopes can achieve 20- to 50-GHz bandwidth with under sampling of stable repetitive signals, but real-time one-shot bandwidth is limited to about 10 to 20 Gsample/s, or 3- to 6-GHz analog bandwidth.

Connecting the measuring instruments to the signals in digital systems is another challenge. The lower density of interconnects and the few, well-defined connection points between subsystems allow RF systems to be equipped with connectors at each major stage, which can be used not only to link the subsequent stages together, but also to connect measuring instruments. The multilayer printed circuit boards and high density of traces do not leave room for RF connectors on high-speed digital systems. Therefore, instead of fixed wideband connectors (fixtures), digital systems

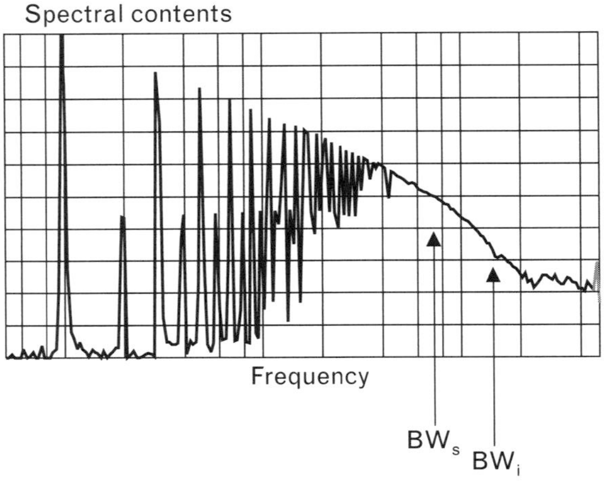

FIGURE 9.30 *Illustration of signal bandwidth (BWs) and instrumentation bandwidth (BWi). The chart shows the spectrum of a periodical clock waveform, measured with a spectrum analyzer. BWs is the bandwidth of the signal, and it marks the frequency where the fall rate of spectral-line magnitude changes from −20 dB/decade to −40 dB/decade. BWi is the bandwidth of instrument, which is minimally required to process the signal with BWs bandwidth.*

rely on probing. As shown in Figure 9.31, lower-frequency, low-cost probes can be hand-held; higher-speed miniature probes often require a large probe station for proper positioning.

As in oscilloscope measurements the time-domain details are of primary interest; the proper bandwidth, and within the required bandwidth, the smooth transfer response or constant group delay are important. To make sure that the probe and oscilloscope bandwidths do not limit the accuracy of measurements, it is a good idea to characterize them before

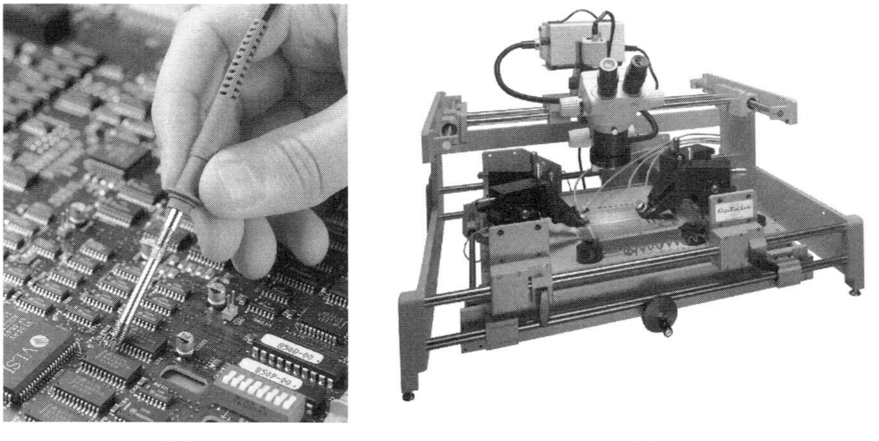

(a) (b)

FIGURE 9.31 *(a) Low-frequency hand-held oscilloscope probe (Courtesy of Tektronix); and (b) large probe station. (Courtesy of GigaTest Labs.)*

starting the measurements, by checking their bandwidth and equivalent rise time. If the measured signals show rise and fall times at least a few times longer than that of the instrumentation, the collected data is not compromised by bandwidth. Otherwise, repeating the measurement with a wider-band setup is necessary. Figures 9.32 shows a situation, where the low bandwidth of the probe distorts the time-domain result. Figure 9.33 illustrates the frequency-domain transfer response and the measured time-domain waveform with sufficiently wide bandwidth.

FIGURE 9.32
(a) Frequency response and (b) time-domain waveform of a low-bandwidth oscilloscope probe, with an input stimulus exceeding the bandwidth of probe.

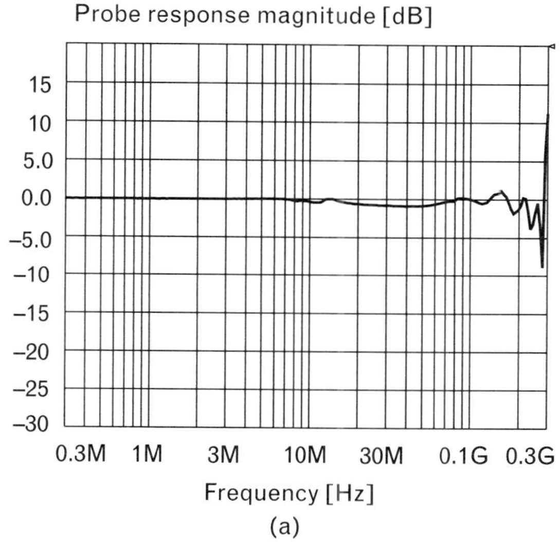

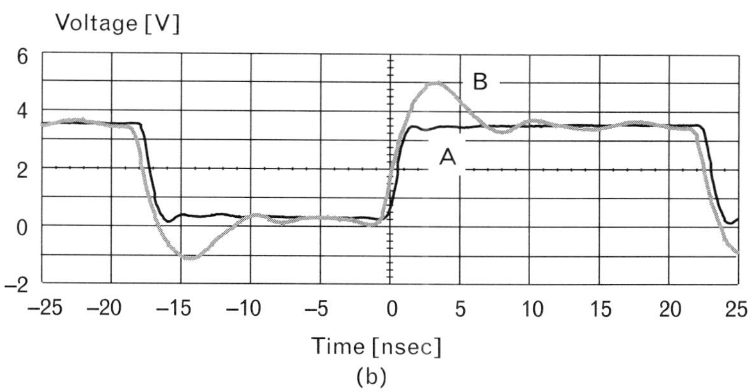

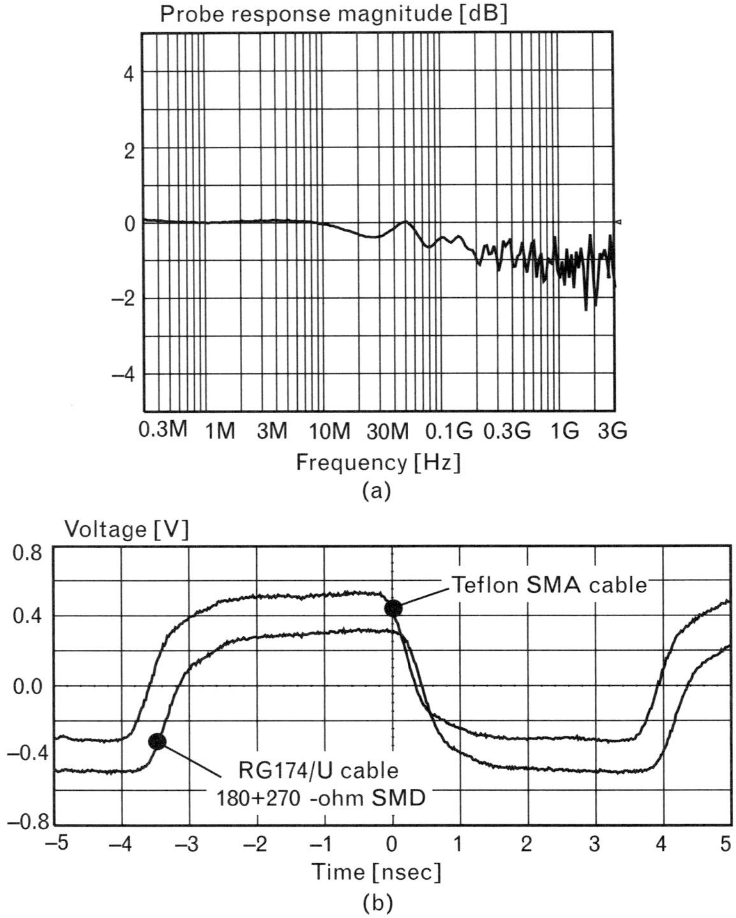

FIGURE 9.33
(a) Frequency response and (b) time-domain waveforms with a wideband oscilloscope probe.

REFERENCES

[1] Cheney, M., *Tesla: Man Out of Time*, New York: Dell Publishing, 1981.

[2] Widmer, A., and P. Franaszek, "A DC-Balanced, Partitioned-Block 8B/10B Transmission Code," *IBM Journal of Research and Development*, Vol. 27, No. 5, September 1983, pp. 440–451.

[3] Feller, A., H. R. Kaupp, and J. J. Digiacomo, "Crosstalk and Reflections in High-Speed Digital Systems," *Proc. of the 1965 Fall Joint Computer Conference*, pp. 511–525.

[4] Collins, H. A., R. E. Nikel, and M. Nagy, "Design of High Speed Source Synchronous Interfaces - DDRS-DRAM Design," *Proc. of DesignCon99, High-Performance Systems Design Conference*, Santa Clara, CA, February 1–4, 1999, pp. 49–58.

[5] Brooks, D., *ESR and Bypass Capacitor Self Resonant Behavior—How to Select Bypass Caps*, Technical Report, Ultracad, http://www.ultracad.com/.

Selected Bibliography

Dally, W. J., and W. J. Poulton, *Digital Systems Engineering*, Cambridge, England: Cambridge University Press, 1998.

Johnson, H. W., and M. Graham, *High-Speed Digital Design*, Englewood Cliffs, NJ: Prentice Hall, 1993.

Paul, C. R., *Introduction to Electromagnetic Compatibility*, New York: Wiley, 1992.

Young, B., *Digital Signal Integrity: Modeling and Simulation with Interconnects and Packages*, Englewood Cliffs, NJ: Prentice Hall, 2001.

Appendix

Summary of Basic Formulas – 1

(Z_0 is the characteristic impedance and Y_0 is the characteristic admittance)

Reactance	$X_L = 6.28 f_{\text{GHz}} L_{\text{nH}}$ $X_C = \dfrac{159}{f_{\text{GHz}} C_{\text{pF}}}$
Conductance	$G = \dfrac{1}{R}$ and $g = \dfrac{1}{r}$
Susceptance	$B = \dfrac{1}{X}$ and $b = \dfrac{1}{x}$
Impedance	$Z = R \pm jX = \dfrac{1}{Y} = Z_0\left(\dfrac{1+\Gamma}{1-\Gamma}\right)$ and $z = \dfrac{Z}{Z_0}$
Admittance	$Y = G \pm jB = \dfrac{1}{Z} = Y_0\left(\dfrac{1-\Gamma}{1+\Gamma}\right)$ and $y = \dfrac{Y}{Y_0}$
Reflection coefficient	$\Gamma = \dfrac{Z - Z_0}{Z + Z_0} = \dfrac{Y_0 - Y}{Y_0 + Y} = \dfrac{VSWR - 1}{VSWR + 1} = \dfrac{z-1}{z+1}$
Voltage standing wave ratio	$VSWR = \dfrac{1+\lvert\Gamma\rvert}{1-\lvert\Gamma\rvert} = \dfrac{R_{\text{LARGER}}}{R_{\text{SMALLER}}}$
Return loss	$RL = -20\log\lvert\Gamma\rvert = -20\log\left\lvert\dfrac{Z-Z_0}{Z+Z_0}\right\rvert$

Mismatch loss $(\Gamma_S = 0, \Gamma_L \neq 0)$

$$ML = -10\log\left(1-|\Gamma_L|^2\right) = -10\log\left(1-\left|\frac{Z_L - Z_0}{Z_L + Z_0}\right|^2\right)$$

$$= -10\log\left[1-\left(\frac{VSWR-1}{VSWR+1}\right)^2\right]$$

Mismatch loss $(\Gamma_S \neq 0, \Gamma_L \neq 0)$

$$ML = -10\log\left[\frac{\left(1-|\Gamma_S|^2\right)\left(1-|\Gamma_L|^2\right)}{|1-\Gamma_L\Gamma_S|}\right]$$

Wavelength in free air $\lambda = \dfrac{c}{f} \approx \dfrac{3\left(10^8\,\text{m}\right)}{f_{\text{Hz}}} \approx \dfrac{30\text{ cm}}{f_{\text{GHz}}} \approx \dfrac{11.8\text{ in}}{f_{\text{GHz}}}$

Conversion to decibels $dB = 20\log\dfrac{v_2}{v_1} = 20\log\dfrac{i_2}{i_1} = 10\log\dfrac{P_2}{P_1}$

Noise factor $F = \dfrac{P_{no}}{G_A P_{ni}} = \dfrac{SNR_{IN}}{SNR_{OUT}} = \dfrac{T_e}{T_o} - 1 \quad [T_0 = 293\text{K}]$

Noise figure $NF = 10\log F$

Cascade noise factor $F = F_1 + \dfrac{F_2 - 1}{G_{A1}} + \dfrac{F_3 - 1}{G_{A1}G_{A2}} + \ldots$

Summary of Basic Formulas – 2

Unnormalized Form		Normalized Form (using $Z_o = 50\Omega$ and $Y_o = 0.02\text{S}$)

Ideal lumped component reactance and susceptance of inductors:

Unnormalized		Normalized
$X_L = 6.283 f_{\text{GHz}} L_{\text{nH}}$	$[+j]$	$x_L = 0.1257 f_{\text{GHz}} L_{\text{nH}}$
$B_L = \dfrac{0.159}{f_{\text{GHz}} L_{\text{nH}}}$	$[-j]$	$b_L = \dfrac{7.96}{f_{\text{GHz}} L_{\text{nH}}}$

Capacitors

Unnormalized		Normalized
$X_C = \dfrac{159}{f_{\text{GHz}} C_{\text{pF}}}$	$[-j]$	$x_C = \dfrac{3.183}{f_{\text{GHz}} C_{\text{pF}}}$
$B_C = 0.006283 f_{\text{GHz}} C_{\text{pF}}$	$[+j]$	$b_C = 0.314 f_{\text{GHz}} C_{\text{pF}}$

Ideal lumped inductance and capacitance in nanohenries and picofarads

Unnormalized	Normalized
$L_{\text{nH}} = \dfrac{159 X_L}{f_{\text{GHz}}} = \dfrac{159}{f_{\text{GHz}} B_L}$	$L_{\text{nH}} = \dfrac{7.96 x_L}{f_{\text{GHz}}} = \dfrac{7.96}{f_{\text{GHz}} b_L}$
$C_{\text{pF}} = \dfrac{0.159}{f_{\text{GHz}} X_C} = \dfrac{0.159 B_C}{f_{\text{GHz}}}$	$C_{\text{pF}} = \dfrac{3.183}{f_{\text{GHz}} x_C} = \dfrac{3.183 b_C}{f_{\text{GHz}}}$

Ideal stub reactance, susceptance, and electrical length

Open stubs

Unnormalized			Normalized	
$X_{OS} = \dfrac{Z_{OS}}{\tan\theta}$	$\theta = \tan^{-1}\left(\dfrac{Z_{OS}}{X_{OS}}\right)$	$[-j]$	$x_{OS} = \dfrac{Z_{OS}}{Z_0 \tan\theta}$	$\theta = \tan^{-1}\left(\dfrac{Z_{OS}}{Z_0 x_{OS}}\right)$
$B_{OS} = \dfrac{\tan\theta}{Z_{OS}}$	$\theta = \tan^{-1}\left(Z_{OS} B_{OS}\right)$	$[+j]$	$b_{OS} = \dfrac{Z_0 \tan\theta}{Z_{OS}}$	$\theta = \tan^{-1}\left(\dfrac{Z_{OS} b_{OS}}{Z_0}\right)$

Shorted stubs

Unnormalized			Normalized	
$X_{SS} = Z_{SS} \tan\theta$	$\theta = \tan^{-1}\left(\dfrac{X_{SS}}{Z_{SS}}\right)$	$[+j]$	$x_{SS} = \dfrac{Z_{SS}}{Z_0} \tan\theta$	$\theta = \tan^{-1}\left(\dfrac{Z_0 x_{SS}}{Z_{SS}}\right)$
$B_{SS} = \dfrac{1}{Z_{SS} \tan\theta}$	$\theta = \tan^{-1}\left(\dfrac{1}{Z_{SS} B_{SS}}\right)$	$[-j]$	$b_{SS} = \dfrac{Z_0}{Z_{SS} \tan\theta}$	$\theta = \tan^{-1}\left(\dfrac{Z_0}{Z_{SS} b_{SS}}\right)$

Input impedance of an ideal transmission line of electrical length θ, terminated with Z_L

$$Z_{IN} = Z_{TL} \frac{Z_L + jZ_{TL} \tan\theta}{Z_{TL} + jZ_L \tan\theta}$$

Impedance and length of a cascade line to match $Z_S = (R_S + jX_S)$ to $Z_L = (R_L + jX_L)$

$$Z_{TL} = \sqrt{\frac{(R_S^2 + X_S^2)R_L - (R_L^2 + X_L^2)R_S}{R_S - R_L}}$$

$$\theta = \tan^{-1}\left[\frac{Z_{TL}(R_L - R_S)}{X_S R_L - X_L R_S}\right]$$

If $X_S = X_L = 0$

$Z_{TL} = \sqrt{R_S R_L}$ and $\theta = 90^\circ, 270^\circ, 450^\circ$

About the Authors

Les Besser is the chairman of Besser Associates, a continuing education organization. He is a Life Fellow of the IEEE, in which he has held various offices and received awards and recognition for past accomplishments. He holds a Ph.D., an M.S., and a B.S. in electrical engineering. Dr. Besser authored the first commercially successful microwave circuit optimization routine, COMPACT, and founded Compact Software (now part of Ansoft), a pioneer group in RF/MW CAE. A master lecturer, he is currently heading an organization dedicated to continuing education through instructor-led and Internet-based short courses, and CD and videotaped presentations. His company has trained nearly 50,000 engineers and managers since the mid-1980s.

Rowan Gilmore is an experienced consulting engineer who introduced the world's first commercial harmonic-balance CAD simulator while he was vice president of engineering at Compact Software. He has held numerous design and management posts in industry, including Central Microwave, Schlumberger, Telstra, and SITA. A senior member of the IEEE, he holds a D.Sc. and an MSEE from Washington University in St. Louis, Missouri, and a B.E. in electrical engineering from the University of Queensland in Brisbane, Australia. He has nearly 15 years of teaching experience with Besser Associates and CEI Europe.

Both authors may be reached at http://www.bessercourse.com.

Index

Recent Titles in the Artech House Microwave Library

Advanced Techniques in RF Power Amplifier Design, Steve C. Cripps

Automated Smith Chart, Version 4.0: Software and User's Manual, Leonard M. Schwab

Behavioral Modeling of Nonlinear RF and Microwave Devices, Thomas R. Turlington

Computer-Aided Analysis of Nonlinear Microwave Circuits, Paulo J. C. Rodrigues

Design of FET Frequency Multipliers and Harmonic Oscillators, Edmar Camargo

Design of Linear RF Outphasing Power Amplifiers, Xuejun Zhang, Lawrence E. Larson, and Peter M. Asbeck

Design of RF and Microwave Amplifiers and Oscillators, Pieter L. D. Abrie

Distortion in RF Power Amplifiers, Joel Vuolevi and Timo Rahkonen

EMPLAN: Electromagnetic Analysis of Printed Structures in Planarly Layered Media, Software and User's Manual, Noyan Kinayman and M. I. Aksun

Feedforward Linear Power Amplifiers, Nick Pothecary

Generalized Filter Design by Computer Optimization, Djuradj Budimir

High-Linearity RF Amplifier Design, Peter B. Kenington

Intermodulation Distortion in Microwave and Wireless Circuits, José Carlos Pedro and Nuno Borges Carvalho

Lumped Elements for RF and Microwave Circuits, Inder Bahl

Microwave Circuit Modeling Using Electromagnetic Field Simulation, Daniel G. Swanson, Jr. and Wolfgang J. R. Hoefer

Microwave Component Mechanics, Harri Eskelinen and Pekka Eskelinen

Microwave Engineers' Handbook, Two Volumes, Theodore Saad, editor

Microwave Filters, Impedance-Matching Networks, and Coupling Structures, George L. Matthaei, Leo Young, and E.M.T. Jones

Microwave Materials and Fabrication Techniques, Third Edition, Thomas S. Laverghetta

Microwave Mixers, Second Edition, Stephen A. Maas

Microwave Radio Transmission Design Guide, Trevor Manning

Microwaves and Wireless Simplified, Thomas S. Laverghetta

Neural Networks for RF and Microwave Design, Q. J. Zhang and K. C. Gupta

Nonlinear Microwave and RF Circuits, Second Edition, Stephen A. Maas

QMATCH: Lumped-Element Impedance Matching, Software and User's Guide, Pieter L. D. Abrie

Practical RF Circuit Design for Modern Wireless Systems, Volume I: Passive Circuits and Systems, Les Besser and Rowan Gilmore

Practical RF Circuit Design for Modern Wireless Systems, Volume II: Active Circuits and Systems, Rowan Gilmore and Les Besser

Radio Frequency Integrated Circuit Design, John Rogers and Calvin Plett

RF Design Guide: Systems, Circuits, and Equations, Peter Vizmuller

RF Measurements of Die and Packages, Scott A. Wartenberg

The RF and Microwave Circuit Design Handbook, Stephen A. Maas

RF and Microwave Coupled-Line Circuits, Rajesh Mongia, Inder Bahl, and Prakash Bhartia

RF and Microwave Oscillator Design, Michal Odyniec, editor

RF Power Amplifiers for Wireless Communications, Steve C. Cripps

RF Systems, Components, and Circuits Handbook, Ferril Losee

Stability Analysis of Nonlinear Microwave Circuits, Almudena Suárez and Raymond Quéré

TRAVIS 2.0: Transmission Line Visualization Software and User's Guide, Version 2.0, Robert G. Kaires and Barton T. Hickman

Understanding Microwave Heating Cavities, Tse V. Chow Ting Chan and Howard C. Reader

For further information on these and other Artech House titles, including previously considered out-of-print books now available through our In-Print-Forever® (IPF®) program, contact:

Artech House Publishers
685 Canton Street
Norwood, MA 02062

Artech House Books
46 Gillingham Street
London SW1V 1AH UK

Phone: 781-769-9750
Fax: 781-769-6334
e-mail: artech@artechhouse.com

Phone: +44 (0)20 7596 8750
Fax: +44 (0)20 7630 0166
e-mail: artech-uk@artechhouse.com

Find us on the World Wide Web at:
www.artechhouse.com